Mathematical Methods for Knowledge Discovery and Data Mining

Giovanni Felici
Consiglio Nazionale delle Ricerche, Rome, Italy

Carlo Vercellis
Politecnico di Milano, Italy

T0324999

Information Science REFERENCE

INFORMATION SCIENCE REFERENCE

Hershey · New York

Acquisitions Editor:	Kristin Klinger
Development Editor:	Kristin Roth
Senior Managing Editor:	Jennifer Neidig
Managing Editor:	Sara Reed
Copy Editor:	Angela Thor
Typesetter:	Jamie Snavely
Cover Design:	Lisa Tosheff
Printed at:	Yurchak Printing Inc.

Published in the United States of America by
Information Science Reference (an imprint of IGI Global)
701 E. Chocolate Avenue, Suite 200
Hershey PA 17033
Tel: 717-533-8845
Fax: 717-533-8661
E-mail: cust@igi-global.com
Web site: http://www.igi-global.com/reference

and in the United Kingdom by
Information Science Reference (an imprint of IGI Global)
3 Henrietta Street
Covent Garden
London WC2E 8LU
Tel: 44 20 7240 0856
Fax: 44 20 7379 0609
Web site: http://www.eurospanonline.com

Library of Congress Cataloging-in-Publication Data

Felici, Giovanni.

Mathematical methods for knowledge discovery and data mining / Giovanni Felici & Carlo Vercellis, editors.

p. cm.

Summary: "This book focuses on the mathematical models and methods that support most data mining applications and solution techniques, covering such topics as association rules; Bayesian methods; data visualization; kernel methods; neural networks; text, speech, and image recognition; an invaluable resource for scholars and practitioners in the fields of biomedicine, engineering, finance, manufacturing, marketing, performance measurement, and telecommunications"--Provided by publisher.

Includes bibliographical references and index.

ISBN 978-1-59904-528-3 (hardcover) -- ISBN 978-1-59904-530-6 (ebook)

1. Data mining. 2. Data mining--Mathematical models. 3. Knowledge acquisition (Expert systems) I. Felici, Giovanni. II. Vercellis, Carlo. III. Title.

QA76.9.D343F46 2007

006.3'12--dc22

2007022228

British Cataloguing in Publication Data
A Cataloguing in Publication record for this book is available from the British Library.

Table of Contents

Foreword xii

Preface xiv

Acknowledgment xx

Chapter I
Discretization of Rational Data / *Jonathan Mugan and Klaus Truemper* 1

Chapter II
Vector DNF for Datasets Classifications: Application to the Financial Timing
Decision Problem / *Massimo Liquori and Andrea Scozzari* 24

Chapter III
Reducing a Class of Machine Learning Algorithms to Logical Commonsense
Reasoning Operations / *Xenia Naidenova* 41

Chapter IV
The Analysis of Service Quality Through Stated Preference Models and
Rule-Based Classification / *Giovanni Felici and Valerio Gatta* 65

Chapter V
Support Vector Machines for Business Applications / *Brian C. Lovell and Christian J. Walder* 82

Chapter VI
Kernel Width Selection for SVM Classification: A Meta-Learning Approach /
Shawkat Ali and Kate A. Smith 101

Chapter VII
Protein Folding Classification Through Multicategory Discrete SVM /
Carlotta Orsenigo and Carlo Vercellis 116

Chapter VIII
Hierarchical Profiling, Scoring, and Applications in Bioinformatics / *Li Liao* 130

Chapter IX
Hierarchical Clustering Using Evolutionary Algorithms / *Monica Chiş* 146

Chapter X
Exploratory Time Series Data Mining by Genetic Clustering / *T. Warren Liao* 157

Chapter XI
Development of Control Signatures with a Hybrid Data Mining and Genetic Algorithm /
Alex Burns, Shital Shah, and Andrew Kusiak 179

Chapter XII
Bayesian Belief Networks for Data Cleaning / *Enrico Fagiuoli, Sara Omerino,*
and Fabio Stella 204

Chapter XIII
A Comparison of Revision Schemes for Cleaning Labeling Noise /*Chuck P. Lam*
and David G. Stork 220

Chapter XIV
Improving Web Clickstream Analysis: Markov Chains Models and Genmax Algorithms /
Paolo Baldini and Paolo Giudici 233

Chapter XV
Advanced Data Mining and Visualization Techniques with Probabilistic Principal Surfaces:
Applications to Astronomy and Genetics / *Antonino Staiano, Lara De Vinco,*
Giuseppe Longo, and Roberto Tagliaferri 244

Chapter XVI
Spatial Navigation Assistance System for Large Virtual Environments:
The Data Mining Approach / *Mehmed Kantardzic, Pedram Sadeghian, and Walaa M. Sheta* 265

Chapter XVII
Using Grids for Distributed Knowledge Discovery / *Antonio Congiusta, Domenico Talia,*
and Paolo Trunfio 284

Chapter XVIII
Fuzzy Miner: Extracting Fuzzy Rules from Numerical Patterns /
Nikos Pelekis, Babis Theodoulidis, Ioannis Kopanakis, and Yannis Theodoridis 299

Chapter XIX
Routing Attribute Data Mining Based on Rough Set Theory / *Yanbing Liu, Menghao Wang,*
and Jong Tang.. 322

Compilation of References 338

About the Contributors 361

Index 368

Detailed Table of Contents

Foreword xii

Preface xiv

Acknowledgment xx

Chapter I

Discretization of Rational Data / *Jonathan Mugan and Klaus Truemper* 1

Frequently, one wants to extend the use of a classification method that, in principle, requires records with True/False values, so that records with rational numbers can be processed. In such cases, the rational numbers must first be replaced by True/False values before the method may be applied. In other cases, a classification method, in principle, can process records with rational numbers directly, but replacement by True/False values improves the performance of the method. The replacement process is usually called discretization or binarization. This chapter describes a recursive discretization process called Cutpoint. The key step of Cutpoint detects points where classification patterns change abruptly. The chapter includes computational results where Cutpoint is compared with entropy-based methods that, to date, have been found to be the best discretization schemes. The results indicate that Cutpoint is preferred by certain classification schemes, while entropy-based methods are better for other classification methods. Thus, one may view Cutpoint to be an additional discretization tool that one may want to consider.

Chapter II

Vector DNF for Datasets Classifications: Application to the Financial Timing Decision Problem / *Massimo Liquori and Andrea Scozzari* 24

Traditional classification approaches consider a dataset formed by an archive of observations classified as positive or negative according to a binary classification rule. In this chapter, we consider the financial timing decision problem, which is the problem of deciding the time when it is profitable for the investor to buy shares or to sell shares or to wait in the stock exchange market. The decision is based on classifying a dataset of observations, represented by a vector containing the values of some financial numerical attributes, according to a ternary classification rule. We propose a new technique based on partially defined vector Boolean functions. We test our technique on different time series of the Mibtel stock exchange market in Italy, and we show that it provides a high classification accuracy, as well as wide applicability for other classification problems where a classification in three or more classes is needed.

Chapter III

Reducing a Class of Machine Learning Algorithms to Logical Commonsense
Reasoning Operations / *Xenia Naidenova* 41

The purpose of this chapter is to demonstrate the possibility of transforming a large class of machine-learning algorithms into commonsense reasoning processes based on using well-known deduction and induction logical rules. The concept of a good classification (diagnostic) test for a given set of positive examples lies in the basis of our approach to the machine-learning problems. The task of inferring all good diagnostic tests is formulated as searching the best approximations of a given classification (a partitioning) on a given set of examples. The lattice theory is used as a mathematical language for constructing good classification tests. The algorithms of good tests inference are decomposed into subtasks and operations that are in accordance with main human commonsense reasoning rules.

Chapter IV

The Analysis of Service Quality Through Stated Preference Models and
Rule-Based Classification / *Giovanni Felici and Valerio Gatta* 65

The analysis of quality of services is an important issue for the planning and the management of many businesses. The ability to address the demands and the relevant needs of the customers of a given service is crucial to determine its success in a competitive environment. Many quantitative tools in the areas of statistics and mathematical modeling have been designed and applied to serve this purpose. Here we consider an application of a well-established statistical technique, the stated preference models (SP), to identify, from a sample of customers, significant weights to attribute to different aspects of the service provided; such aspects may additively compose an overall satisfaction index. In addition, such a weighting system is applied to a larger set of customers, and a comparison is made between the overall satisfaction identified by the SP index and the overall satisfaction directly declared by the customers. Such comparison is performed by two rule-based classification systems, decision trees and the logic data miner Lsquare. The results of these two tools help in identifying the differences between the two measurements from the structural point of view, and provide an improved interpretation of the results. The application considered is related to the customers of a large Italian airport.

Chapter V

Support Vector Machines for Business Applications / *Brian C. Lovell and Christian J. Walder* 82

This chapter discusses the use of support vector machines (SVM) for business applications. It provides a brief historical background on inductive learning and pattern recognition, and then an intuitive motivation for SVM methods. The method is compared to other approaches, and the tools and background theory required to successfully apply SVM to business applications are introduced. The authors hope that the chapter will help practitioners to understand when the SVM should be the method of choice, as well as how to achieve good results in minimal time.

Chapter VI
Kernel Width Selection for SVM Classification: A Meta-Learning /
Shawkat Ali and Kate A. Smith 101

The most critical component of kernel-based learning algorithms is the choice of an appropriate kernel and its optimal parameters. In this chapter, we propose a rule-based metalearning approach for automatic radial basis function (rbf) kernel, and its parameter selection for support vector machine (SVM) classification. First, the best parameter selection is considered on the basis of prior information of the data, with the help of maximum likelihood (ML) method and Nelder-Mead (N-M) simplex method. Then the new rule-based metalearning approach is constructed and tested on different sizes of 112 datasets with binary class, as well as multiclass classification problems. We observe that our rule-based methodology provides significant improvement of computational time, as well as accuracy in some specific cases.

Chapter VII
Protein Folding Classification Through Multicategory Discrete SVM /
Carlotta Orsenigo and Carlo Vercellis 116

In the context of biolife science, predicting the folding structure of a protein plays an important role for investigating its function and discovering new drugs. Protein folding recognition can be naturally cast in the form of a multicategory classification problem, which appears challenging due to the high number of folds classes. Thus, in the last decade, several supervised learning methods have been applied in order to discriminate between proteins characterized by different folds. Recently, discrete support vector machines have been introduced as an effective alternative to traditional support vector machines. Discrete SVM have been shown to outperform other competing classification techniques both on binary and multicategory benchmark datasets. In this chapter, we adopt discrete SVM for protein folding classification. Computational tests performed on benchmark datasets empirically support the effectiveness of discrete SVM, which are able to achieve the highest prediction accuracy.

Chapter VIII
Hierarchical Profiling, Scoring, and Applications in Bioinformatics / *Li Liao* 130

Recently, clustering and classification methods have seen many applications in bioinformatics. Some are simply straightforward applications of existing techniques, but most have been adapted to cope with peculiar features of the biological data. Many biological data take a form of vectors, whose components correspond to attributes characterizing the biological entities being studied. Comparing these vectors, a.k.a. profiles, is a crucial step for most clustering and classification methods. We review the recent developments related to hierarchical profiling where the attributes are not independent, but rather are correlated in a hierarchy. Hierarchical profiling arises in a wide range of bioinformatics problems, including protein homology detection, protein family classification, and metabolic pathway clustering. We discuss in detail several clustering and classification methods where hierarchical correlations are tackled with effective and efficient ways, by incorporation of domain specific knowledge. Relations to other statistical learning methods and more potential applications are also discussed.

Chapter IX

Hierarchical Clustering Using Evolutionary Algorithms / *Monica Chiş* 146

Clustering is an important technique used in discovering some inherent structure present in data. The purpose of cluster analysis is to partition a given data set into a number of groups such that objects in a particular cluster are more similar to each other than objects in different clusters. Hierarchical clustering refers to the formation of a recursive clustering of the data points: a partition into many clusters, each of which is itself hierarchically clustered. Hierarchical structures solve many problems in a large area of interests. In this chapter, a new evolutionary algorithm for detecting the hierarchical structure of an input data set is proposed. The method could be very useful in economy, market segmentation, management, biology taxonomy, and other domains. A new linear representation of the cluster structure within the data set is proposed. An evolutionary algorithm evolves a population of clustering hierarchies. Proposed algorithm uses mutation and crossover as (search) variation operators. The final goal is to present a data clustering representation to quickly find a hierarchical clustering structure.

Chapter X

Exploratory Time Series Data Mining by Genetic Clustering / *T. Warren Liao* 157

In this chapter, we present genetic-algorithm (GA)-based methods developed for clustering univariate time series with equal or unequal length as an exploratory step of data mining. These methods basically implement the k-medoids algorithm. Each chromosome encodes, in binary, the data objects serving as the k-medoids. To compare their performance, both fixed-parameter and adaptive GAs were used. We first employed the synthetic control chart data set to investigate the performance of three fitness functions, two distance measures, and other GA parameters such as population size, crossover rate, and mutation rate. Two more sets of time series with or without known number of clusters were also experimented: one is the cylinder-bell-funnel data and the other is the novel battle simulation data. The clustering results are presented and discussed.

Chapter XI

Development of Control Signatures with a Hybrid Data Mining and Genetic
Algorithm Approach / *Alex Burns, Shital Shah, and Andrew Kusiak* 179

This chapter presents a hybrid approach that integrates a genetic algorithm (GA) and data mining to produce control signatures. The control signatures define the best parameter intervals leading to a desired outcome. This hybrid method integrates multiple rule sets generated by a data-mining algorithm with the fitness function of a GA. The solutions of the GA represent intersections among rules providing tight parameter bounds. The integration of intuitive rules provides an explanation for each generated control setting, and it provides insights into the decision-making process. The ability to analyze parameter trends and the feasible solutions generated by the GA with respect to the outcomes is another benefit of the proposed hybrid method. The presented approach for deriving control signatures is applicable to various domains, such as energy, medical protocols, manufacturing, airline operations, customer service, and so on. Control signatures were developed and tested for control of a power-plant boiler. These signatures discovered insightful relationships among parameters. The results and benefits of the proposed method for the power-plant boiler are discussed in the chapter.

Chapter XII
Bayesian Belief Networks for Data Cleaning / *Enrico Fagiuoli, Sara Omerino,*
and Fabio Stella 204

The importance of data cleaning and data quality is becoming increasingly clear, as evidenced by the surge in software, tools, consulting companies, and seminars addressing data quality issues. In this contribution, the authors present and describe how Bayesian computational techniques can be exploited for data-cleaning purposes to the extent of reducing the time to clean and understand the data. The proposed approach relies on the computational device named Bayesian belief network, which is a general statistical model that allows the efficient description and treatment of joint probability distributions. This work describes the conceptual framework that maps the Bayesian belief network computational device to some of the most difficult tasks in data cleaning, namely imputing missing values, completing truncated datasets, and outliers detection. The proposed framework is described and supported by a set of numerical experiments performed by exploiting the Bayesian belief network programming suite named HUGIN.

Chapter XIII
A Comparison of Revision Schemes for Cleaning Labeling Noise /
Chuck P. Lam and David G. Stork 220

Data quality is an important factor in building effective classifiers. One way to improve data quality is by cleaning labeling noise. Label cleaning can be divided into two stages. The first stage identifies samples with suspicious labels. The second stage processes the suspicious samples using some revision scheme. This chapter examines three such revision schemes: (1) removal of the suspicious samples, (2) automatic replacement of the suspicious labels to what the machine believes to be correct, and (3) escalation of the suspicious samples to a human supervisor for relabeling. Experimental and theoretical analyses show that only escalation is effective when the original labeling noise is very large or very small. Furthermore, for a wide range of situations, removal is better than automatic replacement.

Chapter XIV
Improving Web Clickstream Analysis: Markov Chains Models and Genmax Algorithms /
Paolo Baldini and Paolo Giudici 233

Every time a user links up to a Web site, the server keeps track of all the transactions accomplished, in a log file. What is captured is the "click flow" (clickstream) of the mouse and the keys used by the user during the navigation inside the site. Usually every click of the mouse corresponds to the viewing of a Web page. The objective of this chapter is to show how Web clickstream data can be used to understand the most likely paths of navigation in a Web site, with the aim of predicting, possibly online, which pages will be seen, having seen a specific path of other pages before. Such analysis can be very useful to understand, for instance, what is the probability of seeing a page of interest (such as the buying page in an e-commerce site) coming from another page. Or what is the probability of entering (or exiting) the Web site from any particular page. From a methodological viewpoint, we present two main research contributions. On one hand we show how to improve the efficiency of the Apriori algorithm; on the other hand we show how Markov chain models can be usefully developed and implemented for Web usage mining. In both cases we compare the results obtained with classical association rules algorithms and models.

Chapter XV
Advanced Data Mining and Visualization Techniques with Probabilistic Principal Surfaces:
Applications to Astronomy and Genetics / *Antonino Staiano, Lara De Vinco,*
Giuseppe Longo, and Roberto Tagliaferri 244

Probabilistic principal surfaces (PPS) is a nonlinear latent variable model with very powerful visualization and classification capabilities that seem to be able to overcome most of the shortcomings of other neural tools. PPS builds a probability density function of a given set of patterns lying in a high-dimensional space that can be expressed in terms of a fixed number of latent variables lying in a latent Q-dimensional space. Usually, the Q-space is either two- or three-dimensional and thus, the density function can be used to visualize the data within it. The case in which $Q = 3$ allows to project the patterns on a spherical manifold, which turns out to be optimal when dealing with sparse data. PPS may also be arranged in ensembles to tackle complex classification tasks. As template cases, we discuss the application of PPS to two real- world data sets from astronomy and genetics.

Chapter XVI
Spatial Navigation Assistance System for Large Virtual Environments:
The Data Mining Approach / *Mehmed Kantardzic, Pedram Sadeghian, and Walaa M. Sheta* 265

Advances in computing techniques, as well as the reduction in the cost of technology have made possible the viability and spread of large virtual environments. However, efficient navigation within these environments remains problematic for novice users. Novice users often report being lost, disorientated, and lacking the spatial knowledge to make appropriate decisions concerning navigation tasks. In this chapter, we propose the frequent wayfinding-sequence (FWS) methodology to mine the sequences representing the routes taken by experienced users of a virtual environment in order to derive informative navigation models. The models are used to build a navigation assistance interface. We conducted several experiments using our methodology in simulated virtual environments. The results indicate that our approach is efficient in extracting and formalizing recommend routes of travel from the navigation data of previous users of large virtual environments.

Chapter XVII
Using Grids for Distributed Knowledge Discovery / *Antonio Congiusta, Domenico Talia,*
and Paolo Trunfio 284

Knowledge discovery is a compute- and data-intensive process that allows for finding patterns, trends, and models in large datasets. The grid can be effectively exploited for deploying knowledge discovery applications because of the high performance it can offer, and its distributed infrastructure. For effective use of grids in knowledge discovery, the development of middleware is critical to support data management, data transfer, data mining, and knowledge representation. To such purpose, we designed the knowledge grid, a high-level environment providing for grid-based knowledge discovery tools and services. Such services allow users to create and manage complex knowledge discovery applications, composed as workflows, that integrate data sources and data-mining tools provided as distributed grid services. This chapter describes the knowledge grid architecture, and describes how its components can be used to design and implement distributed knowledge discovery applications. Then, the chapter describes how the knowledge grid services can be made accessible using the open grid services architecture (OGSA) model.

Chapter XVIII

Fuzzy Miner: Extracting Fuzzy Rules from Numerical Patterns / *Nikos Pelekis,*
Babis Theodoulidis, Ioannis Kopanakis, and Yannis Theodoridis... 299

We study the problem of classification as this is presented in the context of data mining. Among the various approaches that are investigated, we focus on the use of fuzzy logic for pattern classification, due to its close relation to human thinking. More specifically, this chapter presents a heuristic fuzzy method for the classification of numerical data, followed by the design and the implementation of its corresponding tool (fuzzy miner). The initial idea comes from the fact that fuzzy systems are universal approximators of any real continuous function. Such an approximation method coming from the domain of fuzzy control is appropriately adjusted into pattern classification, and an « adaptive » procedure is proposed for deriving highly accurate linguistic if-then rules. Extensive simulation tests are performed to demonstrate the performance of fuzzy miner, while a comparison with a neuro-fuzzy classifier of the area is taking place in order to contradict the methodologies and the corresponding outcomes. Finally, new research directions in the context of fuzzy miner are identified, and ideas for its improvement are formulated.

Chapter XIX

Routing Attribute Data Mining Based on Rough Set Theory / *Yanbing Liu, Menghao Wang,*
and Jong Tang... 322

QOSPF (Quality of Service Open Shortest Path First) based on QoS routing has been recognized as a missing piece in the evolution of QoS-based services in the Internet. Data mining has emerged as a tool for data analysis, discovery of new information, and autonomous decision making. This chapter focuses on routing algorithms and their applications for computing QoS routes in OSPF protocol. The proposed approach is based on a data-mining approach using rough set theory, for which the attribute-value system about links of networks is created from network topology. Rough set theory offers a knowledge discovery approach to extracting routing-decisions from attribute set. The extracted rules can then be used to select significant routing-attributes and make routing-selections in routers. A case study is conducted to demonstrate that rough set theory is effective in finding the most significant attribute set. It is shown that the algorithm based on data mining and rough set offers a promising approach to the attribute-selection problem in Internet routing.

Compilation of References 338

About the Contributors 361

Index 368

Foreword

The importance of knowledge discovery and data mining is evident by the great plethora of books and papers dedicated to this subject. Such methods are finding applications in almost any area of human endeavor. This includes applications in engineering, science, business, medicine, humanities, just to name a few. At the same time, however, there is a great confusion about the development and application of such methods. The main reason for this situation is that many, if not most, of the books examine issues on data mining in a narrow manner. Very few books study issues from the mathematical/algorithmic and also the applications point of view simultaneously. Even fewer books present a comprehensive view of all the critical issues involved with the development and application of such methods to many real-life domains. The present book, edited by two world-renowned scholars, Drs. Giovanni Felici and Carlo Vercellis, is a bright example of the most valuable books in this fast emerging field. The emphasis of this book on the mathematical aspects of knowledge discovery and data-mining methods makes the presentations scientifically sound and easy to understand in depth.

The 19 chapters of this book have been written by a number of distinguished scholars, from all over the world, who discuss the most critical subjects in this area. The book starts by discussing an important first step for any application of such methods, that is, how to discretize the data. This step is essential as many methods use binary data, while real-life applications may be associated with nonbinary data. If the analyst is not careful at this step, then it is possible to end up with too many nonrelevant variables that generate computational problems associated with highly dimensional data. A related step is that of cleaning the data before they are used to extract the pertinent models. As before, if this step is not done properly, the validity of the final results may be in jeopardy. Another interesting topic discussed in this book is the development of sophisticated visualization techniques, which are presented in relation with many diverse domains, ranging from astronomy to genetics. The successful application of the proposed visualization techniques to these two highly demanding application areas witnesses the high potential of these methods for a wide spectrum of applications. The high volumes of log data, produced by recording the way people surf the Web, provides an exciting opportunity, amid with interesting algorithmic challenges, for knowledge discovery and data-mining methods. This fascinated topic is also discussed here. Another very interesting subject discussed is how to mine data that come from virtual environments that involve some kind of spatial navigation. Such studies involve the analysis of sequences of routes of actions. A highly promising direction of research seems to be based on the development of methods that attempt to combine characteristics of various approaches. Such methods are known as hybrids, and an interesting development of a new hybrid approach and its applications are presented as well.

No book in this area would be complete without the discussion of logic-based methods that offer some unique algorithmic and application advantages. The relevant discussions are done by some of the most knowledgeable world-renowned scholars on this subject; logic methods are also applied to the analysis of financial data. The potential of using grids for distributed approaches and also parallelism is explained too. Another prominent topic is the use of clustering approaches, which are discussed by developing some specialized evolutionary approaches. Clustering is also used in one of the most promising application areas for the future, the analysis of time series of data. Applications can be found in many domains as one realizes that systems or phenomena of interest usually generate data over time. In such settings, data from one point of time are somehow related to the data of the next point of time. Again, this very fascinating problem is discussed in great depth, and in an easy-to-understand manner by a distinguished expert in this fast-growing field. An extensive treatment of the classification technique, known as support vector machines, is provided by three chapters of the book; altogether with a complete treatment of the main theory of this method and of the related kernel function theory, the new extension of discrete support vector machines is described and applied to bioinformatics. This type of data is also the topic of other applications described in the book. The picture is completed with the description of data-mining approaches for service quality measurement, of fuzzy set and rough set theory applied in different contexts, and of other industrial applications of data mining.

It is quite clear that this book is very valuable to all practitioners and researchers working on different fields but unified by the need to analyze their voluminous and complex data. Therefore, it is strongly recommended to anyone who has an interest in data mining. Furthermore, it is hoped that others will follow the example of this book and present more studies that combine algorithmic developments and applications in the way this edited book by Drs. Felici and Vercellis does so successfully.

Evangelos Triantaphyllou, PhD
Professor
Department of Computer Science
Louisiana State University
Baton Rouge, LA 70803 USA

Preface

The idea of this book was conceived in June 2004, when a small group of researchers in the data-mining field gathered on the shores of the lake of Como, in Italy, to attend a focused conference–MML, *Mathematical Methods for Learning* 2004—having the objective of fostering the interaction among scholars from different countries and with different scientific backgrounds, sharing their research interests in data mining and knowledge discovery. As one of the side effects of that meeting, the conference organizers took on the exciting task of editing high quality scientific publications, where the main contributions presented at the MML conference could find an appropriate place, one next to the other, as they fruitfully did within the conference sessions. Some of the papers presented in Como, sharing a focus on mathematical optimization methods for data mining, found their place in a special issue of the international journal *Computer Optimization and Applications* (*COAP, 38*(1), 2007). Another large group of papers constituted the most appropriate building blocks for an edited book that would span a vast area of data-mining methods and applications, showing, on one hand, the relevance of mathematical methods and algorithms aimed at extracting knowledge from data, and on the other hand, how wide the application domains of data mining are. Shortly later, such project found interest and support by IGI Global, a dynamic publisher very active in promoting research-oriented publications in technological and advanced fields of knowledge. We eventually managed to finalize all the chapters, and moreover, enriched the book with additional research work that, although not presented at the MML conference, appear to have a strong relevance within the scope of the book. Most of the chapters have evolved since they were presented in 2004, and authors had the opportunity to update their work with additional results until the beginning of 2007.

The Motivations of Data Mining

The interest in data mining of researchers and practitioners with different backgrounds has increased steadily year after year. This growth is due to several reasons.

First, data mining plays today a fundamental role in analyzing and understanding the vast amount of information collected by business, government, and scientific applications. The ability to analyze large bodies of data and extract from them relevant knowledge has become a valuable service for most organizations that operate in the highly globalized and competitive business arena. The technical skills required to operate and put to use data-mining techniques are now appreciated, and often required, by the business intelligence units of financial institutions, government agencies, telecommunication companies, service providers, retailers, and distribution operators.

A second reason is to be found in the excellent and constantly improving quality of the methods and tools that are being developed in this field. Advanced mathematical models, state-of-the-art algorithmic techniques, and efficient data management systems, combined with a decreasing cost of computational power and computer memory, are now able to support data analysts with methodologies and tools that were not available a few years ago. Furthermore, such instruments are often available at low cost and with easy-to-use interfaces, integrated into well-established data management systems.

A third reason that is not to be overlooked is connected with the role that data-mining methods are playing in providing support to basic research in many scientific areas. To mention an example, biology and genetics are currently enjoying the results of the application of advanced mining techniques that allow discovery of valuable facts in complex data gathered from experiments in vitro.

Finally, we wish to mention the impulse to methodological research that has been given in many areas by the open problems posed by data-mining applications. The learning and classification problems coming from real-life problems have been exploited through many mathematical theories under different formalizations, and theoretical results of unusual relevance have been reached in optimization theory, computer science, and statistics, also thanks to the many new and stimulating problems.

Data Mining as a Practical Science

Data mining is located at the crossing of different disciplines. Its roots are to be found in the data analysis techniques that were originally the main object of the study of statistics. The fundamental ideas at the basis of estimation theory, classification, clustering, sampling theory, are indeed still one of the major ingredients of data mining. But other methods and techniques have been added to the toolbox of the data analyst, extending the limits of the classical parametric statistics with more complex models, reaching their maturity with the actual state of knowledge on decision trees, neural networks, support vector machines, just to mention a few. In addition, the need to organize and manage large bodies of data has required the deployment of computer science techniques for database management, query optimization, optimal coding of algorithms, and other tasks devoted to the storing of information in the memory of computers and to the efficient execution of algorithms.

A common trademark of the modern approaches is the formalization of estimation and classification problems arising in data mining as mathematical optimization problems, and the use of consistent algorithmic techniques to determine optimal solutions for these problems. Such methodological framework has been strongly supported by applied mathematics and operations research (OR), a scientific discipline characterized by a deep integration of mathematical theory and practical problems. A significant evidence of the role of OR in data mining is the contribution that nonlinear and integer optimization methods have given to the solution of the error minimization functions that need to be optimized to train neural networks and support vector machines. Analogously, integer programming and combinatorial optimization have been largely used to solve problems arising in the identification of synthetic rule-based classification models and in the selection of optimal subsets of features in large datasets.

Despite its strong methodological characterization, data mining cannot be successfully applied without a deep understanding of the semantic of each specific problem, which often requires the customization of existing methods or the development of ad hoc techniques, partially based on already existing algorithms. To some extent, the real challenge that the data mining practitioner has to face is the selection, among many different methods and approaches, of the one that best serves the scope of the task considered, often assessing a compromise between the complexity of the chosen model and its generalization capability.

The Contribution of this Edited Book

This book aims to provide a rich collection of current research on a broad array of topics in data mining, ranging from recent theoretical advancements in the field to relevant applications in diverse domains. Future directions and trends in data mining are also identified in most chapters.

Therefore, this volume should be an excellent guide to researchers, practitioners, and students. Its audience is represented by the research community; business executives and consultants; and senior students in the fields of data mining, information and knowledge creation, optimization, statistics, and computer science.

A Guided Tour of the Chapters

The book is composed of 19 chapters. Each one is authored by a different group of scientists, treats one of the many different theoretical or practical aspects of data mining, and is self contained with respect to the treated subject.

The first four chapters deal, to different degrees, with data-mining problems in logic setting, where the main purpose is to extract rules in logic format from the available data.

In particular, **Chapter I** is written by *Johnathan Mugan* and *Klaus Truemper*, and describes a sophisticated and complete technique to transform a set of data represented in various formats by means of an extended set of logic variables. Such task, often referred to as discretization, or binarization, is a key step in the application of logic-based classification methods to data that is described by rational or nominal variables. The chapter extends the notion of rational variables with the definition of set variables, for example, variables that are represented by their membership functions to one or more sets. The method described is characterized by the fact that the set of logic variables extracted is compact, but strongly aimed at the task of classifying, with high precision, the available data with respect to a given binary target variable. The algorithm that implements the ideas described in the chapter has been implemented and integrated into the logic data mining software Lsquare, made available by the authors as open source.

Chapter II is written by *Massimo Liguori* and *Andrea Scozzari*. Here the subject is the use of another well-known logic data mining technique, the logical analysis of data (LAD), originally developed at the University of Rutgers by the research team led by Peter Hammer. The authors propose an interesting use of this method to treat logic classification where the target variable is of ternary nature (i.e., it can assume one of three possible values). Even more interesting is the application for which the method has been developed: the financial timing decision problem, namely the problem of deciding when to buy and when to sell a given stock to maximize the profit of the trading operations. The results presented in this chapter testify how logic methods can give a significant contribution in a field where classical statistics has always played the main role.

Chapter III, authored by *Xenia Naidenova*, brings to the readers' attention several interesting theoretical aspects of logic deduction and induction that find relevant application in the construction of machine-learning algorithms. The chapter treats extensively the many details connected with this topic, and enlightens many results with simple examples. The author adopts the lattice theory as the basic mathematical tool, and succeeds in proposing a sound integration of inductive and deductive reasoning for learning implicative logic rules. The results described are the basis for the implementation of an algorithm that efficiently infers good maximally redundant tests.

In **Chapter IV**, *Giovanni Felici* and *Valerio Gatta* describe a study where the results of a stated preference model for measuring quality of service is combined with logic-based data mining to gain deeper insight in the system of preferences expressed by the customers of a large airport. The data-mining methods considered are decision trees and the logic miner Lsquare. The results are presented in the form of a set of rules that enables one to understand the similarities and the differences in two different methods to compute a quality of service index.

The topics of the following three chapters evolve around the concept of support vector machines (SVM), a mathematical method for classification and regression emerged in the last decade from statistical learning theory, which quickly attained remarkable results in many applications. SVM are based on optimization methods, particularly in the field of nonlinear programming, and are a vivid example of the contributions that can be given to data mining by state-of-the-art theoretical research in mathematical optimization.

In **Chapter V**, *Brian C. Lovell* and *Christian J. Walder* provide a rich overview of SVM in the context of data mining for business applications. They describe, with high clarity, the basic steps in SVM theory, and then integrate the chapter with several practical considerations on the use of this class of methods, comparing it with other learning approaches in the context of real-life applications.

An important role in SVM is played by kernel functions, which provide an implicit transformation of the representation of the original space of data into a high dimensional space of features. By means of such transformations, SVM can efficiently determine linear transformations in the feature space that correspond to nonlinear separations into the original space.

The identification of the right kernel function is the topic of **Chapter VI**, written by *Shawkat Ali* and *Kate A. Smith*, where they describe the application of a metalearning approach to optimally estimate the parameters that identify the kernel function before SVM is applied. The chapter highlights clearly the role of parameter estimation in the use of learning models, and discusses how the estimation procedure should be able to adapt to the specific dataset under analysis. The experimental analysis provides tests on both binary and multicategory classification problems.

An interesting evolution of SVM is represented by discrete support vector machines, proposed in the last few years by *Carlotta Orsenigo* and *Carlo Vercellis*, authors of **Chapter VII**. According to statistical learning theory, discrete SVM directly face the minimization of the misclassification rate, within the risk functional, instead of replacing it with the misclassification distance as traditional SVM. The problem is then modeled as a mixed-integer programming problem. The method, already successful in other applications, is extended and applied here to protein folding, a very challenging task in multicategory classification. The experiments performed by the authors on benchmark datasets show that the proposed method achieves the highest accuracy in comparison to other techniques.

The use of data-mining methods to extract knowledge from large databases in genetic and biomedical applications is increasing at a fast pace, and **Chapter VIII**, written by *Li Liao*, deals with this topic. Often the data in this context is based on vectors of extremely large dimensions, and specific techniques must be deployed to obtain successful results. Li Liao tackles several of the specific problems related with handling biomedical data, in particular those related with data described by attributes that are correlated with each other and are organized in a hierarchical structure. Clustering and classification methods that exploit the hierarchies in data are considered and compared with statistical learning methods.

Chapters nine and ten both deal with clustering, a fundamental problem in nonsupervised learning. In **Chapter IX**, *Monica Chiş* discusses hierarchical clustering, where the clusters are obtained by recursively separating the data into groups of similar objects. The methods investigated belong to the family of genetic algorithms, where an initial population of chromosomes, corresponding to potential clusters, is evolved at each iteration, generating new chromosomes with the objective of minimizing a fitness function. The genetic operators adopted here are standard mutation and crossover.

Evolving on these concepts, *T. Warren Liao* presents, in **Chapter X**, a method based on genetic algorithms to cluster univariate time series. The study of time series is indeed a very central topic in data analysis, and is often overlooked in standard data-mining applications, where the main attention is addressed to multivariate data. Time series, on the other hand, present several complex aspects linked to autocorrelation and lag parameters, and surely can benefit by the use of the new methods developed in the area of data mining. Using the method of the k-medoids, the author compares the performances of three fitness functions, two distance measures, and other parameters that characterize the genetic algorithms considered. The chapter presents several experiments on data derived from cylinder-bell-funnel data and battle simulation data.

Chapter XI is a sound example of how advanced data-mining techniques can provide relevant information in production systems. *Alex Burns*, *Shital Shah*, and *Andrew Kusiak* describe the implementation of a method that integrates genetic algorithms and data mining. The results of a rule-based data-mining algorithm are evaluated and scored using a fitness function, and the related methods made available in the context of genetic algorithms. Here again we find a strong connection between data analysis and optimization techniques, and we see how certain decision problems can be successfully solved building ad hoc procedures, where methodologies and techniques from different backgrounds are deployed. The authors describe an application of the method to a power-plant boiler and highlight the contribution given to the production process.

Chapter XII is written by *Enrico Fagiuoli*, *Sara Omerino*, and *Fabio Stella*. It is an interesting work that shows how complex models derived from classical statistical techniques can play an important role in the data treatment process. The chapter describes the use of Bayesian belief networks to perform data cleaning, a relevant problem in most data-mining applications where the information available is obtained with noisy, incomplete, or error-prone procedures. Here Bayesian belief networks are used to instantiate missing values of incomplete records, to complete truncated datasets, and to detect outliers. The effectiveness of the approach is supported by numerical experiments.

Chapter XIII deals with a similar topic, data cleaning. Here, *Chuck P. Lam* and *David G. Stork* describe, in a complete and accurate way, the problem of labeling noise, requiring the identification and treatment of records when some of the labels attached to the records are different from the correct value due to some source of noise present in the data collection process. The chapter depicts the two main problems in cleaning labeling noise: the identification of noise and the consequent revision scheme, through removal, replacement, or escalation to human supervision. In particular, the authors examine the k-nearest neighbor method to solve the identification problem, while they use probabilistic arguments to evaluate the alternative revision schemes. The public domain UCI repository is used as a source of datasets where the proposed methods are tested.

In **Chapter XIV** another tool originated in the statistical and stochastic processes environment is used to solve a relevant mining problem, clickstream analysis, which is attracting a growing attention. The problem is generated by the need to investigate the log files produced when users visit a Web site. These log files report the sequence of steps in navigation (clicks) made by the Web users. These applications can be useful for designing Web sites and for related business-oriented analysis. The authors of the chapter, *Paolo Baldini* and *Paolo Giudici*, propose to use Markov chain models to investigate the structure of the most likely navigation path in a Web site, with the objective of predicting the next step made by a Web user, based on the previous ones.

Antonino Staiano, *Lara De Vinco*, *Giuseppe Longo*, and *Roberto Tagliaferro* are the authors of **Chapter XV**. Here the topic is the visualization of complex and multidimensional data for exploration and classification purposes. The method used is based on probabilistic principal surfaces: by means of a density function in the original space, data are projected into a reduced space defined by a set of latent variables. A special case arises when the number of latent variables is equal to three and the projected space is a spherical manifold, particularly indicated to represent sparse data. Besides visualization, such reduced spaces can be used to apply classification algorithms for efficiently determining surfaces that separate groups of data belonging to different classes. Applications of the method for data in the astronomy and genetics domains are discussed.

Chapter XVI presents an unconventional application of data-mining techniques that assists spatial navigation in virtual environments. In this setting, users are able to navigate in a three-dimensional virtual space to accomplish a number of tasks. Such navigation may be difficult or inefficient for nonexperienced users, and the application discusses the use of data-mining techniques to extract knowledge from the navigation patterns of expert users, and create good navigation models. Such process is put into action by a navigation interface that implements a frequent wayfinding-sequence method. The authors, *Mehmed Kantardzic*, *Pedram Sadeghian*, and *Walaa M. Sheta*, have run experiments in simulated virtual environments, extensively discussed in the chapter.

Data-mining algorithms can be very demanding from the point of view of computational requirements, such as speed and memory, especially when large datasets are analyzed. One possible solution to deal with the computational burden is the use of parallel and distributed computing. Such an issue is the topic of **Chapter XVII**, written by *Antonio Congiusta*, *Domenico Talia*, and *Paolo Trunfio*, on the use of grid computing for distributed data mining. An integrated architecture that can properly host all the steps of the data analysis process (data management, data transfer, data mining, knowledge representation) has been designed and is presented in the chapter. The components of this data-mining-oriented middleware, termed *knowledge grid*, are described, explaining how these services can be accessed using the standard open grid architecture model.

The last two chapters of the book are devoted to two knowledge extraction methods that have received large attention in the scientific community. They extend the limits of standard machine learning theory, and can be used to build data-mining applications able to deal with unconventional applications, and to provide information in an original format. **Chapter XVIII** is about the use of fuzzy logic. Here *Nikos Pelekis*, *Babis Theodoulidis*, *Ioannis Kopanakis*, and *Yannis Theodoridis* cover the design of a classification heuristic scheme based on fuzzy methods. The performances of the method are analyzed by means of extensive simulated experiments. The topic of **Chapter XIX** is the method of rough sets. This method presents several noticeable features that originally characterize the rules extracted from the data. The interest of this chapter, written by *Yanbing Liu*, *Menghao Wang*, and *Jong Tang*, is also due to the application of the method to analyze and evaluate network topologies in routing problems. The application of data mining techniques in network problems associated with telecommunication problems is novel, and is likely to represent a relevant object of research in the future.

Acknowledgment

The editors want to express their gratitude to all those colleagues who supported, in many different ways, the pleasant effort of editing this book. First and foremost, the 30 authors that contributed to the writing of the book's chapters; they are the actual creators of the qualified research that is described in the book, and they must be especially thanked for the patience exhibited in the editing process. Among them, special thanks go to Dr. Carlotta Orsenigo, who supported, with her careful and patient touch, many of the steps that led this volume to publication. Speaking of patience, we need not to overlook the contribution of the IGI Globals's publishing editor, Kristin Roth, altogether with all the very efficient staff at IGI Global. A final word of thanks is warmly dedicated to all those who supported our editing work without even knowing it: our families and our friends, who were on our side in the hard and the exciting moments.

Giovanni Felici
Carlo Vercellis
Editors

Chapter I
Discretization of Rational Data

Jonathan Mugan
University of Texas at Austin, USA

Klaus Truemper
University of Texas at Dallas, USA

ABSTRACT

Frequently, one wants to extend the use of a classification method that, in principle, requires records with True/False values, so that records with rational numbers can be processed. In such cases, the rational numbers must first be replaced by True/False values before the method may be applied. In other cases, a classification method in principle can process records with rational numbers directly, but replacement by True/False values improves the performance of the method. The replacement process is usually called discretization or binarization. This chapter describes a recursive discretization process called Cutpoint. The key step of Cutpoint detects points where classification patterns change abruptly. The chapter includes computational results, where Cutpoint is compared with entropy-based methods that, to date, have been found to be the best discretization schemes. The results indicate that Cutpoint is preferred by certain classification schemes, while entropy-based methods are better for other classification methods. Thus, one may view Cutpoint to be an additional discretization tool that one may want to consider.

INTRODUCTION

One often desires to apply classification methods that, in principle, require records with *True/False* values to records that, besides *True/False* values, contain rational numbers. For ease of reference, we call rational number entries *rational data* and refer to *True/False* entries as *logic data*. In such situations, a discretization process must first convert the rational data to logic data. Discretization is also desirable in another setting. Here, a classification method in principle can process records with rational numbers directly, but its performance is improved when the rational data are first converted to logic data.

This chapter describes a method called *Cutpoint* for the discretization task, and compares its effectiveness with that of entropy-based methods, which presently are considered to be the best discretization schemes.

Define *nominal data* to be elements or subsets of a given finite set. In Bartnikowski et al. (Bartnikowski, Granberry, Mugan, & Truemper, 2004), an earlier version of Cutpoint is described and used for the transformation of some cases of nominal data to logic data. Specifically, the nominal data are first converted to rational data, which are then transformed to logic data by Cutpoint.

We focus here on the following case. We are given records of two training classes, A and B, that have been randomly selected from two populations **A** and **B**, respectively. We want to derive a classification scheme from the records of A and B. Later, that scheme is to be applied to records of **A** − A and **B** − B.

For the purpose of a simplified discussion in this section, we assume for the moment that the records have no missing entries. That restriction is removed in the next section.

Abrupt Pattern Changes and Cutpoint

Generally, the discretization may be accomplished by the following, well-known approach. One defines, for a given attribute, $k \geq 1$ breakpoints and encodes each rational number of the attribute by k *True/False* values, where the jth value is *True* if the rational number is greater than the jth breakpoint, and is *False* otherwise. The selection of the k breakpoints requires care if the records of **A** − A and **B** − B are to be classified with good accuracy.

A number of techniques for the selection of the breakpoints have been proposed, and later in this section, we provide a review of those methods. Suffice it to say here that the most effective methods to date are based on the notion of entropy. In these methods, the breakpoints are so selected that the rational numbers of a given

attribute can be most compactly classified by a decision tree as coming from A or B. In contrast, Cutpoint is based on a different goal. Recall that the records of the sets A and B are presumed to be random samples of the populations **A** and **B**. Taking a different viewpoint, we may view each record of **A** − A and **B** − B to be a random variation of some record of A or B, respectively. The goal is then to select the breakpoints so that these random variations largely leave the *True/False* values induced by the breakpoints unchanged.

Cutpoint aims for the stated goal by selecting breakpoints, called *markers*, that correspond to certain abrupt changes in classification patterns, as follows. First, for a given attribute, the rational numbers are sorted. Second, each value is labeled as A or B, depending on whether the value comes from a record of A or B, respectively. For the sake of a simplified discussion, we ignore, for the moment, the case where a rational number occurs in both a record of A and a record of B. Third, each entry with label A (resp. B) is assigned a *class value* of 1 (resp. 0). Fourth, Gaussian convolution is applied to the sequence of class values, and the midpoint between two adjacent entries, where the smoothed class values change by the largest amount, is declared to be a marker.

For example, if the original sorted sequence, with class membership in parentheses, is ..., 10.5(A), 11.7(A), 15.0(A), 16.7(A), 19.5(B), 15.2(B), 24.1(B), 30.8(B),..., then the sequence of class values is ..., 1, 1, 1, 1, 0, 0, 0, 0,.... Note the abrupt transition of the subsequence of 1s to the subsequence of 0s. When a Gaussian convolution with small standard deviation σ is performed on the sequence of class values, a sequence of smoothed values results, which exhibits a relatively large change at the point where the original sequence changes from 1s to 0s. If this is the largest change for the entire sequence of smoothed class values, then the original entries 16.7(A) and 19.5(B), which correspond to that change, produce a marker with value (16.7 + 19.5)/2 = 18.1.

Evidently, a large change of the smoothed class values corresponds in the original sorted sequence of entries to a subsequence of rational numbers, mostly from A, followed by a subsequence of numbers, mostly from B, or *vice versa*. We call such a situation an abrupt pattern change. Thus, markers correspond to *abrupt pattern changes*.

We differentiate between two types of abrupt pattern changes. We assume, reasonably, that an abrupt change, produced by all records of the populations $\mathbf{A}$ and $\mathbf{B}$, signals an important change of behavior and thus should be used to define a *True/False* value. The records of the subsets A and B may exhibit portions of such pattern changes. We say that these pattern changes of the records of A and B are of the *first kind*. The records of A and B may also have additional abrupt pattern changes that do not correspond to abrupt pattern changes in the records of the populations $\mathbf{A}$ and $\mathbf{B}$. This is particularly so if A and B are comparatively small subsets of the populations $\mathbf{A}$ and $\mathbf{B}$, as is typically the case. We say that the latter pattern changes are of the *second kind*.

There is another way to view the two kinds of pattern changes. Suppose we replace records r of $A \cup B$ by records $\tilde{r}$ of $(\mathbf{A} - A) \cup (\mathbf{B} - B)$, respectively, where $\tilde{r}$ is similar to r. Then abrupt pattern changes of the first (resp. second) kind produced by the records r likely (resp. unlikely) are abrupt pattern changes produced by the records $\tilde{r}$.

There is a third interpretation. Suppose we extract from the sorted sequence of numerical values just the A and B labels. For example, the above sequence ..., 10.5(A), 11.7(A), 15.0(A), 16.7(A), 19.5(B), 15.2(B), 24.1(B), 30.8(B),... becomes ..., A, A, A, A, B, B, B, B,.... We call this a *label sequence*. Then for an abrupt pattern change of the first (resp. second) kind, the random substitution of records r by records $\tilde{r}$ is unlikely (resp. likely) to change the label sequence.

Cutpoint relies on the third interpretation in an attempt to distinguish between the two kinds of pattern changes, as follows. The method estimates the probability that A or B is selected in label sequences of abrupt pattern changes of the second kind, by assuming $p = |A|/(|A| + |B|)$ (resp. $q = |B|/(|A|+|B|)$) to be the probability for the label A (resp. B) to occur. Then the standard deviation of the Gaussian convolution process is so selected that the following is assured. Suppose there is at least one abrupt pattern change that, according to the probabilities p and q, has low probability and thus is estimated to be of the first kind. Then the largest change of the smoothed class values and the associated marker tends to correspond to one such abrupt pattern change. Informally, one may say that the standard deviation σ is so selected that marker positions corresponding to abrupt pattern changes of the first kind are favored.

Cutpoint has been added to the version of the *Lsquare* method of Truemper (2004), which is based on prior versions of Felici and Truemper (2002) and Felici et al. (Felici, Sun, & Truemper, 2004). Lsquare computes DNF (disjunctive normal form) logic formulas from logic training data. Cutpoint initially determines one marker for each attribute of the original data, as described previously. Let the transformation of A and B via these markers produce sets A' and B'. If A' and B' cannot be separated by logic formulas, then Cutpoint recursively determines additional markers. The Cutpoint/Lsquare combination is so designed that it does not require user specification of parameters or rules, except for a limit on the maximum number of markers for any attribute. In the tests described later, that maximum was fixed to 6.

Computational Results

To date, we have used Cutpoint in conjunction with Lsquare in a variety of projects such as credit rating, video image analysis, and word sense disambiguation. In each case, Cutpoint has proved to be effective and reliable.

We also have compared the performance of Cutpoint with that of two entropy-based methods that differ by the subdivision selection and termination criterion. In one of the methods, the criterion is the clash condition of Cutpoint introduced later. We refer to this method as Entropy CC. For the other method, the criterion is the minimum description length (MDL) principle (Dougherty, Kohavi, & Sahami, 1995). Accordingly, we refer to that method as Entropy MDL.

For the comparison, we applied Cutpoint and the two entropy-based methods to a number of data sets, and processed the resulting logic data by four classification algorithms. The latter schemes were so chosen that classification by decision trees, naive Bayes methods, support vector machines, and learning logic methods were represented. Note that we do not claim that each of the selected classification methods should use discretization as a preprocessing step. But if one decides to use such preprocessing, then the results indicate the following. Entropy MDL is preferred for decision tree methods and support vector machines, while Cutpoint is preferred for learning logic methods and naive Bayes methods. The performance of Entropy CC lies somewhere between that of Cutpoint and Entropy MDL, and thus is always dominated either by Cutpoint or Entropy MDL. These general conclusions are based on average performance results. For specific data sets, preference can be quite different. Thus, if one needs highest possible accuracy for a given situation, one should try all three schemes and select the one performing best.

Optimized Classification

These results apply only to the situation where costs of obtaining attribute data for records need not be considered. When such costs are present, Cutpoint, Entropy CC and Entropy MDL must be modified. For Cutpoint, the needed adjustments are discussed toward the end of the chapter.

In the remainder of this section, we review prior work on discretization. For a comprehensive survey and a computational comparison of techniques, see Liu et al. (Liu, Hussain, Tan, & Dash, 2002).

Entropy-Based Approaches

The concept of entropy, as used in information theory, measures the purity of an arbitrary collection of examples (Mitchell, 1997). Suppose we have two classes of data, labeled N and P. Let n be the number of N instances, and define p to be the number of P instances. An estimate of the probability that class P occurs in the set is $p/(p + n)$, while an estimate of the probability that class N occurs is $n/(p + n)$. Entropy is then estimated as:

entropy(p,n)

$$= -\frac{p}{p+n}\log_2\frac{p}{p+n} - \frac{n}{p+n}\log_2\frac{n}{p+n} \quad (1)$$

Another value, called *gain*, indicates the value of separating the data records on a particular attribute. Let V be an attribute with two possible values. Define p_1 (resp. n_1) to be the number of P (resp. N) records that contain one of the two values. Similarly, let p_2 (resp. n_2) be the number of P (resp. N) records that contain the second value. Then:

gain = entropy(p,n)

$$-[\frac{p_1+n_1}{p+n}\text{entropy}(p_1,n_1) + \frac{p_2+n_2}{p+n}\text{entropy}(p_2,n_2)]$$

$$(2)$$

In generating decision trees, for example, the attribute with the highest gain value is used to split the tree at each level.

The simplest approach to discretization is as follows. Assume that each record has a rational attribute, V. The records are first sorted according to V, yielding rational values $v_1, v_2,..., v_k$. For each pair of values v_i and v_{i+1}, the average of the two is a potential marker to separate the P records from the N records. For each possible marker, the associated gain is computed. The highest gain indicates the best marker that separates the two classes of data (Quinlan, 1986). The method has been further developed to separate rational data into more than just two classes. In Fayyad and Irani (1992, 1993), a recursive heuristic for that task is described. The multiinterval technique first chooses a marker giving minimal entropy. It then recursively uses the minimum description length (MDL) principle to determine whether additional markers should be introduced.

Another concept, called *minimum splits*, is introduced in Wang and Goh (1997). Minimum splits minimize the overall impurity of the separated intervals with respect to a predefined threshold. Although, theoretically, any impurity measurement could be used, entropy is commonly chosen. Since many minimum splits can be candidates, the optimal split is discovered by searching the minimum splits' space. The candidate split with the smallest product of entropy and number of intervals is elected to be the optimal split.

Entropy-based methods compete well with other data transformation techniques. In Dougherty et al. (1995), it is shown not only that discretization prior to execution of naive Bayes decision algorithms can significantly increase learning performance, but also that recursive minimal entropy partitioning performs best when compared with other discretization methods such as equal width interval binning and Holte's 1R algorithm (Holte, 1993). More comparisons involving entropy-based methods can be found in Kohavi and Sahami (1996), which demonstrates situations in which entropy-based methods using the MDL principle slightly outperform error-minimization methods. The error-minimization methods used in the comparison can be found in Maass (1994) and Auer et al. (Auer, Holte, & Maass, 1995). For information regarding the performance of entropy-based methods for learning classification rules, see An and Cercone (1999).

Bottom-Up Methods

Bottom-up methods initially partition the data set then recombine similar adjacent partitions.

The basic method is introduced in Srikant and Agrawal (1996). Major problems are low speed and bloating of the produced rule set. To offset long execution times, the number of intervals must be reduced. Uninteresting excess rules may be pruned using an interest measure. Data clustering has been used (Miller & Yang, 1997) to generate more meaningful rules. Yet another approach to merging related intervals is used in the contrast set miner (Bay & Pazzani, 1999). The use of one such machine, called STUCCO, is illustrated in Bay (2000).

Other Approaches

Bayes' Law has also been utilized for discretization. Wu (1996) demonstrates one such method. In it, curves are constructed based upon the Bayesian probability of a particular attribute's value in the data set. Markers are placed where leading curves differ on two sides.

A number of investigations have focused on simultaneous analysis of attributes during the transformation process. Dougherty et al. (1995) coin the term *dynamic* to refer to methods that conduct a search through the space of possible k values for all features simultaneously. For an example method, see Gama et al. (Gama, Torgo, & Soares, 1998).

Relatedly, publications tend to use the term *multivariate* with different interpretations. Kwedlo and Krętowski (1999) refer to a multivariate analysis as one that simultaneously searches for threshold values for continuous-valued attributes. They use such an analysis with an evolutionary algorithm geared for decision rule induction. Bay (2000), however, declares that a multivariate test of differences takes as input instances drawn from two probability distributions and determines if the distributions are equivalent. This analysis maintains the integrity of any hidden patterns in the data.

Boros et al. (Boros, Hammer, Ibaraki, & Kogan, 1997) explores several optimization approaches for the selection of breakpoints. In each case, all attributes of the records of the training sets *A* and *B* are considered simultaneously. For example, minimization of the total number of breakpoints is considered. The reference provides polynomial solution algorithms for some of the optimization problems and establishes other problems to be NP-hard. Boros et al. (Boros, Hammer, Ibaraki, Kogan, Mayoraz, & Muchnik, 2000) describes a discretization method that is integrated into the so-called logic analysis of data (LAD) method. In that setting, the discretization requires solution of a potentially large set-covering problem. A heuristic method is employed to solve that problem approximately.

DEFINITIONS

We need a few definitions for the discussion of Cutpoint.

Unknown Values

At times, the records of *A* and *B* may be incomplete. Following Truemper (2004), we consider two values that indicate entries to be unknown. They are *Absent* and *Unavailable*. The value *Absent* means that the value is unknown but could be obtained, while *Unavailable* means that the value cannot be obtained. Of course, there are in-between cases. For example, a diagnostic value could be obtained in principle but is not determined, since the required test would endanger the life of the patient. Here, we force such in-between cases to be classified as *Absent* or *Unavailable*. For the cited diagnostic case, the choice *Unavailable* would be appropriate.

Another way to view *Absent* and *Unavailable* is as follows. Absent means that the value is unknown, and that this fact is, in some sense, independent from the case represented by the given record. On the other hand, *Unavailable* tells that the reason why the value is not known is directly connected with the case of the record.

Thus, *Unavailable* implicitly is information about the case of the record, while *Absent* is not. This way of differentiating between *Absent* and *Unavailable* implies how irrelevant values are handled. That is, if a value is declared to be *irrelevant* or *inapplicable*, then this fact is directly connected with the case of the record, and thus is encoded by the value *Unavailable*.

In prior work, the treatment of unknown values typically does not depend on whether the unknown value could be obtained. For example, the average value of the attribute is often used for missing values (Mitchell, 1997). As another example, database methods such as SQL use NULL to represent unknown entries (Ramakrishnan & Gehrke, 2003). In applications, we have found the distinction between *Absent* and *Unavailable* to be useful. For example, a physician may declare that it is unnecessary that a certain diagnostic value be obtained. In that case, we call the value *irrelevant* and encode it by assigning the value *Unavailable*. Conversely, if a diagnostic value is deemed potentially useful but is not yet attained, we assign the value *Absent*.

It is convenient that we expand the definition of logic data and rational data so that *Absent* and *Unavailable* are allowed. Thus, *logic data* have each entry equal to *True, False, Absent,* or *Unavailable*, while rational data have each entry equal to a rational number, *Absent,* or *Unavailable*.

Records

A *record* contains any mixture of logic data and rational data. There are two sets *A* and *B* of records. Each record of the sets has the same number of entries. For each fixed *j*, the *j*th entries of all records are of the same data type. We want to transform records of *A* and *B* to records containing just logic data, with the objective that logic formulas, determined by any appropriate method, can classify the records correctly as coming from *A* or *B*.

Populations

Typically, the sets A and B come from populations **A** and B, respectively, and we want the transformations and logic formulas derived from A and B to classify the remaining records of $\mathbf{A} - A$ and $\mathbf{B} - B$ with high accuracy.

DNF Formulas

A *literal* is the occurrence of a possibly negated variable in a logic formula. A *disjunctive normal form* (DNF) formula is a disjunction of conjunctions of literals. For example, $(x_1 \wedge \neg x_2) \vee (x_2 \wedge x_3) \vee (x_1 \wedge \neg x_3)$ is a DNF formula. The evaluation of DNF formulas requires the following adjustments when the values *Absent* and *Unavailable* occur. Let D be the DNF formula $D = D_1 \vee D_2 \vee \cdots \vee D_k$, where the D_j are the DNF clauses. For example, we may have $D_j = x \wedge y \wedge \neg z$, where x, y, and $\neg z$ are the literals of logic variables x, y, and z.

The DNF clause D_j evaluates to *True* if the variable of each literal has been assigned a *True/False* value so that the literal evaluates to *True*. For example, $D_j = x \wedge y \wedge \neg z$ evaluates to *True* if $x = y = True$ and $z = False$. The clause D_j evaluates to *False* if, for at least one variable occurring in D_j, the variable has a *True/False* value so that the corresponding literal evaluates to *False*, or if the variable has the value *Unavailable*. For example, $x = False$ or $x = Unavailable$ cause $D_j = x \wedge y \wedge \neg z$ to evaluate to *False*. If one of these cases does not apply, then D_j has the value *Undecided*. Thus, the *Undecided* case occurs if the following three conditions hold: (1) Each variable of D_j has *True*, *False*, or *Absent* as values; (2) there is at least one *Absent* case; and (3) all literals for the *True/False* cases evaluate to *True*. For example, $D_j = x \wedge y \wedge \neg z$ evaluates to *Undecided* if $x = Absent$, $y = True$, and $z = False$.

The DNF formula $D = D_1 \wedge D_2 \wedge \cdots \wedge D_k$ evaluates to *True* if at least one D_j has value *True*, to *False* if all D_j have value *False*, and to *Undecided* otherwise. Thus in the *Undecided* case, each D_j has value *False* or *Undecided*, and there is at least one *Undecided* case.

As an aside, prior rules on the treatment of unknown values effectively treat them as *Absent*. For example, the stated evaluation of DNF formulas for *Absent* values is consistent with the evaluation of logic formulas of SQL for NULL values (Ramakrishnan & Gehrke, 2003).

Clash Condition

Suppose we desire classification by DNF formulas. Specifically, we want two DNF formulas of which one evaluates to *True* on the records derived from A and to *False* on the records derived from B, while the second formula achieves the opposite *True/False* values. We call these formulas *separating*. Note that the outcome *Undecided* is not allowed. That value may occur, however, when a DNF formula evaluates records of $(\mathbf{A} - A) \cup (\mathbf{B} - B)$. Effectively, a formula then votes for membership in **A** or B, or declares the case to be open. We associate with the vote for **A** and **B** a numerical value of 1 or -1, resp., and assign to the *Undecided* case the value 0. This rule is useful when sets of formulas are applied, since then the vote total expresses the strength of belief that a record is in **A** or B.

There is a simple necessary and sufficient condition for the existence of the separating formulas. We call it the *clash condition*. For the description of the condition, we assume for the moment that the records of A and B contain just logic data. We say that an A record and a B record *clash* if the A record has a *True/False* entry for which the corresponding entry of the B record has the opposite *True/False* value or *Unavailable*, and if the B record has a *True/False* entry for which the corresponding entry of the A record has the opposite *True/False* value or *Unavailable*.

For example, let each record of $A \cup B$ have three entries x_1, x_2, and x_3, and suppose that an A record is ($x_1 = True$, $x_2 = Unavailable$, $x_3 = False$) and that a B record is ($x_1 = False$, $x_2 = True$, $x_3 = False$). Then the entry $x_1 = True$ of the A record differs from $x_1 = False$ of the B record, and thus the two records clash. On the other hand, take the same A record, but let the B record be ($x_1 =$ *True*, $x_2 =$ *Unavailable*, $x_3 =$ *Unavailable*). Then there is no *True/False* value in the B record for which the A record has the opposite *True/False* value or *Unavailable*, and thus the two records do not clash.

Define the *clash condition* to be satisfied by sets A and B containing only logic data if every record of A clashes with every record of B. The following theorem links the existence of separating DNF formulas and the clash condition. We omit the straightforward proof.

Theorem 1. *Let sets A and B contain just logic data. Then two separating DNF formulas exist if and only if the clash condition is satisfied.*

Let sets A and B of records be given. Define J to be the set of indices j for which the jth entries of the given records contain rational data.

Cutpoint recursively defines markers for the jth entries of the $j \in J$, where in each pass one marker is defined. It is convenient that we divide the description of the marker selection into two parts, which make up the subsequent two sections. The next section covers the selection of the initial marker for an arbitrary $j \in J$, and the following section deals with the case where an additional marker is to be found.

INITIAL MARKER

Let $j \in J$. We denote the rational numbers in jth position, sorted in increasing order, by $z_1 \leq z_2 \leq \cdots \leq z_N$. For the moment, we ignore all *Absent* and *Unavailable* values that may occur in the jth position.

Class Values

We associate with each z_i a class value v_i that depends on whether z_i is equal to any other z_h, and whether z_i is in a record of set A or B. Specifically, if z_i is unique and thus not equal to any other z_h, then v_i is 1 (resp. 0) if the record

with z_i as jth entry is in A (resp. B). If z_i is not unique, let H be the set of indices h for which $z_h = z_i$. Note that $i \in H$. Let H_A (resp. H_B) be the subset of the $h \in H$ for which z_h is the jth entry of a record in set A (resp. B). If $h \in H_A$ (resp. $h \in H_B$), we say that z_h produces a *local class value* equal to 1 (resp. 0). The class value v_i is then the average of the local class values for the z_h with $h \in H$. Thus, $v_i = [1 \cdot |H_A| + 0 \cdot |H_B|]/|H|$ or, compactly,

$$v_i = |H_A|/|H| \tag{3}$$

The formula also covers the case of unique z_i, since then $H = \{i\}$ and either $H_A = \{i\}$ or $H_A = \varnothing$ depending on whether the record with z_i as jth entry is in A or B, respectively.

For example, suppose $z_1 = 2$, $z_2 = 5$, and $z_5 = 10$ occur in records of set A, and $z_3 = 7$ and $z_4 = 10$ occur in records of set B. Since z_1 and z_2 are unique and occur in records of set A, we have $v_1 = v_2 = 1$. Similarly, uniqueness of z_3 and occurrence in a B record produce $v_3 = 0$. The values z_4 and z_5 are equal and exactly one of them, z_5, occurs in a record of set A. Thus for both z_4 and z_5, we have $H = \{4, 5\}$ and $H_A = \{5\}$, and by (3), $v_4 = v_5 = |H_A|/|H| = 0.5$.

Recall that a marker corresponds to an abrupt change of classification pattern. In terms of class values, a marker is a value c where many if not all z_i close to c and satisfying $z_i < c$ have high class values, while most if not all z_i close to c and satisfying $z_i > c$ have low class values, or *vice versa*. We identify markers following a smoothing of the class values by Gaussian convolution, a much used tool. For example, it is employed in computer vision for the detection of edges in digitized images; see Forsyth and Ponce (2003).

Smoothed Class Values

Gaussian convolution uses the normal distribution with mean equal to 0 for smoothing of data. For completeness, we include the relevant formulas.

For mean 0 and standard deviation $\sigma \geq 0$, the probability density function of the normal distribution is:

$$f(y) = \frac{1}{\sigma\sqrt{2\pi}} e^{-y^2/(2\sigma^2)}, \quad 0 < y < \infty \qquad (4)$$

In our case, we always choose σ to be a positive integer. We cover the selection in a moment. For any integer g and the selected σ, let β_g denote the probability that the random variable defined by $f(y)$ falls into the open interval $(g-0.5, g+0.5)$. Since g is the midpoint of the open unit interval $(g-0.5, g+0.5)$, we have:

$$\beta_g = \int_{g-0.5}^{g+0.5} f(y)dy \cong f(g) \qquad (5)$$

The smoothing process uses the β_g values to derive, from the class values v_i, smoothed values $\bar{v}_i$ by the formula:

$$\bar{v}_i = \sum_{g=-\infty}^{\infty} \beta_g \cdot v_{i+g}, \quad 1 \leq i \leq N \qquad (6)$$

The formula relies on the convention that each v_{i+g} without defined value, that is, with $i+g < 1$ or $i+g > N$, is declared to be 0. For the values of σ of interest and for $|g| \geq 2\sigma + 1$, the β_g are sufficiently small that they can be ignored. That fact and the relation $\beta_g = \beta_{-g}$, for all g, allow us to simplify (6) for each actual computation to:

$$\bar{v}_i = \beta_0 \cdot v_i + \sum_{g=1}^{2\sigma} \beta_g \cdot (v_{i+g} + v_{i-g}), \quad 1 \leq i \leq N \qquad (7)$$

The assumption of $v_i = 0$ outside the known values $v_1, v_2, ..., v_N$ results in biased or, rather,

unusable values $\bar{v}_i$ for $1 \leq i \leq 2\sigma$ and $N - 2\sigma + 1 \leq i \leq N$. As a consequence, we ignore these values and declare the remaining $\bar{v}_i$ values *usable*.

Selection of Standard Deviation

We select the standard deviation σ via an analysis of classification patterns. Suppose we produce sequences made up of the letters A and B. We construct a given sequence by randomly selecting one letter at a time, choosing the letter A with probability p and the letter B with probability $q = 1 - p$. In the construction of a sequence, we begin with the sequence AB. For given $k \geq 1$ and $l \geq 1$, we adjoin $k - 1$ As in front of AB and $l - 1$ Bs behind AB. At this point, we have k As followed by l Bs. Finally, we add a B in front and an A at the end. What is the probability that such a sequence S is constructed from AB when we randomly select letters and add them first in front and then at the end, until a sequence of the described form is achieved? Since the initial sequence AB is given, the probability is:

$$P[S] = p^k q^l \qquad (8)$$

For $m \geq 1$, consider the event E_m where the previously mentioned process constructs any S for which $k \geq m$ or $l \geq m$. We add up the appropriate probabilities of (8) to get the probability α_m that E_m occurs. Using the fact that the sum of the probabilities of all possible cases is 1, that is,

$$\sum_{k \geq 1, \, l \geq 1} p^k q^l = 1 \qquad (9)$$

we compute α_m as shown in equation (10).

Equation 10.

$$
\begin{aligned}
\alpha_m &= \sum_{k \geq m, \, l \geq 1} p^k q^l + \sum_{k \geq 1, \, l \geq m} p^k q^l - \sum_{k \geq m, \, l \geq m} p^k q^l \\
&= p^{m-1} \sum_{k \geq 1, \, l \geq 1} p^k q^l + q^{m-1} \sum_{k \geq 1, \, l \geq 1} p^k q^l - (pq)^{m-1} \sum_{k \geq 1, \, l \geq 1} p^k q^l \\
&= p^{m-1} + q^{m-1} - (pq)^{m-1}
\end{aligned}
$$

Define the length of S to be the number of As and Bs minus 2, which is $k + l$. Effectively, we do not count the initial B of S and the final A of S. The expected length L of S is:

$$
\begin{aligned}
L &= \sum_{k \geq 1, l \geq 1} (k+l) p^k q^l \\
&= \sum_{k \geq 1} k\, p^k \sum_{l \geq 1} q^l + \sum_{k \geq 1} p^k \sum_{l \geq 1} l\, q^l \\
&= \frac{p}{(1-p)^2} \cdot \frac{q}{1-q} + \frac{p}{1-p} \cdot \frac{q}{(1-q)^2} \\
&= \frac{1}{pq}
\end{aligned}
$$

$$(11)$$

Suppose we have a sequence T of N randomly selected As and Bs. What is the expected number of the above sequences S occurring in T? For our purposes, a sufficiently precise estimate is:

$$
N / L = N\, p\, q \tag{12}
$$

Of the expected number of sequences S occurring in T, the fraction of sequences that qualify for being sequences of event E_m is approximately equal to α_m. Thus, a reasonable estimate of the expected number of sequences of E_m occurring in T, which we denote by $K(N, m)$, is:

$$
\begin{aligned}
K(N, m) &= (N/L)\alpha_m \\
&= N\, p\, q[p^{m-1} + q^{m-1} - (pq)^{m-1}]
\end{aligned} \tag{13}
$$

Each S occurring in T is a potential case for a marker that corresponds to the point where k As transition to l Bs. We do not want markers to result from sequences S that likely have been produced by randomness. We try to avoid such choices as follows.

Suppose that, for some $m \geq 1$, $K(N, m)$ is approximately equal to 1. This implies that, on average, there is one sequence S of E_m. Since α_m of (10) decreases geometrically as m increases, such a sequence S of E_m typically has not much more than m As or Bs, and any sequence S' with larger number of As or Bs is very unlikely to

occur. We use this fact as follows. We select a value $m^* \geq 1$ so that $K(N, m^*)$ is as close to 1 as possible. Then we choose the standard deviation σ so that the sequence S with about m^* As or Bs that we can expect to occur does not produce a marker if there is a sequence S' with length significantly larger than m^*.

By these arguments, the latter sequence S' is unlikely to have been produced by randomness, and thus is likely due to a particular behavior of the values of the attribute under consideration. In terms of the discussion in the introduction, we estimate that we have an abrupt pattern change of the first kind.

We achieve the desired effect by selecting $\sigma = m^*$. Indeed, that choice produces significant probabilities β_g for g, $m < g \leq 2m$, and these probabilities tend to smooth out the classification values v_i associated with the As and Bs of all randomly produced sequences S.

When N is not large, certain boundary effects should be addressed. We describe the adjustment and then justify it. Instead of demanding that $K(N, m^*)$ is close to 1, we ignore the first and last m^* As and Bs of the sequence T, and require that $K(N - 2\,m^*, m^*)$ defined from:

$$
K(N - 2m, m) = (N - 2m)\, p\, q[p^{m-1} + q^{m-1} - (pq)^{m-1}] \tag{14}
$$

is as close to 1 as possible. We motivate the adjustment as follows. When Gaussian convolution is performed with $\sigma = m^*$, the first smoothed class value is computed using the values v_i of the $4\sigma + 1$ As and Bs at the beginning of T. Denote that subsequence by T'. If the central $2\sigma + 1$ As and Bs of T' contain an S of some E_m with $m \leq m^*$, then the class values of the As and Bs of any such S tend to be smoothed out. Thus, S is unlikely to result in a marker if a subsequence S' with length greater than m^* exists.

We establish the probability p for (14) and compute m^* as follows. We take p to be the fraction of the number of training records of class A divided by the total number of training records, and we find m^* by dichotomous search.

We note that, due to the symmetry of the formula $K(N - 2m, m)$, the choice of m^* implicitly also considers subsequences in which the roles of A and B are reversed. Table 1 shows σ as a function of N, for $\sigma \leq 10$ and $p = q = 0.5$.

There is an exceptional case where the selected σ must be reduced. As we argue shortly—see the discussion following (16)—we do not consider a marker between z_i and z_{i-1} if $v_i = v_{i-1}$. Thus, no marker can be placed if no i satisfies $2\sigma + 2 \leq i \leq N - 2\sigma$ and $v_i \neq v_{i-1}$. If that case occurs, several corrective actions are possible. We have found that reduction of σ to 1 is a good choice. If for the reduced σ there still is no index i satisfying $2\sigma + 2 \leq i \leq N - 2\sigma$ and $v_i \neq v_{i-1}$, then we declare that no intervals should be created for the jth entry; as a consequence, we delete the jth entry from all records of A and B. Otherwise, we proceed with the reduced $\sigma = 1$.

For example, if $N = 37$, $\sigma = 6$, and $v_{13} = 1$, $v_{14} = v_{15} = \cdots v_{30} = 0$, $v_{31} = 1$, then no i satisfies $2\sigma + 2 = 14 \leq i \leq N - 2\sigma = 25$ and $v_i \neq v_{i-1}$. Thus, σ should be reduced to 1. For that value, both $i = 14$ and $i = 31$, and possibly other values of i, satisfy $2\sigma + 2 = 4 \leq i \leq N - 2\sigma = 35$ and $v_i \neq v_{i-1}$. Thus, $\sigma = 1$ should be used. On the other hand, let $N = 37$ and $\sigma = 6$ as before, but suppose $v_1 = 1$, $v_2 = v_3 = \cdots = v_{35} = 0$, $v_{36} = 1$, $v_{37} = 0$. For $\sigma = 6$, no i satisfies $2\sigma + 2 = 14 \leq i \leq N - 2\sigma = 25$. Reduction of σ to 1 produces the same negative conclusion. Thus, no intervals should be created for the jth entries, and we delete these entries from all records of A and B.

Table 1. σ as function of N for $\sigma \leq 10$ and $p = q = 0.5$

N	σ
3–7	1
8–11	2
12–15	3
16–31	4
32–54	5
55–99	6
100–186	7
187–359	8
360–702	9
703–1387	10

Definition of Marker

Suppose we have selected σ, as described previously, and have computed the smoothed class values $\bar{v}_i$. As we move along the sequence of usable values $\bar{v}_i$, the absolute difference δ_i between adjacent $\bar{v}_{i-1}$ and $\bar{v}_i$,

$$\delta_i = |\bar{v}_i - \bar{v}_{i-1}| \tag{15}$$

measures the abruptness with which class values change. We call δ_i a *difference value*. The largest such value, say δ_{i^*}, produces a marker c between z_{i^*-1} and z_{i^*}. That is,

$$c = (z_{i^*-1} + z_{i^*})/2 \tag{16}$$

The selection rule for c requires a small adjustment due to a quirk that may be introduced by the convolution process. It is possible that, for the selected c, the corresponding original class values v_{i^*-1} and v_{i^*} are equal. In case all z_i are distinct, both values z_{i^*-1} and z_{i^*} separated by c come either from A records or from B records. If several z_i are equal, more complex interpretations are possible. However, all of them reflect unattractive cases.

To rule out all such situations, we restrict the selection of the difference values δ_i by considering δ_i values only if $v_i \neq v_{i-1}$. Thus,

$$\delta_{i^*} = \max_i \{\delta_i \mid v_i, v_{i-1} \in U, v_i \neq v_{i-1}\} \tag{17}$$

where U is the set of usable values. If the maximum is attained by several i^*, we pick one closest to $N/2$, breaking any secondary tie by a random choice.

For example, if $\sigma = 6$ and $N = 60$, then the $\bar{v}_i$ with index i satisfying $2\sigma + 1 = 13 \leq i \leq N - 2\sigma = 48$ are usable. Suppose these values are $\bar{v}_{13} = 0.3214$, $\bar{v}_{14} = 0.3594$, $\bar{v}_{15} = 0.4042$, $\bar{v}_{16} = 0.4439$, $\bar{v}_{17} = 0.4760$, $\bar{v}_{18} = 0.4986,...$, $\bar{v}_{45} = 0.4740$, $\bar{v}_{46} = 0.4410$, $\bar{v}_{47} = 0.4007$, and $\bar{v}_{48} = 0.3612$. For these values, formula (15) produces $\delta_{14} = 0.0380$, $\delta_{15} = 0.0448$, $\delta_{16} = 0.0397$, $\delta_{17} = 0.0321$, $\delta_{18} = 0.0226,...$, $\delta_{46} = 0.0330$, $\delta_{47} = 0.0403$, and $\delta_{48} = $

11

0.0395. Suppose the largest δ_i for which $v_i \neq v_{i-1}$, is unique and is $\delta_{15} = 0.0448$. Thus, $i^* = 15$. If $z_{i^*} = z_{15} = 7$ and $z_{i^*-1} = z_{14} = 5$, the marker c is defined by $c = (z_{i^*-1} + z_{i^*})/2 = (5+7)/2 = 6$.

The next scheme summarizes the computation producing the initial marker c. The scheme also outputs the standard deviation σ of the convolution process since that information is needed later in another application of the algorithm.

ALGORITHM INITIAL MARKER

Input: Rational numbers $z_1 \leq z_2 \leq \cdots \leq z_N$ of the jth attribute of the records of A and B.

Output: Either: Marker c for the jth attribute, standard deviation σ of the convolution process, and the difference value δ_{i^*} associated with the marker. Or: "Marker cannot be determined."

Procedure:

1. (Check if N is too small or if $\sigma = 1$ cannot produce a marker.) If $N \leq 6$ or if, for $\sigma = 1$, there is no index i satisfying $2\sigma + 2 \leq i \leq N - 2\sigma$ and $v_i \neq v_{i-1}$, then output "Marker cannot be determined," and stop. (In that case, one should delete the jth entries from all records of A and B.)

2. (Compute class values.) For $i = 1, 2,..., N$, define $H^i = \{h \mid z_h = z_i\}$, $H_A^i = \{h \in H \mid z_h$ is taken from an A record$\}$, and compute the class value $v_i = |H_A^i|/|H^i|$.

3. (Define p, q, and σ.) Define $p = |A|/(|A| + |B|)$ and $q = 1 - p$. Let m^* be the value $m \geq 1$ for which $K(N - 2m, m)$ is closest to 1. Let $\sigma = m^*$. If there is no index i satisfying $2\sigma + 2 \leq i \leq N - 2\sigma$ and $v_i \neq v_{i-1}$, lower σ to 1.

4. (Compute smoothed class values.) For $i = 1, 2,..., N$, use the class values v_i, the standard deviation σ, and the β_g values of (5) to compute the smoothed class values $\overline{v}_i = \beta_0 \cdot v_i + \sum_{i=1}^{2\sigma} \beta_g (v_{i+g} + v_{i-g})$.

5. (Select marker.) For $i = 2\sigma + 2$, $2\sigma + 3,...$, $N - 2\sigma$, let $\delta_i = |\overline{v}_i - \overline{v}_{i-1}|$. Select i^* so that $\delta_{i^*} = \max_i \{\delta_i \mid v_i \neq v_{i-1}\}$. If i^* is not unique, select an i^* closest to $N/2$ and break any secondary tie by random choice. Define the marker c by $c = (z_{i^*-1} + z_{i^*})/2$. Output the marker c, the standard deviation σ, and the difference value δ_{i^*}.

ADDITIONAL MARKER

This section covers how one additional marker is selected, assuming that a certain collection of markers is already at hand. The procedure is invoked if the sets A' and B', derived from the sets A and B via the markers obtained so far, do not satisfy the clash condition and thus cannot be fully separated.

Critical Interval

The markers on hand define intervals of the rational line for each index $j \in J$, and these markers produce a transformation of A and B to A' and B'. Define such an interval to be critical if a properly chosen subdivision can lead to a transformation of A and B to, say, A'' and B'' such that A'' and B'' have more clashing pairs of records than A' and B'. Clearly, each critical interval is associated with a particular attribute $j \in J$, and all critical intervals are readily determined via the nonclashing pairs of records of A' and B'. We omit the obvious process. For each critical interval, we compute an additional marker using a method virtually identical to Algorithm INITIAL MARKER. Specifically, the input sets A and B of the algorithm are now the subsets $\overline{A} \subseteq A$ and $\overline{B} \subseteq B$ of records for which the values of the associated attribute $j \in J$ falls into the critical interval.

The algorithm either outputs a marker together with the associated standard deviation σ and the difference value δ_{i^*}, or it declares that a marker cannot be found. In the latter case, we do not delete any attribute values from A and B, but instead record that the interval cannot be refined, and thus exclude it from further consideration. When all critical intervals have been processed, two cases are possible. Either we have at least one additional marker, or no additional markers could be determined. In the latter case, the transformation process outputs A', B', the markers on hand, and the warning message "A and B cannot be fully separated," and then stops.

If at least one additional marker has been determined, we select one of them and proceed recursively as described previously. The selection of the marker is based on a measure that considers the attractiveness of pattern change at the point of the marker, and on the number of nonclashing pairs of records of A' and B' that determine the interval to be critical. The latter number is called the *relevance count*. We first discuss the attractiveness of the pattern change.

Attractiveness of Pattern Change

The attractiveness of a pattern change is based on a lower bound ε on the difference values δ_i of (15) for certain label subsequences. Each such subsequence has, for some $n \geq 1$ yet to be specified, $k \geq n + 1$ As followed by $l \geq n + 1$ Bs, and δ_i is the difference value produced by the last A and the first B of the sequence. We establish a lower bound ε for δ_i.

Theorem 2. *Let a label sequence be given for which the original rational numbers z_i are all distinct. For some $n \geq 1$, let a label subsequence have $k \geq n + 1$ As followed by $l \geq n + 1$ Bs. Then $\varepsilon = \beta_0 - 2\beta_{n+1}$ is a lower bound for δ_i of (15).*

Proof: Since $\delta_i \geq 0$, the claim is trivial if $\beta_0 - 2\beta_{n+1} \leq 0$. Hence, we suppose that $\beta_0 > 2\beta_{n+1}$. Using the formula (7) for $\overline{v}_i$ in the definition of δ_i of (15), we have, (see equation (18)).

Consider δ_i produced by the last A and first B of the label sequence. Due to the $k \geq n + 1$ As (resp. $l \geq n + 1$ Bs) in the label subsequence, we have $v_{i-1} = v_{i-2} = \cdots = v_{i-n-1} = 1$ (resp. $v_i = v_{i+1} = \cdots = v_{i+n} = 0$). We use these class values in (18) and simplify to get:

$$\delta_i = \left| \beta_{n+1} - \beta_0 + \sum_{g=n+1}^{\infty} (\beta_g - \beta_{g+1})(v_{i+g} - v_{i-g-1}) \right|$$

(19)

Since $\beta_0 > 2\beta_{n+1}$ and, for all $g \geq 0$, $\beta_g \geq 2\beta_{g+1}$, the right hand side of (19) is minimum if, for all $g \geq n + 1$, we have $v_{i+g} = 1$ and $v_{i-g-1} = 0$. For that case, δ_i becomes $\delta_i = |2\beta_{n+1} - \beta_0| = \beta_0 - 2\beta_{n+1} = \varepsilon$.

If n is sufficiently large, then the label subsequence of Theorem 2 is quite unlikely to be a random occurrence. Thus, if the label subsequence does occur, we estimate that it corresponds to an abrupt pattern change of the first kind. Indeed, as was discussed previously, as n grows beyond σ, this conclusion tends to become valid. For example, $n = \lfloor 1.5\sigma \rfloor$ is large enough for the desired conclusion, and we choose this value of n to compute the lower bound ε. Thus,

$$\varepsilon = \beta_0 - 2\beta_{\lfloor 1.5\sigma \rfloor + 1}$$

(20)

Equation 18.

$$\delta_i = \left| \left[\beta_0 v_i + \sum_{g=1}^{\infty} \beta_g (v_{i+g} + v_{i-g}) \right] - \left[\beta_0 v_{i-1} + \sum_{g=1}^{\infty} \beta_g (v_{i+g-1} + v_{i-g-1}) \right] \right|$$

$$= \left| \sum_{g=0}^{\infty} (\beta_g - \beta_{g+1})(v_{i+g} - v_{i-g-1}) \right|$$

Let c be a marker, and define δ_i to be the change of smoothed class values corresponding to the marker c. To measure how likely the marker c corresponds to a pattern change of the first kind, we compare δ_i with ε. Specifically, if the ratio:

$$\delta_{i^*} / \varepsilon = \delta_{i^*} / (\beta_0 - 2\beta_{\lfloor 1.5\sigma \rfloor + 1}) \qquad (21)$$

is near or above 1, then we estimate that we likely have a pattern change of the first kind. Thus, the ratio δ_i / ε measures the attractiveness of the marker. We say that the marker c has *attractiveness* δ_{i^*}/ε.

Selection of Marker

For each critical interval for which we have determined an additional marker, define the *potential* of the marker to be the product of the relevance count of the interval and the attractiveness of the marker. Letting γ and R denote the potential and relevance count, respectively, we have, for each marker, the potential γ as:

$$\gamma = R\delta_{i^*} / \varepsilon = R\delta_{i^*} / (\beta_0 - 2\beta_{\lfloor 1.5\sigma \rfloor + 1}) \qquad (22)$$

We select the marker with highest potential, add that marker to the list of markers on hand, and proceed recursively as described earlier.

For example, suppose we have two critical intervals. For the first interval, we have $\sigma = 6$, $N = 58$, $\delta_{i^*} = 0.037$, and $R = 12$. For $\sigma = 6$, we have $\varepsilon = (\beta_0^{i} - 2\beta_{\lfloor 1.5\sigma \rfloor + 1}) = 0.035$, and the potential is $\gamma = R\delta_{i^*}/\varepsilon = 12(0.037/0.035) = 12.7$. If the second critical interval has a smaller potential, then we refine the first interval. Suppose that for the first interval we have $z_{i^*} = 17$ and $z_{i^*-1} = 14$. Then the new marker is $p = (z_{i^*-1} + z_{i^*})/2 = (14+17)/2 = 15.5$.

We summarize the selection process.

ALGORITHM ADDITIONAL MARKER

Input: List of critical intervals.
Output: Either: "No critical interval can be refined." or: Additional marker for one critical interval.
Procedure:

1. For each critical interval, do Algorithm INITIAL MARKER where the input sets are the subsets $\bar{A} \subseteq A$ and $\bar{B} \subseteq B$ of records for which the value of the associated attribute $j \in J$ falls into the critical interval. If the algorithm declares that no marker can be determined, remove the interval from the list of candidates.
2. If the list of critical intervals is empty, output "No critical interval can be refined," and stop.
3. For each critical interval, use the value δ_{i^*} and σ determined in Step 1 and the relevance count R to compute the potential $\gamma = R\delta_{i^*} / (\beta_0 - 2\beta_{\lfloor 1.5\sigma \rfloor + 1})$.
4. Select the critical interval with maximum potential. In case of a tie, favor the interval with larger number of z_i values, and break any secondary tie randomly. Using i^* of the associated δ_{i^*}, output the marker $p = (z_{i^*-1} + z_{i^*})/2$ for the selected interval, and stop.

CUTPOINT ALGORITHM

With algorithms INITIAL MARKER and ADDITIONAL MARKER at hand, we are ready to describe the entire algorithm of Cutpoint. We begin the scheme as follows. For each $j \in J$, we carry out algorithm INITIAL MARKER and thus get either a marker, say c_j, or conclude that a marker cannot be obtained. In the latter case, the attribute j is deleted from all records of A and B.

For the reduced sets, which we again denote by A and B, we select a $j \in J$ whose marker has the largest associated difference value δ_j. We apply the transformation implied by that single marker and thus obtain sets A' and B'.

We test if A' and B' satisfy the clash condition. If this is not so, that is, if at least one record of A' and one record of B' do not clash, then we compute one additional marker with algorithm ADDITIONAL MARKER, update the sets A' and B' accordingly, and proceed recursively. That is, we test if the sets A' and B' satisfy the clash condition, and so on.

The process stops either when A' and B' satisfy the clash condition, or when an additional marker cannot be determined. In the implementation of the method, we also stop introducing additional markers for a given $j \in J$ when the number of markers reaches a specified maximum. In the tests described later, that limit was set to six.

As an option, one may also stop the refinement process if the introduction of an additional marker does not reduce the total number of nonclashing pairs of records. Indeed, since that marker had the highest potential among the possible choices and did not reduce the number of nonclashing pairs, one might conjecture that additional markers may not be desirable in any interval, and thus may terminate the refinement process.

When the recursive process terminates, some attributes $j \in J$ may not have received any marker. Of course, one such marker was determined by algorithm INITIAL MARKER in the initial part of Cutpoint, and we now assign that marker. Thus, each $j \in J$ now has at least one marker. At this point, we have the desired collection of markers. We use them for one final update of the sets A' and B'.

We output the collection of markers and the associated sets A' and B'. If these sets do not satisfy the clash condition, we also output the warning message "**A** and **B** cannot be fully separated."

Table 2. Accuracy of C4.5 on Testing Data

Dataset	Cutpoint	Entropy CC	Entropy MDL
heart	78.90	77.98	77.31
australian	85.22	86.23	85.94
hepatitis	78.71	75.48	76.78
horse-colic	84.53	83.46	88.62
boston housing	80.63	77.06	83.59
wisconsin breast	94.98	93.69	95.41
crx	85.07	86.23	86.09
haberman	69.92	70.91	69.26
ionosphere	88.31	90.88	90.88
pima	75.13	75.26	73.56
spectf	74.11	79.77	77.86
Overall	81.41	81.54	82.30

COMPUTATIONAL RESULTS

Cutpoint has been used so far in a variety of projects including credit rating, video image analysis, and word sense disambiguation. In each of the numerous cases, Cutpoint has proved to be effective and reliable.

We also have compared Cutpoint with the two entropy-based methods Entropy CC and Entropy MDL, described in the introduction.

For the comparison, we selected the following data sets of the UC Irvine repository of machine-learning databases: cleveland heart, australian, hepatitis, horse-colic, boston housing, wisconsin breast, crx, haberman, ionosphere, pima, and spectf. Boston housing was run using the attribute median housing value with a cutoff point of $21,000. For both the heart and the crx databases, some attributes were treated as nominal.

In a 5-fold cross-validation approach, we used Cutpoint, Entropy CC, and Entropy MDL for discretization, and finally applied four classification schemes. The latter methods were chosen so that classification by decision trees, naive Bayes methods, support vector machines, and learning logic formulas were represented. For the first three cases we chose the version J4.8 of C4.5, the naive Bayes method, and the SMO support vector machine implemented by Witten and Frank (2000). In each case, we used the default parameters for the runs, except that we carried out 5-fold cross-validation instead of 10-fold cross-validation. For the fourth method, we selected the Lsquare version of Truemper (2004). We emphasize that we do not claim that discretization is needed or even desired for the first three classification methods. We do say that, if one contemplates a discretization preprocessing step followed by application of a classification method of one of the four types, then one may want to consider the results shown in Table 2.

Table 3. Accuracy of Naive Bayes on Testing Data

Dataset	Cutpoint	Entropy CC	Entropy MDL
heart	82.87	81.89	83.87
australian	85.07	85.80	85.36
hepatitis	83.23	82.58	81.29
horse-colic	82.63	82.63	82.34
boston-housing	80.03	79.63	78.64
wisconsin-breast	96.98	96.55	97.13
crx	85.07	86.09	85.80
haberman	69.92	69.92	67.94
ionosphere	90.88	90.88	88.89
pima	76.18	73.83	75.26
spectf	77.48	73.00	72.60
Overall	82.76	82.07	81.74

Prediction Accuracy for Testing Data

For a compact representation and ease of comparison, we have grouped the results for each of the four classification methods. Tables 2–5 show the accuracy established by the 5-fold cross-validation process. The headings contain the term "Testing Data," which may seem superfluous. The term has been added to differentiate the results from a second evaluation, to be discussed shortly. Note that the performance data should only be used for a comparison of Cutpoint, Entropy CC, and Entropy MDL, and not for evaluation of the four classification methods. The reason is that each classification method almost certainly is not the best method of that type. For example, there are better commercial decision-tree methods than C4.5, there are numerous ongoing developments concerning naive Bayes methods and support vector machines, and full implementation of Lsquare, as conceived at present, has not yet been accomplished.

From the results, we conclude the following. Based on the average performance, Cutpoint is the preferred approach for naive Bayes and Lsquare, while Entropy MDL is best for C4.5 and SMO. Furthermore, Entropy CC is dominated by Cutpoint and Entropy MDL.

When one examines the performance for individual data sets, then the preference is not clear-cut. For example, for naive Bayes and Lsquare, Cutpoint is best for 7 of the 11 data sets. In the case of C4.5 and SMO, Entropy MDL is best for 5 of the 11 data sets. Also, there are several cases where Entropy CC is better than Cutpoint and Entropy MDL. Thus, if one needs highest possible accuracy for a given situation, and if sufficient time and data are available to estimate accuracy, then one should try all three discretization schemes and select the one performing best.

Table 4. Accuracy of SMO on Testing Data

Dataset	Cutpoint	Entropy CC	Entropy MDL
heart	82.89	82.94	80.92
australian	85.80	85.94	85.22
hepatitis	81.29	80.00	77.42
horse-colic	85.37	83.99	86.97
boston housing	79.07	80.65	80.83
wisconsin breast	95.69	95.69	95.40
crx	85.36	85.94	85.80
haberman	70.25	70.91	74.84
ionosphere	88.88	91.46	90.02
pima	74.86	74.59	75.90
spectf	76.77	79.73	80.84
Overall	82.38	82.89	83.10

Table 5. Accuracy of Lsqcc on Testing Data

Dataset	Cutpoint	Entropy CC	Entropy MDL
heart	83.25	78.25	81.22
australian	86.67	85.65	83.77
hepatitis	80.65	80.00	76.77
horse-colic	85.62	87.00	88.61
boston housing	83.80	83.99	82.60
wisconsin breast	95.83	95.69	95.83
crx	85.51	84.78	84.20
haberman	66.65	68.29	—*
ionosphere	90.90	91.46	92.88
pima	73.16	76.16	72.12
spectf	82.37	79.35	80.12
Overall	83.13	82.78	82.40

* Entropy MDL failed to produce separable sets. For the computation of average performance, the value 68.29 for Entropy CC was used.

Table 6. Accuracy of C4.5 on Training Data

Dataset	Cutpoint	Entropy CC	Entropy MDL
heart	87.05	88.70	88.21
australian	90.80	90.73	89.06
hepatitis	88.87	89.68	90.48
horse-colic	89.07	89.27	90.90
boston housing	88.78	89.82	90.32
wisconsin breast	96.71	97.03	96.89
crx	89.82	90.07	88.98
haberman	75.98	77.04	75.00
ionosphere	94.59	96.08	95.65
pima	81.32	82.36	81.12
spectf	92.51	92.88	92.50
Overall	88.68	89.42	89.01

Table 7. Accuracy of Naive Bayes on Training Data

Dataset	Cutpoint	Entropy CC	Entropy MDL
heart	85.48	85.56	84.57
australian	87.79	87.90	86.74
hepatitis	87.74	88.55	88.23
horse-colic	83.83	84.03	84.51
boston housing	81.87	82.71	81.67
wisconsin breast	97.28	97.46	97.50
crx	87.18	87.93	87.32
haberman	79.00	79.57	74.67
ionosphere	92.73	93.02	92.17
pima	78.97	80.73	77.83
spectf	82.21	82.57	81.64
Overall	85.83	86.37	85.17

Table 8. Accuracy of SMO on Training Data

Dataset	Cutpoint	Entropy CC	Entropy MDL
heart	88.86	88.04	87.63
australian	89.24	90.11	85.87
hepatitis	92.90	94.19	91.77
horse-colic	90.69	90.83	90.83
boston housing	93.03	94.76	91.05
wisconsin breast	97.71	98.07	97.60
crx	89.06	89.56	88.48
haberman	77.37	78.84	74.84
ionosphere	97.29	98.29	99.64
pima	80.99	83.10	78.26
spectf	95.60	94.94	93.16
Overall	90.25	90.98	89.01

Prediction Accuracy for Training Data

In some applications, one not only desires high accuracy when records of $\mathbf{A} - A$ and $\mathbf{B} - B$ are to be classified, but also wants perfect or near-perfect accuracy for the training sets A and B. For example, if a diagnostic system for certain diseases is trained on sets A and B, then a physician may demand that, at a minimum, the system handles all training cases correctly. As a second example, consider the situation where the training data are obtained via a simulation process, and where the classification method is to produce compact rules that replace a complex decision mechanism of the simulation process. The user of the classification rules then may demand that all training data produced from simulation runs must be correctly classified.

One may evaluate the performance of Cutpoint, Entropy CC, and Entropy MDL and of the four classification methods on training data as follows. In each case of the 5-fold cross-validation process, 80% of the given data sets A and B are used for training, while 20% are used for testing. One applies the classification rules derived from the training sets to these very same sets, and thus obtains for each pair of sets A and B five accuracy estimates. The average of these five figures is an estimate of the accuracy of the classification method for training data. Tables 6–9 contain these results in the format of Tables 2–5.

A first conclusion from Tables 6–9 is that three of the four methods fail to achieve perfect or near-perfect accuracy for the training data. The exception is Lsquare. Indeed, according to Table 9, the average accuracy for the Cutpoint/Lsquare combination is 98.97%. That figure includes the comparatively poor accuracy of 88.6% of the haberman data set that, with only three attributes

Table 9. Accuracy of Lsqcc on Training Data

Dataset	Cutpoint	Entropy CC	Entropy MDL
heart	100.00	100.00	99.34
australian	99.89	99.78	96.60
hepatitis	99.68	99.68	99.19
horse-colic	99.66	99.73	98.91
boston housing	100.00	99.85	96.25
wisconsin breast	100.00	100.00	99.57
crx	99.86	99.82	98.77
haberman	89.62	85.54	*__
ionosphere	100.00	100.00	100.00
pima	100.00	99.77	83.73
spectf	100.00	100.00	98.88
Overall	98.97	98.56	96.07

* Entropy MDL failed to produce separable sets. For the computation of average performance, the value 85.54 for Entropy CC was used.

and 306 records defies full separation. When that set is excluded from consideration, the average accuracy of Cutpoint/Lsquare is 99.91%.

For the other three methods, the highest accuracy figures are as follows: 89.42% for C4.5, 86.37% for naive Bayes, and 90.98% for SMO. Curiously, each of these numbers is produced with discretization by Entropy CC. When one removes the results for the haberman data set and selects the highest accuracy among the three discretization methods, one gets the following figures: 90.66% for C4.5, 87.05% for naive Bayes, and 92.19% for SMO.

One may argue, quite properly so, that classification methods such as C4.5, naive Bayes, and SMO are designed to give highest accuracy for testing data, and that they are not optimized for accuracy on training data. Thus, it would be interesting to see how the methods may be modified so that simultaneously perfect or near-perfect accuracy for training data and high accuracy for testing data are achieved. Some condition needs to be imposed on the construction to rule out trivial modifications such as making the training data part of the output of the classification method, and enlarging the classification rule so that one first checks if the record is part of the training set. For example, one may impose the restriction that the size of the encoding of the entire classification rule must be logarithmic in the size of the encoding of the training sets.

The additional requirement of perfect or near-perfect accuracy for training data likely influences the choice of the discretization method. Thus, for a revised classification method, it is quite possible that Cutpoint or Entropy CC lead to better results than Entropy MDL.

Optimized Classification

The results implicitly assume that costs for obtaining attribute values of records need not be considered. When such costs are important, the entire classification approach must be reconsidered. For the sake of discussion, we assume the setting described in Truemper (2004), which is as follows.

Tests $T_1, T_2, ..., T_m$ are available to obtain attribute values. In particular, when test T_j is performed, then a specified subset of attribute values is obtained. Each test T_j carries a certain cost. In the general case, a test may produce rational, nominal, or logic data, except that the value *Absent* is not possible. On the other hand, the value *Unavailable* is allowed. A test produces that value if it has been determined that the attribute cannot or should not be obtained.

Classification of a record into one of the populations **A** and **B** is done as follows. Initially, some entries are given, and the remaining entries are equal to *Absent*. The classification method recursively decides if it should declare the record to be in **A** or **B**, or if it should carry out one of the tests to get additional entries. In the first case, the methods stops with the declaration that the record is classified into **A** or **B**. In the second case, the method selects a test, requests that the test be carried out, adds the new values to the record, and invokes recursion.

The goal is classification with an accuracy that is above a given lower bound, and that, subject to that condition, involves minimum or close-to-minimum total cost of tests. Depending on the setting, one may also impose the additional condition of perfect or near-perfect accuracy on the training data. We call any scheme that carries out this recursive process and that achieves the desired goal an *optimized classification process*. At this time, it is largely open how the various classification methods in existence should be modified so that they can carry out optimized classification. For Lsquare, the solution via so-called optimized formulas is given in Truemper (2004). We omit details here, but cover the adjustment needed for Cutpoint. Instead of enforcing at least one marker per attribute, we now require at least k markers, where k is a small positive integer. Then Lsquare is applied with the so-called optimized formula option. Typically, $1 \leq k \leq 6$ is appropriate. The specific choice may be obtained by trying several values and selecting one so that the classification of testing data is sufficiently accurate and can be done at low total test costs. If such a trial-and-error process is not possible, $k = 2$ or $k = 3$ is likely to work well. Regardless of the choice of k, the method always achieves perfect accuracy on the training data, assuming that Cutpoint produces fully separable sets of logic data.

CONCLUSION

The chapter introduces the Cutpoint method for discretization of rational data and compares the scheme with two entropy-based methods called Entropy CC and Entropy MDL. According to tests, Cutpoint seems best when the classification is done by naive Bayes methods or learning logic methods, while Entropy MDL appears to be best for decision tree methods and support vector machines. The performance differences are fairly small so that, for specific cases, one may want to apply each of the three methods and select the one giving best results. The chapter also discusses the choice of discretization methods when perfect or near-perfect accuracy is desired for training data, or when costs of obtaining data are to be considered in so-called optimized classification processes.

ACKNOWLEDGMENT

This research was supported in part by the Technical SupportWorking Group (TSWG) under contract N41756-03-C-4045.

REFERENCES

An, A., & Cercone, N. (1999). Discretization of continuous attributes for learning classification rules. In *Proceedings of the Third Pacific-Asia Conference on Methodologies for Knowledge Discovery and Data Mining* (pp. 509-514).

Auer, P., Holte, R.C., & Maass, W. (1995). Theory and applications of agnostic PAC-learning with small decision trees. In *Proceedings of the Eighth European Conference on Machine Learning* (pp. 21-29).

Bartnikowski, S., Granberry, M., Mugan, J., & Truemper, K. (2006). Transformation of rational and set data to logic data. In E. Triantaphyllou & G. Felici (Eds.), *Data mining and knowledge discovery approaches based on rule induction techniques*. Berlin: Springer-Verlag.

Bay, S.D. (2000). Multivariate discretization of continuous variables for set mining. In *Proceedings of the Sixth ACM SIGKDD International Conference on Knowledge Discovery and Data Mining* (pp. 315-319).

Bay, S.D., & Pazzani, M.J. (1999). Detecting change in categorical data: Mining contrast sets. In *Proceedings of the Fifth ACM SIGKDD International Conference on Knowledge Discovery and Data Mining* (pp. 302-306).

Boros, E., Hammer, P.L., Ibaraki, T., & Kogan, A. (1997). A logical analysis of numerical data. *Mathematical Programming, 79*, 163-190.

Boros, E., Hammer, P.L., Ibaraki, T., Kogan, A., Mayoraz, E., & Muchnik, I. (2000). An implementation of logical analysis of data. *IEEE Transactions on Knowledge and Data Engineering,12*, 292-306.

Dougherty, J., Kohavi, R., & Sahami, M. (1995). Supervised and unsupervised discretization of continuous features. In *Machine Learning: Proceedings of the Twelfth International Conference* (pp. 194-202).

Fayyad, U.M. & Irani, K.B. (1992). On the Handling of Continuous-Valued Attributes in Decision Tree Generation. *Machine Learning 8*, 87-102.

Fayyad, U.M., & Irani, K.B. (1993). Multi-interval discretization of continuous-valued attributes for classification learning. In *Proceedings of the Thir-*

teenth International Joint Conference on Artificial Intelligence (pp. 1022-1027).

Felici, G., Sun, F., & Truemper, K. (2004). Learning logic formulas and related error distributions. In E. Triantaphyllou & G. Felici (Eds.), *Data mining and knowledge discovery approaches based on rule induction techniques*. Berlin: Springer-Verlag.

Felici, G., & Truemper, K. (2002). A MINSAT approach for learning in logic domains. *INFORMS Journal on Computing, 14*, 20-36.

Forsyth, D.A., & Ponce, J. (2003). *Computer vision: A modern approach*. Englewood Cliffs, NJ: Prentice Hall.

Gama, J., Torgo, L., & Soares, C. (1998). Dynamic discretization of continuous attributes. In *Proceedings of the Sixth Ibero-American Conference on Artificial Intelligence* (pp. 160-169).

Holte, R.C. (1993). Very simple classification rules perform well on most commonly used datasets. *Machine Learning, 11*, 63-91.

Kohavi, R., & Sahami, M. (1996). Error-based and entropy-based discretization of continuous features. In *Proceedings of the Second International Conference on Knowledge Discovery and Data Mining* (pp. 114-119).

Kwedlo, W., & Krętowski, M. (1999). An evolutionary algorithm using multivariate discretization for decision rule induction. In *Proceedings of the European Conference on Principles of Data Mining and Knowledge Discovery* (pp. 392-397).

Liu, H., Hussain, F., Tan, C.L., & Dash, M. (2002). Discretization: An enabling technique. *Data Mining and Knowledge Discovery, 6*, 393-423.

Maass, W. (1994). Efficient agnostic PAC-learning with simple hypotheses. In *Proceedings of the Seventh Annual ACM Conference on Computerized Learning Theory* (pp. 67-75).

Miller, R.J., & Yang, Y. (1997). Association rules over interval data. In *Proceedings of the ACM SIGMOD International Conference on Management of Data* (pp. 452-461).

Mitchell, T. (1997). *Machine learning*. Boston, MA: McGraw Hill.

Quinlan, J.R. (1986). Induction of decision trees. *Machine Learning, 1*, 81-106.

Ramakrishnan, R., & Gehrke, J. (2003). *Database management systems* (3rd ed.). New York: McGraw-Hill.

Srikant, R., & Agrawal, R. (1996). Mining quantitative association rules in large relational tables. In *Proceedings of the ACM SIGMOD International Conference on Management of Data* (pp. 1-12).

Truemper, K. (2004). *Design of logic-based intelligent systems*. New York: Wiley.

Wang, K., & Goh, H.C. (1997). Minimum splits based discretization for continuous features. In *Proceedings of the Fifteenth International Joint Conference on Artificial Intelligence* (pp. 942-951).

Witten, I.H., & Frank, E. (2000). *Data mining*. San Diego: Academic Press.

Wu, X. (1996). A Bayesian discretizer for real-valued attributes. *The Computer Journal, 39*, 688-691.

Chapter II
Vector DNF for
Datasets Classifications:
Application to the Financial Timing
Decision Problem

Massimo Liquori
Università di Roma "La Sapienza", Italy

Andrea Scozzari
Università di Roma "La Sapienza", Italy

ABSTRACT

Traditional classification approaches consider a dataset formed by an archive of observations classified as positive or negative according to a binary classification rule. In this chapter, we consider the financial timing decision problem, which is the problem of deciding the time when it is profitable for the investor to buy shares or to sell shares or to wait in the stock exchange market. The decision is based on classifying a dataset of observations, represented by a vector containing the values of some financial numerical attributes, according to a ternary classification rule. We propose a new technique based on partially defined vector Boolean functions. We test our technique on different time series of the Mibtel stock exchange market in Italy, and we show that it provides a high classification accuracy, as well as wide applicability for other classification problems where a classification in three or more classes is needed.

INTRODUCTION

In the area of knowledge-based expert systems, the aim is to detect structural information from large datasets in order to extract salient features for identifying differences that separate one set of data from another. Classification methods developed in the literature try to classify the given observations and, in addition, to classify new observations in a way consistent with past classifications. Such structural information can provide powerful means for the solution of a variety of problems, including classification, automated knowledge acquisition for expert systems, development of pattern-based decision support systems, detection of inconsistencies in databases, feature selection, medical diagnosis, marketing, and numerous aspects of etiology.

Several approaches coming from different fields have been proposed in the literature to tackle the classification problem. One of the best-known methods is support vector machines (SVM) that has proved highly successful in a number of classification studies. Although the subject traces its origin to the seminal work of Vapnik and Lerner (1963), it is only now receiving a growing attention. In the simplest case, given a set of observations classified into two classes, the aim is to construct a function to discriminate between classes. This can be done via a mathematical programming approach. A linear programming-based approach, stemming from the multisurface method of Mangasarian (1965, 1968), has been used for a breast cancer diagnosis system (Mangasarian, Setiono, & Wolberg, 1990; Mangasarian,, Street, & Wolberg, 1995; Wolberg, & Mangasarian, 1990). Another approach is the quadratic programming method based on Vapnik's Statistical Learning Theory (Cortes & Vapnik, 1995; Vapnik1995). See Burges (1998) for a tutorial on classification via SVMs. Bredensteiner and Bennett (1999) show how the linear programming and quadratic programming methods can be combined to yield two new approaches for the multiclass problem. Other mathematical programming techniques, based on the minimization of some function measuring the classification error (Freed & Glover, 1986; Glover, 1990; Kamath, Karmarkar, Ramakrishnan, & Resende, 1992; Triantaphyllou, Allen, Soyster, & Kumara, 1994), have been used in classification problems. A MINSAT approach for learning logic relationship that correctly classify a given

dataset has been recently proposed in Felici and Truemper (2000).

Decision trees are another popular technique for classification. The main reason behind their popularity seems to be their relative advantage in terms of interpretability. There are several efficient and simple implementations of decision trees (Quinlan, 1993). In a recent work, Street (2004) presents an algorithm based on nonlinear programming for multicategory decision trees. Unfortunately, one of the limitations of most decision trees is that they are known to be unstable, especially when dealing with large data sets (Fu, Golden, Lele, Raghavan, & Wasil, 2003). In the literature, there are several papers that provide heuristics and metaheuristics for the problem of finding an optimal decision tree, which is known to be an NP-complete problem (Fu, et al., 2003; Niimi & Tazaki, 2000).

Naive Bayes method is another simple but effective classifier (Jefferys & Berger, 1992; Yeung, 1993). The attributes, observed in the training set, are assumed to be conditionally independent, given the value of the class attribute. In order to derive a good classification rule, and considering the independence assumption made, the marginal probabilities of each attribute must be estimated. In Lin (2002), it is shown that the asymptotic target of support vector machines is some interesting classification functions that are directly related to the Bayes rule. Actually, the independence assumption is unrealistic, thus, Bayesian networks have been introduced that explicitly model dependencies between attributes (Pernkopf, 2005). Thus, given a set of observations, the problem is to find a network that best matches the training set. The search for the best network is based on a scoring function that evaluates each network with respect to the training data (Heckerman, Geiger, & Chickering, 1995; Lam & Bacchus, 1994).

Several classification problems can also be formulated as an artificial neural network problem. An artificial neural network (ANN) can be thought of as a mathematical paradigm that models the biological neural system. The theory and the design of ANNs has significantly developed during the last 20 years. The increasing interest for ANNs is mainly due to their ability to learn both from supervised and unsupervised datasets. ANNs are very well suited for solving large classification problems (Archer & Wang, 1993; Bishop, 2004; Boulle, Chandramohan, & Weller, 2001; Coakley & Brown, 2000). There exist several architectures for dealing with real applications, and different algorithms and methods are used for training a neural network. For instance, the backpropagation algorithm (Archer & Wang, 1993; Lawrence & Giles, 2000; Ooyen & Nienhuis, 1992) and the multisurface method of Mangasarian are often used for the training phase.

In this chapter, we propose a new classification technique, based on combinatorics, optimization, and partially defined *vector* Boolean functions, which is suitable when the archive of observations is, in particular, classified in more than two classes. Our method relates to the logical analysis of data (LAD), that is, a classification method proposed in (Hammer, 1986). LAD was first applied to the classification of binary datasets when a binary classification rule was adopted (Boros, Hammer, Ibaraki, Kogan, Mayoraz, & Muchnick, 2000). This method assumes that the archive of observations can be naturally represented by a partially defined Boolean function. The goal of LAD is to obtain an extension of the partially defined Boolean function, that is, a completely defined Boolean function that represents the classification of all the vectors in the sample space. The central concepts used by LAD are those of prime implicants, which are special logical conjunctions of literals imposed on the values of the attributes in the dataset. The aim is to generate a set of prime implicants for finding a suitable minimal *disjunctive normal form* (DNF) (Crama, Hammer, & Ibaraki, 1988) representation of a Boolean function that allows both to describe the archive, and to correctly classify all known and most new observations. Such a minimal DNF

provides an extension of the partially defined Boolean function. A straightforward extension of the LAD technique when an archive S is classified, for instance, into three classes, S_1, S_2, and S_3, can be obtained by first finding a minimal DNF that separates the class S_1 from $S_2 \cap S_3$, and then by referring to a second minimal DNF for classifying a new observation either in S_2 or in S_3. The disadvantage of this approach is that the two DNF so obtained may not capture all the relationships between the attributes and those salient features that separate one set from another.

We apply our classification technique to the *financial timing decision problem*, which is the problem of deciding, at each time period, if it is profitable for the investor to buy shares, or to sell shares, or to wait in the stock exchange market. The decision is based on finding a ternary classification of a dataset of observations of the stock exchange course. Each observation in the dataset is represented by a vector containing the values of some financial numerical attributes (Murphy, 1997) like, for instance, the relative strength index (RSI), the rate of change (ROC), the stochastic oscillators (SO), and so forth. More precisely, we classify as *positive* those observations that refer to periods when it is profitable for the investor to buy shares, *negative* when it is profitable to sell shares, and *null* when it is a waiting period. In this chapter, we develop an approach that produces a *vector disjunctive normal form* (VDNF) representation that allows one to classify a new observation directly in one of the sets S_1, S_2, or S_3. In fact, our archive S can be represented by a partially defined *vector* Boolean function ψ: $S \rightarrow \{0,1\}^2$ in such a way that for a given observation $s \in S$, we have $s \in S_1$, $s \in S_2$ and $s \in S_3$ if and only if $\psi(s)=(1,1)$, $\psi(s)=(0,0)$ and $\psi(s)=(1,0)$ or $\psi(s)=(0,1)$, respectively. We then proceed to find a simple VDNF representation consistent to this classification, that is, to find a simple extension of ψ. Here, the simple requirement means that we want to find a VDNF as short as possible. The VDNF embeds the structural information of the dataset so that we can correctly classify new observations and find out indications about the interpretation of the financial phenomenon under study. Both exact and heuristic procedures are used for generating a simple VDNF. Our technique has a wide applicability for classification problems where a classification in three or more classes is needed.

We provide the effectiveness of our method by means of numerical experiments. We use three time series of the Mibtel stock exchange market in Italy over a period of 1 year. Each time series reports the *open*, *close*, *high*, and *low* daily prices. We generate the attributes of the financial timing decision problem by computing four technical oscillators (Murphy, 1997): the Relative Strength Index, the Difference on Average based on 2 days moving average, the Rate of Change, and the Difference on Average based on a period of 3 days. By means of graphical and numerical analysis, we divide the set of observations into three classes: S_1, the set of the daily oscillators values suggesting to buy shares (positive observations); S_2, the set of values suggesting to sell shares (negative observations); and S_3, the set of values indicating a waiting period (null observations). We show that our method provides a high classification accuracy. We also compare our technique with some standard classification methods. In particular, we compare it with support vector machines, decision trees, naive Bayes, neural networks, and linear and quadratic discriminant analysis.

The remainder of the chapter is organized as follows: Section 2 provides some notation and the main results of the chapter, Section 3 presents an example and describes the application considered in this chapter. Section 4 reports some experimental results along with the comparison analysis with other classification methods, and finally, in Section 5, conclusions and further research issues are discussed. In the Appendix, we provide a short description of the financial attributes considered in our application.

NOTATION AND MAIN RESULTS

The input information to be classified is represented by an archive S of financial observations. We assume that each observation is represented by a d-dimensional vector containing the values of some financial numerical attributes (technical oscillators), where d is the number of attributes considered. We propose to have an initial classification of S into three classes, and we denote by S_1 the *positive* class, by S_2 the *negative* class, and by S_3 the *null* class. In financial problems, these classes refer to periods when it is profitable to buy shares, or to sell shares, or to wait in the stock exchange market. That is, an observation that refers to a particular day, and that is classified, for instance, in S_1, indicates that it is profitable to buy shares on that day because the next day the same operation will not be as profitable as the day before, that is, there could be an increase in the stocks' prices. Hence, each daily observation forecasts what will happen in the stock exchange market.

Our classification technique is based on a binary representation of the attributes. In fact, a *binarization* procedure, consisting of the transformation of numerical (real valued) data to binary (0,1) ones, can always be implemented. This procedure can be performed by referring to a *single cut-point* method, or to an *interval cut-point* method (see (Boros, Hammer, Ibaraki, & Kogan, 1997; Boros, *et al.*, 2000) for details). The cut points will be chosen in a way that allows one to distinguish between positive, negative, and null observations. A set of cut-points is *consistent* if in the resulting binarized archive $S_1 \cap S_2 = \varnothing$, $S_1 \cap S_3 = \varnothing$, and $S_2 \cap S_3 = \varnothing$. Hence, our binarized archive S can be represented by a partially defined *vector* Boolean function (pdVBf) $\psi: S \to \{0,1\}^2$ in such a way that for a given observation $s \in S$, we have $s \in S_1$, $s \in S_2$ and $s \in S_3$ if and only if $\psi(s)=(1,1)$, $\psi(s)=(0,0)$, and $\psi(s)=(1,0)$ or $\psi(s)=(0,1)$, respectively. When we introduce the largest set of cut-points in the binarization process, the resulting binarized archive S is called *Master* pdVBf.

Partially Defined Vector Boolean Functions and Their Extension

A *vector Boolean function* is a mapping $\phi:\{0,1\}^n \to \{0,1\}^m$ where $x \in \{0,1\}^n$ is a Boolean vector. In the sequel we will consider mostly the case $m=2$. The extension to the general case is straightforward, but it requires heavier notation. Following the notation introduced previously, we denote by $S_1(\phi)$ the set of all Boolean vectors x such that $\phi(x)=(1,1)$, $S_2(\phi)$ the set of all Boolean vectors x such that $\phi(x)=(0,0)$, and $S_3(\phi)$ the set of all Boolean vectors x such that $\phi(x)=(1,0)$, or $\phi(x)=(0,1)$. A partially defined *vector* Boolean function (pdVBf) ψ is defined by a triple of sets (S_1, S_2, S_3) such that S_1 denotes a set of positive examples, S_2 denotes a set of negative examples, and S_3 denotes a set of null examples, that is, ψ is defined on $S_1 \cup S_2 \cup S_3$ and $\psi(x)=(1,1)$ if $x \in S_1$, $\psi(x)=(0,0)$ if $x \in S_2$, and $\psi(x)=(1,0)$ or $(0,1)$ if $x \in S_3$.

We call a function ϕ an extension of the pdVBf $\psi(S_1, S_2, S_3)$ if $S_1 \in S_1(\phi)$, $S_2 \in S_2(\phi)$, and $S_3 \in S_3(\phi)$. Referring to our archive of observations $\psi(S_1, S_2, S_3)$, our goal is the determination of an extension ϕ that agrees with the classification in the archive S and that represents the classification of all the vectors in the sample space.

Referring to the theory of the Boolean functions, given a partially defined *scalar* Boolean function $\varphi: S \subseteq \{0,1\}^n \to \{0,1\}$, necessary and sufficient conditions for the existence of an extension $f: \{0,1\}^n \to \{0,1\}$ in different subclasses of *scalar* Boolean functions are provided in Boros, *et al.* (Boros, Ibaraki, & Mikino, 1998). We will denote by f a *scalar* Boolean function and by ϕ a *vector* Boolean function.

A simple extension of the arguments in Boros, *et al.* (1998) provides the following necessary and sufficient condition for the existence of an extension ϕ of a partially defined *vector* Bool-

ean function ψ in the class of all *vector* Boolean functions.

Fact. A partially defined *vector* Boolean function $\psi(S_1, S_2, S_3)$ has an extension in the class of all *vector* Boolean functions if and only if $(S_1 \cap S_2) = \varnothing$, $(S_1 \cap S_3) = \varnothing$, and $(S_2 \cap S_3) = \varnothing$.

Other classes of Boolean functions are of interest. For instance, given a *scalar* Boolean function $f: \{0,1\}^n \to \{0,1\}$, f is positive if $x \leq y$ always implies $f(x) \leq f(y)$ (Boros *et al.*, 1998). In the case of a *vector* Boolean function $\phi:\{0,1\}^n \to \{0,1\}^m$, ϕ is positive if $x \leq y$ implies $\phi(x) \leq \phi(y)$, that is, $\phi_i(x) \leq \phi_i(y)$, for all $i=1,\dots,m$. Therefore, a positive *vector* Boolean function ϕ has all the *scalar* Boolean functions ϕ_i, $i=1,\dots,m$, positive. As in Boros, *et al.* (1998), we give necessary and sufficient conditions for the existence of an extension ϕ of a pdVBf $\psi(S_1, S_2, S_3)$ in the class of positive *vector* Boolean functions:

Theorem 1. A pdVBf $\psi(S_1, S_2, S_3)$, has an extension ϕ in the class of positive *vector* Boolean functions if and only if for all $x \in S_1$, $y \in S_2$, and $z \in S_3$, we have $x \hbar y$, $x \hbar z$, and $z \hbar y$.

Proof: Necessity. Suppose that there exists a pair $x \leq y$ with $x \in S_1$ and $y \in S_2$ (or $x \leq z$ with $z \in S_3$ or $z \leq y$), then $\phi(x) \leq \phi(y)$, by positivity of ϕ, contradicting $\phi(x)=(1,1)$ and $\phi(y)=(0,0)$.

Sufficiency. Define the following *vector* Boolean function ϕ:

$$S_1(\phi) = \{c \in B^n \mid c \geq x \text{ for some } x \in S_1\}$$
$$S_2(\phi) = \{b \in B^n \mid b \leq y \text{ for some } y \in S_2\}$$
$$S_3(\phi) = B^n - S_1(\phi) - S_2(\phi).$$

The function ϕ is positive. Moreover, ϕ is a positive extension of ψ. From the positivity of ϕ, it follows immediately that $S_1 \subseteq S_1(\phi)$, and $S_2 \subseteq S_2(\phi)$. In order to complete the proof, it only remains to show that $(S_3 \cap (S_1(\phi) \cap S_2(\phi))) = \varnothing$. Indeed, if there exists $z \in (S_1(\phi) \cap S_3)$, then we have $z \in S_3$, and there exists $x \in S_1$ such that $x \leq z$, contradicting the assumption $x \not\geq z$. A similar argument shows that $S_3 \cap S_2(\phi) = \varnothing$. Hence, $S_3 \subseteq S_3(\phi)$, and this completes the proof.

In this chapter, given our archive of observations, represented as a pdVBf, we consider the problem of finding an extension ϕ as short as possible in the class of all *vector* Boolean functions. We first need to introduce some classical definitions from Boolean algebra.

The Boolean variables $x_1, x_2,..., x_n$ and their complements $x'_1, x'_2,..., x'_n$ are called literals. A *term T* is a conjunction of literals such that at most, one of x_i and x'_i appears for each variable. A *minterm* is a maximum length conjunction term, that is, it contains all variables normal (uncomplemented) or complemented (Schneeweiss, 1989). A disjunction of conjunctions T defines a *disjunctive normal form* (DNF). In the case of *scalar* Boolean functions, a DNF defines a function, and it is well-known that every *scalar* Boolean function f can be represented as a DNF (Schneeweiss, 1989); however, such a representation may not be unique. The goal is to find the minimal DNF representation of a *scalar* Boolean function f. To make the minimality requirement more precise, we need to introduce some further definitions.

A term T is an *implicant* for a given *scalar* Boolean function f if $T = 1$ implies $f = 1$. An *implicant T* is a *prime implicant* for the *scalar* Boolean function f if any term obtained by dropping a literal from it is not an implicant (Boros *et al.*, 1997).

Prime implicants are the fundamental blocks for generating an extension of a given partially *scalar* Boolean function. Indeed, the minimal DNF representation of a *scalar* Boolean function is obtained when its terms T are prime implicants (Crama & Hammer, 2002). The concepts of implicants and prime implicants can be generalized in the case of *vector* Boolean functions.

Definition 1. (McCluskey, 1986). Given a *vector* Boolean function $\phi:\{0,1\}^n \rightarrow \{0,1\}^m$, of components $\phi_1, \phi_2,..., \phi_m$, a term T is a *multiple implicant* (resp. *multiple prime implicant*) of ϕ if:

1. It is either an implicant (resp. prime implicant) of one of the functions ϕ_i, $i=1,...,m$ or;
2. It is an implicant (resp. prime implicant) of one of the product (conjunction) function $\phi_{i1} \cdot \phi_{i2} \cdots \phi_{ik}$, $1 \leq i_k \leq m$.

Here we introduce the concept of a *vector disjunctive normal form* VDNF as a vector of m components, each corresponding to a single DNF. A *vector* Boolean function may be represented by a VDNF. Notice that, by considering the class of positive *scalar* Boolean functions introduced above, it is well-known that a *scalar* Boolean function f is positive if and only if f can be represented by a DNF in which all the literals of each term are uncomplemented (Boros, *et al.*, 1998). Hence, by referring to the definition of positive *vector* Boolean functions, a *vector* Boolean function ϕ is positive if and only if it can be represented by a VDNF whose components are represented by DNFs in which all the literals of each term are uncomplemented. We want to find a short representation of a *vector* Boolean function ϕ in the sense described by the following theorem.

Theorem 2. (Existence of a short VDNF (McCluskey, 1986)). There exists a short representation of a *vector* Boolean function ϕ, in which each component ϕ_i is the disjunction of multiple prime implicants of ϕ, such that, all the terms that occur only in the expression for ϕ_i are prime implicants of ϕ_i; all the terms that occur in both the expressions for ϕ_i and ϕ_j with $i \neq j$ but in no other expressions are prime implicants for $\phi_i \cdot \phi_j$; and so forth.

Multiple Logic Minimization Problem

It was already noticed that a *scalar* Boolean function $f :\{0,1\}^n \rightarrow \{0,1\}$ may have numerous DNF representations (Schneeweiss, 1989). In many applications a short DNF representation of f is preferred over a longer one (Crama & Hammer, 2002). The problem of constructing a DNF representation as short as possible is referred to as the *logic minimization* problem. The same problem arises for a *vector* Boolean function $\phi:\{0,1\}^n \rightarrow \{0,1\}^m$. That is, we search for a short VDNF representation of ϕ with $m=2$ in our application. Perhaps, the most obvious technique is to find a short representation of every component (DNF) of the VDNF. Unfortunately, some simple examples show that this method does not necessarily lead to a short VDNF representation (McCluskey, 1986). The problem of finding a short VDNF is usually referred to as the *multiple logic minimization* problem. In our work, we refer to the *Quine-McCluskey* approach for solving it (McCluskey, 1986). This approach is divided into two phases. In the first phase, a *prime implicant table* is generated. For each (binarized) observation s in the Master pdVBf $\psi(S_1, S_2, S_3)$, along with its classification $\psi(s)=(\psi_1, \psi_2)$, a minterm and a multiple prime implicant for $\psi(s)=(\psi_1, \psi_2)$ are found. That is, for each observation s, a minterm for ψ_1 and ψ_2, and a prime implicant for ψ_1, ψ_2 and $\psi_1 \cdot \psi_2$ are generated. Each row of the *prime implicant table* corresponds to a multiple prime implicant, while each column corresponds to a minterm.

In the second phase, the Quine-McCluskey approach solves a set-covering problem by finding the minimum number of rows that covers all the columns. It is well-known that the set-covering problem is NP-complete, and therefore, this approach to multiple logic minimization does not provide a polynomial algorithm. In order to solve the set-covering problem, a reduction of the table is performed. For this, let us first recall the following definitions from McCluskey (1986):

Definition 2. A multiple prime implicant of a *vector* Boolean function $\phi = (\phi_1, \phi_2,..., \phi_m)$ is *essential* for a function ϕ_i if there is a minterm in the representation of ϕ_i that is included in only one multiple prime implicant.

Definition 3. Let $\phi = (\phi_1, \phi_2,..., \phi_m)$ be a *vector* Boolean function and let $P_1, P_2,..., P_t$ be the corresponding set of multiple prime implicants. Then, a minterm for a function ϕ_i is a *distinguished minterm* if and only if it is included in only one conjunction of literals that is a multiple prime implicant of ϕ_i, or of any of the product (conjunction) of functions involving ϕ_i.

Definition 4. A multiple prime implicant of a *vector* Boolean function $\phi = (\phi_1, \phi_2,..., \phi_m)$ is *essential* for a given function ϕ_i if and only if it includes a *distinguished minterm* of ϕ_i.

In the prime implicant table, rows and columns corresponding to essential multiple prime implicant and distinguished minterm are called *essential rows* and *distinguished columns*, respectively. The reduction of the table is first performed by deleting all the essential rows and dstinguished columns (McCluskey, 1986). The table so obtained can be further reduced by applying the *essential reducing algorithm* (ERA) described in Crama and Hammer (2002). The final reduced table A^* is the matrix of the corresponding set-covering problem. Given a set-covering matrix A, in Crama and Hammer (2002), it is proved that the ERA algorithm applied to A preserves the solutions of a set-covering problem. For further details about the algorithm for finding a short VDNF representation of a *vector* Boolean function ϕ, the interested reader can refer to McCluskey (1986). However, in the next section we provide a numerical example for finding an extension of a pdVBf.

A NUMERICAL EXAMPLE

In this section, we provide a numerical example for the computation of a VDNF. Let us consider the Master pdVBf given in Table 1, where s_i is a binarized observation, $i=1,...,14$ and $x_j, j=1,2,3,4$ are the cut-points.

Our aim is to find an extension of the above Master pdVBf as short as possible. By applying the first step of the Quine-McCluskey procedure, we can find an extension of the pdVBf of Table 1 by generating all the prime implicants both for ψ_1 and ψ_2 and for the function $\psi_1 \times \psi_2$ (see Table 2).

The resulting extension is:

$$\psi_1 = P_1 \vee P_2 \vee P_3 \vee P_5 \vee P_6 \vee P_7 =$$
$$= (x_1 \cdot x'_2) \vee (x_1 \cdot x_4) \vee (x_2 \cdot x_4) \vee (x'_1 \cdot x_2 \cdot x_4) \vee (x_3 \cdot x_4) \vee (x'_2 \cdot x_3)$$

$$\psi_2 = P_4 \vee P_5 \vee P_6 \vee P_7 = x_3 \vee (x'_1 \cdot x_2 \cdot x_4) \vee (x_3 \cdot x_4) \vee (x'_2 \cdot x_3).$$

This extension may not be a short one, so then, we apply the second step of the Quine-McCluskey procedure in order to eliminate all the redundancies. The first step is to select all the *essential* prime implicants (see Definitions 2-4).

Table 1. Partially defined vector Boolean function

	x_1	x_2	x_3	x_4	ψ_1	ψ_2
s_1	0	0	0	0	0	0
s_2	0	0	1	0	1	1
s_3	0	0	1	1	1	1
s_4	0	0	0	1	0	0
s_5	0	1	0	1	1	1
s_6	0	1	1	0	0	1
s_7	0	1	1	1	1	1
s_8	1	0	0	0	1	0
s_9	1	0	0	1	1	0
s_{10}	1	0	1	0	1	1
s_{11}	1	0	1	1	1	1
s_{12}	0	1	0	0	0	0
s_{13}	1	1	0	1	1	0
s_{14}	1	1	1	0	0	1

Table 2. Prime implicants

Prime Implicants for ψ_1	Prime Implicants for ψ_2	Prime Implicants for $\psi_1 \times \psi_2$
$P_1 = (x_1 \times x'_2)$	$P_4 = x_3$	$P_5 = (x'_1 \times x_2 \times x_4)$
$P_2 = (x_2 \times x_4)$		$P_6 = (x'_2 \times x_3)$
$P_3 = (x_1 \times x_4)$		$P_7 = (x_3 \times x_4)$

Table 3. The covering table

	s_2	s_3	s_5	s_7	s_8	s_9	s_{10}	s_{11}	s_{13}	s_2	s_3	s_5	s_6	s_7	s_{10}	s_{11}	s_{14}	
P_1					x	x	x	x										
P_2			x	x					x									ψ_1
P_3						x		x	x									
P_4										x	x		x	x	x	x	x	ψ_2
P_5			x	x									x	x				
P_6	x	x			x	x				x	x				x	x		$\psi_1 \times \psi_2$
P_7		x		x						x				x		x		
	ψ_1									ψ_2								

Table 4. The second covering table

	s_5	s_7	s_{13}	
P_2	x	x	x	
P_3			x	ψ_1
P_5	x	x		
P_7		x		$\psi_1 \times \psi_2$
		ψ_1		

For ψ_1 they are P_1 and P_6, while for ψ_2 they are P_4 and P_5. We delete from Table 2 all the minterms covered by the essential prime implicants. The remaining minterms to be covered are shown in Table 4.

The prime implicant P_2 in Table 4, dominates the other ones for ψ_1, that is, it covers all the minterms covered by $P_3, P_5, P_7,$ and other minterms not covered by P_3, P_5, P_7. It also completes, with the essential prime implicants P_1 and P_6, the covering of the component ψ_1 of the table.

The resulting extension is:

$$\psi_1 = P_1 \vee P_2 \vee P_6 = (x_1 \cdot x'_2) \vee (x_2 \cdot x_4) \vee (x'_2 \cdot x_3)$$

$$\psi_2 = P_4 \vee P_5 = x_3 \vee (x'_1 \cdot x_2 \cdot x_4)$$

which is shorter than the first extension found.

In fact, ψ_1 contains three terms less than before, while ψ_2 contains two terms less than before. Alternatively, a short extension can be obtained by solving the following set-covering problem:

$$min \; s_1 + s_2 + s_3 + s_4$$
s.t.
$$s_1 + s_3 \geq 1$$
$$s_1 + s_3 + s_4 \geq 1$$
$$s_1 + s_2 \geq 1$$
$$s_i = \{0,1\}, \; i=1,2,3$$

whose optimal solution is (1, 0, 0, 0) which refers to the same dominating prime implicant P_2.

THE APPLICATION

In our application, we refer to three time series of the Mibtel stock exchange market for the years 1999-2001. We considered these time series because they do not present a well-defined *primary trend* and, for each attribute, they show a considerable presence of lateral movements, or the so called *sideways*.

Each observation is related to the *open, close, high,* and *low* daily prices of the Mibtel stock exchange market. We use both graphical and algorithmic tools of the financial technical analysis to generate numerical attributes or *technical indicators*. The interpretation of these indicators allows us to classify our dataset into the three classes $S_1, S_2,$ and S_3. These indicators are obtained by computing the moving averages defined on some combinations of the open, close, high, and low prices. There are several indicators that can be considered by a financial expert to predict future movements in the market, and there is not a standard criterium for choosing a given subset of indicators. In the present work, we use a particular type of technical indicator called *Oscillators*. The oscillators are particularly useful when the market does not present a well-defined trend that can be used to identify situations of *oversold* and *overbought*. An overbought or oversold condition merely indicates that there is a high probability of a reaction in the market. These conditions suggest that there could be an opportunity to buy or sell securities. Just one oscillator is not able to provide information about the market situation. In the Appendix, we describe the characteristics as well as the meaning of the oscillators we used in our application.

EXPERIMENTAL RESULTS

This section reports on numerical experiments with the use of the VDNF technique for classification. We consider the three time series (1999-2001) as the initial classified archive. It is composed of 747 observations, where $|S_1| = 53$, $|S_2| = 66$, and $|S_3| = 379$ are the cardinalities of the three classes, respectively. In our experiments, we used as training sets both the 1999 time series only, representing 33% of the observations of the entire archive, and the 1999-2000 time series, representing 66% of the whole archive. Then, the accuracy of our method was tested on the complement of the two training sets. Since most of the observations in the archive are in the S_3 class, we did not use larger training sets. Otherwise, we would have obtained a test set with very few observations in one of the two classes S_1 and S_2, and this would have resulted in a bad classifica-

tion accuracy. Moreover, in contrast with other classification applications, we did not estimate the effectiveness of our technique by randomly selecting a subset of the dataset as training set. This is due to our particular application. In fact, for financial applications, the aim is to predict whether the market will continue to go up or down, that is, whether it is in an oversold or in an overbought condition. Thus, the data included in a training set must represent a continuum of observations for a given time period. However, determining the appropriate time period for historical inputs is one of the biggest challenges in financial forecasting.

We present a comparison of our VDNF approach with some standard classification methods. In particular, we compared our technique with support vector machines, decision trees, and naïve Bayes. For this, we use the implementation of these techniques contained in the Weka software (Witten & Frank, 2005). Moreover, classification results provided by neural networks and linear and quadratic discriminant analysis are reported as well. These last methods were implemented in the Matlab 7.0 environment. All the experiments were performed on a PC AMD Athlon 2500+ GHz. More details on the implementation of the classification methods compared are given as follows:

Support vector machines (SVM): We obtained support vector classifiers by using both a radial basis function (RBF) and a polynomial function as *kernels*. Furthermore, parameter-selection strategies have also been considered. That is, we searched for the appropriate value of a parameter C, which controls the trade-off between the classifier capacity and the training errors, and of parameters γ and d in the two kernel functions. For instance, in the case of RBF, we find the best values of the parameters (C, γ) to be used in the classification algorithm by searching on the following two grids of possible points for (C, γ).

We first consider the grid $[1 \cdot 10^{-4}, 1 \cdot 10^{-3}, ..., 1 \cdot 10^{4}] \times [1 \cdot 10^{-3}, 1 \cdot 10^{-2}, ..., 1 \cdot 10^{3}]$. Let (C_0, γ_0) be the pair associated with the best classification value. Then, we use the grid $[0.2C_0, ..., 8C_0] \times [0.2\gamma_0, ..., 8\gamma_0]$ to select the best pair. Similarly, for the polynomial kernel, the best classification values were obtained by searching the parameters (C, d) sequentially in the two grids $[1 \cdot 10^{-4}, 1 \cdot 10^{-3}, ..., 1 \cdot 10^{4}]$ ‰ $[1, 10, 50, 100]$ and $[0.2C_0, ..., 8C_0] \times [0.2 d_0, 0.4 d_0, 0.6 d_0, 0.8d_0, 2d_0, 3d_0, 4d_0]$. These grids were considered both when the training set consisted of 33% of the dataset, and when it consisted of 66% of the dataset. For the RBF, by referring to the two percentages of the training sets, the pairs that gave the best classification values were $C'=600$ and $\gamma'=6$, and $C'=20$ and $\gamma'=60$, respectively. For the polynomial kernel, the best pairs were $C'=600$ and $d'=20$, and $C'=4000$ and $d'=8$, respectively. In Table 5, we report the percentages of well-classified instances obtained by using the two kernel functions.

Decision trees: We referred to the popular *C4.5* algorithm implemented in the Weka software for constructing decision trees. We tested a set of parameters in the range [1,...,20], indicating the minimum number of instances to specify in a leaf of the tree. For both the percentages of the training sets, the best classification was obtained with the parameter set to the value 2. Moreover,

we observed that a pruning strategy did not yield better percentages of correctly classified instances than an unpruned strategy. In Table 5, we report the best values of well-classified instances found.

Naïve Bayes: The popular Naïve Bayes method is another simple but yet effective classifier. This method learns the conditional probability of each attribute, given the class label from the training data. Classification is then done by applying Bayes rule to compute the probability of a class value, given the particular instance, and predicting the class value with the highest probability. A strong independence assumption is made, that is, all the attributes are assumed conditionally independent, given the value of the class attribute. Numerical values of the attributes are usually handled by assuming that they have a Gaussian probability distribution. Since this assumption may be incorrect, we also implemented the method referring to the kernel density estimation that does not assume any particular distribution for the attribute values.

Neural networks: For training a feedforward neural network, we used the backpropagation algorithm implemented in Matlab 7.0 with the regularized mean square error (MSE regularized) as performance function. Although there are many variants of the backpropagation algorithm

Table 5. Percentage of well-classified instances with standard classification methods

Classifier	33% Training	66% Training
Support Vector Machine with RBF	95.38	96.78
Support Vector Machine with Polynomial Kernel	95.58	96.38
Decision Trees	94.97	95.79
Naïve Bayes	91.76	86.34
Naïve Bayes with kernel estimation	93.17	86.54
Neural Networks	74.83	79.52
Linear Discriminant Analysis	82.73	70.60
Quadratic Discriminant Analysis	52.81	75.10

in Matlab, we adopted the resilient backpropagation training algorithm that is able to eliminate the harmful effects of the magnitudes of the partial derivatives and is generally much faster than other procedures. A validation set was also introduced in order to check the progress of training. This set was formed by 3-months observations from December 2000 to February 2001. Both with a training set formed by 33% of the observations and with one formed by 66% of the observations, we trained six single-layer and four double-layer networks with [50,100,150,200,250,300] neurons and [10,15,25,50] neurons per layer, respectively. Each network was trained five times with randomly generated input weights. In Table 5, we report only the best classification percentages obtained.

Discriminant analysis: We used the tools for the linear and quadratic discriminant analysis of Matlab 7.0 that implement Fisher's Method.

We now consider the VDNF approach. Given the archive of observations, we derive three different binarized tables, each referring to three different binarization strategies. Namely, the three binarized archives are obtained by first applying only a *single cut-point* method, then only an *interval cut-point* method, and finally, by using both a *single cut-point* and an *interval cut-point* method (Boros, *et al.*, 1997; Boros, *et al.*, 2000). We report the classification accuracy of our technique by referring to each Master pdVBf obtained. In general, regardless of the method used, a binary encoding of a dataset of observations generates a great number of Boolean variables (cut-points). In fact, many of the Boolean variables introduced by a binarization procedure may not be needed to explain the phenomenon. Hence, a size reduction is actually necessary in order to prevent insurmountable computational difficulties at the VDNF generation stage. Following Boros *et al.* (2000), we reduce the dimension of the Master pdVBf by deleting the redundant variables. In order to generate the VDNF with the Quine-McCluskey procedure, we used two softwares:

1. **ESPRESSO II:** It generates the prime implicant table, as described before, and then applies a reduction process to this table.

Table 6. Classification accuracy with 33% training

	Single Cut Point		Interval Cut-Point		Single-Interval Cut Point	
	Dimension	Classified	Dimension	Classified	Dimension	Classified
Espresso II	249×155	94.98	249×146	88.55	249×288	84.54
Boom 2.3	249×155	95.18	249×146	94.77	249×288	95.18

Table 7. Classification accuracy with 66% training

	Single Cut Point		Interval Cut-Point		Single-Interval Cut Point	
	Dimension	Classified	Dimension	Classified	Dimension	Classified
Espresso II	498×191	98.99	498×181	98.99	498×346	95.58
Boom 2.3	498×191	98.99	498×181	98.99	498×346	97.98

Finally, depending on the dimension of the resulting reduced table, either an exact algorithm or a greedy-like procedure is implemented for solving the set-covering problem.

2. **BOOM v2.3:** It implements a stochastic greedy-like procedure for generating the prime implicant table. After applying a reduction process to this table, a stochastic greedy-like procedure is implemented for solving the resulting set-covering problem.

In Tables 6 and 7, we report the results when the training set consists of 33% of the dataset and of 66% of the dataset, respectively. For each software used, we present the Master pdVBf dimension (i.e., number of observations times number of cut-points used) and the percentage of the well-classified points in the test set.

Recalling the definition of the three classes, S_1, S_2 and S_3, we observed that the percentage of well-classified observations in the test set represents how many times we were able to correctly classify the observations in S_1, S_2 and S_3, and therefore, how many times we were able to make a profitable operation (to buy, to sell or to wait) in the stock exchange market.

At the beginning of this section we pointed out that, from an application point of view, in financial forecasting problems it make no sense to apply a classification method by randomly selecting a subset of the data set as training set. Nevertheless, we next report some experimental results on randomly generated training sets with the aim of evaluating the efficiency of our technique.

Let us consider the 1999 time series only as dataset. It has 249 observations where $|S_1| = 34$, $|S_2| = 27$, and $|S_3| = 188$ are the cardinalities of the three classes in which the dataset is subdivided, respectively. We tested the classification accuracy of our method by extracting from the 1999 series two different samples that were used as training sets, each containing about 47% of the observations in the archive (119 observations). The observations in the samples were chosen according to this rule: (1) for each sample, the number r_1, r_2, and r_3 of the observations that must be classified in each class were firstly decided; (2) from the 1999 time series we extract, at random, r_1 units from the observations in S_1, r_2 units from the observations in S_2, and r_3 units from the observations in S_3. We choose this rule since, by extracting at random 119 observations directly from the 1999 series, it may result in a sample in which one of the r_i, $i=1,2,3$ could be zero. Also, a training set (a sample) with very few observations in one class may give a bad classification accuracy.

In our application, for the first sample we considered r_1=19, r_2=12 and r_3=88, while for the second sample we considered r_1=19, r_2=15 and r_3=85. As test set, we used the rest of the 1999 time series composed of 130 observations. The results are reported in Table 8. For each sample, we provide the dimension of the Master pdVBf generated, and the percentages of the well-classified observations in the test set for both software. We used only the simple cut-point method for the binarization of the numerical dataset.

Table 8. Dataset composed by the 1999 time series

	Master pdVBf dimension	Well-Classified	
		Espresso II	Boom 2.3
First sample	119×101	90.77	91.54
Second sample	119×112	86.15	92.31

CONCLUSION AND FURTHER RESEARCH

In this chapter, we proposed a new classification technique that is based on combinatorics, optimization, and partially defined *vector* Boolean functions. Our technique is concerned with classification problems, where the goal is to extract salient features from an archive of observations in order to separate one set of observations from another. We developed a method that can be efficiently used when the observations in the archive are divided into three (or more) classes according to a ternary classification rule. In particular, in this chapter, we applied our classification method to a financial problem, the *financial timing decision* problem. The classification performance of our technique was tested on three financial time series. We compared our technique with some standard classification approaches. In particular, we compared it with support vector machines, decision trees, naïve Bayes, neural networks, and linear and quadratic discriminant analysis, obtaining encouraging results. Our technique seems to provide a high classification accuracy, as well as a wide applicability for classification problems where a classification in three or more classes is needed. In fact, it outperforms almost all the standard classification methods, and compares favorably with the SVM classifiers that have proved highly successful in a number of classification studies. Moreover, it reveals a good explanatory power of the phenomenon under study, since a VDNF, which makes use of multiple prime implicants, better captures the combination between the attributes considered.

Of course we are conscious that further research is needed for a better understanding of the mathematical and computational aspect of this technique, and also further classification problems need to be considered in order to better understand the domain of applicability of our method.

REFERENCES

Archer, N., & Wang, S. (1993). Application of the backpropagation neural network algorithm with monotonicity constraints for two-group classification problems. *Decision Sciences, 24*, 60-75.

Bennett, K., & Mangasarian, O.L. (1992). Robust linear programming discrimination of two linearly inseparable sets. *Optimization Methods and Software, 1*, 23-34.

Bishop, C.M. (Ed.). (2004). *Neural network in pattern recognition*. New York: Oxford University Press.

Boom 2.3. Retrieved from http://vlsicad.eecs. umich.edu

Boros, E., Hammer, P. L., Ibaraki, T., & Kogan, A. (1997). A logical analysis of numerical data. *Mathematical Programming, 79*, 163-190.

Boros, E., Hammer, P. L., Ibaraki, T., Kogan, A., Mayoraz, E., & Muchnick, I. (2000). An implementation of Logical Analysis of Data. *IEEE Transactions on Knowledge and Data Engineering, 12*, 292-306.

Boros, E., Ibaraki, T., & Mikino, K. (1998). Error-free and best-fit extensions of partially defined Boolean Functions. *Information and Computation, 140*, 254-283.

Boulle, A., Chandramohan, D., & Weller, O. (2001). A case study of using artificial neural networks for classifying cause of death from autopsy. *International Journal of Epidemiology, 30*, 515-520.

Bredensteiner, E.J., & Bennett, K. P. (1999). Multicategory classification by support vector machines. *Computational Optimization and Application, 12*, 53-79.

Burges, C.J.C. (1998). A tutorial on support vector machines for pattern recognition. *Data Mining and Knowledge Discovery, 2*, 121-167.

Coakley, J., & Brown, C. (2000). Artificial neural networks in accounting and finance: Modeling issues. *International Journal of Intelligent Systems in Accounting, Finance and Management, 9*, 119-144.

Cortes, C., & Vapnik, V. (1995). Support vector networks. *Machine Learning, 20*, 273-297.

Crama, Y., & Hammer, P. L. (2002). *Boolean functions—Theory, algorithms, and applications.* Retrieved from http://www.rogp.hec.ulg.ac.be/Crama/

Crama, Y., Hammer, P.L., & Ibaraki, T. (1988). Cause-effect relationships and partially defined Boolean functions. *Annals of Operations Research. 16*, 299-326.

Espresso II. Retrieved from http://www-cad.eecs.berkeley.edu

Felici, G., & Truemper, K. (2000). A MINSAT approach for learning in logic domains. *INFORMS Journal on Computing, 13*, 1-17.

Freed, N., & Glover, F. (1986). Evaluating alternative linear programming models to solve the two-group discriminant problem. *Decision Sciences, 17*, 151-162.

Fu, Z., Golden, B., Lele, S., Raghavan, S., & Wasil, E. (2003). A genetic algorithm-based approach for building accurate decision trees. *Informs Journal of Computing, 5*, 3-22.

Glover, F. (1990). Improved linear programming models for discriminant analysis. *Decision Sciences, 21*, 771-785.

Hammer, P.L. (1986). *Partially defined Boolean functions and cause-effect relationships.* Paper presented at the International Conference on Multi-Attribute Decision Making Via OR-Based Expert Sytems, University of Passau, Passau Germany.

Heckerman, D., Geiger, D., & Chickering, D. M. (1995). Learning Bayesian networks: The combination of knowledge and statistical data. *Machine Learnin,. 20*, 197-243.

Kamath, A.P., Karmarkar, N.K., Ramakrishnan, K.J., & Resende, M.G.C. (1992). A continuous approach to inductive inference. *Mathematical Programming, 57*, 215-238.

Jang, G.S., Lai, F., & Parng, T.M. (1993). Intelligent stock trading decision support system using dual adaptive-structure neural networks. *Journal of Information Science and Engineering, 9*, 271-297.

Jefferys, W.H., & Berger, J.O. (1992). Ockhams razor and Bayesian analysis. *American Science, 80*, 64-72.

Lam W., & Bacchus, F. (1994). Learning bayesian belief networks. An approach based on the MDL principle. *Computational Intelligence, 10*, 269-293.

Lawrence, S., & Giles, L. (2000). Overfitting and neural networks: Conjugate gradient and backpropagation. *IEEE Computer Society,* 114-119.

Lin, Y. (2002). Support vector machines and the bayes rule in classification. *Data Mining and Knowledge Discovery, 6*, 259-275.

Mangasarian, O.L. (1965). Linear and nonlinear separation of patterns by linear programming. *Operations Research, 13*, 444-452.

Mangasarian, O.L. (1968). Multi-surface method of pattern separation. *IEEE Transactions on Information Theory, 14*, 801–807.

Mangasarian, O.L., Street, W.N., & Wolberg, W.H. (1995). Breast cancer diagnosis and prognosis via linear programming. *Operations Research, 43*, 570-577.

Mangasarian, O.L., Setiono, R., & Wolberg, W. H. (1990). Pattern recognition via linear programming: Theory and application to medical diagnosis. In T. H. Coleman & Y. Li (Eds.), *Large-scale numerical optimization* (pp. 22-30). Philadelphia: SIAM Publication.

McCluskey, E. J. (Ed.). (1986). *Logic design principle.* NJ: Prentice Hall, Inc.

Murphy, J. J. (Ed.). (1997). *Analisi Tecnica dei Mercati Finanziari.* New York Institute of Finance.

Niimi, A., & Tazaki, E. (2000). *Genetic programming combined with association rule algorithm for decision tree construction.* Paper presented at the fourth international conference on knowledge-based intelligent engeneering systems & allied technologies, Brighton, UK.

Ooyen, A., & Nienhuis, B. (1992). Improving the convergence of the back-propagation algorithm. *Neural Networks, 5,* 465-471.

Pernkopf, F. (2005). Bayesian network classifiers versus selective k-NN classifier. *Pattern Recognition, 38,* 1-10.

Quinlan, J. R. (Ed.). (1993). *C4.5: Programs for machine learning.* San Mateo, CA: Morgan-Kaufmann.

Schneeweiss, W.G. (Ed.). (1989). *Boolean funcions with engineering applications and computer programs.* Berlin: Springer-Verlag.

Street, W.N. (in press). Multicategory decision trees using nonlinear programming. *Informs Journal on Computing.*

Triantaphyllou, E., Allen, L., Soyster, L., & Kumara, S.R.T. (1994). Generating logical expressions from positive and negative examples via a branch-and-bound approach. *Computer and Operations Research, 21,* 185-197.

Vapnik, V., & Lerner, A. (1963). Pattern recognition using generalized portrait method. *Automation and Remote Control, 24,* 774-780.

Vapnik, V. (Ed.). (1995). *The nature of statistical learning theory.* Springer-Verlag.

Witten, I.H., & Frank, E. (Ed.). (2005). *Data mining: Practical machine learning tools and techniques* (2nd ed.). San Francisco: Morgan Kaufmann.

Wolberg, W.H., & Mangasarian, O.L. (1990). Multisurface method of pattern separation for medical diagnosis applied to breast cytology. In *Proceedings of the National Academy of Sciences USA, 87,* 9193-9196.

Yeung, D.Y. (1993). Constructive neural networks as estimators of Bayesian discriminant functions. *Pattern Recognition, 26,* 189-204.

APPENDIX: THE TECHNICAL OSCILLATORS

Relative Strength Index (RSI)

The relative strength index is a well-known oscillator developed by Welles Wilder Jr. RSI measures the relative changes between high and low closing prices, and provides an insight of overbought and oversold conditions. The term relative strength generally implies a comparison between two different markets or indices. RSI provides early "warning signals" (buy or sell signals) if it is used in conjunction with other indicators. The relative strength index values can be plotted on a vertical scale ranging from 0 to 100. The 70 and 30 values are refered to as warning signals. An RSI value above 70 is related to an overbought condition, indicating a (probably) selling period, while a value below 30 refers to an oversold condition, indicating a (probably) buying period. The values 80 and 20 are often preferred by some traders. The information provided by the RSI depends upon the time interval on which it is computed. The shorter the interval, the more sensitive is the information provided by the index. Time intervals of 9, 10, and 25 days are often considered. Extending the time period makes the oscillator smoother and narrower in amplitude. RSI signals should always be used in conjunction with trend-reversal indicator prices.

Rate of Change (ROC)

ROC measures the "speed" of the prices in the market. Indeed, ROC is sometimes referred to as the price rate of change (PROC). Growing values of ROC indicate a *bullish* prices increasing period, while falling values of ROC indicate a *bearish* prices decreasing period. The ROC index displays the amount of price changes over a given time period. ROC can be represented as a *wave*. When the wave is above an equilibrium line, usually the zero-line, we assume to be in a buying period. When the wave falls below the equilibrium line, we assume to be in a selling period. When the wave starts growing from below the equilibrium line, we have an indication of a coming bullish period. The symmetric configuration is considered a forthcoming bearish period. Like the RSI, the ROC can also be computed referring to different time periods. If ROC is computed referring to a 10- or 12-day interval, it is a good short-term price indicator.

Moving Average Convergence/Divergence Trading Method (MACD)

The MACD method, developed by Gerald Appel, is a trend indicator, telling us whether a stock is in an *uptrend* or in a *downtrend* (Murphy, 1997). The direction of a long-term trend is the first assessment one should consider in any market. An *uptrend* is preferred and indicates a buying period, while a *downtrend* indicates a selling period. The simplest representation of this indicator is composed of two lines: the MACD line, which is the difference between two exponential moving averages (EMAs), and a signal line, which is an EMA of the MACD line itself. The signal or trigger line is plotted on top of the MACD to show buy or sell opportunities. Gerald Appel's MACD method uses a 26-day and 12-day EMA, based on the daily close prices, and a 9-day EMA for the signal line. The basic MACD trading rule is to buy when the MACD rises above its signal line. Similarly, a sell signal occurs when the MACD crosses below its signal line. If the MACD line is above the signal line, it denotes the beginning of a trend. An uptrend typically stops when the MACD line falls below the signal line.

Chapter III
Reducing a Class of Machine Learning Algorithms to Logical Commonsense Reasoning Operations

Xenia Naidenova
Military Medical Academy, Russia

ABSTRACT

The purpose of this chapter is to demonstrate the possibility of transforming a large class of machine-learning algorithms into commonsense reasoning processes based on using well-known deduction and induction logical rules. The concept of a good classification (diagnostic) test for a given set of positive examples lies in the basis of our approach to the machine-learning problems. The task of inferring all good diagnostic tests is formulated as searching the best approximations of a given classification (a partitioning) on a given set of examples. The lattice theory is used as a mathematical language for constructing good classification tests. The algorithms of good tests inference are decomposed into subtasks and operations that are in accordance with main human commonsense reasoning rules.

INTRODUCTION

The development of a full online computer model for integrating deductive and inductive reasoning is of great interest in machine learning. The main tendency of integration is to combine, into a whole system, some already well-known models of learning (inductive reasoning) and deductive reasoning. For instance, the idea of combining inductive learning from examples with prior knowledge and default reasoning has been advanced in Giraud-Carrier and Martinez (1994).

Obviously, this way leads to a lot of difficulties in knowledge representation because deductive reasoning tasks are often expressed in the classical first-order logic language (FOL), but machine-learning tasks use a variant of simbolic-valued attribute language (AVL).

The principe of "aggregating" different models of human thinking for constructing intelligent computer systems leads to dividing the whole process into two separate modes: learning and execution or deductive reasoning. This division is used, for example, in Zakrevskij (1982, 1987; Zakrevskij & Vasylkova, 1997). This approach is based on using finite spaces of Boolean or mul-tivalued attributes for modeling natural subject areas. It combines inductive inference used for extracting knowledge from data with deductive inference (the type of theorem proving) for solving pattern recognition problems. The inductive inference is reduced to looking for empty (forbidden) intervals of Boolean space of attributes describing a given set of positive examples. The deductive inference relates to the situation when an object is contemplated with known values of some attributes and unknown values of some others, including a goal attribute. The possible values of the latter ones are to be calculated on the base of implicative regularities in the Boolean space of attributes. In Zakrevskij (2001), the results of prolonged research conducted in that direction at the Institute of Engineering Cybernetics in Minsk are given.

The fundamental unified model for combining inductive reasoning with deductive reasoning is developed in the framework of inductive logic programming (ILP). ILP is a discipline that investigates the inductive construction of first-order clausal theories from examples and background knowledge. ILP has the same goal as machine learning, namely, to develop tools and techniques to induce hypotheses from examples and to obtain new knowledge from experience; but, the traditional theoretical basis of ILP is in the framework of first-order predicate calculus.

Inductive inference in ILP is based on inverting deductive inference rules; for example, inverting resolution (rules of absorption, identification, intraconstruction, and interconstruction), inverting implication (inductive inference under θ-subsumption).

There is a distinction between concept learning and program synthesis. Concept learning and classification problems, in general, are inherently object oriented. It is difficult to interpret concepts as subsets of domain examples in the frameworks of ILP. One of the ways to overcome this difficulty has been realized in a transformation approach: an ILP task is transformed into an equivalent learning task in different representation formalism. This approach is realized in LINUS (Lavrač & Džeroski, 1994; Lavrač, Gamberger, & Jovanoski, 1999), which is an ILP learner-inducing hypotheses in the form of constrained deductive hierarchical database (DHDB) clauses. The main idea of LINUS is to transform the problem of learning relational DHDB descriptions into the attribute-value learning task. This is achieved by the so-called DHDB interface. The interface transforms the training examples from the DHDB form into the form of attribute-value tuples. Some well-known attribute-value learners can then be used to induce "if-then" rules. Finally, the induced rules are transformed back into the form of DHDB clauses. The LINUS uses already-known algorithms, for example, the decision tree induction system ASSISTANT, and two rule induction systems: an ancestor of AQ15 named NEWGEM, and CN2.

A simple form of predicate invention through first-order feature construction is proposed by Lavrač and Flash (2000). The constructed features are used then for propositional learning.

Another way for combining ILP with an attribute-value learner has been developed in Lisi and Malerba (2004). In this work, a novel ILP setting is proposed. This setting adopts AL-log as a knowledge representation language. It allows a unified treatment of both the relational and structural features of data. This setting has been implemented in SPADA, an ILP system developed for mining multilevel association rules in spatial databases and applied to geographic data mining.

AL-log is a hybrid knowledge representation system that integrates the description logic ALC (Schmidt-Schauss & Smolka, 1991) and the deductive database language DATALOG (Ceri, Gotlob, & Tanca, 1990). Therefore, it embodies two subsystems, called structural and relational, respectively.

The description logic ALC allows for the specification of structural knowledge in terms of concepts, roles, and individuals. Individuals represent objects in the domain of interest. Concepts represent classes of these objects, while roles represent binary relations between concepts. Complex concepts can be defined from primitive concepts and roles by applying constructors such as $\cap$ (conjunction), $\cup$ (disjunction), and $\neg$ (negation).

ALC knowledge bases have an intensional part and an extensional part. In the intensional part, relations between concepts are syntactically expressed as inclusion statements of the form $C \subseteq D$ where C and D are two arbitrary concepts. As for the extensional part, it is possible to specify instances of relations between individuals and concepts. Relations are expressed as membership assertions, for example, concept assertions of the form $a : C$ ("a belongs to C").

The formal model of conceptual reasoning, based on an algebraic lattice, has been obtained in two independent ways. One way goes back to the works of the great psychologist J. Piaget, who introduced the concept of grouping (1959) to explain methods of object classification used mainly by 7- to 11-year-old children.

The idea of concepts' classification as a lattice arose from practical tasks of developing information retrieval and pattern recognition systems. In 1974, Shreider described the classification algebra as idempotent semigroup with the unit element.

In the same year, Boldyrev (1974) advanced the formalization of pattern recognition system as algebra with two binary operations of refinement and generalization defined by an axiom system, including lattice axioms. The ideas of Boldyrev have been used often for minimization of Boolean partial functions with a large number of "Don't Care" conditions, but we have been interested, from the beginning of our investigation, in applying the lattice theory for feature extraction and classification of attribute-value's tuples, and later, of concepts (symbols, names...).

The formal concept analysis (FCA), based on the concept lattice, has been advanced by Wille (1992). The problems of the FCA have been extensively studied by Stumme et al. (Stumme, Taouil, Bastide, Pasquier, & Lakhal, 2000), Dowling (1993), Salzberg (1991). Some algorithms for building concept lattices are considered in Nourine and Raynaud (1999), Ganter (1984), Kuznetsov (1993), and Kuznetsov and Obiedkov (2001).

A lot of experience has been obtained on the application of algebraic lattices in machine learning. From this point of view, the JSM-method of reasoning (Finn, 1984, 1988, 1991, 1999) is interesting.

The JSM-method of hypotheses' automatic generation formalizes a special class of plausible reasoning. The technique of this method is a synthesis of several cognitive procedures: empirical induction based on modeling John S. Mill's joint rule of similarity-distinction (Mill, 1900), causal analogy, and Charles S. Peirce's abduction.

Similarity in the JSM-method is both a relation and an operation that is idempotent, commutative and associative (i.e., it induces a semilattice on objects' descriptions and their generalizations). Being described in algebraic terms, the JSM-method can be implemented in the procedural programming languages.

In Galitsky et al. (Galitsky, Kuznetsov, & Vinogradov, 2005), the system JASMINE, based on the JSM-method, is presented. The system extends this methodology by implementing (1) a combination of abductive, inductive, and analogical reasoning for hypotheses generation, and (2) multivalued logic-based deductive reasoning for verification of their consistency. Formally, all the above components can be represented as deductive inference via logic programming (Anshakov, Finn, & Skvortsov, 1989; Finn, 1999). In fact, JASMINE is based on the logic programming implementation (Vinogradov, 1999).

The idea of using algebraic lattices for knowledge or data representation is realized by a lot of researchers. We can mention some of them: the works of the French group (Ganascia, 1989); the work on conceptual clustering (Carpineto & Romano, 1996); the works related to conceptual knowledge discovery (Mephu & Njiwoua, 1998; Stumme, Wille, & Wille, 1998). The following works are devoted to the application of algebraic lattices for extracting functional and implicative dependencies from data: Demetrovics and Vu (1993), Mannila and Räihä (1992), Mannila and Räihä (1994), Huntala et al. (Huntala, Karkkainen, Porkka, & Toivonen, 1999), Cosmadakis et al. (Cosmadakis, Kanellakis, & Spyratos, 1986), Naidenova and Polegaeva (1986), Megretskaya (1988), Naidenova et al. (Naidenova, Polegaeva, & Iserlis, 1995a), Naidenova et al. (Naidenova, Plaksin, & Shagalov, 1995b), Naidenova (1992, 2001.

An advantage of the algebraic lattices approach is based on the fact that an algebraic lattice can be defined both as an algebraic structure that is declarative, and as a system of dual operations with the use of which the elements of this lattice and the links between them can be generated.

Our approach to machine-learning problems is based on the concept of a good diagnostic (classification) test. We have chosen the lattice theory as a model for inferring good diagnostic tests from examples from the very beginning of our work in this direction. This concept has been advanced firstly in the framework of inferring functional and implicative dependencies from relations (Naidenova & Polegaeva, 1986). But

later, the fact has been revealed that the task of inferring all good diagnostic tests for a given set of positive and negative examples can be formulated as the search of the best approximation of a given classification on a given set of examples, and that it is this task that some well-known machine-learning problems can be reduced to (Naidenova, 1996): finding keys and functional dependencies in database relations, finding association rules, finding implicative dependencies, inferring logical rules (if-then rules, rough sets, "ripple down" rules), decision tree construction, learning by discovering concept hierarchies, eliminating irrelevant features from the set of exhaustively generated features.

In this chapter, we would like to demonstrate the possibility of transforming a large class of machine-learning algorithms for inferring good classification tests into the commonsense reasoning processes based on using well-known logical reasoning rules.

In this chapter, we describe the forms of an expert's rules (rules of the first type). The rules of the first type can be represented with the use of only one class of logical rules based on implicative dependencies between concepts (names). Then we describe commonsense reasoning operations (deductive and inductive) or rules of the second type. The concept of a good diagnostic test is introduced, and the problem of inferring all good diagnostic tests for a given classification on a given set of examples is formulated. We give the description of the mathematical model underlying algorithms of inferring good tests from examples. This model allows one to demonstrate that the inferring good tests entails applying deductive and inductive commonsense reasoning rules of the second type. We propose a decomposition of learning algorithms into operations and subtasks with the use of which good diagnostic tests inferring is transformed into an incremental process. The concepts of an essential value and an essential example are also introduced. We describe an incremental learning algorithm DIAGaRa and an approach to incremental inferring good diagnostic tests. The chapter ends with a brief summary section.

THE LOGICAL REASONING RULES

We need the following three types of rules in order to realize logical inference (deductive and inductive):

- *INSTANCES* or relationships between objects or facts really observed. Instance can be considered as a logical rule with the least degree of generalization. On the other hand, instances can serve as a source of a training set of positive and negative examples for inductive inference of generalized rules.
- *RULES OF THE FIRST TYPE,* or logical rules. These rules describe regular relationships between objects and their properties and between properties of different objects. The rules of the first type can be given explicitly by an expert, or derived automatically from examples with the help of some learning process. These rules are represented in the form of "if-then" assertions.
- *RULES OF THE SECOND TYPE* or inference rules with the help of which rules of the first type are used, updated, and inferred from data (instances). The rules of the second type embrace both inductive and deductive reasoning rules.

The Rules of the First Type

The rules of the first type can be represented with the use of only one class of logical statements; namely, the statements based on implicative dependencies between names. Names are used for designating concepts, things, events, situations, or any evidences. They can be considered as attributes' values in the formal representations of logical rules. In our further consideration, the letters A, B, C, D, a, b, c, d ...will be used as attributes' values in logical rules.

We consider the following rules of the first type:

- **Implication:** $a, b, c \rightarrow d$. This rule means that if the values standing on the left side of the rule are simultaneously true, then the value on the right side of the rule is always true.
- **Interdiction or forbidden rule:** (a special case of implication) $a, b, c \rightarrow false$ (*never*). This rule interdicts a combination of values enumerated on the left side of the rule. The rule of interdiction can be transformed into several implications such as $a, b \rightarrow$ not c; a, $c \rightarrow$ not b; b, $c \rightarrow$ not a.
- **Compatibility:** $a, b, c \rightarrow rarely$; $a, b, c \rightarrow frequently$. This rule says that the values enumerated on the left side of the rule can simultaneously occur rarely (frequently). The rule of compatibility presents the most frequently observed combination of values that is different from a law or regularity, with only one or two exceptions. Compatibility is equivalent to a collection of assertions as follows:

$a, b \rightarrow c$ *rarely* (*frequently*)
$a, c \rightarrow b$ *rarely* (*frequently*)
$b, c \rightarrow b$ *rarely* (*frequently*)

- **Diagnostic rule:** $x, d \rightarrow a$; $x, b \rightarrow$ not a; d, $b \rightarrow false$. For example, d and b can be two values of the same attribute. This rule works when the truth of "x" has been proven and it is necessary to determine whether "a" is true or not. If "x & d" is true, then "a" is true, but if "x & b" is true, then "a" is false.
- **Rule of alternatives:** a or $b \rightarrow true$ (*always*); $a, b \rightarrow false$. This rule says that a and b cannot be simultaneously true; either a or b can be true, but not both. This rule is a variant of interdiction.

Deductive Reasoning Rules of the Second Type

Deductive steps of commonsense reasoning consist of inferring consequences from some observed facts with the use of statements of the form "if-then" (i.e., knowledge). For this goal, deductive rules of reasoning are applied, the main forms of which are modus ponens, modus tollens, modus ponendo tollens, and modus tollendo ponens.

Let x be a collection of true values of some attributes (or evidences), observed simultaneously.

- **Using implication:** Let r be an implication, left(r) be the left part of r and right(r) be the right part of r. If left(r) $\subseteq x$, then x can be extended by right(r): $x \leftarrow x \cup$ right(r). Using implication is based on modus ponens: if A, then B; A; hence B.
- **Using interdiction:** Let r be an implication $y \rightarrow$ not k. If left(r) $\subseteq x$, then k is a forbidden value for all the extensions of x. Using interdiction is based on modus ponendo tollens: either A or B (A, B – alternatives); A; hence not B; either A or B; B; hence not A.
- **Using compatibility:** Let r = "$a, b, c \rightarrow k$, *rarely* (*frequently*)," where *rarely, frequently* are the values of a special attribute (SA). If left(r) $\subseteq x$, then k can be used for an extension of x with the value of SA equal to "*rarely*" ("*frequently*"). The application of several rules of compatibility leads to the appearance of several values "*rarely*" and/or "*frequently*" in the extension of x.

Computing the value of SA for the extension of x requires special consideration. In any case, the appearance of at least one value "*rarely*" means that the total result of the extension will have the value of SA equal to "*rarely*." Two values equal to "*frequently*" lead to the result "*less frequently*," three values equal to "*frequently*" lead to the

result "*less less frequently*," and hence the values "*rarely*" and "*frequently*" must have the ordering scale of measuring.

Using compatibility is based on modus ponens.

- **Using diagnostic rules:** Let r be a diagnostic rule such as "$x, d \rightarrow a$; $x, b \rightarrow$ not a," where "x" is true, and "a," "not a" are hypotheses or possible values of some attribute. Using a diagnostic rule is based on modus ponens and modus ponendo tollens.
  There are several ways for refuting one of the hypotheses:
  1. To infer either d or b with the use of one's knowledge;
  2. To involve new known facts and/or statements for inferring (with the use of inductive reasoning rules of the second type) new rules of the first type for distinguishing the hypotheses "a" and "not a"; to apply these new rules;
  3. To get, from an observation, which of the values d or b is true?
- **Using rule of alternatives**: Let "a" and "b" be two alternative hypotheses about the value of some attribute. If one of these hypotheses is inferred with the help of reasoning operations, then the other one is rejected. Using a rule of alternatives is based on modus tollendo ponens: either A or B (A, B – alternatives); not A; hence B; either A or B; not B; hence A.

The operations enumerated can be named as "forward reasoning" rules.

Experts also use implicative assertions in a different way. This way can be named as "backward reasoning."

- **Generating hypothesis or abduction rule:** Let r be an implication $y \rightarrow k$. Then the following hypothesis is generated "if k is true, then it is possible that y is true."

- **Using modus tollens:** Let r be an implication $y \rightarrow k$. If "not k" is inferred, then "not y" is also inferred.

Natural diagnostic reasoning is not any method of proving the truth. It has another goal: to infer all possible hypotheses about the value of some target attribute. These hypotheses must not contradict with the expert's knowledge and the situation under consideration. The process of inferring hypotheses is reduced to extending maximally a collection x of attribute values such that none of the forbidden pairs of values would belong to the extension of x.

Inductive Reasoning Rules of the Second Type

Inductive steps of common sense reasoning consist of using already known facts and statements, observations, and experience for inferring new logical rules of the first type or correcting those that turn out to be false.

For this goal, inductive reasoning rules are applied. The main forms of induction are the canons of induction that have been formulated by English logician Mill (1900). These canons are known as the five induction methods of reasoning: method of only similarity, method of only distinction, joint method of similarity-distinction, method of concomitant changes, and method of residuum.

- **The method of only similarity:** This rule means that if the previous events (values) A, B, C lead to the events (values) a, b, c and the events (values) A, D, E lead to the events (values) a, d, e, then A is a plausible reason of a.
- **The method of only distinction:** This rule means that if the previous events (values) A, B, C lead to (or give rise to) the events (values) a, b, c and the events (values) B, C lead to the events (values) b, c, then A is a plausible reason of a.

- **The joint method of similarity-distinction:** This method consists of applying two previous methods simultaneously.
- **The method of concomitant changes:** This rule means that if the change of a previous event (value) A is accompanied by the change of an event (value) a, and all the other previous events (values) do not change, then A is a plausible reason of a.
- **The method of residuum:** Let U be a complex phenomenon $abcd$, and we know that A is the reason of a, B is the reason of b, and C is the reason of c. Then it is possible to suppose that there is an event D that is a reason of d.

THE CONCEPT OF A GOOD CLASSIFICATION TEST

Our approach to machine-learning problems is based on the concept of a good diagnostic (classification) test. A good classification test can be understood as an approximation of a given classification on a given set of examples (Naidenova, 1996; Naidenova & Polegaeva, 1986).

A good diagnostic test is defined as follows. Let R be a set of examples and $S = \{1, 2, \ldots i, \ldots, n\}$ be the set of indices of examples, where n is the number of examples of R. Let $R(+)$ and $S(+)$ be the set of positive examples and the set of indices of positive examples, respectively. Let $R(-) = R/R(+)$ denote the set of negative examples. Let U be the set of attributes and T be the set of attributes values (values, for short), each of which appears at least in one of the examples of R.

Denote by $s(A)$, $A \in T$ the subset $\{i \in S: A$ appears in $t_i, t_i \in R\}$, where $S = \{1, 2, \ldots, n\}$. Following Cosmadakis et al. (1986), we call $s(A)$ the interpretation of $A \in T$ in R. The definition of $s(A)$ can be extended to the definition of $s(t)$ for any collection $t, t \subseteq T$ of values as follows: if $t = A_1 A_2 \ldots A_m$, then $s(t) = s(A_1) \cap s(A_2) \cap \ldots \cap s(A_m)$.

Definition 1. A collection $t \subseteq T (s(t) \neq \varnothing)$ of values is a diagnostic test for the set $R(+)$ of examples if and only if the following condition is satisfied: $t \not\subset t^*, \forall t^*, t^* \in R(-)$ (the equivalent condition is $s(t) \subseteq S(+)$).

Let k be the name of a set $R(k)$ of examples. To say that a collection t of values is a diagnostic test for $R(k)$ is equivalent to say that it does not cover any example $t^*, t^* \notin R(k)$. At the same time, the condition $s(t) \subseteq S(k)$ implies that the following implicative dependency is true: "if t, then k." Thus a diagnostic test, as a collection of values, makes up the left side of a rule of the first type.

It is clear that the set of all diagnostic tests for a given set $R(+)$ of examples (call it "$DT(+)$") is the set of all the collections t of values for which the condition $s(t) \subseteq S(+)$ is true. For any pair of diagnostic tests t_i, t_j from $DT(+)$, only one of the following relations is true: $s(t_i) \subseteq s(t_j), s(t_i) \supseteq s(t_j), s(t_i) \approx s(t_j)$, where the last one means that $s(t_i)$ and $s(t_j)$ are incomparable, that is, $s(t_i) \not\subset s(t_j)$ and $s(t_j) \not\subset s(t_i)$. This consideration leads to the concept of a good diagnostic test.

Definition 2. A collection $t \subseteq T (s(t) \neq \varnothing)$ of values is a good test for the set $R(+)$ of examples if and only if $s(t) \subseteq S(+)$ and, simultaneously, the condition $s(t) \subset s(t^*) \subseteq S(+)$ is not satisfied for any $t^*, t^* \subseteq T$, such that $t^* \neq t$.

Now we shall give the following definitions.

Definition 3. A collection t of values is irredundant if for any value $v \in t$ the following condition is satisfied: $s(t) \subset s(t/v)$.

If a collection t of values is a good test for $R(+)$ and, simultaneously, it is an irredundant collection of values, then any proper subset of t is not a test for $R(+)$.

Definition 4. A collection $t \subseteq T$ of values is maximally redundant if for any implicative de-

pendency $X \to v$, which is satisfied in R, the fact that t contains X implies that t also contains v.

If t is a maximally redundant collection of values, then for any value $v \notin t, v \in T$ the following condition is satisfied: $s(t) \supset s(t \cup v)$. In other words, a maximally redundant collection t of values covers the number of examples greater than any collection $(t \cup v)$ of values, where $v \notin t$.

If a diagnostic test t for a given set $R(+)$ of examples is a good one and it is a maximally redundant collection of values, then for any value $v \notin t, v \in T$ the following condition is satisfied: $(t \cup v)$ is not a good test for $R(+)$.

Any example t in R is a maximally redundant collection of values because for any value $v \notin t$, $v \in T s(t \cup v)$ is equal to $\varnothing$.

For example, in Table 1, the collection "*Blond Bleu*" is a good irredundant test for class 1 and, simultaneously, it is maximally redundant collection of values. The collection "*Blond Embrown*" is a test for class 2 but it is not good and, simultaneously, it is maximally redundant collection of values.

The collection "*Embrown*" is a good irredundant test for class 2. The collection "*Red*" is a good irredundant test for class 1. The collection "*Tall Red Bleu*" is a good maximally redundant test for class 1.

It is clear that the best tests for pattern recognition problems must be good irredundant tests. These tests allow constructing the shortest rules of the first type with the highest degree of generalization.

One of the possible ways for searching for good irredundant tests for a given class of positive examples is the following: first, find all good maximally redundant tests; second, for each good maximally redundant test, find all good irredundant tests contained in it. This is a convenient strategy as each good irredundant test belongs to one and only one good maximally redundant test with the same interpretation (Naidenova, 1999).

Table 1. Example 1of Data Classification (This example is adopted from (Ganascia, 1989))

Index of example	Height	Color of hair	Color of eyes	Class
1	Low	Blond	Bleu	1
2	Low	Brown	Bleu	2
3	Tall	Brown	Embrown	2
4	Tall	Blond	Embrown	2
5	Tall	Brown	Bleu	2
6	Low	Blond	Embrown	2
7	Tall	Red	Bleu	1
8	Tall	Blond	Bleu	1

.*Note to Table 1 and all the following tables: the values of attributes must not be considered as the words of English language, they are the abstract symbols only*

THE DUALITY OF GOOD DIAGNOSTIC TESTS

In our definition of good tests, we used, implicitly, correspondences of Galois G on $S \times T$ and two relations $S \to T, T \to S$ (Ore, 1944; Riguet, 1948). Let $s \subseteq S, t \subseteq T$. We define the relations as follows: $S \to T$: $t(s) = \{$intersection of all t_i: $t_i \subseteq T, i \in s\}$ and $T \to S$: $s(t) = \{i: i \in S, t \subseteq t_i\}$.

Extending s by an index j^* of some new example leads to receiving a more general feature of examples:

$$(t \cup A) \supseteq t \text{ implies } s(t \cup A) \subseteq s(t).$$

Extending t by a new value A leads to decreasing the number of examples possessing the general feature "tA" in comparison with the number of examples possessing the general feature "t":

$$(t \cup A) \supseteq t \text{ implies } s(t \cup A) \subseteq s(t).$$

Now we shall introduce the following generalization operations (functions):

generalization_of(t) = t' = $t(s(t))$; generalization_of(s) = s' = $s(t(s))$.

As a result of the generalization of s, the sequence of operations $s \to t(s) \to s(t(s))$ gives that $s(t(s)) \supseteq s$. This generalization operation gives the maximal set of examples possessing the feature $t(s)$.

As a result of the generalization of t, the sequence of operations $t \to s(t) \to t(s(t))$ gives that $t(s(t)) \supseteq t$. This generalization operation gives the maximal general feature for examples the indices of which are in $s(t)$.

These generalization operations are not artificially constructed operations. One can perform, mentally, a lot of such operations during a short period of time. We give some examples of these operations. Suppose that somebody has seen two films (s) with the participation of Gerard Depardieu ($t(s)$). After that he tries to know all the films with his participation ($s(t(s))$). One can know that Gerard Depardieu acts with Pierre Richard (t) in several films ($s(t)$). After that he can discover that these films are the films of the same producer, Francis Veber $t(s(t))$.

Namely, these generalization operations are used for searching for good diagnostic tests.

Now we define a diagnostic test as a dual object, that is, as a pair (SL, TA), $SL \subseteq S, TA \subseteq T$, $SL = s(TA)$ and $TA = t(SL)$.

Definition 5. Let $PM = \{s_1, s_2, ..., s_m\}$ be a family of subsets of some set M. Then PM is a Sperner system (Sperner, 1928) if the following condition is satisfied: $s_i \not\subset s_j$ and $s_j \not\subset s_i$, $\forall(i,j)$, $i \neq j$, $i, j = 1, ..., m$.

Let R, S, $R(+)$, $R(-)$, $S(+)$ be defined as before.

Definition 6. To find all *good maximally redundant tests* (GMRTs) for a given class $R(+)$ of examples means to construct a family PS of subsets $s_1, s_2, ..., s_j, ..., s_{np}$ of the set $S(+)$ such that:

1. PS is a Sperner system.
2. Each s_j is a maximal set in the sense that adding to it the index i, such that $i \notin s_j$, $i \in S(+)$, implies $s(t(s_j \cup i)) \not\subset S(+)$. Putting it in another way, $t(s_j \cup i)$ is not a test for the class $R(+)$.

The set *TGOOD* of all GMRTs is determined as follows: $\{t: t(s_j), s_j \in PS, \forall j = 1, ..., np\}$.

Definition 7. To find all *good irredundant tests* (GIRTs) for a given class $R(+)$ of examples means to construct a family PRT of subsets $t_1, t_2, ..., t_j, ..., t_{nq}$ of the set T such that:

1. $\forall t_j, j = 1, ..., nq, t_j \not\subset t, \forall t, t \in R(-)$ and, simultaneously, $\forall t_j, j = 1, ..., nq, s(t_j) \neq \varnothing$ there does not exist a collection $s^* \neq s(t_j)$, $s^* \subseteq S$ of indices such that $s(t_j) \subset s^* \subseteq S(+)$.
2. PRT is a Sperner system.
3. Each t_j is a minimal set in the sense that removing from it any value A belonging to it implies $s(t_j$ without $A) \not\subset S(+)$.

GENERATION OF DUAL OBJECTS WITH THE USE OF LATTICE OPERATIONS

Let R be a table of examples and S, T are defined as before. Let MUT be the set of all dual objects, that is, the set of all pairs (s, t), $s \subseteq S$, $t \subseteq T$, $s = s(t)$, and $t = t(s)$. This set is partially ordered by

the relation $\leq$, where $(s, t) \leq (s^*, t^*)$ is satisfied if and only if $s \subseteq s^*$ and $t \supseteq t^*$.

The set $\Psi = (MUT, \cup, \cap)$ is an algebraic lattice, where operations $\cup$, $\cap$ are defined for all pairs (s^*, t^*), $(s, t) \in MUT$ in the following way (Wille, 1992):

$$(s^*, t^*) \cup (s, t) = ((s^* \cup s), (t^* \cap t)),$$
$$(s^*, t^*) \cap (s, t) = ((s^* \cap s), (t^* \cup t)).$$

The unit element and the zero element are $(S, \varnothing)$ and $(\varnothing, T)$, respectively.

Inferring good tests is reduced to inferring, for any element $(s^*, t^*) \in MUT$, all the elements nearest to it in the lattice with respect to the ordering $\leq$, that is, inferring all (s, t), that $(s^*, t^*) \leq (s, t)$, and there does not exist any (s^{**}, t^{**}) such that $(s^*, t^*) \leq (s^{**}, t^{**}) \leq (s, t)$, or inferring all (s, t), that $(s^*, t^*) \geq (s, t)$, and there does not exist any (s^{**}, t^{**}) such that $(s^*, t^*) \geq (s^{**}, t^{**}) \geq (s, t)$.

Inferring the chains of lattice elements ordered by the inclusion relation lies in the foundation of generating all types of diagnostic tests:

$$s_0 \subseteq ... \subseteq s_i \subseteq s_{i+1} \subseteq ... \subseteq s_m \ (t(s_0) \supseteq t(s_1) \supseteq ... \supseteq t(s_i) \supseteq t(s_{i+1}) \supseteq ... \supseteq t(s_m)),$$
$$(2) \ t_0 \subseteq ... \subseteq t_i \subseteq t_{i+1} \subseteq ... \subseteq t_m \ (s(t_0) \supseteq s(t_1) \supseteq ... \supseteq s(t_i) \supseteq s(t_{i+1}) \supseteq ... \supseteq s(t_m)). \quad (1)$$

Inductive Rules for Constructing Elements of a Dual Lattice

We use the following variants of inductive transition from one element of a chain to its nearest element in the lattice:

1. from $s_q = (i_1, i_2, ..., i_q)$ to $s_{q+1} = (i_1, i_2, ..., i_{q+1})$;
2. from $t_q = (A_1, A_2, ..., A_q)$ to $t_{q+1} = (A_1, A_2, ..., A_{q+1})$;
3. from $s_q = (i_1, i_2, ..., i_q)$ to $s_{q-1} = (i_1, i_2, ..., i_{q-1})$;
4. from $t_q = (A_1, A_2, ..., A_q)$ to $t_{q-1} = (A_1, A_2, ..., A_{q-1})$.

We need the special rules for realizing these inductive transitions.

The Generalization Rule

The generalization rule is used to get all the collections of indices $s_{q+1} = \{i_1, i_2, \ldots i_q, i_{q+1}\}$ from a collection $s_q = \{i_1, i_2, \ldots i_q\}$ such that $t(s_q)$ and $t(s_{q+1})$ are tests for a given class of positive examples.

The termination condition for constructing a chain of generalizations is: for all the extension s_{q+1} of s_q, $t(s_{q+1})$ is not a test for a given class of positive examples.

Consider some of the possible realizations of this rule for inferring GMRTs.

The first variant of generalization rule. Let $S(test)$ be the partially ordered set of elements $s = \{i_1, i_2, \ldots i_q\}$, $q = 1, 2, \ldots, nt - 1$ obtained as a result of generalizations and satisfying the following condition: $t(s)$ is a test for $R(+)$. Here nt denotes the number of positive examples. Let $STGOOD$ be the partially ordered set of elements s satisfying the following condition: $t(s)$ is a GMRT for $R(+)$.

Next we use an inductive rule for extending elements of $S(test)$ and constructing $\{i_1, i_2, \ldots i_{q+1}\}$ from $\{i_1, i_2, \ldots i_q\}$, $q = 1, 2, \ldots, nt - 1$. This rule relies on the following consideration: if the set $\{i_1, i_2, \ldots i_{q+1}\}$ corresponds to a test for $R(+)$, then all its proper subsets must correspond to tests too and, consequently, they must be in $S(test)$. Having constructed a set $s_{q+1} = \{i_1, i_2, \ldots i_{q+1}\}$, we determine whether it corresponds to the test or not. If $t(s_{q+1})$ is not a test, then s_{q+1} is deleted, otherwise it is inserted in $S(test)$. If all the extensions of s do not correspond to tests, then s corresponds to a GMRT and it is inserted in $STGOOD$.

The function to_be_test(t) is defined as follows: if $s(t) \cap S(+) = s(t)$ then *true* else *false*.

This variant of generalization rule is used in an algorithm of inferring GMRTs given in Naidenova and Polegaeva (1991). An analogous inductive extension of items' collections is also used in two algorithms, *Apriory* and *AprioryTid*, proposed in Agrawal and Srikant (1994) for min-

ing association rules between items in a large database of sales transactions.

The second variant of generalization rule. This rule allows for each element s the following:

- To avoid constructing the set of all its subsets.
- To avoid the repetitive generation of it.

Consider a way for choosing indices admissible for extending s_q.

Suppose that $S(test)$ and $STGOOD$ are not empty and $s \in S(test)$. Construct the set V:

$$V = \{\cup s', s \subseteq s', s' \in \{S(test) \cup STGOOD\}\}.$$

The set V is the union of all the collections of indices in $S(test)$ and $STGOOD$ containing s; hence, s is in the intersection of these collections. If we want an extension of s not to be included in any element of $\{S(test) \cup STGOOD\}$, we must use, for extending s, the indices not appearing simultaneously with s in the set V. The set of indices, candidates for extending s, is the set:

$$CAND(s) = nts/V, \text{ where } nts = \{\cup s, s \in S(test)\}.$$

An index $j^* \in CAND(s)$ is not admissible for extending s if, at least for one index $i \in s$, the pair $\{i, j^*\}$ either does not correspond to a test or it corresponds to a good test (it belongs to $STGOOD$). Let Q be the set of forbidden pairs of indices for extending s: $Q = \{\{i, j\} \subseteq S(+): t(\{i, j\})$ is not a test for $R(+)\}$. Then the set of admissible indices is $select(s) = \{i, i \in CAND(s): (\forall j) (j \in s), \{i, j\} \notin \{STGOOD$ or $Q\}\}$.

The set Q can be generated in the beginning of searching all GMRTs for $R(+)$.

The procedure EXTENSION(s) takes $select(s)$ and returns the set $ext(s)$ of all possible extensions of s in the form $snew = (s \cup j)$, $j \in select(s)$ and $snew$ corresponds to a test for $R(+)$. This procedure executes the function generalization_of($snew$)

for each element *snew* $\in$ *ext*(*s*) (for this function, please see the introduction to Section 4).

If *ext*(*s*) and the set *V* are empty, then *s* corresponds to a GMRT for *R*(+) and *s* is transferred from *S*(*test*) to *STGOOD*. If *ext*(*s*) contains one and only one element, then this element corresponds to a GMRT, it is inserted in *STGOOD* and *s* is deleted from *S*(*test*). In all other cases, the set *ext*(*s*) substitutes *s* in *S*(*test*).

This variant of generalization rule is a complex process in which both deductive and inductive reasoning rules of the second type are performed (please, see Table 2). The knowledge acquired during the process of generalization (the sets *Q*, *S*(*test*), *STGOOD*) is used for pruning the search in the domain space.

The generalization rule realizes the joint method of similarity-distinction. The extending of *s* results in obtaining the subsets of positive examples of more and more power with more and more generalized features (set of values). This operation is analogous to the generalization rule applied for star generation under conceptual clustering (Michalski, 1983).

An algorithm, NIAGaRa, based on this variant of generalization rule is used in Naidenova (2001), for inferring GMRTs.

The Specification Rule

The specification rule is used to get all the collections of values $t_{q+1} = \{A_1, A_2, ..., A_{q+1}\}$ from a collection $t_q = \{A_1, A_2, ..., A_q\}$ such that t_q and t_{q+1} are irredundant collections of values, and they are not tests for a given set of positive examples.

The termination condition for constructing a chain of specifications is: for all the extensions t_{q+1} of t_q, t_{q+1} is either a redundant collection of values or a test for a given set of positive examples.

This rule is used for inferring GIRTs.

The first variant of specification rule. Let *TGOOD* be the partially ordered set of elements *t* satisfying the following condition: *t* is a good irredundant test for *R*(+). We denote by *SAFE* the set of elements *t* such that *t* is an irredundant collection of values but not a test for *R*(+).

Next we use an inductive rule for extending elements of *SAFE* and constructing $t_{q+1} = (A_1, A_2, ..., A_{q+1})$ from $t_q = (A_1, A_2, ..., A_q)$ $q = 1, 2, .., na - 1$, where *na* is the number of values in the set *T*. This rule relies on the following consideration: if the collection of values $\{A_1, A_2, ... A_{q+1}\}$ is an irredundant one, then all its proper subsets must be irredundant collections of values too and, consequently, they must be in *SAFE*. Having constructed a set $t_{q+1} = \{A_1, A_2, ... A_{q+1}\}$, we determine whether it is the irredundant collection of values or not. If the collection t_{q+1} is redundant, then it is deleted from *SAFE*. If it is the test for *R*(+), then it is transferred from *SAFE* to *TGOOD*. If t_{q+1} is irredundant but not a test for *R*(+), then it is a candidate for extension.

Table 2. Using deductive and inductive rules of the second type

Inductive rules	Process	Deductive and inductive rules of the second type
Generalization rule		
	Forming *Q*	Generating forbidden rules
	Forming *CAND*(*s*)	The joint method of similarity-distinction
	Forming *select*(*s*)	Using forbidden rules
	Forming *ext*(*s*)	The method of only similarity
	Function_to_be test(*t*)	Using implication
	Generalization_of(*snew*)	Lattice operations

We use the function to_be_irredundant$(t) =$ if $(\forall A_i)\,(A_i \in t)\,s(t) \neq s(t/A_i)$ then *true* else *false*

It is easy to see that this variant of specification rule is algorithmically equivalent to the first variant of the generalization rule. This rule is used in an algorithm given in Megretskaya (1988) for inferring GIRTs.

The second variant of specification rule. It is an inductive extension rule containing a method for choosing admissible values for extending t in case of t is not a test, but its extension is a test for $R(+)$. We extend t by choosing admissible values as follows: these values appear simultaneously with t in the examples of $R(+)$, and do not appear with t in any example of $R(-)$. These values are to be said essential ones. To get them, we construct two sets $V(-)$ and $V(+)$ as follows: $V(-) = \{\cup\ t': t \subseteq t',\ t' \in R(-)\}$; $V(+) = \{\cup\ t': t \subseteq t',\ t' \in R(+)\}$. The set $ess(t)$ of essential values for t is equal to $V(+)/V(-)$. Thus searching essential values requires a special reasoning operation, *a diagnostic induction reasoning rule*.

The Inductive Diagnostic Rule

The inductive diagnostic rule is used to get a collection of values $t_{q+1} = \{A_1, A_2, ..., A_{q+1}\}$ from a collection $t_q = \{A_1, A_2, ... A_q\}$ such that t_q is not a test, but t_{q+1} is a test for a given set of positive examples.

We extend t_q by choosing values that appear simultaneously with it in the examples of $R(+)$, and do not appear in any example of $R(-)$. These values are to be said essential ones.

Definition 8. Let t be a collection of values that is a test for a given set of positive examples. We say that the value A in t is essential if (t/A) is not a test for a given set of positive examples.

Generally, we are interested in finding the maximal subset $sbmax(t) \subset t$ such that t is a test, but $sbmax(t)$ is not a test for a given set of posi-

tive examples. Then $sbmin(t) = t/\,sbmax(t)$ is the minimal set of essential values in t.

This inductive rule generates diagnostic rules of the first type. It is based on the inductive method of only distinction. We see that the diagnostic rules of the first type obtained with the use of inductive diagnostic rules are used immediately in the process of good tests construction.

An analogous rule is defined in Michalski (1983) and Michalski and Larson (1978). If a newly presented training example contradicts an already constructed concept description, the specialization rule is applied to generate a new consistent concept description.

The Dual Inductive Diagnostic Rule

The dual inductive diagnostic rule is used to get a collection of indices $s_{q-1} = (i_1, i_2, ..., i_{q-1})$ from a collection $s_q = (i_1, i_2, ..., i_q)$ such that $t(s_{q-1})$ is a test, but $t(s_q)$ is not a test for a given set of positive examples. This rule uses a method for choosing indices admissible for deleting from s_q. By analogy with an essential value, we define an essential example.

Definition 9. Let s be a subset of indices of positive examples; assume also that $t(s)$ is not a test. The example $t_j,\ j \in s$ is to be said an essential one if $t(s/j)$ proves to be a test for a given set of positive examples.

Generally, we are interested in finding the maximal subset $sbmax(s) \subset s$ such that $t(s)$ is not a test, but $t' = t(sbmax(s))$ is a test for a given set of positive examples. Then $sbmin(s) = s/sbmax(s)$ is the minimal set of indices of essential examples in s.

The dual inductive diagnostic rule is used for inferring compatibility rules of the first type. The number of indices in $sbmax(s)$ can be understood as a measure of "carrying-out" for an acquired rule related to $sbmax(s)$, namely, $t(sbmax(s)) \rightarrow$

$k(R(+))$ *frequently,* where $k(R(+))$ is the name of the set $R(+)$.

Assume s^* is a collection of indices of positive examples such that $t(s^*)$ is not a test. Next we describe the procedure with the use of which a quasi-maximal subset $qsbmax(s^*) \subset s^*$ is obtained such that $t(qsbmax(s^*))$ is a test for a given set of positive examples.

We begin with the first index i_1 of s^*, then we take the next index i_2 of s^* and evaluate the function to_be_test $(t(\{i_1, i_2\}))$. If the value of this function is "*true,*" then we take the next index i_3 of s^* and evaluate the function to_be_test $(t(\{i_1, i_2, i_3\}))$. If the value of the function to_be_test $(t(\{i_1, i_2\}))$ is "*false,*" then the index i_2 of s^* is skipped and the function to_be_test $(t(\{i_1, i_3\}))$ is evaluated. We continue this process until we achieve the last index of s^*.

The dual inductive diagnostic rule is based on the inductive method of only distinction.

We see that the compatibility rules of the first type, obtained with the use of dual inductive diagnostic rule, are used immediately in the process of good tests construction.

The rules for constructing diagnostic tests as elements of dual lattice generate logical rules of the first type, as shown in Table 3.

THE DECOMPOSITION OF INFERRING GOOD DIAGNOSTIC TESTS INTO SUBTASKS

To transform good diagnostic tests inferring into an incremental process, we introduce two kinds of subtasks (Naidenova & Ermakov, 2001):

For a given set of positive examples:

1.   Given a positive example t, find all GMRTs contained in t.
2.   Given a nonempty collection of values X (maybe only one value) such that it is not a test, find all GMRTs containing X.

The subtask of the first kind. We introduce the concept of an example's projection proj$(R)[t]$ of a given positive example t on a given set $R(+)$ of positive examples. The proj$(R)[t]$ is the set $Z = \{z: (z \text{ is nonempty intersection of } t \text{ and } t') \& (t' \in R(+)) \& (z \text{ is a test for a given class of positive examples})\}$.

If the proj$(R)[t]$ is not empty and contains more than one element, then it is a subtask for inferring all GMRTs that are in t. If the projection contains one and only one element equal to t, then t is a GMRT.

Table 3. Deductive rules of the first type obtained with the use of inductive rules for inferring diagnostic tests

Inductive rules	Action	Inferring deductive rules of the first type
Generalization rule	Extending s (narrowing t)	Implications
Specification rule	Extending t (narrowing s)	Implications
Inductive diagnostic rule	Searching for essential values	Diagnostic rules
Dual inductive diagnostic rule	Searching for essential examples	Compatibility rules (approximate implications)

The subtask of the second kind. We introduce the concept of an attributive projection proj(R)[A] of a given value A on a given set R(+) of positive examples.

The projection proj(R)[A] = {t: (t ∈ R(+)) & (A appears in t)}. Another way to define this projection is: proj(R)[A] = {t_i: i ∈ (s(A) ∩ S(+))}. If the attributive projection is not empty and contains more than one element, then it is a subtask of inferring all GMRTs containing a given value A. If A appears in one and only one example, then A does not belong to any GMRT different from this example.

Forming the projection of A makes sense if A is not a test and the intersection of all positive examples in which A appears is not a test too, that is, s(A) ⊄ S(+) and t' = t(s(A) ∩ s(+)) is not a test for a given set of positive examples.

The decomposition of good classification tests inferring into subtasks of the first and second kinds implies introducing a set of special rules to realize the following operations: choosing an example (value) for a subtask, forming a subtask, deleting values or examples from a subtask, and some other rules controlling the process of inferring good tests.

The following theorem gives the foundation for reducing projections both of the first and the second kind. The proof of this theorem can be found in Naidenova et al. (1995b).

THEOREM 1

Let A be a value from T, X be a maximally redundant test for a given set R(+) of positive examples, and s(A) ⊆ s(X). Then A does not belong to any maximally redundant good test for R(+) different from X.

It is convenient to choose essential values in an example and essential examples in a projection for the decomposition of inferring GMRTs into the subtasks of the first or second kind.

Table 4. Example 2 of data classification

Index of example	Height	Color of hair	Color of eyes	Class
1	Low	Blond	Bleu	1
2	Low	Brown	Bleu	1
3	Tall	Brown	Embrown	1
4	Tall	Blond	Embrown	2
5	Tall	Brown	Bleu	2
6	Low	Blond	Embrown	2
7	Tall	Red	Bleu	2
8	Tall	Blond	Bleu	2

Table 5. The subtask for the value "Low"

Index of example	Height	Color of hair	Color of eyes	Class
1	Low	Blond	Bleu	1
2	Low	Brown	Bleu	1
6	Low	Blond	Embrown	2

We give a small example for inferring all the GMRTs for the instances of class 1 presented in Table 4.

In Table 4, we have: $S(+) = \{1,2,3\}$, $s(Low) \to \{1,2,6\}$, $s(Brown) \to \{2,3,5\}$, $s(Bleu) \to \{1,2,5,7,8\}$, $s(Tall) \to \{3,4,5,7,8\}$, $s(Embrown) \to \{3,4,6\}$, and $s(Blond) \to \{1,4,6,8\}$.

We discover that the value "*Low*" is essential in lines 1 and 2. Then it is convenient to form the subtask of the second kind for this value as shown in Table 5.

In Table 5, we have: $S(+) = \{1,2\}$, $s(Low) \to \{1,2,6\}$, $s(Brown) \to \{2\}$, $s(Bleu) \to \{1,2\}$, and $s(Blond) \to \{1,6\}$.

We have: $s(Bleu) = \{1,2\} \subseteq S(+)$. It means that the collection of values "*Low Bleu*" is a test for class 1. Analogously, for the value "*Brown,*" we have: $s(Brown) = \{2\} \subseteq S(+)$. It means that the collection of values "*Low Brown*" is a test for class 1 but not a good one because of $s(Brown) = \{2\} \subseteq s(Bleu)$.

It is clear that these values cannot belong to any test different from the tests already obtained. We delete "*Brown*" and "*Bleu*" from further consideration in this subtask. But after deleting these values, line 1 and 2 are not tests for class 1. Hence, the subtask is over.

Return to the main problem. Now we can delete the value "*Low*" from further consideration because we have gotten all good tests containing this value for class 1. But we know that the value "*Low*" is essential in lines 1 and 2; this fact means that these lines are not tests for class 1 after deleting this value.

The following step may be the inference of all irredundant tests contained in line 3 (covering only one line 3) for class 1. In our case, the collection of values "*Brown Embrown*" is a GIRT contained in line 3.

A recursive procedure, based on using attributive subtasks for inferring GMRTs, has been described in Naidenova et al. (1995b). In the following part of this chapter, we give an algorithm based on the subtasks of the first kind combined with searching essential examples. This algorithm is used only for inferring GMRTs.

An Algorithm for Inferring GMRTs with the use of the Subtask of the First Kind

The algorithm DIAGaRa is the basic recursive algorithm for solving a subtask of the first kind.

The initial information for the algorithm of finding all the GMRTs contained in a positive example is the projection of this example on the current set $R(+)$. Essentially the projection is simply a subset of examples defined on a certain restricted subset t^* of values. Let s^* be the subset of indices of positive examples producing the projection.

It is useful to introduce the characteristic $W(t)$ of any collection t of values named by the weight of t in the projection: $W(t) = \|s(t) \cap s^*\|$ is the number of positive examples of the projection containing t. Let WMIN be the minimal permissible value of the weight.

Let *STGOOD* be the partially ordered set of elements s satisfying the condition that $t(s)$ is a good test for $R(+)$.

The basic algorithm consists of applying the sequence of the following steps:

- **Step 1:** Check whether the intersection of all the elements of projection is a test and if so, then s^* is stored in *STGOOD* if s^* corresponds to a good test at the current step; in this case, the subtask is over. Otherwise the next step is performed (we use the function to_be_test(t): if $s(t) \cap S(+) = s(t)$ $(s(t) \subseteq S(+))$ then *true* else *false*).

- **Step 2:** For each value A in the projection, the set $splus(A) = \{s^* \cap s(A)\}$ and the weight $W(A) = \|splus(A)\|$ are determined, and if the weight is less than the minimum permissible weight WMIN, then the value A is deleted from the projection. We can also delete the value A if $W(A)$ is equal to WMIN and

$t(splus(A))$ is not a test; in this case A will not appear in a maximally redundant test t with $W(t)$ equal to or greater than WMIN.

- **Step 3:** The generalization operation is performed: $t' = t(splus(A))$, $A \in t^*$; if t' is a test, then the value A is deleted from the projection and $splus(A)$ is stored in *STGOOD* if $splus(A)$ corresponds to a good test at the current step.

- **Step 4:** The value A can be deleted from the projection if $splus(A) \subseteq s'$ for some s' $\in$ *STGOOD*.

- **Step 5:** If at least one value has been deleted from the projection, then the reduction of the projection is necessary. The reduction consists of deleting the elements of projection that are not tests (as a result of previous eliminating values). If, under reduction, at least one element has been deleted from the projection, then Step 2, Step 3, Step 4, and Step 5 are repeated.

- **Step 6:** Check whether the subtask is over or not. The subtask is over when either the projection is empty or the intersection of all elements of the projection corresponds to a test (see Step 1). If the subtask is not over, then the choice of an essential example in this projection is performed and the new subtask is formed with the use of this essential example. The new subsets s^* and t^* are constructed and the basic algorithm runs recursively. The important part of the basic algorithm is how to form the set *STGOOD*.

We give in the Appendix an example of the work of the algorithm DIAGaRa.

An approach for forming the set *STGOOD*. Let $L(S)$ be the set of all subsets of the set S. $L(S)$ is the set lattice (Rasiova, 1974). The ordering determined in the set lattice coincides with the set-theoretical inclusion. It will be said that subset s_1 is absorbed by subset s_2, that is, $s_1 \leq s_2$, if and only if the inclusion relation is hold between them,

that is, $s_1 \subseteq s_2$. Under formation of *STGOOD*, a collection s of indices is stored in *STGOOD* if and only if it is not absorbed by any collection of this set. It is necessary also to delete from *STGOOD* all the collections of indices that are absorbed by s if s is stored in *STGOOD*. Thus, when the algorithm is over, the set *STGOOD* contains all the collections of indices that correspond to GMRTs and only such collections. Essentially, the process of forming *STGOOD* is an incremental procedure of finding all maximal elements of a partially ordered set. The set *TGOOD* of all the GMRTs is obtained as follows: $TGOOD = \{t: t = t(s), (\forall s) (s \in STGOOD)\}$.

An Approach to Incremental Inferring Good Diagnostic Tests

Incremental learning is necessary when a new portion of observations or examples becomes available over time. Suppose that each new example comes with the indication of its class membership. The following actions are necessary with the arrival of a new example:

- Check whether it is possible to perform generalization of some existing GMRTs for the class to which the new example belongs (class of positive examples), that is, whether it is possible to extend the set of examples covered by some existing GMRTs or not.
- Infer all the GMRTs contained in the new example.
- Check the validity of the existing GMRTs for negative examples, and if it is necessary:
- Modify tests that are not valid (test for negative examples is not valid if it is included in a positive example, that is, in other words, it accepts an example of positive class).

Thus the process of inferring all the GMRTs is divided into the subtasks that conform to three acts of reasoning:

- Pattern recognition or using already known rules (tests) for determining the class membership of a new positive example and generalization of these rules that recognize it correctly (deductive reasoning and increasing the power of already existing inductive knowledge).
- Inferring new rules (tests) that are generated by a new positive example (inductive reasoning a new knowledge).
- Correcting rules (tests) of alternative (negative) classes that accept a new positive example (these rules do not permit to distinguish a new positive example from some negative examples) (deductive and inductive diagnostic reasoning to modify knowledge).

The first act reveals the known rules satisfied with a new example, the induction base of these rules can be enlarged.

The second act can be reduced to the subtask of the first kind.

The third act can be reduced either to the inductive diagnostic rule and the subtask of the first or to the subtask of the second kind.

CONCLUSION

This work is an attempt to transform a large class of machine-learning tasks into a commonsense reasoning process based on using well-known deduction and induction logical rules.

For this goal, we have chosen the task of inferring good classification (diagnostic) tests for a given partitioning on a given training set of examples because a lot of well-known machine-learning problems, such as inferring functional, implicative, and associative dependencies from data, are reduced to this task.

We proposed a unified model for combining inductive reasoning with deductive reasoning in the framework of inferring and using implicative

logical rules. The key concept of our approach is the concept of a good diagnostic test. We define a good diagnostic test as the best approximation of a given classification on a given set of examples.

We have used the lattice theory as the mathematical model for constructing good classification tests. We define a diagnostic test as a dual object, that is, as an element of the concept lattice introduced in the formal concept analysis.

The links between dual elements of concept lattice reflect both inclusion relations between concepts (structural knowledge) and implicative relations between concept descriptions (deductive knowledge).

Inferring the chains of lattice elements ordered by the inclusion relation lies in the foundation of generating all types of diagnostic tests. We considered four variants of inductive transition from one element of a chain to its nearest element in the lattice. We have constructed the special rules for realizing these inductive transitions: the generalization rule, the specification rule, the inductive diagnostic rule, and the dual inductive diagnostic rule.

We have divided commonsense reasoning rules in two classes: rules of the first type and rules of the second type. The rules of the first type are represented with the use of implicative logical statements. The rules of the second type or reasoning rules (deductive and inductive) are rules with the help of which rules of the first type used, updated, and inferred from data. The deductive reasoning rules of the second type are modus ponens, modus ponendo tollens, modus tollendo ponens, and modus tollens. The inductive reasoning rules of the second type are the following ones: the method of only similarity, the method of only distinction, the joint method of similarity-distinction, and some others. The analysis of the inference for lattice construction allows demonstrating that this inference engages both inductive and deductive reasoning rules of the second type. During the lattice construction, the rules of the first type (implications, interdictions, rules of compatibility) are generated and used immediately.

We have introduced the decomposition of inferring good tests for a given set of positive examples into operations and subtasks that are in accordance with human commonsense reasoning operations. This decomposition allows, in principle, to transform the process of inferring good tests into a "step by step" commonsense reasoning process.

We have given also the algorithm DIAGaRa for inferring good maximally redundant tests, and an approach to incrementally inferring good diagnostic tests.

ACKNOWLEDGMENT

The author is very grateful to Professor Evangelos Triantaphyllou (Louisiana State University), who inspired and supported this work, to Dr. Giovanni Felici (IASI – Italian National Research Council), for his invaluable critical remarks, and to Prof. Carlo Vercellis (Milan Polytechnic Institute) for his invariable attention to the author. None of these people bear any responsibility for the content as presented, of course.

Appendix

An example of using algorithm DIAGaRa. The data to be processed are in Table 6 (the set of positive examples) and in Table 7 (the set of negative examples).

We begin with $s^* = S(+) = \{\{1\}, \{2\}, …, \{14\}\}$, $t^* = T = \{A_1, A_2, ….., A_{26}\}$, $SPLUS = \{splus(A_i): A_i \in t^*\}$ (see $SPLUS$ in Table 8). In Tables 8, 9, A_* denotes the collection of values $\{A_8, A_9\}$ and A_+ denotes the collection of values $\{A_{14}, A_{15}\}$ because $splus(A_8) = splus(A_9)$ and $splus(A_{14}) = splus(A_{15})$.

We use the algorithm DIAGaRa for inferring all the GMRTs having a weight equal to or greater than WMIN = 4 for the training set of the positive examples represented in Table 6.

Please observe that $splus(A_{12}) = \{2,3,4,7\}$ and $t(\{2,3,4,7\})$ is a test; therefore, A_{12} is deleted from t^* and $splus(A_{12})$ is inserted into $STGOOD$.

Then $W(A_*)$, $W(A_{13})$, and $W(A_{16})$ are less than WMIN; hence, we can delete A_*, A_{13}, and A_{16} from t^*. Now t_{10} is not a test and can be deleted.

Table 6. The set of positive Examples R(+)

Index of example	R(+)
1	$A_1 A_2 A_5 A_6 A_{21} A_{23} A_{24} A_{26}$
2	$A_4 A_7 A_8 A_9 A_{12} A_{14} A_{15} A_{22} A_{23} A_{24} A_{26}$
3	$A_3 A_4 A_7 A_{12} A_{13} A_{14} A_{15} A_{18} A_{19} A_{24} A_{26}$
4	$A_1 A_4 A_5 A_6 A_7 A_{12} A_{14} A_{15} A_{16} A_{20} A_{21} A_{24} A_{26}$
5	$A_2 A_6 A_{23} A_{24}$
6	$A_7 A_{20} A_{21} A_{26}$
7	$A_3 A_4 A_5 A_6 A_{12} A_{14} A_{15} A_{20} A_{22} A_{24} A_{26}$
8	$A_3 A_6 A_7 A_8 A_9 A_{13} A_{14} A_{15} A_{19} A_{20} A_{21} A_{22}$
9	$A_{16} A_{18} A_{19} A_{20} A_{21} A_{22} A_{26}$
10	$A_2 A_3 A_4 A_5 A_6 A_8 A_9 A_{13} A_{18} A_{20} A_{21} A_{26}$
11	$A_1 A_2 A_3 A_7 A_{19} A_{20} A_{21} A_{22} A_{26}$
12	$A_2 A_3 A_{16} A_{20} A_{21} A_{23} A_{24} A_{26}$
13	$A_1 A_4 A_{18} A_{19} A_{23} A_{26}$
14	$A_{23} A_{24} A_{26}$

Table 7. The set of negative Examples R(-)

Index of example	R(-)	Index of example	R(-)
15	$A_3 A_8 A_{16} A_{23} A_{24}$	32	$A_1 A_2 A_3 A_7 A_9 A_{13} A_{18}$
16	$A_7 A_8 A_9 A_{16} A_{18}$	33	$A_1 A_5 A_6 A_8 A_9 A_{19} A_{20} A_{22}$
17	$A_1 A_{21} A_{22} A_{24} A_{26}$	34	$A_2 A_8 A_9 A_{18} A_{20} A_{21} A_{22} A_{23} A_{26}$
18	$A_1 A_7 A_8 A_9 A_{13} A_{16}$	35	$A_1 A_2 A_4 A_5 A_6 A_7 A_9 A_{13} A_{16}$
19	$A_2 A_6 A_7 A_9 A_{21} A_{23}$	36	$A_1 A_2 A_6 A_7 A_8 A_{13} A_{16} A_{18}$
20	$A_{10} A_{19} A_{20} A_{21} A_{22} A_{24}$	37	$A_1 A_2 A_3 A_4 A_5 A_6 A_7 A_{12} A_{14} A_{15} A_{16}$
21	$A_1 A_{20} A_{21} A_{22} A_{23} A_{24}$	38	$A_1 A_2 A_3 A_4 A_5 A_6 A_9 A_{12} A_{13} A_{16}$
22	$A_1 A_3 A_6 A_7 A_9 A_{16}$	39	$A_1 A_2 A_3 A_4 A_5 A_6 A_{14} A_{15} A_{19} A_{20} A_{23} A_{26}$
23	$A_2 A_6 A_8 A_9 A_{14} A_{15} A_{16}$	40	$A_2 A_3 A_4 A_5 A_6 A_7 A_{12} A_{13} A_{14} A_{15} A_{16}$
24	$A_1 A_4 A_5 A_6 A_7 A_8 A_{16}$	41	$A_2 A_3 A_4 A_5 A_6 A_7 A_9 A_{12} A_{13} A_{14} A_{15} A_{19}$
25	$A_7 A_{13} A_{19} A_{20} A_{22} A_{26}$	42	$A_1 A_2 A_3 A_4 A_5 A_6 A_{12} A_{16} A_{18} A_{19} A_{20} A_{21} A_{26}$
26	$A_1 A_2 A_3 A_6 A_7 A_{16}$	43	$A_4 A_5 A_6 A_7 A_8 A_9 A_{12} A_{13} A_{14} A_{15} A_{16}$
27	$A_1 A_2 A_3 A_5 A_6 A_{13} A_{16}$	44	$A_3 A_4 A_5 A_6 A_8 A_9 A_{12} A_{13} A_{14} A_{15} A_{18} A_{19}$
28	$A_1 A_3 A_7 A_{13} A_{19} A_{21}$	45	$A_1 A_2 A_3 A_4 A_5 A_6 A_7 A_8 A_9 A_{12} A_{13} A_{14} A_{15}$
29	$A_1 A_4 A_5 A_6 A_7 A_8 A_{13} A_{16}$	46	$A_1 A_3 A_4 A_5 A_6 A_7 A_{12} A_{13} A_{14} A_{15} A_{16} A_{23} A_{24}$
30	$A_1 A_2 A_3 A_6 A_{12} A_{14} A_{15} A_{16}$	47	$A_1 A_2 A_3 A_4 A_5 A_6 A_8 A_9 A_{12} A_{14} A_{15} A_{18} A_{22}$
31	$A_1 A_2 A_5 A_6 A_{14} A_{15} A_{16} A_{26}$	48	$A_2 A_8 A_9 A_{12} A_{14} A_{15} A_{16}$

Table 8. The set SPLUS of the collection splus(A) for all A in Tables 6 and 7

SPLUS = {splus(A_i): s(A_i) ∩ s(+), A_i ∈ T}:	
$splus(A_*) \to \{2,8,10\}$	$splus(A_{22}) \to \{2,7,8,9,11\}$
$splus(A_{13}) \to \{3,8,10\}$	$splus(A_{23}) \to \{1,2,5,12,13,14\}$
$splus(A_{16}) \to \{4,9,12\}$	$splus(A_3) \to \{3,7,8,10,11,12\}$
$splus(A_1) \to \{1,4,11,13\}$	$splus(A_4) \to \{2,3,4,7,10,13\}$
$splus(A_5) \to \{1,4,7,10\}$	$splus(A_6) \to \{1,4,5,7,8,10\}$
$splus(A_{12}) \to \{2,3,4,7\}$	$splus(A_7) \to \{2,3,4,6,8,11\}$
$splus(A_{18}) \to \{3,9,10,13\}$	$splus(A_{24}) \to \{1,2,3,4,5,7,12,14\}$
$splus(A_2) \to \{1,5,10,11,12\}$	$splus(A_{20}) \to \{4,6,7,8,9,10,11,12\}$
$splus(A_.) \to \{2,3,4,7,8\}$	$splus(A_{21}) \to \{1,4,6,8,9,10,11,12\}$
$splus(A_{19}) \to \{3,8,9,11,13\}$	$splus(A_{26}) \to \{1,2,3,4,6,7,9,10,11,12,13,14\}$

After modifying $splus(A)$ for A_5, A_{18}, A_2, A_3, A_4, A_6, A_{20}, A_{21}, and A_{26} we find that $W(A_5) = 3$, therefore, A_5 is deleted from t^* .

Then $W(A_{18})$ turns out to be less than WMIN and we delete A_{18}; this implies deleting t_{13}. Next we modify $splus(A)$ for A_1, A_{19}, A_{23}, A_4, A_{26} and find that $splus(A_4) = \{2,3,4,7\}$. A_4 is deleted from t^*. Finally, $W(A_1)$ turns out to be less than WMIN and we delete A_1.

We can delete also the values A_2, A_{19} because $W(A_2)$, $W(A_{19}) = 4$, $t(splus(A_2))$, $t(splus(A_{19}))$ are not tests and, therefore, these values will not appear in a maximally redundant test t with $W(t)$ equal to or greater than 4.

After deleting these values, we can delete the examples t_9, t_5 because A_{19} is essential in t_9, and A_2 is essential in t_5. Next we can observe that $splus(A_{23}) = \{1,2,12,14\}$ and $t(\{1,2,12,14\})$ is a test; thus, A_{23} is deleted from t^* and $splus(A_{23})$ is inserted into STGOOD. We can delete the values A_{22} and A_6 because $W(A_{22})$ and $W(A_6)$ are now equal to 4, $t(splus(A_{22}))$ and $t(splus(A_6))$ are not tests, and these values will not appear in a maximally redundant test with weight equal to or greater than 4. Now t_{14} and t_1 are not tests and can be deleted.

Choose t_{12} as a subtask because $t(splus(A_{21})/\{12\})$ and $t(splus(A_{24})/\{12\})$ will be tests. By resolving this subtask, we find that t_{12} does not produce a new test. We delete it. Then $splus(A_{21})$ is equal to $\{4,6,8,11\}$, $t(\{4,6,8,11\})$ is a test, thus A_{21} is deleted from t^* and $splus(A_{21})$ is inserted into STGOOD. We can also delete the value A_{24} because $t(splus(A_{24}))$ is the GMRTs already obtained.

We can delete the value A_3 because $W(A_3)$ is now equal to 4, $t(splus(A_3))$ is not a test, and this value will not appear in a maximally redundant test with weight equal to or greater than 4. We can delete t_6 because now this example is not a test. Then we can delete the value A_{20} because $t(splus(A_{20}))$ is the GMRTs already obtained.

These deletions imply that all of the remaining rows t_2, t_3, t_4, t_7, t_8, and t_{11} are not tests.

The list of the GMRTs with the weight equal to or greater than WMIN = 4 is given in Table 9.

REFERENCES

Agraval, R., & Srikant, R. (1994). Fast algorithms for mining association rules. In *Proceedings of the 20th VLDB Conference*. Santiago, Chile.

Anshakov, O.M., Finn, V.K., & Skvortsov, D.P. (1989). On axiomatization of many-valued logics associated with formalization of plausible reasoning. *Studia Logica*, *42*(4), 423-447.

Boldyrev, N.G. (1974). Minimization of Boolean partial functions with a large number of "Don't Care" conditions and the problem of feature extraction. *Proceedings of International Symposium "Discrete Systems"* (pp.101-109). Riga, Latvia.

Carpineto, C., & Romano, G. (1996). A lattice conceptual clustering system and its application to browsing retrieval. *Machine Learning, 24*, 95-122.

Ceri, C., Gotlob, G., & Tanca, L. (1990). *Logic programming and databases*. Springer.

Cosmadakis, S., Kanellakis, P.C., & Spyratos, N. (1986). Partition semantics for relations. *Journal of Computer and System Sciences*, *33*(2), 203-233.

Demetrovics J., & Vu, D.T. (1993). Generating Armstrong relation schemes and inferring functional dependencies from relations. *International Journal on Information Theory & Applications*, *1*(4), 3-12.

Table 9. The Sets STGOOD and TGOOD for the Examples of Tables 6 and 7.

№	STGOOD	TGOOD
1	{2,3,4,7}	$A_4 A_{12} A_+ A_{24} A_{26}$
2	{1,2,12,14}	$A_{23} A_{24} A_{26}$
3	{4,6,8,11}	$A_7 A_{20} A_{21}$

Dowling, C.E. (1993). On the irredundant generation of knowledge spaces. *Journal of Math. Psych.*, *37*(1), 49-62.

Finn, V. K. (1984). Inductive models of knowledge representation in man-machine and robotics systems. *Proceedings of VINITI, Vol. A*, 58-76.

Finn, V. K. (1988). Commonsense inference and commonsense reasoning. *Review of Science and Technique* (*Itogi Nauki i Tekhniki*), *Series "The Theory of Probability. Mathematical Statistics. Technical Cybernetics,"* 28, 3-84.

Finn, V. K. (1991). Plausible reasoning in systems of JSM type. *Review of Science and Technique* (*Itogi Nauki i Tekhniki*), *Series "Informatika,"* 15, 54-101.

Finn, V. K. (1999). The synthesis of cognitive procedures and the problem of induction. *NTI, Series 2*(1-2), 8-44. Moscow, Russia: VINITI.

Galitsky, B. A., Kuznetsov, S. O., & Vinogradov, D. V. (2005). JASMINE: A hybrid reasoning tool for discovering causal links in biological data. Retrieved from http://www.dcs.bbk.ac.uk/~galitsky/Jasmine

Ganascia, J. - Gabriel. (1989). EKAW - 89 tutorial notes: Machine learning. *Third European Workshop on Knowledge Acquisition for Knowledge-Based Systems* (pp. 287-296). Paris, France.

Ganter, B. (1984). *Two basic algorithms in concepts* analysis (FB4-Preprint, No. 831). TH Darmstadt.

Giraud-Carrier, C., & Martinez, T. (1994). An incremental learning model for commonsense reasoning. *Proceedings of the Seventh International Symposium on Artificial Intelligence* (*ISAI*'94), *ITESM* (pp. 134-141).

Huntala, Y., Karkkainen, J., Porkka, P., & Toivonen, H. (1999). TANE: An efficient algorithm for discovering functional and approximate dependencies. *The Computer Journal*, *42*(2), 100-111.

Kuznetsov, S. O. (1993). Fast algorithm of constructing all the intersections of finite semi-lattice objects. *NTI, Series 2*(1), 17-20. Moscow, Russia: VINITI.

Kuznetsov, S. O., & Obiedkov, S. A. (2001). Comparing performance of algorithms for generating concept lattices. *J. Exp. Theor. Artif. Intell.* 14(2-3), 183-216.

Lavraĉ, N., & Džeroski, S. (1994). *Inductive logic programming: Techniques and applications*. Chichester: Ellis Horwood.

Lavraĉ, N., & Flash, P. (2000). *An extended transformation approach to inductive logic programming. CSTR-00-002*, March, 2000 (pp. 1-42). University of Bristol, Department of Computer Science.

Lavraĉ, N., Gamberger, D., & Jovanoski, V. (1999). A study of relevance for learning in deductive databases. *Journal of Logic Programming*, *40*(2/3), 215-249.

Lisi, F., & Malerba, D. (2004). Inducing multi-level association rules from multiple relations. *Machine Leaning*, *55*, 175-210.

Mannila, H., & Räihä, K.-J. (1992). On the complexity of inferring functional dependencies. *Discrete Applied Mathematics*, *40*, 237-243.

Mannila, H., & Räihä, K.-J. (1994). Algorithm for inferring functional dependencies. *Data & Knowledge Engineering*, *12*, 83-99.

Megretskaya, I. A. (1988). Construction of natural classification tests for knowledge base generation. In *The Problem of the Expert System Application in the National Economy* (pp. 89-93). Kishinev, Moldavia.

Mephu Nguifo, E., & Njiwoua, P. (1998). Using lattice based framework as a tool for feature extraction. In H. Lui & H. Motoda (Eds.), *Feature extraction, construction, and selection: A data mining perspective*. Kluwer.

Michalski, R. S. (1983). A theory and methodology of inductive learning. *Artificial Intelligence,* *20,* 111-161.

Michalski, R. S., & Larsen, I. B. (1978). *Selection of most representative training examples and incremental generation of VL1 Hypotheses: The Underlying methodology and the description of programs ESEL and AQII.* (Report No. 78-867). Dep. of Comp. Science, Univ. of Illinois at Urbana-Champaign, IL, USA.

Michalski, R. S., & Ram, A. (1995). Learning as goal-driven inference. In A. Ram & D. B. Leake (Eds), *Goal-driven learning.* Cambridge, MA: MIT Press/Bradford Books.

Mill, J. S. (1900). *The system of logic.* Moscow, Russia: Russian Publishing Company "Book Affair."

Naidenova, X. A. (1992). Machine learning as a diagnostic task. In I. Arefiev (Ed.), *Knowledge-Dialogue-Solution* (pp. 26-36). Materials of the short-term scientific seminar. Saint-Petersburg, Russia.

Naidenova, X. A. (1996). Reducing machine learning tasks to the approximation of a given classification on a given set of examples. In *Proceedings of the 5th National Conference at Artificial Intelligence (Kazan, Tatarstan), 1,* 275-279.

Naidenova, X. A. (1999). The data-knowledge transformation. *Text processing and cognitive technologies (Pushchino, Russia), 3,* 130-151.

Naidenova, X. A. (2001). Inferring good diagnostic tests as a model of common sense reasoning. *Proceedings of the International Conference "Knowledge-Dialog-Solution", 2,* 501-506. Saint-Petersburg, Russia: State North-West Technical University, Publishing House "Lan".

Naidenova, X. A., & Ermakov, A. E. (2001). The decomposition of algorithms of inferring good diagnostic tests. In A. Zakrevskij (Ed.), *Proceedings of the 4th International Conference*

"Computer-Aided Design of Discrete Devices (CAD DD'2001)" (Vol. 3, pp. 61-69), Institute of Engineering Cybernetics, National Academy of Sciences of Belarus. Minsk, Belarus.

Naidenova, X. A., Plaksin, M. V., & Shagalov, V. L. (1995b). Inductive inferring all good classification tests. *Proceedings of International Conference "Knowledge-Dialog-Solution" (Jalta, Ukraine), 1,* 79-84.

Naidenova, X. A., & Polegaeva, J. G. (1986). An algorithm of finding the best diagnostic tests. In G. E. Mintz & P. P. Lorents (Eds), *The 4th All Union Conference "Application of mathematical logic methods"* (pp. 63-67). Tallinn, Estonia: Institute of Cybernetics, National Acad. of Sciences of Estonia.

Naidenova, X. A., & Polegaeva, J. G. (1991). SISIF—The system of knowledge acquisition from experimental facts. In J. L. Alty & L. I. Mikulich (Eds.), *Proceedings of the IFIP TC5/WG5.3 Conference "Industrial applications of artificial intelligence"* (pp. 87-92). North-Holland, Amsterdam, the Netherlands.

Naidenova, X. A., Polegaeva, J. G., & Iserlis, J. E. (1995a). The system of knowledge acquisition based on constructing the best diagnostic classification tests. In *Proceedings of International Conference "Knowledge-Dialog-Solution" (Jalta, Ukraine), 1,* 85-95.

Nourine, L., & Raynaud, O. (1999). A fast algorithm for building lattices. *Information Processing Letters, 71,* 199-204.

Ore, O. (1944). Galois connexions. *Trans. Amer. Math. Society, 55*(1), 493-513.

Piaget, J. (1959). *La genèse des structures logiques elémentaires.* Neuchâtel.

Rasiova, H. (1974). *An algebraic approach to non-classical logic (Studies in Logic, Vol. 78).* Amsterdam; London: North-Holland Publishing Company.

Riguet, J. (1948). Relations binaires, fermetures, correspondences de Galois. *Bull. Soc. Math.*, *76*(3), 114-155.

Salzberg, S. (1991). A nearest hyper rectangle learning method. *Machine Learning, 6,* 277-309.

Schmidt-Schauss, M., & Smolka, G. (1991). Attributive concept descriptions with complements. *Artificial Intelligence, 48*(*1*), 1-26.

Shreider, J. (1974). Algebra of classification. *Proceedings of VINITI, Series 2*(9), 3-6.

Sperner, E. (1928). Eine Satz uber untermengen einer endlichen menge. *Mat. Z, 27*(11), 544-548.

Stumme, G., Taouil, R., Bastide, Y., Pasquier, N., & Lakhal, L. (2000). Fast computation of concept lattices using data mining techniques. In *Proceeding the 7th International Workshop on Knowledge Representation Meets Databases* (*KRDB* 2000) (pp. 129-139).

Stumme, G., Wille, R., & Wille, U. (1998). Conceptual knowledge discovery in databases using formal concept analysis methods. *Proceeding the 2nd European Symposium on Principles of Data Mining and Knowledge Discovery* (*PKDD*'98).

Vinogradov, D. V. (1999). Logic programs for quasi-axiomatic theories, *NTI, Series 2*(1-2), 61-64. Moscow, Russia: VINITI.

Wille, R. (1992). Concept lattices and conceptual knowledge system. *Computer Math. Appl., 23*(6-9), 493-515.

Zakrevskij, A. D. (1987). Implicative regularities in formal cognition models. *LMPS*'87 *Abstracts, 1,* 373-375.

Zakrevskij, A. D. (1982). Revealing implicative regularities in the Boolean space of attributes and pattern recognition, *Kibernetika, 1,* 1-6.

Zakrevskij, A. D. (2001). A logical approach to the pattern recognition problem. *Proceedings of the International Conference "Knowledge-Dialog-Solution"* (*KDS*'2001), *2,* 238-245. *Saint-Petersburg, Russia: State North-West Technical University, Publishing House "Lan"*).

Zakrevskij, A. D., & Vasylkova, I, V. (1997). Inductive inference systems in logical recognition in case of partial data. *Proceedings of the Fourth International Conference on Pattern Recognition and Information Processing* (*Minsk-Szczecin*), *1,* 322-326.

Chapter IV
The Analysis of Service Quality Through Stated Preference Models and Rule–Based Classification

Giovanni Felici
Consiglio Nazionale delle Ricerche, Italy

Valerio Gatta
Sapienza Università de Roma, Italy

ABSTRACT

The analysis of quality of services is an important issue for the planning and the management of many businesses. The ability to address the demands and the relevant needs of the customers of a given service is crucial to determine its success in a competitive environment. Many quantitative tools in the areas of statistics and mathematical modeling have been designed and applied to serve this purpose. Here we consider an application of a well-established statistical technique, the stated preference models (SP), to identify, from a sample of customers, significant weights to attribute to different aspects of the service provided; such aspects may additively compose an overall satisfaction index. In addition, such a weighting system is applied to a larger set of customers and a comparison is made between the overall satisfaction identified by the SP index and the overall satisfaction directly declared by the customers. Such a comparison is performed by two rule-based classification systems, decision trees, and the logic data miner Lsquare. The results of these two tools help in identifying the differences between the two measurements from the structural point of view, and provide an improved interpretation of the results. The application considered is related to the customers of a large Italian airport.

INTRODUCTION

Although quality is recognized as a key tool in the management of services, its measurement still remains a fairly subjective concept. The range of definitions used is vast and spreads from "the conformity of the specific or requisites" through to "the suitability for use" arriving at the ample sphere of "client satisfaction" (Franceschini, 2001; Negro, 1995). Many statistical and data analysis techniques have been proposed to measure the effective and perceived quality of the customers of a given service. Despite such efforts, some aspects of the issue still remain unsettled and the decision maker is faced with a number of choices to make when he/she has to plan a quality measurement campaign. In this chapter, we try to extend the range of tools usually deployed in this setting, integrating the results of consolidated techniques for quality surveys, the Stated P reference models (SP), with the application of rule-based classification algorithms. Such algorithms are used to analyze the results obtained by SP and to compare them with the satisfaction level directly declared by the users of the service under study. Our intention is to show the appropriateness of such advanced data analysis tools, typical of the area of data mining, to perform a deeper analysis of the survey data and to better understand the structure of the different methods available to measure service quality and customer satisfaction. The data considered for this application is derived from a survey on airport customers conducted on a large Italian airport, where some of the variables have been appropriately coded, as part of the results obtained are to be considered confidential. The results presented are not to be considered for interpretation purposes.

The chapter is organized as follows. The next section analyzes in more detail the issue of measuring the quality of a service through interviews to service users. The main techniques available are briefly introduced and described. Following, we explain, with a larger degree of detail, the main

concepts behind SP, how such types of surveys are built, and what statistical and inferential tools are typically used to put such models to work. Then, we describe some partial results obtained from the realization of an SP survey on airport users. Such results are used to infer a factor-weighting system for a larger customer satisfaction survey. The comparison between the quality index obtained by the SP model and the one detected directly in the survey is the topic of the last two sections of this chapter. In one we propose the use of decision trees to compare the classification models for both quality indices that are obtained from a set of explanatory variables; in the other, we use a logic-based data-mining system, *Lsquare*, to derive explanations that link, through logic formulas, the overall quality index, and the preference level attributed by the customers to five relevant factors. Finally, some conclusions are drawn.

MEASURING SERVICE QUALITY

In marketing literature, the study of service quality has focused on its evaluation by the customers. When a consumer is put in a central position as the final judge of the quality, the typical customer satisfaction survey (CSS) is based on the compilation of assessment by the clients regarding the diverse characteristics of the services through suitable scales, to which specific graduation techniques are applied (Edwards, 1957). Above all, customer satisfaction market research is used by means of questionnaires and verbal scales, which the people interviewed use to express judgement about the aspects that influence the quality of the said service. These scales are usually made up of five or seven levels pinpointed by adjectives, labels, or graduated segments. In such a way, the person interviewed is able to agree or disagree with each item. Each individual identifies an association between his own feelings and one of the categories in the scale that is offered to him/her. The most common instrument for measuring service quality is the *Servqual* scale, a method that takes inspiration from the *disconfirmation theory*, based on the difference between the quality conceived and that expected by the client (Parasuraman Zeithaml, & Berry, 1988). Servqual is a two-part questionnaire containing several statements: one part to measure what the client would expect from a general firm in the sector to which the service under examination refers, the other to assess how the client has perceived the service offered. Despite Servqual has been applied across a broad variety of service contexts, it has been criticized on methodological and psychometric grounds by many researchers: the *Servperf* model (Cronin & Taylor, 1992), the *Evaluated Performance* (Teas, 1993), and the *Retail Service Quality Scale* (Dabholkar, Thorpe, & Rentz, 1996) are additional examples of how the subject has been treated in literature. According to these approaches, the analysis of the data is achieved through multivariate statistical techniques such as factorial analysis, hierarchical, and multidimensional model.

Often, the global service-quality index is simply computed as the average of the clients' responses on the overall service evaluation and then, through the relationship among the latter and the judgements on each service dimension (attribute), the importance weights of the single service characteristics are calculated. Sometimes, the importance of each dimension is obtained by directly asking respondents to allocate a certain amount of points across the dimensions. It is to stress that these procedures may lead to partial or biased measures. With the aim to overcome this problem, we consider an SP survey combined with CSS. By doing so, we are able to get the relative importance measures of the attributes, jointly evaluated, that is, based on an explicit trade-off between attributes, and we use the latter information to calculate the service quality indicator (SQI in the following).

STATED PREFERENCE MODELS

SP methods refers to a family of techniques that foresee interviewing individuals concerning their preferences regarding a set of different options to estimate utility functions. The options are none other than descriptions of goods or services that differentiate for the characteristics they hold. They mainly deal with hypothetical situations made up *ad hoc* by the researcher. By their nature, SP methods require purpose-designed surveys for their collection of data. Such methods were originally developed in the marketing research field in the early 1970s, and have become widely used since 1978, with the objective of identifying the customers' preferences structure for products available or not yet available on the market. The flexibility of these techniques and the rich information that can be extracted allow their application also in transport, environmental, and medical fields.

A preference can be expressed in three different ways: respondent may give a rank between options (no metric valuation); they may rate a set of alternatives; or they may choose the best scenario in a given set. The latter is less informative but easier and faster for individuals than the other tasks, and it is the one that they make in reality, by comparing a set of situations and selecting one. Furthermore, this method does not require any assumptions to be made about order or cardinality measurement (Louviere, 1988). We therefore concentrate, in this chapter, on choice-based conjoint analysis, whose seminal precursory paper was written by Mc Fadden (1974).

The formation of a preference and the decisional process are, however, two very delicate aspects of the theory of the behavior. The huge complexity which stems from their analysis, a series of simplified measures, as well as the knowledge of the theory of the process that leads the individual to give certain answers (Louviere, Hensher, & Swait, 2000).

The theoretical basis is represented by the microeconomic theory of choice and by the random utility theory (RUT). The first maintains that each decision maker possesses a preference relation, $(\succsim)$, amidst the range of possible choices that satisfy a rational axiom. Such rationality is guaranteed by the completeness and transitivity properties (Mas-Colell, Whinston, & Green, 1995), which guarantees the representability of the structure of the individual's preferences through the mathematical function U, called utility function, which has an ordinal worth. Consider two alternatives, i and j (which can be goods or services), belonging to a set of choices C, meaning a collection of available alternatives from which the individual is asked to choose, we get:

$$i \succsim j \Leftrightarrow U_i > U_j \qquad (1)$$

In this context, utility is defined as the capacity of the object in question (goods or service) to satisfy the needs and meet the preferences of the decision maker. The choices will be carried out in order to guarantee the highest level of utility possible. Utility maximization, as a decisional rule, implies that the alternative i would be chosen if:

$$U_i > U_j, \quad \forall j \neq i \in C \qquad (2)$$

A first extension of the microeconomic theory for individual choice is suggested by Lancaster (1966). Here utility is defined in terms of different attributes. The decision, then, would directly stem from the utility that springs from the attributes and, consequently, the preference towards any certain product or service would only be indirect. This hypothesis allows one to represent the choice between alternatives as between attributes.

A coherent approach with the above-mentioned measures is RUT, originally proposed by Thurstone (1927), by which the decision maker has a perfect discrimination capacity, while the analyst has incomplete information mainly caused by the impossibility to consider all the factors that influence the preference of the individual. That implies that utility is not an exact known factor and must be treated as a random variable, made up of a systematic component with a margin of error. Utility of alternative i perceived by individual q can be represented as the sum of both a systematic component and a random one:

$$U_{iq} = V_{iq} + \varepsilon_{iq} \qquad (3)$$

The systematic component is a function, linear in its parameters, of the fundamental attributes:

$$V_i = \bar{\beta}\bar{X}_i \qquad (4)$$

where $\bar{\beta}$ is the vector of the coefficients associated to the vector $\bar{X}$ of explanatory variables associated with alternative i. The random component is then included as it is envisaged that some factors that influence the choice of the decision maker are not measurable. Manski (1973) identified four fonts of randomness due to incomplete information: important attributes not taken into consideration, preferences not detected that differed between individuals, measurement errors, errors gone unnoticed. In synthesis it is assumed that the decision maker is fully informed, has rational preferences, can observe the alternatives with ease and without cost, and choose in a rewarding way that which offers the greatest utility.

In the case of choice between two or more alternatives, equation (2) becomes:

$$\forall j \neq i \in C, \quad U_{iq} > U_{jq} \Leftrightarrow (V_{iq} - V_{jq}) > (\varepsilon_{jq} - \varepsilon_{iq}) \qquad (5)$$

According to RUT, the analyst, not being able to observe the difference to the right-hand member of the last equation, is not able to indicate, with a deterministic concept, when such an inequality is valid and, therefore, turns to a probabilistic approach. Then, the probability that the individual

q chooses the alternative i from the set of choices C is given by:

$$\forall j \neq i, \quad P_q(i \mid C) = P\left[(\varepsilon_{jq} - \varepsilon_{iq}) < (V_{iq} - V_{jq})\right]$$

(6)

In order to calculate such a probability, it is enough to define the statistical distribution of the random term, and equation (6) could be rewritten as follows:

$$\forall j \neq i,$$
$$P_q(i \mid C) = \int_\varepsilon I[(\varepsilon_{jq} - \varepsilon_{iq}) < (V_{iq} - V_{jq})]f(\varepsilon_q)d\varepsilon_q$$

(7)

where $f(\varepsilon_q)$ is the density function of the random vector $\varepsilon_q = (\varepsilon_{1q}, ..., \varepsilon_{Jq})$, while $I(\cdot)$ is the indicator function that assumes value 1 when the expression in brackets is true, and 0 otherwise. The probability that each random term $(\varepsilon_{jq} - \varepsilon_{iq})$ is below the observed quantity $(V_{iq} - V_{jq})$ is none other than a cumulated distribution that can be rewritten in terms of multidimensional integral over the density of the unobserved portion of utility. Different specifications of density, meaning different assumptions about the distribution of the error term, generate various discrete choice models that can be used to analyze the gathered choice data with the purpose of estimating the β-parameters and calculating an SQI.

The most popular models are the Logit and the Probit (see (Train, 2003) for reference). The first derives from the Gumbel distribution, the latter from the Normal distribution. The latter has the disadvantage of presenting yet another complex calculation; in fact, they do not have a closed-form expression for the integral in (7), quite the opposite to that which happens for the Logit models that are much easier to use.

The most popular model is the multinominal logit (ML), which is expressed as: (see equation (8)).

Parameters of this model are estimated using maximun likelihood, thus determining the set of coefficients that, when inserted in the deterministic part of the utility function, maximize the joint probablity across all the observations of the choices actually made.

AN APPLICATION TO AIRPORT DATA

In this section, we illustrate some results extracted from a survey conducted by the statistics department of the Sapienza Università de Rome, for one of the major Italian airports. The aim of the project was to identify the relative importance weights of the dimensions that still characterize their own customer satisfaction surveys, and use these measures to properly calculate an SQI; as stated, we want to link SP methods with CSS. The survey was conducted over a period of 9 months, and dealt with many aspects related to customer satisfaction and perceived service quality.

While the main survey was taking place, we also conducted a parallel survey using the SP method. One of the first steps in designing a conjoint study is to fix a set of attributes and corresponding attribute levels that need to be evaluated by the respondents. The identification of relevant attributes is usually done through

Equation 8.

$$P_q(i \mid C) = \int [\prod_{j \neq i} \exp(-e^{-(\varepsilon_{iq} + V_{iq} - V_{jq})})]e^{-\varepsilon_{iq}} \exp(-e^{-\varepsilon_{iq}})d\varepsilon_{iq} = \frac{e^{\overline{\beta}\overline{X}_{iq}}}{\sum_{j=1}^{J} e^{\overline{\beta}\overline{X}_{jq}}}$$

literature reviews, focus group discussions, or direct questioning. Given our objective, in the actual study, attribute levels were simply selected according to the items; for these were used verbal scales to evaluate five different elements linked with the airport service (e.g., airport enviroment, waiting, time, and others).

As anticipated, part of the information is confidential to the client of this study and thus, from now on, we confine the description of the five factors to the coded names *F1, F2, F3, F4, F5* with six qualitative levels from *Excellent* to *Very poor*, mapped into the values 6, 5, ...1, respectively. At the same time the results presented are to be considered just an example of the methods adopted and by no means the true conclusions of the study.

Statistical design theory is used to combine the levels of the attributes into a number of alternatives to be presented to respondents. The total number of options is a function both of the number of attributes and of the number of attribute levels. Here the total number of possible combinations was 7776; however, respondents can only evaluate a fairly limited number of options because of cognitive burden and fatigue. Through a for-

mal experimental design, we constructed three choice sets per interview. One of these choice sets had a control function; it was formed by two fixed-design alternatives. Dominance refers to a situation where one option is superior to another on every attribute, so that no trade-offs are involved in selecting the alternatives. In the final analysis, we ignored all the interviews in which the agents failed to correctly answer the control choice exercise. To allow for a rich variation in the combination of attribute levels we used a block design and we prepared 200 different versions of the survey form. The interview was composed of two sections: in the first one, respondents were asked to fill in the form in Table 1; in the second one, the questions were put in a behavioral choice context and the interviewee had to make repeated choices between two alternatives. An example of a choice set is shown in Table 2.

Overall, 1,000 face-to-face interviews were obtained at the airport station, according to a random sampling strategy; such sample size guaranteed the desired level of accuracy on the estimated probabilities. Table 3 provides information about the first section of the interview. The frequencies distributions of the judgments are

Table 1. Items and verbal scales in CSS

What is your general judgement on the airport?	Excellent (6)	Good(5)	Fairly good (4)	Barely Satisfactory (3)	Poor (2)	Very poor (1)
Overall evaluation						
What is your opinion on:	Excellent (6)	Good(5)	Fairly good (4)	Barely Satisfactory (3)	Poor (2)	Very poor (1)
F1						
F2						
F3						
F4						
F5						

very similar between the airport dimensions. The most representative class is *Good*, about 60% of the sample for the overall evaluation and for the other attributes, except for F5 where it is 36%. If we just assign to the six categories the values from 1 to 6, we will see, on average, that F3 is the best evaluated attribute while F5 is the worst.

Before getting into the econometric analysis, SP data need to be correctly organized. For all of the considered attributes there is no continuous scale; more than two levels are specified. We need to use an effect coding scheme. This creates (l-1) variables that can take the values 1, 0, -1, where l is the number of levels. We decided to exclude the lower category. For example, for F1 we have the situation in Table 4.

Now, we turn our attention to the issue of parameter estimation. We may obtain information about the relative importance of the attribute levels by using discrete choice models. In particular, since we have only two alternatives per choice set, binary logit is used. The estimation results are reported in Table 5. The preferred model is the one in which the F3 attribute is recoded as having two categories instead of six, named "High" (Fairly good, Good, or Excellent) and "Low" (Very poor, poor, Barely satisfactory). Therefore, a new single dummy variable is created (F3_h) that takes value 1 when the judgment on F3 is "High" and 0 otherwise. In this final model, we included all the variables that have significant parameter.

Table 2. Example of a Choice set used in the study

If you were to have these alternatives available to you, which one would you choose?.		
Factors	Airport A judgement	Airport B judgement
F1	*Good*	*Fairly good*
F2	*Barely satisfactory*	*Very poor*
F3	*Very poor*	*Good*
F4	*Poor*	*Excellent*
F5	*Barely satisfactory*	*Fairly good*
	☐	☐

Table 3. Frequencies of the judgements on the airport dimensions

	Very poor	Poor	Barely Satisfactory	Fairly good	Good	Excellent	Mean
Overall evaluation	0,3%	1,2%	3,0%	21,3%	66,7%	7,5%	4,75
F1	0,4%	1,8%	6,1%	25,4%	59,0%	7,3%	4,63
F2	0,3%	1,8%	4,8%	24,6%	60,2%	8,3%	4,67
F3	0,6%	1,7%	3,5%	22,1%	56,3%	15,9%	4,80
F4	0,4%	2,1%	4,3%	22,4%	60,1%	10,7%	4,72
F5	4,2%	6,2%	12,1%	33,7%	36,1%	7,7%	4,15

Table 4. Effect coding for factor F1

LEVELS	VARIABLES				
	F1_p	F2_bs	F1_fg	F1_g	F1_e
Very poor	-1	-1	-1	-1	-1
Poor	1	0	0	0	0
Barely Satisfactory	0	1	0	0	0
Fairly good	0	0	1	0	0
Good	0	0	0	1	0
Excellent	0	0	0	0	1

Table 5. Estimation results of binary logit model

Discrete choice (binary logit) model
Maximum Likelihood Estimates
Log likelihood function = -964.3959
Pseudo-R2=0.48923

Variable	Coefficient	Std. Err.	|b/St.Er	P[|Z|>z]
F1_p	-0.5214	0.0784	-6.6530	0.0000
F1_bs	-0.2341	0.0718	-3.2600	0.0011
F1_fg	0.2332	0.0777	3.0010	0.0027
F1_g	0.5543	0.0762	7.2750	0.0000
F1_e	0.7119	0.0804	8.8530	0.0000
F2_p	-0.3031	0.0722	-4.1960	0.0000
F2_g	0.2994	0.0712	4.2040	0.0000
F2_e	0.5167	0.0696	7.4190	0.0000
F3_h	0.2200	0.0676	3.2540	0.0011
F4_p	-0.5079	0.0740	-6.8650	0.0000
F4_bs	-0.1945	0.0771	-2.5220	0.0117
F4_fg	0.1885	0.0731	2.5780	0.0099
F4_g	0.3607	0.0792	4.5540	0.0000
F4_e	0.6781	0.0837	8.1000	0.0000
F5_p	-0.5678	0.0780	-7.2800	0.0000
F5_fg	0.2202	0.0689	3.1940	0.0014
F5_g	0.5194	0.0748	6.9480	0.0000
F5_e	0.7129	0.0740	9.6310	0.0000

The overall explanatory power of this nonlinear model is very good; in fact, a pseudo-R^2 of 0.5 is equivalent to about 0.8-0.9 for a linear model (Domencich & Mc Fadden, 1975). In the last two columns, the Wald test is reported.

The relative importance weights of the attribute levels are summarized in Table 6. The magnitude and the signs are consistent with our *a-priori*. In fact, for each attribute, the effect on utility increases, moving from the lowest category *Very poor* to the highest category *Excellent*. However, it should be noticed that this growth is not linear.

Based on the information presented in Table 6 and those gathered from the first section of the

Table 6. Relative importance weights of the attribute levels

ATTRIBUTES	IMPORTANCE WEIGHTS					
	Very poor	Poor	Barely	Fairly	Good	Excellent
F1	-0.7438	-0.5214	-0.2341	0.2332	0.5543	0.7119
F2	-0.5130	-0.3031	0.0000	0.0000	0.2994	0.5167
F3	0.0000	0.0000	0.0000	0.2200	0.2200	0.2200
F4	-0.5249	-0.5079	-0.1945	0.1885	0.3607	0.6781
F5	-0.8847	-0.5678	0.0000	0.2202	0.5194	0.7129

interview, SQI can be computed through the following formula:

$$SQI_q = \sum_{k=1}^{5} \sum_{l=1}^{6} \beta_{kl} X_{klq} \qquad (10)$$

where β_{kl} is the parameter of the SP model corresponding to the l-th value of the k-th factor, and X_{klq} has value 1 if the judgement expressed by user q for factor k is at level l. Therefore, the SQI for user q is obtained by simply adding up the importance weights of the attribute levels relevant to the judgment expressed by user q. Then the overall SQI is measured by taking the individual SQI average for the sampled users. Table 7 shows the overall SQI and the contributions of each attribute.

The greatest contribution to the actual SQI is given by F1 attribute, while the smaller one is given by F3 attribute. In order to obtain a relative measure of SQI, we normalized the index in this way:

$$0 < SQI' = \frac{SQI - SQI_{min}}{SQI_{max} - SQI_{min}} < 1 \qquad (11)$$

EXPLANATORY MODELS WITH DECISION TREES

In this section, we consider the construction of explanatory models for the satisfaction indices based on *decision trees*. Decision trees are a widely used technique to extract knowledge from data. They are based on an iterative and hierarchic partition of the training set in subsets of decreasing *entropy*, where the entropy is computed on the frequency distribution of the nominal variable that is to be classified. Given the tree-shaped hierarchic nature of the subset identification, the final subsets are called *leaves*. The variables used to split each subset into its child are then used to build, in a leaf-to-root path, the rule that identifies the subset of the training set that represents that leaf. The decision trees thus represent a particular type of rule-based classification system that partitions the training data, and associates with each element of the partition a single class of the variable that is to be classified (such variable is often referred, in the data-mining jargon, as *target variable*).

Extensive variants of such techniques have been proposed and refined since the seminal work

Table 7. Service quality index and its attributes contributions

	SQI	Contributions				
		F1	F2	F3	F4	F5
Mean	1.390	0.411	0.213	0.207	0.312	0.247
Minimum	-2.667	-0.744	-0.513	0	-0.525	-0.885
Maximum	2.840	0.712	0.517	0.22	0.678	0.713

of Breiman et al. (Breiman, Friedman, Olshen, & Stone, 1984) on classification and regression trees. For a detailed description of this method, we address the interested reader to the large body of available literature.

Here we adopt a very flexible and user-friendly data mining tool that implements several variants of decision trees, the open source software WEKA (see Witten & Frank, 2005). WEKA is a large software project developed at the University of Waikito, New Zealand, that puts together a large collection of classification and regression tools (among others, neural network support vector machines, logistic regressions, associative rules) in a common experimental environment where the user can edit and preprocess the data files.

The choice of rule-based models expressed as a decision tree is driven by the objective of understanding and interpreting the relations between the satisfaction and other characteristics expressed by the customers in the survey. In particular, we adopt the J48 algorithm, a recent implementation of the classical Quinlan's C4.5 (Quinlan, 1993). J48 allows one to control the dimension of the tree, thus avoiding potential *overfitting* from data, by two alternative parameters: a pruning process, controlled by a confidence factor, and a lower bound on the minimum number of elements that can be associated with a leaf. While the latter parameter is very straightforward and has no relation with the characteristics of the data analyzed, the former is based on probabilistic models associated with each leaf's data to reduce the size of the tree without losing predictive power, and may exhibit a more consistent behavior. In the following experiments, we tested different levels of confidence, maintaining a very small value of the lower bound on the minimum number of elements per leaf.

In the previous sections, we have examined the process to build a consistent SQI using SP, and how this can provide additional information about the customer's preference structure. When computing the SP-based SQI to a larger amount of interviews (in this work we present partial results obtained on a subset of approx. 5,400 interviews), we can then compare, for each interview, the service quality provided directly by the customer in the interview (referred to as SAT index, for *overall satisfaction*) and the SQI computed from the judgment expressed by the same customer on the five factors that have been considered (namely, F1, F2, F3, F4, and F5 introduced in the previous section). The two indices (the

Table 8. Description of Variables for Decision Trees

Variable Name	Description
MON	month of year
TIME	time slot of flight
SEX	sex of passenger
AGE	age of passenger (classes)
OCC	occupation of passenger
NAT	nationality of passenger
FLYER	flying frequency (Heavy, Light)
USER	using airport frequency (Heavy, Light)
TERM	Terminal
FLIGHT	type of flight (national, international)
REAS	reason for travel
SAT	overall satisfaction declared by passenger
SQI	satisfaction computed by Stated Preference Models

SQI satisfaction index and the SAT satisfaction index) have a low degree of linear correlation; not surprisingly, also the correlation among the SAT index and the satisfaction indices associated with the service components used to compute the SQI index appears to be low.

Here we construct decision trees, where the target variables are, in turn, SAT and SQI, while the explicatory variables are chosen among the set of variables measured by the survey. In Table 8, we report the complete list of the variables that was submitted to the classification algorithm.

As anticipated, the SAT index is determined on a scale from 1 to 6 (1= minimum, 6=maximum). In order to compare with SQI, we rescaled both measures from 0 to 1, and then created for both

indices a dichotomous variable with a threshold value of 0.8, obtaining a binary version of the two variables (referred to as SAT[b] and SQI[b,] respectively). The values of the two binary variables are indicated with GOOD (passengers with satisfaction of at least 80%) and BAD (passengers with satisfaction below 80%). The main interest of the study is towards the analysis of the extreme values, as one wants to understand what are the elements that separate the highly satisfied passengers from the rest. In particular, here we want to check whether the classification model is able to explain why the two indices are different and for what type of customers.

The first experiments reported are related with the J48 decision tree, where the target variable is

Table 9. Performance of Decision Trees J48 for target variable SAT[b]

Confidence Level	Lower Bound on Leaves	Percent Correct	True Positive Rate	True Negative Rate	Number of Leaves	Training Time (secs)
0.5	2	95.37%	94.30%	96.10%	71	0.09
0.4	2	95.51%	94.20%	96.40%	42	0.09
0.3	2	95.61%	93.80%	96.80%	42	0.08
0.25	2	95.70%	93.80%	97.00%	15	0.08
0.2	2	95.90%	93.50%	97.50%	15	0.09
0.1	2	95.90%	93.50%	97.50%	15	0.09

Figure 1. Decision Tree for SAT[b]

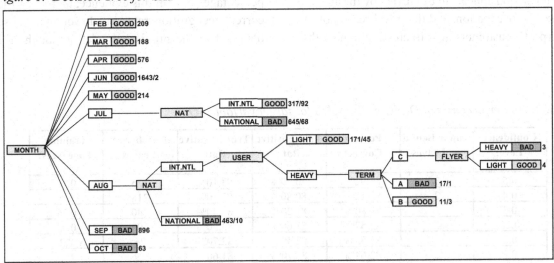

the binarized SAT[b] index. In Table 9, we see the performance of the algorithm with different values of confidence factor. The table reports overall correct recognition rate, false positive and false negative rate, number of leaves in the corresponding tree, and time spent in the training phase. The data set was composed of 5,422 valid records, and straightforward 10-fold cross validation was used to produce the results described.

The performance of the model is of very good quality, and we see a high correct recognition rate (above 95%) without relevant difference between the true positive and the true negative rates. We then analyze the structure of the tree that obtains the best recognition rate, which, in this case, corresponds to that with the lowest number of leaves. The tree is obtained with a confidence value of 0.20 or lower, and is depicted in Figure 1; attached to each leaf is the number of elements of the complete training set that are associated to that leaf (the first figure refers to the elements in the class that characterized the leaf, the second figure to those of the other class; some leaves still exhibit a nonnegligible degree of entropy, due to the effect of pruning). At first glance we note that the relevant variables for the classification are the month of the interview, the nationality of the passenger (national or international), the flying frequency and the terminal. No significant information appears to be carried by the age, the sex, the occupation, and the travel reason of the airport's customers. It is interesting to note that

there is a strong relation between the target variable and the month of the interview; months from February to May present only GOOD level of satisfaction, while September and October are composed only of BAD records. On the other hand, we note that for the months of July and August, the national passengers are dissatisfied (even if with some degree of imprecision, as expressed by the frequency attached to the related leaves). For international passengers during the month of August, a more articulated behavior is brought to evidence: those who do not fly very often from the analyzed airport declare satisfaction (i.e., GOOD value of the target variable), while those who use the airport often are significantly dissatisfied; in particular, those leaving and arriving at one of the three terminals (A) and those that use terminal C and fly frequently (although the small amount of records in these leaves does not provide a strong significance to these leaves).

The same analysis is repeated, substituting the target variable SAT[b] with SQI[b], maintaining the same set of explanatory variables. In Table 10, we report the results of the J48 algorithm for different values of the confidence factor. In this case, the separation problem appears to be slightly more difficult, as for the same level of confidence, the algorithm requires a larger number of leaves for convergence. Training time and recognition percentage confirm this evidence; moreover, the correct recognition rates are lower than those obtained in the previous model, although still

Table 10. Performance of Decision Trees J48 for target variable SQI[b]

Confidence Level	Lower Bound on Leaves	Percent Correct	True Positive Rate	True Negative Rate	Number of Leaves	Training Time (secs)
0.5	2	81.92%	80.70%	83.10%	325	0.44
0.4	2	82.12%	80.80%	83.40%	218	0.2
0.3	2	80.80%	81.20%	83.40%	140	0.24
0.25	2	82.57%	81.60%	83.50%	100	0.22
0.2	2	82.55%	82.40%	82.70%	26	0.2
0.1	2	82.64%	83.70%	81.60%	19	0.55

balanced between true positive and true negative. Nevertheless, the best performance, obtained for a small confidence level (0.1), is above 82%, and thus the information provided by the model can definitely be considered interesting.

The analysis of the tree (Figure 2) highlights several aspects of interest for the comparison of the two indices SAT and SQI in their binarized versions. At first glance we see that the important role of month is maintained in this second model; such variable is still selected by the algorithm to perform the first, and most relevant split in the training data. But here we see that the month of April turns out to be strongly characterized by dissatisfied clients (BAD value of the SQI[b] target variable) differently from what happened with SAT[b]; at the same time, the month of May gets a split on the time of the day variable, where one branch, associated with the morning time slot, sees dissatisfied customers in terminals A and C and satisfied customers in terminal B. The other branch, associated with afternoon and nighttime slots, shows an unusual pattern of dissatisfied international customers and satisfied national ones. As in the SAT[b] model, in the month of July the national passengers are dissatisfied. The

rest of the tree is substantially equivalent to that derived for the SAT[b] model; the month of August is split in the same nodes, as well as September and October.

The comparison of the two trees highlights, with a certain precision, the few structural differences between the two indices. The main conclusion is that the differences are restricted to the months of April and May. In these 2 months it appears that the direct evaluation of the satisfaction given by the customers is somewhat optimistic with respect to the more refined index obtained by the SP method. Such results could be properly interpreted by the users of the survey; one possible interpretation may be related with the way the questionnaires were submitted in those periods, or with the particular type of traffic there present. Of more interest is the coherent structure of both trees for the months of July and August, where the airport traffic is typically characterized by a larger percentage of leisure travelers and international traffic. In both months, the dissatisfaction expressed by national travelers emerges with strong evidence. Analogously, we record the negative characterization of September and October; although, here we see that while the

Figure 2. Decision Tree for SAT[b]

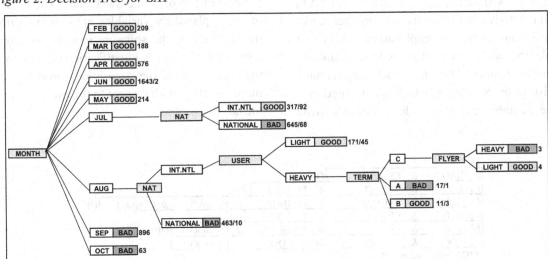

direct index (SAT[b]) reports that all customers are not in the high satisfaction class, the second index (SQI[b]) presents a large proportion of satisfied customers in the month of September (282 compared to 896).

A LOGIC MODEL FOR THE RELATION OF SPECIFIC AND SYNTHETIC INDICES

As already stated in the previous sections, one of the main concerns in the measurement of customer satisfaction is how to express the overall satisfaction stated by customers in terms of the satisfaction declared about specific aspects of the service. In the application here described, we have, at hand, an overall satisfaction index expressed by customers and the satisfaction index expressed, by the same customers, on five elements that characterize the specific service (here coded as F1, F2, F3, F4, F5). The deployment of an SP-based model has indicated a set of weights to attribute to each one of these elements in order to derive a more reliable index of the general satisfaction on the service. Such an index can be used to assess the effect on the overall satisfaction of specific actions on the different elements. Here we intend to analyze the relations between the direct satisfaction index (refered to as SAT) and the derived index (SQI) from another angle. We intend to find an explanatory model that links the value of overall index to the values of the five elements. Obviously, such a model must exist for the SQI index, which is computed by a linear combination of the values attributed to the five elements; on the other hand, we know that the linear correlation between the SAT index and the level of satisfaction expressed for the five elements is moderately poor (0.609). We restrict our attention to the binarized version of the two indices, defining two classes (below and above 80% of the maximum value of satisfaction), as in the previous section. Such simplification may reduce the intrinsic variance of the indices; on the other hand, it forces the focus of the analysis on a marked difference in the judgments expressed by the customers.

For this purpose we use the logic data miner *Lsquare*. *Lsquare* is a classification method based on a logic programming formulation of the classification problem, where the problem of finding a set of logic rules that separates the training data is made equivalent to the solution of a finite sequence of minimum cost satisfiability problems (MINSAT). A description of the system is available in Felici and Truemper (2002, 2005) and Felici, Sun, Truemper K. (2006). A more detailed discussion of the methods and mathematical tools on which the system is based can be found in Truemper (2004).

For the derivation of the logic model we use, in turn, the SAT[b] and the SQI[b] indices as target variables, and as explanatory variables, the satisfaction level expressed by the customers, on a scale from 1 to 6, on the five factors. The values of the explanatory variables are automatically transformed by the software into one or more logic variables by the identification of cutpoints that best express the separation capability of these variables with respect to the two values of the target variable.

IF

F1=BAD	& F5=BAD	OR			
F1=GOOD	& F3=BAD	& F5=BAD	OR		
F1=GOOD	& F3=GOOD	& F3=BAD	& F4=BAD	& F5=BAD	OR
F1=BAD	& F4=BAD	& F5=BAD	OR		
F2=BAD	& F3=BAD	& F4=GOOD	& F5=GOOD	OR	
F1=GOOD	& F3=BAD	& F1=GOOD	& F5=GOOD		

THEN satisfaction = BAD

The first set of results that we examine is related with the model where SAT[b] plays the role of the target variable. We split the available data into a training set composed of 700 records, 400 from those having GOOD satisfaction levels and 300 from the those having BAD (the training set is extracted at random from the 5,422 available records according to the uniform distribution). The remaining data is used for testing purposes.

The system identifies a single cutpoint for each variable, located at the value 4.5; each satisfaction judgment, from 1 to 6, is then mapped into a logic variable that has value *true* if the value is equal to 5 or 6, and *false* otherwise. Interestingly, the binary mapping computed by *Lsquare* strongly resembles the partition of the overall satisfaction index adopted ex-ante for the target variable. For symmetry, we will indicate with GOOD and BAD also the binary value of the five explanatory variables; for example, with *F3 = GOOD* we intend that the satisfaction level expressed on this factor is greater or equal to 5.

The best formula obtained is composed of six logic clauses, has value *true* for BAD records, and has an overall precision of 75.3%, divided into 66.1% on GOOD records and 89.3% on BAD records (such percentages are computed on the 4,722 records kept aside for testing). Thus, such formula holds true for 89.3% of the records with BAD overall satisfaction in the test set, and holds false for 66.1% of records with GOOD overall satisfaction in the test set. Next, we report a description of the formula. With a deeper analysis of the formula produced by *Lsquare,* we can identify a much more compact formula that still exhibits a good behavior. This very simple, one-clause formula holds true for 58% of records with GOOD and for 93.7% of BAD records out of the 4,722 records used for testing, resulting in an overall correctness of 72.1%:

The results just described highlight the difficulty to predict, in a consistent way and with high precision, the general satisfaction expressed by the customers from the satisfaction on the five specific aspects of the service. They show some additional information: while it is somehow difficult to predict the GOOD satisfaction level, it is much easier to predict the BAD. In other words, if a customer satisfies this rule, we can assume that he/she has a GOOD satisfaction level with mild confidence; but, on the other hand, a customer that does not satisfy it can very likely be assigned a BAD satisfaction level.

The same analysis is conducted, adopting SQI[b] as the target variable. The entire set is split into training and test data exactly in the same way adopted for the previous model. The results are as follows: The best rule identified is composed of seven clauses and exhibits an overall correct classification rate of 88.2%; the next formula reported holds *true* for 80.2% of records with GOOD and *false* for 95.9% of BAD records out of the 4,722 records used for testing.

As done in the previous case, we can select the best one-clause formula from the ones available. Such formula has exactly the same form as the one-clause formula derived in the SAT[b] model: the general satisfaction is GOOD when the satisfaction levels on F1, F2, and F4 are, simultaneously, GOOD. The formula has an overall precision on the test set of 83.8%, obtained from a correct recognition of 70.8% on GOOD records and of 96.5% on BAD records.

IF

F1=GOOD	& F3=GOOD	& F4=GOOD

THEN satisfaction = GOOD

IF

F1=BAD	& F4=BAD	OR
F2=BAD	& F5=BAD	OR
F1=BAD	& F5=BAD	OR
F4=BAD	& F5=BAD	OR
F1=BAD	& F3=BAD	OR
F2=BAD	& F3=BAD	OR
F1=BAD	& F3=GOOD	& F3=GOOD

THEN satisfaction = BAD

From these results we can again conclude that the system of preferences that supports the customers' satisfaction does not exhibit a clear structure; even for the second index (that, we recall, is obtained by a linear combination of the five satisfaction levels expressed on the detailed services) the best logic-based model fails to identify, with high precision, the records with GOOD general satisfaction; on the other hand, we have identified very simple logic rules that can be considered extremely reliable when declaring that a record is in the BAD class. Besides, we see that the most compact rule is the same for both models.

CONCLUSION

In this chapter we have considered the problem of service quality measurement using an SP model to assess the contribution of different factors to the overall value of service quality. To analyze the results obtained, we have used two rule-based classification techniques, and have exploited some characteristics of the results obtained, comparing the differential information provided by the SP model and from a direct customer survey. Data from a survey on a large Italian airport were used. The analysis show that the apparent difference between the two measures can be restricted to some particular subsets of the available data and thus, the two indices compared can be better understood and put in relation with specific events and customer types. The use of rule-based data mining techniques allows, to put in clear evidence, the information contained in the data, and to orient the decision makers in the interpretation of the results.

REFERENCES

Breiman, L., Friedman J.H., Olshen R.A., & Stone, C.J. (1984). *Classification and regression trees.* Wadsworth International.

Cronin, J.J,. & Taylor, A. (1992). Measuring service quality: A reexamination and extension. *Journal of Marketing, 56*(3), 55-68.

Dabholkar, P.A., Thorpe, D.I., & Rentz, J.O. (1996). A measure of service quality for retail stores: Scale development and validation. *Journal of the Academy of Marketing Science, 24*(1), 3-16.

Domencich, T., & Mc Fadden, D. (1975). *Urban travel demand, a behavioural analysis.* Amsterdam, Holland: North-Holland Publishing Company.

Edwards, A.L. (1957). *Techniques of attitude scale construction.* New York: Appleton-Century-Crofts Inc.

Felici, G., & Truemper, K. (2002). A minsat approach for learning in logic domains, *INFORMS Journal on Computing, 14*(1), 20-36.

Felici, G., & Truemper K. (2005). The Lsquare system for mining logic data. In D. Wang (Ed.) *Encyclopedia of Data Warehousing and Mining* (vol. 2.) Hershey, PA: Idea Group Reference.

Felici, G., Sun, K-S., & Truemper. K. (2006). Learning logic formulas and related error distributions. In G. Felici and E. Trintaphyllou (Eds.), *Data mining and knowledge discovery approaches based on rule induction techniques.* Springer.

Franceschini, F. (2001). *Dai prodotti ai servizi. Le nuove frontiere per la misura della qualità.* Torino, Italy: UTET.

Lancaster, K. (1966). A new approach to consumer theory. *Journal of Political Economy, 74*(2), 132-157.

Louviere, J. (1988). Conjoint analysis modelling of stated preferences. A review of theory, methods, recent developments and external validity. *Journal of Transport Economics ad Policy, 22*(1), 93-119.

Louviere, J.J., Hensher D.A., & Swait J. (2000). *Stated choice methods. Analysis and application.*

Cambridge, England: Cambridge University Press.

Manski, C. (1973). *The analysis of quantitative choice*. Unpublished doctoral dissertation, Department of Economics, Massachusetts Institute of Technology, Cambridge.

Mas-Colell, A., Whinston M.D., & Green J.R. (1995). *Microeconomic theory*. New York: Oxford University Press.

Mc Fadden, D. (1974). Conditional logit analysis of qualitative choice behaviour. In P. Zarembka (Ed.), *Frontiers in econometrics* (pp. 105-142). New York: Academic Press.

Negro, G. (1995). *Organizzare la qualità nei servizi*. Milano, Italy: Edizioni del Sole 24 Ore.

Parasuraman, A., Zeithaml, V.A., & Berry, L.L. (1988). SERVQUAL: A multiple item scale for measuring consumer perceptions of service quality. *Journal of Retailing, 64*(1), 12-37.

Quinlan, R. (1993). *C4.5: Programs for machine learning*. San Mateo, CA: Morgan Kaufmann Publishers.

Teas, R.K. (1993). Expectations, performance evaluation, and consumers' perceptions of quality. *Journal of Marketing, 57*(4), 18-34.

Thurstone, L. (1927). A law of comparative judgment. *Psychological Review, 34*, 273-286.

Train, K. (2003). *Discrete choice methods with simulation*. UK: Cambridge University Press.

Truemper, K. (1996). *The Leibniz system for logic programming*. Version 4.2, Leibniz Plano, Texas 75023, U.S.A.

Truemper, K. (2004). *Design of logic-based intelligent systems*. New York: Wiley.

Witten, I.H., & Frank E. (2005). *Data mining: Practical machine learning tools and techniques*. Morgan Kaufmann.

Chapter V
Support Vector Machines for Business Applications

Brian C. Lovell
NICTA & The University of Queensland, Australia

Christian J. Walder
Max Planck Institute for Biological Cybernetics, Germany

ABSTRACT

This chapter discusses the use of support vector machines (SVM) for business applications. It provides a brief historical background on inductive learning and pattern recognition, and then an intuitive motivation for SVM methods. The method is compared to other approaches, and the tools and background theory required to successfully apply SVM to business applications are introduced. The authors hope that the chapter will help practitioners to understand when the SVM should be the method of choice, as well as how to achieve good results in minimal time.

INTRODUCTION

Recent years have seen an explosive growth in computing power and data storage within business organisations. From a business perspective, this means that most companies now have massive archives of customer and product data and more often than not these archives are far too large for human analysis. An obvious question has therefore arisen, "How can one turn these immense corporate data archives to commercial advantage?" To this end, a number of common applications have arisen, from predicting which products a customer is most likely to purchase, to designing the perfect product based on responses to questionnaires. The theory and development of these processes has grown into a discipline of its own, known as Data Mining, which draws heavily on the related fields of Machine Learning, Pattern Recognition, and Mathematical Statistics.

The Data Mining discipline is still developing, however, and a great deal of sub-optimal and ad hoc analysis is being done. This is partly due to the complexity of the problems, but is also due to the vast number of available techniques. Even the most fundamental task in Data Mining, that of inductive inference, or making predictions based on examples, can be tackled by a great many different techniques. Some of these techniques are very difficult to tailor to a specific problem and require highly skilled human design. Others are more generic in application and can be treated more like the proverbial "black box." One particularly generic and powerful method, known as the Support Vector Machine (SVM) has proven to be both easy to apply and capable of producing results that range from good to excellent in comparison to other methods. While application of the method is relatively straightforward, the practitioner can still benefit greatly from a basic understanding of the underlying machinery.

Unfortunately most available tutorials on SVMs require a very solid mathematical background, so we have written this chapter to make SVM accessible to a wider community. This chapter comprises a basic background on the problem of induction, followed by the main sections. In the first section we introduce the concepts and equations on which the SVM is based, in an intuitive manner, and identify the relationship between the SVM and some of the other popular analysis methods. In the second section we survey some interesting applications of SVMs on practical real-world problems. Finally, the third section provides a set of guidelines and rules of thumb for applying the tool, with a pedagogical example that is designed to demonstrate everything that the SVM newcomer requires in order to immediately apply the tool to a specific problem domain. The chapter is intended as a brief introduction to the field that introduces the ideas, methodologies, as well as a hands-on introduction to freely available software, allowing the reader to rapidly determine the effectiveness of SVMs for their specific domain.

BACKGROUND

SVMs are most commonly applied to the problem of inductive inference, or making predictions based on previously seen examples. To illustrate what is meant by this, let us consider the data presented in Tables 1 and 2. We see here an example of the problem of inductive inference, more specifically that of supervised learning. In supervised learning we are given a set of input data along with their corresponding labels. The input data comprises a number of examples about which several attributes are known (in this case, age, income, etc.). The label indicates which class a particular example belongs to. In the example above, the label tells us whether or not a given person has a broadband Internet connection to their home. This is called a binary classification problem because there are only two possible classes. In the second table, we are given the attributes for a different set of consumers, for whom the true class labels are unknown. Our goal is to infer from the first table the most likely labels for the people in the second table, that is, whether or not they have a broadband Internet connection to their home.

In the field of data mining, we often refer to these sets by the terms test set, training set, validation set, and so on, but there is some confusion in the literature about the exact definitions of these terms. For this reason we avoid this nomenclature, with the exception of the term training set. For our purposes, the training set shall be all that is given to us in order to infer some general correspondence between the input data and labels. We will refer to the set of data for which we would like to predict the labels as the unlabelled set.

Table 1. Training or labelled set

Age	Income	Years of Education	Gender	Broadband Home Internet Connection?
30	$56,000 / yr	16	male	Yes
50	$60,000 / yr	12	female	Yes
16	$2,000 / yr	11	male	No
35	$30,000 / yr	12	male	No

The dataset in Table 1 contains demographic information for four randomly selected people. These people were surveyed to determine whether or not they had a broadband home internet connection.

Table 2. Unlabelled set

Age	Income	Years of Education	Gender	Broadband Home Internet Connection?
40	$48,000 / yr	17	male	unknown
29	$60,000 / yr	18	female	unknown

The dataset in Table 2 contains demographic information for people who may or may not be good candidates for broadband internet connection advertising. The question arising is, "Which of these people is likely to have broadband internet connection at home?"

Figure 1. Inductive inference process in schematic form (Based on a particular training set of examples with labels, the learning algorithm constructs a decision rule which can then be used to predict the labels of new unlabelled examples.)

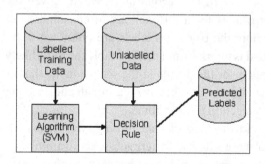

A schematic diagram for the above process is provided in Figure 1. In the case of the SVM classifier (and most other learning algorithms for that matter), there are a number of parameters which must be chosen by the user. These parameters control various aspects of the algorithm, and in order to yield the best possible performance, it is necessary to make the right choices. The process of choosing parameters that yield good performance is often referred to as model selection. In order to understand this process, we have to consider what it is that we are aiming for in terms of classifier performance. From the point of view of the practitioner, the hope is that the algorithm will be able to make true predictions about unseen cases. Here the true values we are trying to predict are the class labels of the unlabelled data. From this perspective it is natural to measure the performance of a classifier by the probability of its misclassifying an unseen example.

It is here that things become somewhat less straightforward, however, due to the following dilemma. In order to estimate the probability of a misclassification, we need to know the true underlying probability distributions of the data that we are dealing with. If we actually knew this, however, we wouldn't have needed to perform inductive inference in the first place! Indeed knowledge of the true probability distributions allows us to calculate the theoretically best possible decision rule corresponding to the so-called Bayesian classifier (Duda, Hart, & Stork, 2001).

In recent years, a great deal of research effort has gone into developing sophisticated theories that make statements about the probability of a particular classifier making errors on new un-labelled cases—these statements are typically referred to as generalization bounds. It turns out, however, that the research has a long way to go, and in practice one is usually forced to determine the parameters of the learning algorithm by much more pragmatic means. Perhaps the most straightforward of these methods involves estimating the probability of misclassification using a set of real data for which the class labels are known— to do this one simply compares the labels predicted by the learning algorithm to the true known labels. The estimate of misclassification probability is then given by the number of examples for which the algorithm made an error (that is, predicted a label other than the true known label) divided by the number of examples that were tested in this manner.

Some care needs to be taken, however, in how this procedure is conducted. A common pitfall for the inexperienced analyst involves making this estimate of misclassification probability using the training set from which the decision rule itself was inferred. The problem with this approach is easily seen from the following simple decision rule example. Imagine a decision rule that makes label predictions by way of the following procedure (sometimes referred to as the notebook classifier):

The notebook classifier decision rule: We wish to predict the label of the example X. If X is present in the training set, make the prediction that its label is the same as the corresponding label in the training set. Otherwise, toss a coin to determine the label.

For this method, while the estimated probability of misclassification on the training set will be zero, it is clear that for most real-world problems the algorithm will perform no better than tossing a coin! The notebook classifier is a commonly used example to illustrate the phenomenon of overfitting—which refers to situations where the decision rule fits the training set well, but does not *generalize* well to previously unseen cases. What we are really aiming for is a decision rule that generalizes as well as possible, even if this means that it cannot perform as well on the training set.

Cross-validation: So it seems that we need a more sophisticated means of estimating the generalization performance of our inferred decision rules if we are to successfully guide the model selection process. Fortunately there is a more effective means of estimating the generalization performance based on the training set. This procedure, which is referred to as *cross-validation* or more specifically *n-fold cross-validation*, proceeds in the following manner (Duda et al., 2001):

1. Split the training set into n equally sized and disjoint subsets (partitions), numbered 1 to n.
2. Construct a decision function using a conglomerate of all the data from subsets 2 to n.
3. Use this decision function to predict the labels of the examples in subset number 1.
4. Compare the predicted labels to the known labels in subset number 1.
5. Repeat steps 1 through 4 a further (n-1) times, each time testing on a different subset, and always excluding that subset from training.

Having done this, we can once again divide the number of misclassifications by the total number of training examples to get an estimate of the true generalization performance. The point is that since we have avoided checking the performance of the classifier on examples that the algorithm had already "seen," we have calculated a far more meaningful measure of classifier quality. Commonly used values for n are 3 and 10 leading to so called *3-fold* and *10-fold* cross-validation.

Now, while it is nice to have some idea of how well our decision function will generalize, we really want to use this measure to guide the model selection process. If there are only, say, two parameters to choose for the classification algorithm, it is common to simply evaluate the generalization performance (using cross validation) for all combinations of the two parameters, over some reasonable range. As the number of parameters increases, however, this soon becomes infeasible due to the excessive number of parameter combinations. Fortunately one can often get away with just two parameters for the SVM algorithm, making this relatively straightforward model selection methodology widely applicable and quite effective on real-world problems.

Now that we have a basic understanding of what supervised learning algorithms can do, as well as roughly how they should be used and evaluated, it is time to take a peek under the hood of one in particular, the SVM. While the main underlying idea of the SVM is quite intuitive, it will be necessary to delve into some mathematical details in order to better appreciate why the method has been so successful.

MAIN THRUST OF THE CHAPTER

The SVM is a supervised learning algorithm that infers from a set of labeled examples a function that takes new examples as input, and produces predicted labels as output. As such the output of the algorithm is a mathematical function that is defined on the space from which our examples are taken. It takes on one of two values at all points in the space, corresponding to the two class labels that are considered in binary classification. One of the theoretically appealing things about the SVM is that the key underlying idea is in fact extremely simple. Indeed, the standard derivation of the SVM algorithm begins with possibly the simplest class of decision functions: linear ones. To illustrate what is meant by this, Figure 2 consists of three linear decision functions that happen to be correctly classifying some simple 2D training sets.

Linear decision functions consist of a decision boundary that is a hyperplane (a line in 2D, plane in 3D, etc.) separating the two different regions of the space. Such a decision function can be expressed by a mathematical function of an input vector **x**, the value of which is the predicted label

Figure 2. A simple 2D classification task, to separate the black dots from the circles (Three feasible but different linear decision functions are depicted, whereby the classifier predicts that any new samples in the gray region are black dots, and those in the white region are circles. Which is the best decision function and why?)

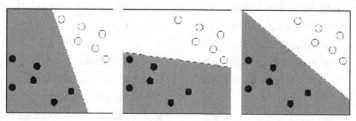

for **x** (either +1 or -1). The linear classifier can therefore be written as:

$$g(\mathbf{x}) = sign(f(\mathbf{x}))$$

where $f(\mathbf{x}) = <\mathbf{w}, \mathbf{x}> + b$.

In this way we have parameterized the function by the weight vector **w** and the scalar b. The notation $<\mathbf{w},\mathbf{x}>$ denotes the inner or scalar product of **w** and **x**, defined by:

$$<\mathbf{w}, \mathbf{x}> = \sum_{i=1}^{d} w_i x_i$$

where d is the dimensionality, and w_i is the i-th component of **w**, where **w** is of the form $(w_1, w_2, \dots w_d)$. Having formalized our decision function, we can now formalize the problem that the linear SVM addresses:

Given a training set of vectors $\mathbf{x}_1, \mathbf{x}_2, \dots \mathbf{x}_n$ with corresponding class membership labels $y_1, y_2, \dots y_n$ that take on the values +1 or -1, choose parameters w and b of the linear decision function that generalizes well to unseen examples.

Perceptron Algorithm: Probably the first algorithm to tackle this problem was the Perceptron algorithm (Rosenblatt, 1958). The Perceptron algorithm simply used an iterative procedure to incrementally adjust **w** and b until the decision boundary was able to separate the two classes of the training data. As such, the Perceptron algorithm would give no preference between the three feasible solutions in Figure 2—any one of the three could result. This seems rather unsatisfactory, as most people would agree that the rightmost decision function is the superior one. Moreover, this intuitive preference can be justified in various ways, for example by considering the effect of measurement noise on the data — small perturbations of the data could easily change the predicted labels of the training set in the first two examples, whereas the third is far more robust in this respect. In order to make use of this intuition,

it is necessary to state more precisely why we prefer the third classifier:

We prefer decision boundaries that not only correctly separate two classes in the training set, but lie as far from the training examples as possible.

This simple intuition is all that is required to lead to the linear SVM classifier, which chooses the hyperplane that separates the two classes with the maximum *margin*. The margin is just the distance from the hyperplane to the nearest training example. Before we continue, it is important to note that while the above example shows a 2D data set, which can be conveniently represented by points in a plane, in fact we will typically be dealing with higher dimensional data. For example, the example data in Table 1 could easily be represented as points in four dimensions as follows:

$\mathbf{x}_1 = [\ 3056000\ 16\ 0\ 1]$; $y_1 = +1$
$\mathbf{x}_2 = [50\ 60000\ 12\ 1\ 0]$; $y_2 = +1$
$\mathbf{x}_3 = [16\ 2000\ 110\ 1]$; $y_3 = -1$
$\mathbf{x}_4 = [\ 35\ 30000\ 12\ 0\ 1]$; $y_4 = -1$

Actually, there are some design decisions to be made by the practitioner when translating attributes into the above type of numerical format, which we shall touch on in the next section. For example here we have mapped the male/female column into two new numerical indicators. For now, just note that we have also listed the labels y_1 to y_4 which take on the value +1 or –1, in order to indicate the class membership of the examples (that is, $y_i = 1$ means that $\mathbf{x}_i$ has a broadband home Internet connection).

In order to easily find the maximum margin hyperplane for a given data set using a computer, we would like to write the task as an *optimization problem*. Optimization problems consist of an objective function, which we typically want to find the maximum or minimum value of, along

Figure 3. Linearly separable classification problem

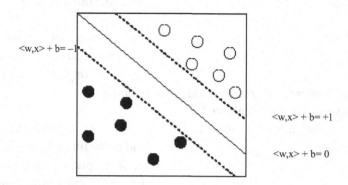

with a set of constraints, which are conditions that we must satisfy while finding the best value of the objective function. A simple example is to minimize x^2 subject to the constraint that $1 \leq x \leq 2$. The solution to this example optimization problem happens to be $x = 1$. To see how to compactly formulate the maximum margin hyperplane problem as an optimization problem, take a look at Figure 3.

The figure shows some 2D data drawn as circles and black dots, having labels +1 and –1 respectively. As before, we have parameterized our decision function by the vector **w** and the scalar b, which means that, in order for our hyperplane to correctly separate the two classes, we need to satisfy the following constraints:

$$< \mathbf{w}, \mathbf{x}_i > +b > 0, \text{for all } y_i = 1$$
$$< \mathbf{w}, \mathbf{x}_i > +b < 0, \text{for all } y_i = -1$$

To aid understanding, the first constraint above may be expressed as: "$< \mathbf{w}, \mathbf{x}_i > +$ b must be greater than zero, *whenever* y_i *is equal to one.*" It is easy to check that the two sets of constraints above can be combined into the following single set of constraints: $(< \mathbf{w}, \mathbf{x}_i > +b)y_i > 0, i = 1...n$

However meeting this constraint is not enough to separate the two classes optimally — we need to do so with the maximum margin. An easy way to see how to do this is the following. First note that we have plotted the decision surface as

a solid line in Figure 3, which is the set satisfying: $< \mathbf{w}, \mathbf{x} > +b = 0$.

The set of constraints that we have so far is equivalent to saying that these data must lie on the correct side (according to class label) of this decision surface. Next notice that we have also plotted as dotted lines two other hyperplanes, which are the hyperplanes where the function $<\mathbf{w},\mathbf{x}> +$ b is equal to -1 (on the lower left) and +1 (on the upper right). Now, in order to find the maximum margin hyperplane, we can see intuitively that we should keep the dotted lines parallel and equidistant to the decision surface, and maximize their distance from one another, while satisfying the constraint that the data lie on the correct side of the dotted lines associated with that class. In mathematical form, the final clause of this sentence (the constraints) can be written as: $y_i(< \mathbf{w}, \mathbf{x}_i > +b) > 1, i = 1...n$.

All we need to do then is to maximize the distance between the dotted lines subject to the constraint set above. To aid in understanding, one commonly used analogy is to think of these data points as nails partially driven into a board. Now we successively place thicker and thicker pieces of timber between the nails representing the two classes until the timber just fits—the centreline of the timber now represents the optimal decision boundary. It turns out that this distance is equal to $2/\sqrt{<\mathbf{w},\mathbf{w}>}$, and since maximizin $2/\sqrt{<\mathbf{w},\mathbf{w}>}$ is the same as minimizing $<\mathbf{w},\mathbf{w}>$, we end up with the following optimization problem, the solution of which yields the parameters of the maximum margin hyperplane. The term ½ in the objective function below can be ignored as it simply makes things neater from a certain mathematical point of view:

$$\min_{\mathbf{w},b} \frac{1}{2}\mathbf{w}\cdot\mathbf{w}$$

such that $y_i(\mathbf{w}\cdot\mathbf{x}_i + b) \geq 1$ (1)

for all $i = 1,2,...n$

The previous problem is quite simple, but it encompasses the key philosophy behind the SVM—maximum margin data separation. If the above problem had been scribbled onto a cocktail napkin and handed to the pioneers of the Perceptron back in the 1960s, then the Machine Learning discipline would probably have progressed a great deal further than it has to date! We cannot relax just yet, however, as there is a major problem with the above method: What if these data are not *linearly separable*? That is if it is not possible to find a hyperplane that separates all of the examples in each class from all of the examples in the other class? In this case there would be no combination of **w** and b that could ever satisfy the set of constraints above, let alone do so with maximum margin. This situation is depicted in Figure 4, where it becomes apparent that we need

to *soften* the constraint that these data lie on the correct side of the +1 and -1 hyperplanes, that is we need to allow some, but not too many data points to violate these constraints by a preferably small amount. This alternative approach turns out to be very useful not only for data sets that are not linearly separable, but also, and perhaps more importantly, in allowing improvements in generalization.

Usually when we start talking about vague concepts such as "not too many" and "a small amount," we need to introduce a parameter into our problem, which we can vary in order to balance between various goals and objectives. The following optimization problem, known as the 1-norm *soft margin* SVM, is probably the one most commonly used to balance the goals of maximum margin separation, and correctness of the training set classification. It achieves various trade-offs between these goals for various values of the parameter C, which is usually chosen by cross-validation on a training set as discussed earlier.

$$\min_{\mathbf{w},b,} \frac{1}{2}\mathbf{w}\cdot\mathbf{w} + C\sum_{i=1}^{m}\mathbf{x}_i$$

such that $y_i(\mathbf{w}\cdot\mathbf{x}_i + b) + \mathbf{x}_i \geq 1$

*   (2)

for all $i = 1,2,...n$.

Figure 4. Linearly inseparable classification problem

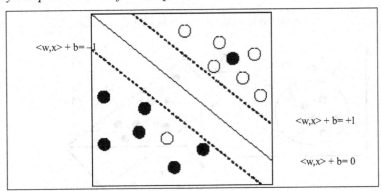

The easiest way to understand this problem is by comparison with the previous formulation that we gave, which is known as the *hard margin* SVM, in reference to the fact that the margin constraints are "hard," and are not allowed to be violated at all. First note that we have an extra term in our objective function that is equal to the sum of the ξ_i's. Since we are minimizing the objective function, it is safe to say that we are looking for a solution that keeps the ξ_i values small. Moreover, since the ξ term is added to the original objective function after multiplication by C, we can say that as C increases we care less about the size of the margin, and more about keeping the ξ's small. The true meaning of the ξ_i's can only be seen from the constraint set, however. Here, instead of constraining the function $y_i(\langle \mathbf{w}, \mathbf{x}_i \rangle + b)$ to be greater than 1, we constrain it to be greater than $1 - \xi_i$. That is, we allow the point $\mathbf{x}_i$ to violate the margin by an amount ξ_i. Thus, the value of C trades between how large of a margin we would prefer, as opposed to how many of the training set examples violate this margin (and by how much).

So far, we have seen that the maximally separating hyperplane is a good starting point for linear classifiers. We have also seen how to write down the problem of finding this hyperplane as an optimization problem consisting of an objective function and constraints. After this we saw a way of dealing with data that is not linearly separable, by allowing some training points to violate the margin somewhat. The next limitation we will address is in the form of solutions available. So far we have only considered very simple linear classifiers, and as such we can only expect to succeed in very simple cases. Fortunately it is possible to extend the previous analysis in an intuitive manner, to more complex classes of decision functions. The basic idea is illustrated in Figure 5.

The example in Figure 5 shows on the left a data set that is not linearly separable. In fact, the data is not even close to linearly separable, and one could never do very well with a linear classifier for the training set given. In spite of this, it is easy for a person to look at the data and suggest a simple elliptical decision surface that ought to generalize well. Imagine, however, that there is a mapping Φ, which transforms these data to some new, possibly higher dimensional space, in which the data is linearly separable. If we knew Φ then we could map all of the data to the *feature space*, and perform normal SVM classification in this space. If we can achieve a reasonable margin in the feature space, then we can expect a reasonably good generalization performance, in spite of a possible increase in dimensionality.

The last sentence of the previous paragraph is far deeper than it may first appear. For some time, Machine Learning researchers have feared the *curse of dimensionality*, a name given to the widely-held belief that if the dimension of the feature space is large in comparison to the number

Figure 5. An example of a mapping Φ to a feature space in which the data become linearly separable

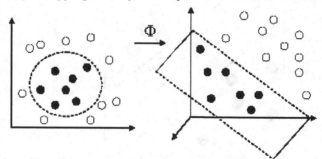

of training examples, then it is difficult to find a classifier that generalizes well. It took the theory of Vapnik and Chervonenkis (Vapnik, 1998) to put a serious dent in this belief. In a nutshell, they formalized and proved the last sentence of the previous paragraph, and thereby paved the way for methods that map data to *very* high dimensional feature spaces where they then perform maximum margin linear separation. Actually, a tricky practical issue also had to be overcome before the approach could flourish: if we map to a feature space that is too high in dimension, then it will become impossible to perform the required calculations (that is, to find **w** and b)—that is, it would take too long on a computer. It is not obvious how to overcome this difficulty, and it took until 1995 for researchers to notice the following elegant and quite remarkable possibility.

The usual way of proceeding is to take the original soft margin SVM, and convert it to an equivalent *Lagrangian dual problem*. The derivation is not especially enlightening, however, so we will skip to the result, which is that the solution to the following *dual* or equivalent problem gives us the solution to the original SVM problem. The dual problem, which is to be solved by varying the α_i's, is as follows (Vapnik, 1998):

$$\min_{\alpha} \frac{1}{2} \sum_{i,j=1}^{m} y_i y_j \alpha_i \alpha_j (\mathbf{x}_i \cdot \mathbf{x}_j) - \sum_{i=1}^{m} \alpha_i$$

$$\text{such that } \sum_{i=1}^{m} y_i \alpha_i = 0$$

$$0 \le \alpha_i \le C, \ i = 1, 2, ..., m. \tag{3}$$

The α_i's are known as the *dual variables*, and they define the corresponding *primal variables* **w** and b by the following relationships:

$$\mathbf{w} = \sum_{i=1}^{m} \alpha_i y_i \mathbf{x}_i$$

$$\alpha_i (y_i (< \mathbf{w}, \mathbf{x}_i > +b) - 1) = 0$$

Note that by the linearity of the inner product (that is, the fact that $<\mathbf{a}+\mathbf{b},\mathbf{c}> = <\mathbf{a},\mathbf{c}> + <\mathbf{b},\mathbf{c}>$), we can write the decision function in the following form: $f(\mathbf{x}) = <\mathbf{w}, \mathbf{x}> +b = \sum_{i=1}^{m} \alpha_i y_i <\mathbf{x}_i, \mathbf{x}> +b$

Recall that it is the sign of f(**x**) that gives us the predicted label of **x**. A quite remarkable thing is that in order to determine the optimal values of the α_i's and b, and also to calculate f(**x**), we do not actually need to know any of the training or testing vectors, we only need to know the scalar value of their inner product with one another. This can be seen by noting that the vectors only ever appear by way of their inner product with one another. The elegant thing is that rather than explicitly mapping all of the data to the new space and performing linear SVM classification, we can operate in the original space, provided we can find a so-called *kernel* function k(.,.) which is equal to the inner product of the mapped data. That is, we need a kernel function k(.,.) satisfying:

$$k(\mathbf{x},\mathbf{y}) = <\Phi(\mathbf{x}), \Phi(\mathbf{y})>$$

In practice, the practitioner need not concern him or herself with the exact nature of the mapping Φ. In fact, it is usually more intuitive to concentrate on properties of the kernel functions anyway, and the prevailing wisdom states that the function k(**x**,**y**) should be a good measure of the similarity of the vectors **x** and **y**. Moreover, not just any function k can be used—it must also satisfy certain technical conditions, known as Mercer's conditions. This procedure of implicitly mapping the data via the function *k* is typically often called the *kernel trick* and has found wide application after being popularized by the success of the SVM (Schölkopf & Smola, 2002). The two most widely used kernel functions are the following.

Polynomial Kernel

$$k(x,y) = (<x,y> + 1)^d$$

The polynomial kernel is valid for all positive integers $d \geq 1$. The kernel corresponds to a mapping Φ that computes all degree d monomial terms of the individual vector components of the original space. The polynomial kernel has been used to great effect on digit recognition problems.

Gaussian Kernel

$$k(x,y) = \exp\left(-\frac{||\mathbf{x} - \mathbf{y}||^2}{\sigma^2}\right)$$

The Gaussian kernel, which is similar to the Gaussian probability distribution from which it gets its name, is one of a group of kernel functions known as radial basis functions (RBFs). RBFs are kernel functions that depend only on the geometric distance between $\mathbf{x}$ and $\mathbf{y}$. The kernel is valid for all non-zero values of the kernel width σ, and corresponds to a mapping Φ into an infinite dimensional and therefore somewhat less interpretable, feature space. Nonetheless, the Gaussian is probably the most useful and commonly used kernel function.

Now that we know the form of the SVM dual problem, as well as how to generalize it using kernel functions, the only thing left is to see is how to actually solve the optimization problem, in order to find the α_i's. The optimization problem is one example of a class of problems known as Quadratic Programs (QPs). The term program, as it is used here, is somewhat antiquated and in fact means a "mathematical optimization problem," not a computer program. Fortunately there are many computer programs that can solve QP's such as this, these computer programs being known as Quadratic Program (QP) solvers. An important factor to note here is that there is considerable structure in the QP that arises in SVM training, and while it would be possible to use almost any QP solver on the problem, there are a number of sophisticated software packages tailored to take advantage of this structure, in order to decrease the requirements of computer time and memory.

One property of the SVM QP that can be taken advantage of is its sparcity—the fact that in many cases, at the optimal solution most of the α_i's will equal zero. It is interesting to see what this means in terms of the decision function f(**x**): those vectors with $\alpha_i = 0$ do not actually enter into the final form of the solution. In fact, it can be shown that one can remove all of the corresponding training vectors before training even commences, and get the same final result. The vectors with non-zero values of α_i are known as the *Support Vectors*, a term that has its root in the theory of convex sets. As it turns out, the Support Vectors are the "hard" cases—the training examples that are most difficult to classify correctly (and that lie closest to the decision boundary). In our previous practical analogy, the support vectors are literally the nails that support the block of wood! Now that we have an understanding of the machinery underlying it, we will soon proceed to solve a practical problem using the freely available SVM software package libSVM written by Hsu and Lin (Chang & Lin, 2001).

Relationship to Other Methods

We noted in the introduction that the SVM is an especially easy-to-use method that typically produces good results even when treated as a processing "black box." This is indeed the case, and to better understand this it is necessary to consider what is involved in using some other methods. We will focus in detail on the extremely prevalent class of algorithms known as artificial neural networks, but first we provide a brief overview of some other related methods.

Linear discriminant analysis (Hand, 1981; Weiss & Kulikowski, 1991) is widely used in business and marketing applications, can work in multiple dimensions, and is well-grounded in the mathematical literature. It nonetheless has two major drawbacks. The first is that linear discriminant functions, as the name implies, can only successfully classify linearly separable data thus limiting their application to relatively simple problems. If we extend the method to higher-order functions such as quadratic discriminators, generalization suffers. Indeed such degradation in performance with increased numbers of parameters corroborated the belief in the "curse of dimensionality" finally disproved by Vapnik (1998). The second problem is simply that generalization performance on real problems is usually significantly worse than either decision trees or artificial neural networks (e.g., see the comparisons in Weiss & Kulikowski, 1991).

Decision trees are commonly used in classification problems with categorical data (Quinlan, 1993), although it is possible to derive categorical data from ordinal data by introducing binary valued features such as "age is less than 20." Decision trees construct a tree of questions to be asked of a given example in order to determine the class membership by way of class labels associated with leaf nodes of the decision tree. This approach is simple and has the advantage that it produces decision rules that can be interpreted by a human as well as a machine. However the SVM is more appropriate for complex problems with many ordinal features.

Nearest neighbor methods are very simple and therefore suitable for extremely large data sets. These methods simply search the training data set for the k examples that are closest (by the criteria of Euclidean distance for example) to the given input. The most common class label that associated with these k is then assigned to the given query example. When the training and testing computation times are not so important, however, the discriminative nature of the SVM will usually yield significantly improved results.

Artificial neural network (ANN) algorithms have become extremely widespread in the area of data mining and pattern recognition (Bishop, 1995). These methods were originally inspired by the neural connections that comprise the human brain—the basic idea being that in the human brain many simple units (neurons) are connected together in a manner that produces complex, powerful behavior. To simulate this phenomenon, neurons are modeled by units whose output y is related to the input x by some *activation function* g by the relationship $y = g(x)$. These units are then connected together in various *architectures*, whereby the output of a given unit is multiplied by some constant *weight* and then fed forward as input to the next unit, possibly in summation with a similarly scaled output from some other unit(s). Ultimately all of the inputs are fed to one single final unit, the output of which is typically compared to some threshold in order to produce a class membership prediction. This is a very general framework that provides many avenues for customization:

- Choice of activation function.
- Choice of network architecture (number of units and the manner in which they are connected).
- Choice of the "weights" by which the output of a given unit is multiplied to produce the input of another unit.
- Algorithm for determining the weights given the training data.

In comparison to the SVM, both the strength and weakness of the ANN lies in its flexibility—typically a considerable amount of experimentation is required in order to achieve good results, and moreover since the optimization problems that are typically used to find the weights of the chosen network are non-convex, many numerical tricks are required in order to find a good solution to the problem. Nonetheless, given sufficient skill and effort in engineering a solution with an ANN, one can often tailor the algorithm very specifically to a given problem in a process that is likely to eventually yield superior results to the SVM. Having said this, there are cases, for example in handwritten digit recognition, in which SVM performance is on par with highly engineered ANN solutions (DeCoste & Schölkopf, 2002). By way of comparison, the SVM approach is likely to yield a very good solution with far less effort than is required for a good ANN solution.

Practical Application of the SVM

As we have seen, the theoretical underpinnings of the SVM are very compelling, especially since the algorithm involves very little trial and error, and is easy to apply. Nonetheless, the usefulness of the algorithm can only be borne out by practical experience, and so in this sub-section we survey a number of studies that use the SVM algorithm in practical problems. Before we mention such specific cases, we first identify the general characteristics of those problems to which the SVM is particularly well-suited. One key consideration is that in its basic form the SVM has limited capacity to deal with large training data sets. Typically the SVM can only handle problems of up to approximately 100,000 training examples before approximations must be made in order to yield reasonable training times. Having said this, the training times depend only marginally on the dimensionality of the features—it is often said that SVM can often defy the so-called *curse of dimensionality*—the difficulty that often occurs when the dimensionality is high in comparison with the number of training samples. It should also be noted that, with the exception of the string kernel case, the SVM is most naturally suited to ordinal features rather than categorical ones, although as we shall see in the next section, it is possible to handle both cases.

Before turning to some specific business and marketing cases, it is important to note that some of the most successful applications of the SVM have been in image processing—in particular handwritten digit recognition (DeCoste & Schölkopf, 2002) and face recognition (Osuna, Freund & Girosi, 1997). In these areas, a common theme of the application of SVM is not so much increased accuracy, but rather a greatly simplified design and implementation process. As such, when considering popular areas such as face recognition, it is important to understand that very simple SVM implementations are often competitive with the complex and highly tuned systems that were developed over a longer period prior to the advent of the SVM. Another interesting application area for SVM is on string data, for example in text mining or the analysis of genome sequences (Joachims, 2002). The key reason for the great success of SVM in this area is the existence of "string kernels"—these are kernel functions defined on strings that elegantly avoid many of the combinatoric problems associated with other methods, whilst having the advantage over generative probability models such as the Hidden Markov Model that the SVM learns to *discriminate* between the two classes via the maximization of the margin. The practical use of text categorization systems is extremely widespread, with most large enterprises relying on such analysis of their customer interactions in order to provide automated response systems that are nonetheless tailored to the individual. Furthermore, the SVM has been successfully used in a study of text and data mining for direct marketing applications (Cheung, Kwok, Law, & Tsui, 2003) in which relatively limited customer information was automatically supplanted with the preferences of a larger population, in order to determine effective marketing strategies. SVMs have enjoyed success in a number of other business related applications, including credit rating analysis (Huang, Chen, Hsu, Chen, & Wu, 2004) and electricity price forecasting (Sansom, Downs, & Saha, 2002). To conclude this survey note that while the majority of the marketing teams do not publish their methodologies, since many of the important data mining software packages (e.g., Oracle Data Mining and SAS Enterprise Miner) have incorporated the SVM, it is likely that there will be a significant and increasing use of the SVM in industrial settings.

A WORKED EXAMPLE

In "A Practical Guide to Support Vector Classification" (Hsu, Chang, & Lin, 2003) a simple procedure for applying the SVM classifier is provided for inexperienced practitioners of the SVM classifier. The procedure is intended to be easy to follow, quick, and capable of producing reasonable generalization performance. The steps they advocate can be paraphrased as follows:

1. Convert the data to the input format of the SVM software you intend to use.
2. Scale the individual components of the data into a common range.
3. Use the Gaussian kernel function.
4. Use cross-validation to find the best parameters C (margin softness) and σ (Gaussian width).
5. With the values of C and σ determined by cross-validation, retrain on the entire training set.

The above tasks are easily accomplished using, for example, the free libSVM software package, as we will demonstrate in detail in this section. We have chosen this tool because it is free, easy to use and of a high quality, although the majority of our discussion applies equally well to other SVM software packages wherein the same steps will necessarily be required. The point of this chapter, then, is to illustrate in a concrete fashion the process of applying an SVM. The libSVM software package with which we do this consists of three main command-line tools, as well as a helper script in the python language. The basic functions of these tools are summarized here:

- **svm-scale:** This simple program simply rescales the data as in step 2 above. The input is a data set, and the output is a new data set that has been rescaled.
- **grid.py:** This function can be used to assist in the cross validation parameter selection process. It simply calculates a cross validation estimate of generalization performance for a range of values of C and the Gaussian kernel width σ. The results are then illustrated as a two dimensional contour plot of generalization performance versus C and σ.
- **svm-train:** This is the most sophisticated part of libSVM, which takes as input a file containing the training examples, and outputs a "model file"—a list of Support Vectors and corresponding α_i's, as well as the bias term and kernel parameters. The program also takes a number of input arguments that are used to specify the type of kernel function and margin softness parameter. As well as some more technical options, the program also has the option (used by grid.py) of computing an n-fold cross-validation estimate of the generalization performance.
- **svm-predict:** Having run svm-train, svm-predict can be used to predict the class labels of a new set of unseen data. The input to the program is a model file and a data set, and the output is a file containing the predicted labels, sign(f(**x**)), for the given data set.

Detailed instructions for installing the software can be found on the libSVM Web site (Chang & Lin, 2001). We will now demonstrate these three steps using the example data set at the beginning of the chapter, in order to predict which customers are likely to be home broadband Internet users. To make the procedure clear, we will give details of all the required input files (containing the labelled and unlabelled data), the output file (containing the learned decision function), and the command line statements required to produce and process these files.

Preprocessing (svm-scale)

All of our discussions so far have considered the input training examples as numerical vectors. In fact this is not necessary as it is possible to define kernels on discrete quantities, but we will not worry about that here. Instead, notice that in our example training data in Table 1, each training example has several individual features, both numerical and categorical. There are three numerical features (age, income and years of education), and one categorical feature (gender). In constructing training vectors for the SVM from these training examples, the numerical features are directly assigned to individual components of the training vectors

Categorical features, however, must be dealt with slightly differently. Typically, if the categorical feature belongs to one of *m* different categories (here the categories are male and female so that our *m* is 2), then we map this single categorical feature into *m* individual binary valued numerical features. A training vector whose categorical feature corresponds to feature *n* (the ordering is irrelevant), will have all zero values for these into binary valued numerical features, except for the *n*-th one, which we set to 1. This is a simple way of indicating that the features are not related to one another by relative magnitudes. Once again, the data in Table 1 would be represented by these four vectors, with corresponding class labels y_i:

$\mathbf{x}_1 = [\ 30\ 56000\ \ 16\ 0\ 1];\ y_1 = +1$
$\mathbf{x}_2 = [\ 50\ 60000\ \ 12\ 1\ 0];\ y_2 = +1$
$\mathbf{x}_3 = [16\ 2000\ 11 0\ 1];\ y_3 = -1$
$\mathbf{x}_4 = [\ 35\ 30000\ \ 12\ 0\ 1];\ y_4 = -1$

In order to use the libSVM software, we must represent the above data in a file that is formatted according to the libSVM standard. The format is very simple, and best described with an example.

The above data would be represented by a single file that looks like this:

```
+1 1:30 2:56000 3:16 5:1
+1 1:50 2:60000 3:12 4:1
-1 1:16 2:2000 3:11 5:1
-1 1:35 2:30000 3:12 5:1
```

Each line of the training file represents one training example, and begins with the class label (+1 or -1), followed by a space and then an arbitrary number of index:value pairs. There should be no spaces between the colons and the indexes or values, only between the individual index: value pairs. Note that if a feature takes on the value zero, it need not be included as an index: value pair, allowing data with many zeros to be represented by a smaller file.

Now that we have our training data file, we are ready to run **svm_scale**. As we discovered in the first section, ultimately all our data will be represented by the kernel function evaluation between individual vectors. The purpose of this program is to make some very simple adjustments to the data in order for it to be better represented by these kernel evaluations. In accordance with step 3 above we will be using the Gaussian kernel, which can be expressed by:

$$k(x,y) = \exp\left(-\frac{\lVert \mathbf{x}-\mathbf{y} \rVert^2}{\sigma^2}\right) = \exp\left(-\sum_{d=1}^{D}\frac{(x_d - y_d)^2}{\sigma^2}\right).$$

Here we have written out the *D* individual components of the vectors **x** and **y**, which correspond to the (*D* = 5) individual numerical features of our training examples. It is clear from the summation on the right, that if a given feature has a much larger range of variation than another feature it will dominate the sum, and the feature with the smaller range of variation will essentially be ignored. For our example, this means that the income feature, which has the largest range of values, will receive an undue amount of attention from the SVM algorithm. Clearly this is a problem, and while the Machine Learning community has yet to give the final word on how to deal with it in an optimal manner, many practitioners simply rescale the data so that each feature falls in the same range, for example between zero and one. This can be easily achieved using **svm_scale**, which takes as input a data file in libSVM format, and outputs both a rescaled data file and a set of scaling parameters. The rescaled data should then be used to train the model, and the same scaling (as stored in the scaling parameters file) should be applied to any unlabelled data before applying the learnt decision function. The format of the command is as follows:

svm-scale –s scaling_parameters_file training_data_file > rescaled_training_data_file

In order to apply the same scaling transformation to the unlabelled set, svm_scale must be executed again with the following arguments:

svm-scale –r scaling_parameters_file unlabelled_data_file >
rescaled_unlabelled_data_file

Here the file unlabelled_data_file contains the unlabelled data, and has an identical format to the training file, aside from the fact that the labels +1 and -1 are optional, and will be ignored if they exist.

Parameter Selection (grid.py)

The parameter selection process is without doubt the most difficult step in applying an SVM. Fortunately the simplistic method we prescribe here is not only relatively straightforward, but also usually quite effective. Our goal is to choose the C and σ values for our SVM. Following the previous discussion about parameter or model selection, our basic method of tackling this problem is to make a cross-validation estimate of the generalization performance for a range of values of C and σ, and

examine the results visually. Given the outcome of this step, we may either choose values for C and σ, or conduct a further search based on the results we have already seen.

The following command will construct a plot of the cross-validation performance for our scaled data set:

grid.py –log2c -5,5,1 –log2g -20,0,1 –v 10 rescaled_training_data_file

The search range of the C and σ values are specified by the –log2c and –log2g commands respectively. In both cases the numbers that follow take the form begin,end,stepsize to indicate that we wish to search logarithmically using the values $2^{begin}, 2^{begin+stepsize}...2^{end}$.

Specifying "-v n" indicates that we wish to do n-fold cross-validation (in the above command n = 10), and the last argument to the command indicates which data file to use. The output of the program is a contour plot, saved in an image file of the name rescaled_training_data_file.png. The output image for the above command is depicted in Figure 6.

The contour plot indicates with various line colors the cross-validation accuracy of the classifier, as a function of C and σ—this is measured as a percentage of correct classifications, so that we prefer large values. Note that óð is in fact referred to as "gamma" by the libSVM software—the variable name is of course arbitrary, but we choose to refer to it as σ for compatibility with the majority of SVM literature.

Given such a contour plot of performance, as stated previously there are generally two conclusions to be reached:

1. The optimal (or at least satisfactory) values of C and σ are contained within the plotting region.
2. It is necessary to continue the search for C and σ, over a different range than that of the plot, in order to achieve better performance.

Figure 6. A contour plot of cross-validation accuracy for a given training set as produced by grid.py

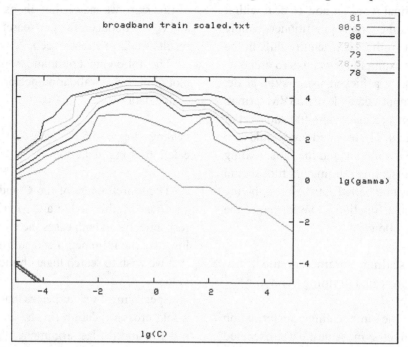

In the first case, we can read the optimal values of C and óð from the output of the program on the command window. Each line of output indicates the best parameters that have been encountered up to that point, and so we can take the last line as our operating parameters.

In the second case, we must choose which direction to continue the search. From Figure 6 it seems feasible to keep searching over a range of smaller σ and larger C. This whole procedure is usually quite effective, however, there can be no denying that the search for the correct parameters is still something of a black art. Given this, we invite interested readers to experiment for themselves, in order to get a basic feel for how things behave. For our purposes, we shall assume that a good choice is $C = 2^{-2} = 0.25$ and $σ = 2^{-2} = 0.25$, and proceed to the next step.

Training (svm-train)

As we have seen, the cross-validation process does not use all of the data for training—at each itera-

tion some of the training data must be excluded for evaluation purposes. For this reason it is still necessary to do a final training run on the entire training set, using the parameters that we have determined in the previous parameter selection process. The command to train is:

svm-train –g 0.25 –c 0.25 rescaled_training_data_ file model_file

This command sets C and σ using the –c and –g switches, respectively. The other two arguments are the name of the training data, and finally the file name for the learnt decision function or model.

Prediction (svm-predict)

The final step is very simple. Now that we have a decision function, stored in the file model_file as well as a properly scaled set of unlabelled data, we can compute the predicted label of each of the examples in the set of unlabelled data by executing the command:

svm-predict rescaled_unlabelled_data_file model_file predictions_file

After executing this command, we will have a new file of the name predictions_file, Each line of this file will contain either "+1" or "–1" depending on the predicted label of the corresponding entry in the file rescaled_unlabelled_data.

CONCLUSION

The general problem of induction is an important one, and can add a great deal of value to large corporate databases. Analyzing this data is not always simple, however, and it is fortunate that methods that are both easy to apply and effective have finally arisen, such as the support vector machine.

The basic concept underlying the support vector machine is quite simple and intuitive, and involves separating our two classes of data from one another using a linear function that is the maximum possible distance from the data. This basic idea becomes a powerful learning algorithm, when one overcomes the issue of linear separability (by allowing margin errors), and implicit mapping to more descriptive feature spaces (through the use of kernel functions).

Moreover, there exist free and easy to use software packages, such as libSVM, that allow one to obtain good results with a minimum of effort. The continued uptake of these tools is inevitable, but is often impeded by the poor results obtained by novices. We hope that this chapter is a useful aid in avoiding this problem, as it quickly affords a basic understanding of both the theory and practice of the SVM.

REFERENCES

Bishop, C. (1995). *Neural networks for pattern recognition*. Oxford: Oxford University.

Chang, C.-C., & Lin, C.-J. (2001). *LIBSVM: A library for support vector machines* [Computer software and manual]. Retrieved May 11, 2005, from http://www.csie.ntu.edu.tw/~cjlin/libsvm

Cheung, K.-W., Kwok, J. T., Law, M. H., & Tsui, K.-C. (2003). Mining customer product ratings for personalized marketing. *Decision Support Systems, 35*(2), 231-243.

DeCoste, D., & Schölkopf, B. (2002). Training invariant support vector machines. *Machine Learning, 45*(1-3), 161-290.

Duda, R. O., Hart, P. E., & Stork, D. G. (2001). *Pattern classification*. New York: John Wiley.

Hand, D. J. (1981). *Discrimination and classification*. New York: John Wiley.

Hsu, C.-W., Chang, C.-C., & Lin, C.-J. (2003). *A practical guide to support vector classification*. Retrieved May 11, 2005, from http://www.csie.ntu.edu.tw/~cjlin/papers/guide/guide.pdf

Huang, Z., Chen, H., Hsu, C.-J., Chen, W.-H., & Wu, S. (2004). Credit rating analysis with support vector machines and neural networks: A market comparative study. *Decision Support Systems, 37*(4), 543-558.

Joachims, T. (2002). *Learning to classify text using support vector machines: Methods, theory and algorithms*. Norwell, MA: Kluwer Academic.

Osuna, E., Freund, R., & Girosi, F. (1997). Training support vector machines: An application to face detection. In *Proceedings of the 1997 Conference on Computer Vision and Pattern Recognition—CVPR'97* (pp. 130-138). Washington, DC: IEEE Computer Society.

Quinlan, R. (1993). *C4.5: Programs for machine learning*. San Francisco: Morgan Kaufmann.

Rosenblatt, F. (1958, November). The Perceptron: A probabilistic model for information storage and organization in the brain. *Psychological Review, 65*, 386-408.

Sansom, D. C., Downs, T., & Saha, T. K. (2002). Evaluation of support vector machine based forecasting tool in electricity price forecasting for Australian National Electricity Market participants. *Journal of Electrical and Electronics Engineering Australia, 22*(3), 227-233.

Schölkopf, B., & Smola, A. (2002). *Learning with kernels: Support vector machines, regularization, optimization, and beyond.* Cambridge, MA: MIT.

Vapnik, V. N. (1998). *Statistical learning theory.* New York: Wiley.

Weiss, S. A., & Kulikowski, C. A. (1991). *Computer systems that learn: Classification and prediction methods from statistics, neural nets, machine learning, and expert systems.* San Mateo, CA: Morgan Kaufmann.

This work was previously published in Business Applications and Computatuional Intelligence, edited by K.E. Vogues and N.K.L. Pope, pp. 267-290, copyright 2006 by Idea Group Publishing (an imprint of IGI Global).

Chapter VI
Kernel Width Selection for SVM Classification:
A Meta–Learning Approach

Shawkat Ali
Monash University, Australia

Kate A. Smith
Monash University, Australia

INTRODUCTION

Support Vector Machines (SVMs) (Boser et al., 1992; Cortes & Vapnik, 1995; Vapnik, 1995) are a relatively new statistical supervised learning method first introduced by Vapnik and his co-worker for binary classification problems. After that, it has been extended to multi-class problems, regression tasks, and novelty detection. These statistical learning algorithms are gaining rapid popularity due to quite a large number of attractive performance results in areas including bioinformatics (Guyon et al., 2002), text mining (Paab et al., 2002), fraud detection (Hyun-Chul et al., 2002), speaker identification (Wan & Renals, 2002), and database marketing (Bennett et al., 1999), among many others.

SVMs adopt the structural risk minimization (SRM) principle, as opposed to the empirical risk minimization (ERM) approach most commonly employed within statistical, neural, and rule-based learning methods. This SRM principle has made SVMs an excellent tool for improved generalization. A kernel transforms the data points from input space to higher dimensional feature space by generating the dot product. The feature space theoretically could be of infinite dimension, where linear discrimination is possible by constructing the optimal hyperplane. This is another significant speciality of SVMs compared to other traditional learning algorithms.

The polynomial and radial basis function (RBF) kernels are the most popular classical SVM kernel. According to Ou et al., RBF kernel is more suitable than others SVM kernels (Ou et al., 2003). Hsu et al. suggested that in general RBF is a reasonable first choice for SVM classification. Up to now a good number of kernels have been proposed by researchers, but there is no any unique kernel that performs best for all problems. The performance of the SVM method depends on the suitable selection of a kernel. The most common procedure for SVM best kernel selection is the trial-and-error approach. Joachims argues SVMs are universal learners with a simple "plug-in" of an appropriate kernel function to learn the problems (Joachims, 1998). This is a very lengthy procedure due to a vast range of kernel function available. Onoda et al. argued selection of the suitable kernel for SVM is an important research issue for real world applications (Onoda, et al. 2002). A priori kernel selection for SVM is a difficult task for the user though (Amari, and Wu 1999; Parrado-Hernandez, et al. 2003). Clearly, automatic kernel selection is a key issue for SVM given the number of kernels available rather than the current trial-and-error nature of selecting the best kernel for a given problem. We found in SVM literature (Joachims, 1998; Morik et al., 1999), manually feeding the parametric kernel parameter is a traditional approach for SVM user. The RBF parameter (width) could be selected by optimization approach (Chapelle et al., 2002). Muller et al. (Muller et al., 2001) suggest RBF width could be selected by following cross-validation procedure. This is the most common way of the RBF width selection method. Carlos et al. argue both optimization and cross-validation methods are computationally very expensive and they suggest selecting the RBF width within a range for regression using meta-learning (Carlos et al., 2004). Schölkopf (Schölkopf, 2003) suggested searching the RBF width between 0.2 and 1. Up to still this is the most popular way to feed RBF kernel parameter for SVM. Therefore, it is a research issue how to choose automatically the

most suitable RBF kernel function and its optimum width for SVM classification.

Our methodology seeks to understand the characteristics of the data (classification problem), understand RBF kernel perform well on which types of problems, and generate rules to assist in the automatic selection of RBF kernel for SVMs. First we classify a wide range of classification problems with different kernels and then identify the dataset characteristics matrix by statistical measures as we have done in some previous related work (Smith et al. 2002; Smith et al. 2001). We then build models for 112 classification problems (see Appendix A) from the UCI Repository (Blake and Merz, 2002) and Knowledge Discovery Central (Lim, 2002) database using SVM with six different kernels. Finally we use the induction algorithm C5.0 (Windows version See5, http://www.rule-quest.com/see5-info.html) to generate the rules to describe RBF kernel is suitable for which type of problem, given the dataset characteristics and the performance of RBF kernel on each dataset. We also examine the rules by 10 Fold Cross Validation (10FCV) performances. Therefore, we estimate the RBF width by maximum likelihood (ML) method and Nelder-Mead (N-M) simplex method. Based on both RBF width estimation methods performance we repeat the rule generation procedure and select the best rule for width estimation methods with 10 fold cross validation performance.

Our chapter is organized as follows: First, we provide some theoretical brief review framework regarding SVM and rules for RBF kernel selection with evaluation. Then we explain the formulation for best RBF width selection and its performance evaluation with statistical significance test results. All statistical measures to identify the dataset characteristics matrix are summarized next. After that, a brief review on rule-based learning algorithm C5.0 and the experimental results post processing method and introduction to the rules for best width selection methods with evaluation are presented. Finally, we conclude our research.

SUPPORT VECTOR MACHINE

Let us consider a dataset D of l independently identically distributed (i.i.d) samples: $(x_1, y_1), \cdots, (x_l, y_l)$. Each sample is a set of feature vectors of length m, $\mathbf{x}_i = \langle x_1, \cdots, x_m \rangle$ and the target value $y_i \in \{1, \cdots, k\}$ that represents the multi-class membership. Now, the machine-learning task is to learn the classes for each pattern by finding a classifier with decision functions $f(\mathbf{x}_i, \alpha_i)$, where $f(\mathbf{x}_i, \alpha_i) = y_i$, $\alpha_i \in \Lambda$, $\forall \langle x_i, y_i \rangle \in D$, and Λ is a set of abstract parameters. We consider SVM to learn this problem. SVM learns the problem to estimate the learning parameter by solving the quadratic optimization as follows (Weston & Watkins, 1999):

$$\min_{\omega, \xi} \phi(\omega, \xi) = \frac{1}{2} \sum_{m=1}^{k} (\omega_m \cdot \omega_m) + C \sum_{i=1}^{l} \sum_{m \neq y_i} \xi_i^m \tag{1}$$

subject to:

$$(\omega_{y_i} \cdot \mathbf{x}_i) + b_{y_i} \geq (\omega_m \cdot \mathbf{x}_i) + b_m + 2 - \xi_i^m$$

$$\xi_i^m \geq 0, i = 1, \cdots, l\, m \in \{1, \cdots, k\} \setminus y_i$$

where ω = weight, ξ = slack variable, C = upper limit, and b = bias.

Now, we can solve this optimization problem by finding the saddle point of the Lagrangian:

$$L(\omega, b, \xi, \alpha, \beta) = \frac{1}{2} \sum_{m=1}^{k} (\omega_m \cdot \omega_m) + C \sum_{i=1}^{\ell} \sum_{m=1}^{k} \xi_i^m -$$

$$\sum_{i=1}^{\ell} \sum_{m=1}^{k} \alpha_i^m [((\omega_{y_i} - \omega_m) \cdot \mathbf{x}_i) + b_{y_i} - b_m - 2 + \xi_i^m]$$

$$- \sum_{i=1}^{\ell} \sum_{m=1}^{k} \beta_i^m \xi_i^m \tag{2}$$

with the dummy variables:

$$\alpha_i^{y_i} = 0, \quad \beta_i^{y_i} = 0, \quad \xi_i^{y_i} = 0$$

subject to:

$$\alpha_i^m \geq 0, \beta_i^m \geq 0, \xi_i^m \geq 0,$$

$$i = 1, \cdots, \ell \text{ and } \{c \in 1, \cdots, k\} \setminus y_i$$

which is maximized with respect to α and β and minimized with respect to ω and ξ by considering the notation:

$$c_i^n = \begin{cases} 1 & \text{if } y_i = n \\ 0 & \text{if } y_i \neq n \end{cases}$$

and

$$A_i = \sum_{m=1}^{k} \alpha_i^m \tag{3}$$

After getting the differentiation, the optimal α is obtained as follows:

$$\alpha_i^o = 2 \sum_{i,m} \alpha_i^m +$$

$$\sum_{i,j,m} [-\frac{1}{2} c_j^{y_i} A_i A_j + \alpha_i^m \alpha_j^{y_i} - \frac{1}{2} \alpha_i^m \alpha_j^{y_i}] (\mathbf{x}_i \cdot \mathbf{x}_j) \tag{4}$$

Finally the decision function for multi-class SVM is:

$$\hat{f}(\mathbf{x})$$

$$= \arg\max_n [\sum_{i:y_i=n} A_i (\mathbf{x}_i \cdot \mathbf{x}) - \sum_{i:y_i \neq n} \alpha_i^n (\mathbf{x}_i \cdot \mathbf{x}) + b_n] \tag{5}$$

The inner product $(\mathbf{x}i \cdot \mathbf{x})$ can be replaced by the convolution inner product $K(\mathbf{x}_i, \mathbf{x}_j)$, also known as the kernel function. Some commonly used SVM kernels with their mathematical expressions are listed in Table 1.

A graphical comparison among different parameter values' effect on RBF kernel is explained in Figure 1 for a binary class synthetic problem. The rectangular and cross signs indicate the two different classes of the problem.

The RBF kernel with width 0.8 and 1 classifies all patterns perfectly with a single optimal hyper-

Table 1. The common uses SVM kernel functions.

Kernels Name	Kernel Functions
linear kernel (Vapnik, 1995)	$K(x_i, x_j) = \left\langle x_i^T x_j \right\rangle$
polynomial kernel (Vapnik, 1995)	$K(x_i, x_j) = \left\langle x_i^T x_j \right\rangle^d$ or $K(x_i, x_j) = (\left\langle x_i^T x_j \right\rangle + 1)^d$
rbf kernel (Vapnik, 1995)	$K(x_i, x_j) = \exp\left(-\dfrac{\left\| x_i - x_j \right\|^2}{2h^2} \right)$ where $h > 0$
multiquadratic kernel (Evgeniou, et al., 2000)	$K(x_i, x_j) = \left(\left\| x_i - x_j \right\|^2 + \tau^2 \right)^{\frac{1}{2}}$ where $\tau > 0$
spline kernel (Gunn, 1998)	$K(x_i, x_j) = 1 + (x_i^T x_j) + \dfrac{1}{2}(x_i^T x_j)\min(x_i^T x_j)^2 - \dfrac{1}{6}\min(x_i^T x_j)^3$
sigmoidal kernel (Evgeniou, et al., 2000)	$K(x_i, x_j) = \tanh(\eta(x_i^T x_j) + \theta)$ where η and θ parameter
Laplace kernel (Ali and Smith, 2004a)	$K(x_i, x_j) = \exp\left(-\dfrac{\left| x_i - x_j \right|}{h} \right)$ where h is the kernel smoothing parameter.

plane. But the other parameters for RBF kernel construct several optimal boundaries to classify all the patterns. It is interesting to observe from Figure 1 how each RBF kernel width generates the optimal hyperplane and how certain kernel parameters are limited in their ability to find the optimal hyperplane for highly non-separable data.

In our previous study, we found that the best rule for RBF kernel and the rule evaluation performance based on 10FCV is presented in Table 2. For other kernel rules, see Ali and Smith (2004b); the summarized kernels' performance are mentioned in Appendix B.

The best rules for RBF kernel are generated with c = 70% and m = 2 as follows: (see Figure 1).

Rule # 1. IF (range > 9 AND normal cdf > 7.2957) OR (discrete uniform cdf <= 2.8185),

THEN we should choose RBF kernel for SVM classification.

In the following sections, we attempt to select the RBF kernel parameter based on data set properties. The rule for this kernel is highly acceptable due to higher accuracy rating. We found that RBF kernel showed best classification performance for 44.64% data sets. Now that we have found two issues for RBF kernel, first, should we search the best width between 0.2 to 1.2 and how we can estimate the best RBF kernel width? We will examine two different RBF width estimation methods (i.e., maximum likelihood (ML) and Nelder-Mead (N-M) simplex method), present comparative performance results, and then attempt to gain insight into which method should be used for certain datasets.

Figure 1. A pictorial view of the rbf kernel performance is shown on an artificial dataset with different width effect. The cross and rectangular sign indicate the two classes of data. The middle continuous lines (for width 1) of the above graphs represent the optimal hyperplane for classification.

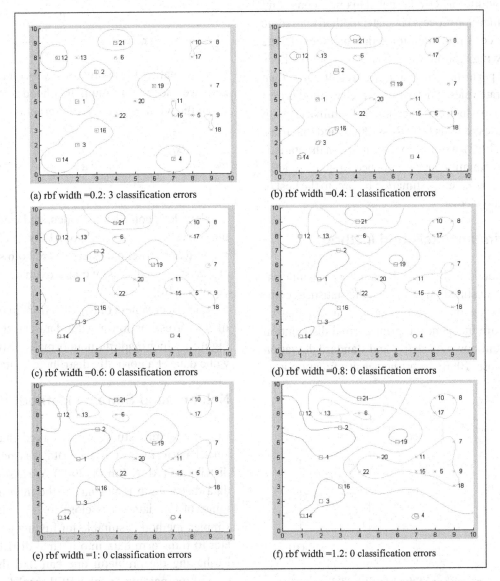

(a) rbf width =0.2: 3 classification errors

(b) rbf width =0.4: 1 classification errors

(c) rbf width =0.6: 0 classification errors

(d) rbf width =0.8: 0 classification errors

(e) rbf width =1: 0 classification errors

(f) rbf width =1.2: 0 classification errors

OPTIMUM RBF KERNEL WIDTH SELECTION

We consider ML and N-M simplex method to estimate best RBF kernel width. The ML method estimates the variance of the data set, and then, the normalized variance is considered as the width

Table 2. Confusion matrix based on 10FCV results for the RBF kernel selection rule

		rbf Kernel Best	
		Yes (Y)	No (N)
Data Condition Satisfied	Y	2.9	0.7
	N	0.7	6.7

Accuracy = 87.27%

of the RBF kernel. The N-M simplex method searches for the appropriate variance from the transformed data and then selects this value as the best width for RBF kernel. This non-constraint optimization process is a faster method than some other constrained optimization methods. In the following section, we will explain both methods with performance evaluation results and then generate rules to determine which method is best suited to which classification. We summarize both of these methods from Ross (2000) and Lagarias, et al. (1998), which are well studied in the statistical community but have not been applied to RBF kernel parameter estimation before, to the best of our knowledge.

Maximum Likelihood Method

Likelihood method has been a very popular parameter estimation method in the statistical community for many years. Let us consider that x_1, $\cdots$, x_n are independent, prior normal distribution with unknown mean (μ), and standard deviation (σ) (Ross, 2000). Now, the density function is as follows:

$$f(x_1,\cdots,x_n \mid \mu,\sigma) = \prod_{i=1}^{n} \frac{1}{\sqrt{2\pi}\sigma} \exp\left[\frac{-(x_i-\mu)^2}{2\sigma^2}\right]$$

$$= \left(\frac{1}{2\pi}\right)^{n/2} \frac{1}{\sigma^n} \exp\left[\frac{-\sum_{1}^{n}(x_i-\mu)^2}{2\sigma^2}\right] \quad (6)$$

The logarithm of the likelihood density function is given by:

$$\log f(x_1,\cdots,x_n \mid \mu,\sigma) = -\frac{n}{2}\log(2\pi) - n\log\sigma - \frac{\sum_{1}^{n}(x_i-\mu)^2}{2\sigma^2} \quad (7)$$

Now, after the differentiating with respect to μ and σ, we can write

$$\frac{d}{d\mu}\log f(x_1,\cdots,x_n \mid \mu,\sigma) = \frac{\sum_{1}^{n}(x_i-\mu)}{\sigma^2} \quad (8)$$

$$\frac{d}{d\sigma}\log f(x_1,\cdots,x_n \mid \mu,\sigma) = -\frac{n}{\sigma} + \frac{\sum_{1}^{n}(x_i-\mu)^2}{\sigma^3} \quad (9)$$

By equating these previous two equations to zero, we find the maximum likelihood is obtained when the width σ of the RBF kernel is

$$\sigma = \left(\frac{\sum_{n=1}^{n}(x_i-\hat{\mu})^2}{n}\right)^{1/2} \quad \text{where } \hat{\mu} = \frac{\sum_{i=1}^{n}x_i}{n} \quad (10)$$

Now, let's consider the calculated σ value as the smoothing parameter h of RBF kernel. Since σ and RBF kernel parameter h are both serving as variance measures, we approximate h using equation (10).

The effectiveness of σ ($= 0.1$) from maximum likelihood estimation on wdbc dataset is shown in Figure 2, which made the nonnormally distributed wdbc dataset a normally distrubuted dataset.

Nelder-Mead Simplex Method

In our previous study, we observed that RBF kernel perform best if the data was normally distributed (Ali & Smith, 2003). We assume the data distribution is normal, if the interquantile range of the data set is close to 1.3. The N-M simplex method is suitable for finding a parameter to transform data into normal distribution. Reshaping the problem, one can find the best smoothing parameter for RBF kernel h so that the data is effectively transformed.

The N-M simplex method for unconstrained optimization has been used extensively to solve parameter estimation problems over a few decades. It is still the method of choice in statistics, engineering, and the physical and medical sciences, due to its ease of use. This method does not require derivatives and is often claimed to be

Figure 2. The effect of (σ=0.1) on wdbc dataset. The suitable σ value makes the non normally distributed wdbc dataset as a normally distributed dataset.

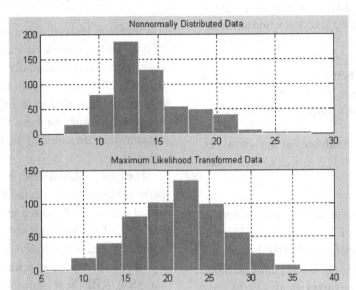

robust for problems with discontinuous attribute values (Lagarias et al. 1998). First, we transform the data by following a Box-Cox transformation (Gentle, 2002) in order to produce data that follows a normal distribution more closely than the original data:

$$\tau(X;\lambda) = \begin{cases} \left(X^{\lambda}-1\right)/\lambda & \text{if } \lambda \neq 0 \\ \log X & \text{if } \lambda = 0 \end{cases} \qquad (11)$$

This transformation can be used only for positive response variables. Box and Cox (Gentle, 2002) suggested the transformation for negative elements variable as follows:

$$\tau(X;\lambda) = \begin{cases} \dfrac{(X+\delta)^{\lambda}-1}{\lambda} & \text{if } \lambda \neq 0 \\ \log(X+\delta) & \text{if } \lambda = 0 \end{cases}$$

$$(12)$$

Now, our aim is to find the appropriate value of λ, which can be considered as similar to the width of the RBF kernel, since they are both measures of variance. We use N-M simplex method to find out the best value of λ.

Each iteration of the N-M method begins with a simplex, specified by its $n+1$ vertices and the associate function values. One or several test points are computed in correspondence to their function values, and then the iteration terminates with a new simplex such that the function values at their vertices satisfy some form of descent condition compared with previous simplex.

One iteration of the N-M simplex algorithm consists of the following steps:

1. **Order:** Order and relabel the $n+1$ vertices as $x_1, \cdots, x_n$ such that $f(x_1) \leq, \cdots, \leq f(x_{n+1})$. Since we want to maximize, we refer to x_1 as the *best* vertex or point, to x_{n+1} as the *worst* point and to x_n as the *next worst* point. Let $\bar{X}$ refer to the centroid of the n best points in the vertex.

2. **Reflect:** Compute the *reflection* point x_r, $x_r = \bar{X} + \rho(\bar{X} - x_{n+1})$, where ρ is a parameter. Evaluate $f(x_r)$. If $f_1 \leq f_r < f_n$, accept the reflected point x_r and terminate the iteration.

3. **Expand:** If $f_r < f_l$, compute the *expansion* point x_e, $x_e = \bar{X} + x(x_r - \bar{X})$, where x is a parameter. If $f_e < f_r$ accept x_e and terminate the iteration; otherwise (i.e., if $f_e \geq f_r$), accept x_r and terminate the iteration.

4. **Contract:** If $f_r \geq f_n$, perform a contraction between $\bar{X}$ and the better of x_{n+1} and x_r, $x_c = \bar{X} - \gamma(\bar{X} - x_{n+1})$, where γ is a parameter. If $f_c < f_{n+1}$, accept x_c and terminate the iteration.

5. **Shrink Simplex:** Evaluate f at the n new vertices for $i = 1, \cdots, n$. $v_i = x_1 + \varsigma(x_i - x_1)$, where ς is a parameter.

Now the highest vertices point considers as the optimum value of λ and also considers as the smoothing parameter h of RBF kernel. For the four coefficients (ρ, χ, γ, and ς), the standard values reported in Lagarias et al. (1998) have been adopted.

The effectiveness of λ with Box-Cox transformation on wine dataset is shown in Figure 3. The suitable value of λ is 0.5, which made the nonnormally distributed wine dataset a normally distributed dataset.

RBF Width Estimation Algorithms Performance: Accuracy

The average test set classification performance of RBF kernel with parameter 0.2-1.2, RBF_best (best performance manually selected from width 0.2-1.2), best width approximation by ML and NM methods is shown in Figure 4. The results are presented only for those 58 of the 112 original datasets that are suited to the RBF kernel (satisfy rule #1).

The RBF_ML and RBF_N-M methods showed close performance with the best RBF accuracy found through exhaustive search of width range 0.2 to 1.2. Both methods showed average higher accuracy than some individual RBF width performance. For large datasets (more than 1,000 samples) RBF_best showed average accuracy 77.52%; RBF_ML and RBF_N-M methods showed 75.94% and 71.41%. The RBF_ML showed better performance than RBF_N-M method. On the other hand for small datasets (less than 1,000 samples) RBF_best showed average accuracy 69.76%, and RBF_N-M methods showed 67.26% and 64.98%.

Figure 3. The effect of Box-Cox transformation on wine dataset ($\lambda = 0.5$). The suitable λ value makes the non nonnormally distributed wine dataset as a normally distributed dataset.

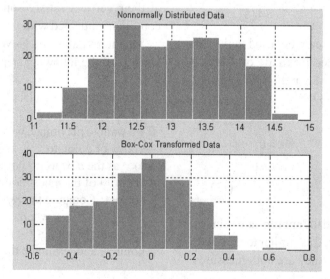

Figure 4. Average test set accuracy for different rbf kernel parameter fitting methods for problems satisfying rule # 1 (58 datasets).

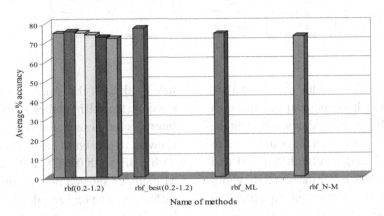

Figure 5. Average test set accuracy for different rbf kernel parameter fitting methods for problems not satisfying rule # 1 (54 datasets).

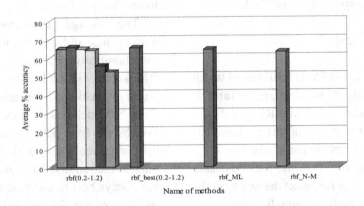

The RBF_ML again showed better performance than RBF_N-M method for small dataset. The RBF_ML method predicted the best width for RBF kernel for 29.31% of the datasets, where RBF kernel is expected to be best. On the other hand, RBF_N-M method predicted the best width for 31.03% of the datasets. We observed that 24.13% of the datasets have best width outside the range of 0.2 to 1.2. For many of the datasets, RBF_ML and RBF_N-M methods predicted the same RBF width among the 112 problems. The RBF kernel

performance with datasets better suited to other (non-RBF) kernels is showed in Figure 5.

RBF Width Estimation Algorithms Performance: Computational Time

The computational performance to determine the best RBF width using the three methods: RBF_best (exhaustive search of width 0.2-1.2), estimation by RBF_ML, and RBF_N-M methods, as shown in Table 3.

Table 3. Average computational performance of the different rbf kernel width estimation methods.

Average Computational Time in Sec.	rbf_Best r	bf_ML	rbf_N-M
	1269.39 0	.5336 5	.1350

The exhaustive best width search method needed extremely higher computational time than RBF_ML and RBF_N-M methods. It selects the RBF width one by one from a range of 0.2-1.2 to train the SVM RBF model. But both RBF_ML and RBF_N-M methods estimate the best RBF width for SVM by simply implementing equation (10) and a simple iteration of equation (12), respectively, that estimates the likely performance of the SVM model without the need to build such models. Therefore, RBF_ML and RBF_N-M methods show superior computational performance compared to the exhaustive search method.

Significance Test

The t-test results for different RBF kernel width estimation methods are summarized in Table 4. We considered the base method as RBF_best. The test input was the percentage of correct classification for all width estimation methods.

The outputs of $H = 0$ in the previous table indicated that we may not reject the null hypothesis that both methods are equally significant. Alternatively, $H = 1$ means we may reject the null hypothesis. The RBF_best with RBF width 1-1.2 showed significant performance difference. The lower values of the significance level suggested rejecting the null hypothesis. The 95% confidence intervals for these methods are highly positively skewed, as shown in Table 4. But the RBF_best with RBF width 0.2-0.8, RBF_ML and RBF_N-M methods showed no significant performance difference. The higher values of the significance level suggested accepting the null hypothesis. The 95% confidence intervals for these kernels are highly balance skewed, as shown in Table 4. The RBF_ML and RBF_N-M methods give results comparable to exhaustive search but are much faster to implement.

The average percentage of classification performance and significance testing has shown that classification accuracy depends on particular RBF kernel width selection. A detailed RBF best width estimation performance by RBF_ML and RBF_N-M methods is represented in Appendix C. We observe from both of these best RBF widths estimation methods that the best width could be outside the range 0.2-1.2. Any single method is not always best to estimate the best RBF width for all problems. So, we need a method to provide a priori information about which best width estimation method is suitable for which classification problem with SVM.

Table 4. Results of the t-test for all methods of rbf width selection.

Algorithms H	ypothesis H	Significance δ	Confidence Interval CI	
rbf_best vs rbf_0.2	0	0.4539	-7.6381 3	.4257
rbf_best vs rbf_0.4	0	0.4368 -	7.8229	3.3904
rbf_best vs rbf_0.6	0	0.2093 -	9.4083 2	.0726
rbf_best vs rbf_0.8	0	0.0732	-11.1674 0	.5058
rbf_best vs rbf_1	1	0.0346 -	12.2829	-0.4664
rbf_best vs rbf_1.2	1	0.0171	-13.1628 -	1.3012
rbf_best vs rbf_ML 0	0	.4064	-7.8649	3.1961
rbf_best vs rbf_NM 0	0	.0746	-11.2024	0.5342

In the following section, we describe the methodology we use to assist in the appropriate selection of the best width estimation method for a given dataset. First, each dataset is described by a set of measurable meta characteristics; we then combine this information with the performance results; and finally, use a rule-based induction method to provide rules describing when each best parameter estimation method for RBF kernel is likely to perform well.

Dataset Characteristics Measurement

Each dataset can be described by simple-, distance-, and distribution-based statistical measures (Smith et al., 2001, 2002). These three sets of measures characterize the datasets in different ways. First, the simple classical statistical measures identify the data characteristics, based on variable to variable comparisons. Then, the distance-based measures identify the data characteristics based on sample-to-sample comparisons. Finally, the density-based measures consider the single data point from a matrix to identify the dataset characteristics. We average most statistical measures with all the variables and take these as global measures of the dataset characteristics.

Simple Statistical Measures

Descriptive statistics can be used to summarize any large dataset into a few numbers that contain most of the relevant characteristics of that dataset. The following table lists the statistical measures used in this work, as provided by the Matlab Statistics Toolbox and some other different sources (Mandenhall & Sincich, 1995; Tamhane & Dunlop, 2000) as follows:

Meta Attribute Names	Meta Attribute Names
Geometric Mean	Max. and Min. Eigenvalue
Harmonic Mean	Skewness
Trim Mean	Kurtosis
Standard Deviation	Correlation Coefficient
Interquantile Range	Prctile

Distance-Based Measures

Distance-based measures calculate the dissimilarity between samples. We measure the euclidean, city block, and mahalanobis distance between each pair of observations for each dataset as follows:

Meta Attribute Names	Meta Attribute Names
Euclidean Distance	Mahalanobis Distance
City Block Distance	

Distribution-Based Measures

The probability distribution of a random variable describes how the probabilities are distributed over the various values that the random variable can take. We measure the probability density function (PDF) and cumulative distribution function (CDF) for all datasets by considering different types of distributions as follows:

Meta Attribute Names	Meta Attribute Names
Chi-Square PDF	Chi-Square CDF
Normal PDF	Normal CDF
Binomial PDF	Discrete Uniform CDF
Exponential PDF	F PDF
Gamma PDF	Hypergeometric CDF
Lognormal PDF	Poisson PDF
Rayleigh PDF	Student's t PDF

These measures are calculated for each of the datasets to produce a dataset characteristics matrix. Finally, by combining this matrix with the performance results in Appendix C, we can derive rules to suggest when certain best width estimation methods are appropriate.

Rule Generation

The trial-and-error approach is a very common procedure to select the best width for RBF kernel. It is a computationally complex task to find the best width by following this procedure. If we are interested in applying a specific method to a particular problem, we have to consider which method is more suitable for which problem. The suitability test can be done from rules developed with the help of the data characteristics properties.

Rule-based learning algorithms, especially decision trees (also called classification trees or hierarchical classifiers) are a divide-and-conquer approach or a top-down induction method, which has been studied with interest in the machine learning community. Quinlan (1993) introduced the C4.5 and C5.0 algorithms to solve classification problems. C5.0 works in three main steps. First, the root node at the top node of the tree considers all samples and passes them through to the second node called *branch node*. The branch node generates rules for a group of samples based on an entropy measure. In this stage, C5.0 constructs a very big tree by considering all attribute values and finalizes the decision rule by pruning. It uses a heuristic approach for pruning based on statistical significance of splits. After fixing the best rule, the branch nodes send the final class value in the last node, called the *leaf node* (Duin, 1996; Quinlan, 1993). C5.0 has two parameters: the first one is called the pruning confidence factor (c), and the second one represents the minimum number of branches at each split (m). The pruning factor has an effect on error estimation and, hence, the severity of pruning the decision tree. The smaller value of c produces more pruning of the generated tree and a higher value results in less pruning. The minimum branches m indicates the degree to which the initial tree can fit the data. Every branch point in the tree should contain at least two branches (so, a minimum number of $m = 2$). For detail formulations, see Quinlan (1993).

Now that the characteristics of each dataset can be quantitatively measured, we can combine this information with the empirical evaluation of RBF width estimation performance and construct the dataset characteristics matrix. Thus, the result of the jth width selection method on the ith dataset is calculated as:

$$R_{ij} = 1 - \frac{e_{ij} - \max(e_i)}{\min(e_i) - \max(e_i)} \qquad (13)$$

where e_{ij} is the percentage of correct classification for the jth method on dataset i, and e_i is a vector of accuracy for dataset i. The class values in the matrix are assigned based on the performance best rank. The best rank is defined as 1, and the worst is 0. For example, if RBF_ML method shows the ranking performance 1 for the dataset A, then the class in the matrix for problem A is RBF_ML. Based on the 112 classification problems, we then can train a rule-based classifier (C5.0) to learn the relationship between dataset characteristics and width selection method performance. We split the matrix 90% to construct the model tree. The process then is repeated using a 10-fold cross validation approach so that 10 trees are constructed. From these 10 trees, the best rules are found for each best width selection method based on the best test set results. The generalization of these rules is then tested by applying each of the randomly extracted test sets and calculating the average accuracy of the rules, as discussed below in Tables 5 and 6. We found the suitable parameter value by tuning for global pruning factor; c is 70-90%, and the number of minimum branches m is 2.

We have demonstrated the rules for RBF kernel in the second section. Now, if any dataset satisfies the RBF kernel rule, then we need to find best RBF width estimation method. So, in the following section, we will generate the rules describing when to choose the RBF_ML and RBF_N-M methods for best RBF width estimation.

Rules for RBF_ML Method

The best rules for RBF_ML method are generated with c = 85% and m = 2 as follows:

Rule # 2. IF ($md_{rs} <= 4.3713$ and $y_{bio\_pdf} > 2.7036$) or (mean > 49.0052 and s > -1.7911, and $y_{gama\_pdf} <= 0.00030981$) or ($p_{chi\_cdf} <= 30.3236$), THEN we

Table 5. Confusion matrix based on 10FCV results for the RBF_ML method selection rule

		rbf_Ml Method Best	
		Y	N
Data Condition Satisfied	Y	2.7	0.2
	N	0.3	2.6

Accuracy = 91.38%

Table 6. Confusion matrix based on 10FCV results for the RBF_N-M method selection rule

		RBF_N-M Method Best	
		Y	N
Data Condition Satisfied	Y	2.9	0.2
	N	0.4	2.3

Accuracy = 89.66%

should choose RBF_ML method for RBF kernel width estimation.

Rules for RBF_N-M Method

The best rules for RBF_NM method are generated with c = 80% and m = 2 as follows:

Rule # 3. IF (s > -0.4284 and $y_{norm\_pdf} \leq 0.23362$ and $y_{ray\_pdf} \leq 2.3497$) or ($y_{bio\_pdf} > 2.7036$) THEN we should choose RBF_NM method for RBF kernel width estimation.

The generated rules show around 90% accuracy. On average, we observed that RBF_ML approximation method showed slightly better performance than RBF_N-M approximation. However, which method was best for individual datasets has been shown to be quite data dependent. These rules might be useful to determine which RBF width

approximation method is most appropriate for which problem.

CONCLUSION

In this research, we have widely investigated empirically how to select RBF kernel and its best width for SVM. We proposed a simple rule for RBF kernel, based on data set information. This method is faster than trial-and-error-based kernel selection with SVM. We observed that the suitable RBF width could be out of the range 0.2 and 1.2, which is the commonly tested range in the literature. The estimated higher width increased the kernel performance accuracies for some specific cases. The RBF_ML and RBF_N-M methods are very fast to estimate the best RBF width. We examined the generated rules 10-fold cross validation evaluation. All generated rules showed higher efficiency rating. The main benefit of our methodology based on meta-learning is that we can achieve higher accuracy for some classification problems and significant savings in time for at least similar accuracy for all problems.

REFERENCES

Ali, S., & Smith, K.A. (2003). Automatic parameter selection for polynomial kernel. In *Proceedings of the IEEE International Conference on Information Reuse and Integration*, USA (pp. 243-249).

Ali, S., & Smith, K.A. (2004a). Laplace kernel with automatic smoothing parameter estimation for support vector machine. *Computational Management Science (submitted)*.

Ali, S., & Smith, K.A. (2004b). Automatic kernel selection for support vector machines. *Neurocomputing (submitted)*.

Amari, S.-I., & Wu, S. (1999). Improving support vector machine classifiers by modifying kernel functions. *Neural Networks, 12*, 783-789.

Bennett, K.P., Wu, S., & Auslender, L. (1999). On support vector decision trees for database marketing. In *Proceedings of the IEEE International Joint Conference on Neural Networks, IJCNN'99* (pp. 904-909).

Blake, C., & Merz, C.J. (2002). UCI repository of machine learning databases. Retrieved from http://www.ics.uci.edu/~mlearn/MLRepository.html

Boser, B.E., Guyon, I., & Vapnik, V.N. (1992). A training algorithm for optimal margin classifiers. In *Proceedings of the Fifth Annual Workshop of Computational Learning Theory*, Pittsburgh, Pennsylvania (pp. 144-152).

Carlos, S., Pavel, B., & Brazdil, P. (2004). A meta-learning method to select the kernel width in support vector regression. *Machine Learning, 54*, 195-209.

Chapelle, O., Vapnik, V., Bousquet, O., & Mukherjee, S. (2002). Choosing multiple parameters for support vector machines. *Machine Learning, 46*(1), 131-159.

Cortes C., & Vapnik, V. (1995). Support vector networks. *Machine Learning, 20*, 273-297.

Duin, R.P.W. (1996). A note on comparing classifier. *Pattern Recognition Letters, 1*, 529-536.

Evgeniou, T., Pontil, M., & Poggio, T. (2000). Regularization networks and support vector machines. *Advances in Computational Mathematics, 13*(1), 1-50.

Gentle, J.E. (2002). *Elements of computational statistics*. New York: Springer-Verlag.

Gunn, S.R. (1998). Support vector machines for classification and regression. Southampton, UK: University of Southampton.

Guyon, I., Weston, J., Barnhill, S., & Vapnik, V. (2002). Gene selection for cancer classification using support vector machines. *Machine Learning, 46*(1/3), 389-422.

Hsu, C.-W., Chang, C.-C., & Lin, C.-J. (2004). A practical guide to support vector classification (technical report). Taiwan. National Taiwan University.

Hyun-Chul, K., Shaoning, P., Hong-Mo, J., Daijin, K., & Sung-Yang, B. (2002). Pattern classification using support vector machine ensemble. In *Proceedings of the IEEE 16th International Conference on Pattern Recognition* (pp. 160-163).

Joachims, T. (1998). Text categorization with support vector machines: Learning with many relevant features. In *Proceedings of the ECML'98, 10th European Conference on Machine Learning* (pp. 137-142).

Lagarias, J.C., Reeds, J.A., Wright, M.H., & Wright, P.E. (1998). Convergence properties of the Nelder-Mead simplex method in low dimensions. *SIAM Journal of Optimisation, 9*, 112-147.

Lim, T.-S. (2002). Knowledge discovery central, datasets. Retrieved from *http://www.KDCentral.com*

Mandenhall, W., & Sincich, T. (1995). *Statistics for engineering and the sciences* (4th ed.). Prentice Hall.

Morik, K., Brockhausen, P., & Joachims, T. (1999). Combining statistical learning with a knowledge-based approach: A case study in intensive care monitoring. In *Proceedings of the 16th International Conference on Machine Learning* (pp. 268-277).

Muller, K.-R., Mika, S., Ratsch, G., Tsuda, K., & Scholkopf, B. (2001). An introduction to kernel-based learning algorithms. *IEEE Transactions on Neural Networks, 12*(2), 181-201.

Onoda, T., Murata, H., Ratsch, G., & Muller, K.-R. (2002). Experimental analysis of support vector machines with different kernels based on non-intrusive monitoring data. In *Proceedings of the IEEE International Joint Conference on Neural Networks* (pp. 2186-2191).

Ou, Y.-Y., Chen, C.-Y., Hwang, S.-C., & Oyang, Y.-J. (2003). Expediting model selection for support vector machines based on data reduction. In *Proceedings of the IEEE International Conference on Systems, Man and Cybernetics* (pp. 786-791).

Paab, G., Leopold, E., Larson, M., Kindermann, J., & Eickeler, S. (2002). SVM classification using sequences of phonemes and syllables. In *Proceedings of the European Conference on Machine Learning, ECML*, Helsinki.

Parrado-Hernandez, E., Mora-Jimenez, I., Arenas-Garca, J., Figueiras-Vidal, A.R., & Navia-Vazquez, A. (2003). Growing support vector classifiers with controlled complexity. *Pattern Recognition, 36*, 1479-1488.

Quinlan, R. (1993). *C4.5: Programs for machine learning*. San Mateo, CA: Morgan Kaufman Publishers.

Ross, S.M. (2000). *Introduction to probability and statistics for engineers and scientists* (2nd ed.). London: Academic Press.

Schölkopf, B. (personal communication, 2003).

Smith, K.A., Woo, F., Ciesielski, V., & Ibrahim, R. (2001). Modelling the relationship between problem characteristics and data mining algorithm performance using neural networks. In C. Dagli et al. (Eds.), *Smart engineering system design: Neural networks, fuzzy logic, evolutionary programming, data mining, and complex systems* (pp. 357-362). ASME Press.

Smith, K.A., Woo, F., Ciesielski, V., & Ibrahim, R. (2002). Matching data mining algorithm suitability to data characteristics using a self-organising map. In A. Abraham & M. Koppen (Eds.), *Hybrid information systems* (pp. 169-180). Heidelberg: Physica-Verlag.

Tamhane, A.C., & Dunlop, D.D. (2000). *Statistics and data analysis*. Prentice Hall.

Vapnik, V. (1995). *The nature of the statistical learning theory*. New York: Springer-Verlag.

Wan, V., & Renals, S. (2002). Evaluation of kernel methods for speaker verification and identification. In *Proceedings of IEEE International Conference on Acoustics, Speech, and Signal Processing, ICASSP'02*, (pp. 669-672).

Weston, J., & Watkins, C. (1999). Multi-class support vector machines. In M. Verleysen (Ed.), In *Proceedings of the Seventh European Symposium on Artificial Neural Networks*, Bruges, Belgium.

This work was previously published in International Journal of Data Warehousing and Mining, Vol. 1, No.4, edited by D. Taniar, pp. 78-97, copyright 2005 by Idea Group Publishing (an imprint of IGI Global).

Chapter VII
Protein Folding Classification Through Multicategory Discrete SVM

Carlotta Orsenigo
Università degli Studi di Milano, Italy

Carlo Vercellis
Università degli Studi di Milano, Italy

ABSTRACT

In the context of biolife science, predicting the folding structure of a protein plays an important role for investigating its function and discovering new drugs. Protein folding recognition can be naturally cast in the form of a multicategory classification problem that appears challenging due to the high number of folds classes. Thus, in the last decade, several supervised learning methods have been applied in order to discriminate between proteins characterized by different folds. Recently, discrete support vector machines have been introduced as an effective alternative to traditional support vector machines. Discrete SVM have shown to outperform other competing classification techniques both on binary and multicategory benchmark datasets. In this chapter, we adopt discrete SVM for protein folding classification. Computational tests performed on benchmark datasets empirically support the effectiveness of discrete SVM, which are able to achieve the highest prediction accuracy.

INTRODUCTION

Proteins are sequences of amino acids organized into three-dimensional structures that largely influence their function and evolution. The prediction of the three-dimensional structure of a protein is a challenging problem that has many applications in discovering new drugs and therapies, and for which different approaches have been proposed. The early efforts were aimed at predicting the function based on sequence similarity comparison (Holm & Sander, 1999). However, it has been observed that this approach may fail since sometimes proteins with similar functions can differ substantially in terms of primary sequence structure.

The alternative taxonometric approach relies on predicting the protein *fold*, which can be defined as a common three-dimensional pattern with the same major secondary structure elements in the same arrangements and with the same topological connections (Craven, Mural, Hauser, & Uberbacher, 1995). Hence, the problem can be cast in the form of a multicategory classification task in the context of learning from data, where one has to determine an explanatory relationship between protein folding and the underlying primary structure (Baldi & Brunak, 1998; Durbin, Eddy, Krogh, & Mitchison, 1998).

Notice that multicategory classification has proven to be a much more complex task in comparison to its binary counterpart, particularly for protein folding recognition, due to the large number of distinct folds, which can be in the order of hundreds. For this reason, algorithms for protein folding classification are usually characterized by a low degree of prediction accuracy and require a high computational effort. This complexity has been partially mitigated by confining the attention only to the most populated classes of folds. Sometimes, folds are grouped into four *structural classes* that correspond to a higher concept, with respect to the folds, in the hierarchical representation of proteins. Hence, the protein folding recognition appears much harder to accomplish than the prediction of the structural class of a protein.

In general, we can distinguish multicategory classifiers into two main groups. Some techniques directly address the multicategory nature of the classification task, such as classification trees (Breiman, Friedman, Olshen, & Stone, 1984;

Murthy, 1998; Quinlan, 1993). Other methods, which appear more effective, are based on a sequence of binary classification problems. Referring to protein folding, the classification of a new protein is then performed by assigning the fold that is closest to the predictions of the binary classifiers according to a suitable metric. To some extent, the approaches developed are not tied to a specific binary classifier. A common scheme for deriving a multicategory classifier is based on the *one-against-all* framework, in which binary classification problems are obtained with the aim of discriminating between examples of one class and all the remaining examples (Dietterich & Bakiri, 1995; Guruswami & Sahai, 1999). If D denotes the number of different folds, in this scheme one has to train only D binary classifiers, although each of them is rather complex due to the heterogeneity of the class collecting the remaining folds. A different scheme, termed *pairwise decomposition* or *round-robin*, has been devised by letting the binary problems to discriminate among all pairs of folds (Kreßel, 1999). In this case, $D(D-1)/2$ binary classifiers have to be trained, where each of them is composed by a small number of proteins belonging to one of two homogeneous fold classes. Finally, other hybrid schemes can be devised, as the *unique-one-against-all* proposed in Ding and Dubchak (2001), where a cascading combination of the two previous schemes is adopted.

Discrete support vector machines, originally introduced in Orsenigo and Vercellis (2003, 2004), are a successful alternative to SVM that is based on the idea of accurately evaluating the number of misclassified examples instead of measuring their distance from the separating hyperplane. Starting from the original formulation, discrete SVM have been effectively extended in several directions, to deal with multiclass problems (Orsenigo & Vercellis, 2007a) or to learn from a small number of training examples (Orsenigo & Vercellis, 2007b).

In this chapter, we perform protein folding classification by means of discrete SVM. In order to assess the usefulness of the proposed method, some computational tests have been performed on benchmark datasets composed of proteins grouped into 27 different folds. The performance exhibited by the proposed method appears superior to that achieved by the best alternative classification techniques.

The chapter is organized as follows. In the next section, we provide a description of SVM and discrete SVM. Then, the different schemes for multicategory classification, based on discrete SVM, are introduced in the subsequent section. Finally, computational tests are illustrated in the last section.

SVM AND DISCRETE SVM

In a classification problem, we are required to discriminate among examples belonging to different classes. Formally, a training set S_m composed by *m examples* (x_i, y_i), $i \in M = \{1,2,...,m\}$, in the $(n+1)$–dimensional real space $\Re^{n+1}$ is given, where $\mathbf{x}_i \in \Re^n$ is a vector of *attributes* or *features* and y_i is a scalar representing the *label* or *class* of $\mathbf{x}_i$. Let $\mathcal{D} = \{1,2,...,D\}$ be the set of distinct class values. Each component x_{ij} of an example $\mathbf{x}_i$ is assumed to be a realization of a random variable $\mathbf{A}_j$, $j \in N = \{1,2,...,n\}$, representing the *j*-th attribute of S_m.

Let $\mathcal{H}$ denote a set of functions $f(\mathbf{x}): \Re^n \mapsto \mathcal{D}$ that represent hypothetical relationships between $\mathbf{x}_i$ and y_i. A classification problem consists of defining an appropriate hypotheses space $\mathcal{H}$ and a function $f^* \in \mathcal{H}$ that optimally describes the relationship between the examples and their class values. A function $f^* \in \mathcal{H}$ can be considered optimal according to different criteria that should take into account the minimization of the empirical classification error on the training set and the maximization of the generalization capability on new data.

To assess the accuracy of $f \in \mathcal{H}$, the whole set of examples is usually partitioned into two disjoint subsets, denoted respectively as *training* and *test* set. For a given classifier, the discriminant function is learned using the examples from the training set, and then applied to predict the class of the examples in the test set. In the remainder of this section, we will refer to *binary* classification problems for which the class attribute y_i takes only two different values, which may be labeled as -1 and 1 without loss of generality.

Most binary classifiers actually generate as output a function $g: \Re^n \to \Re$, termed *score function* or *margin*, whose sign discriminates between the two classes, so that $f(\mathbf{x}) = \text{sgn}(g(\mathbf{x}))$. Moreover, for these binary margin classifiers, the magnitude of the score $g(\mathbf{x})$ can be viewed as a measure of confidence in the class assignment, therefore expressing the likelihood that the example $\mathbf{x}$ belongs to the predicted class $f(\mathbf{x}) = \text{sgn}(g(\mathbf{x}))$.

Linear separating score functions have been widely used for binary classification, often as a building block within more complex schemes for pattern recognition. In this case, $g(\mathbf{x}) = \mathbf{w'x} - b$ is a hyperplane and $f(\mathbf{x}) = \text{sgn}(\mathbf{w'x} - b)$. If the training examples are linearly separable, it is possible to find a pair $(\mathbf{w}, b)$ such that $f(\mathbf{x}_i) = \text{sgn}(\mathbf{w'x}_i - b) = y_i, i \in M$, by solving a linear optimization problem. Conversely, when the examples are not linearly separable, for every choice of the parameters $(\mathbf{w}, b)$ there exists a subset $M' \subseteq M$ such that $f(\mathbf{x}_i) = \text{sgn}(\mathbf{w'x}_i - b) \neq y_i, i \in M'$. Hence, in order to derive the optimal separating hyperplane, one is led to define a suitable loss function whose expectation with respect to the unknown distribution of the examples has to be minimized.

The structural risk minimization (SRM) principle, developed in the context of statistical learning theory (Vapnik, 1995, 1998), establishes the concept of reducing the empirical classification error as well as the generalization error in order to achieve a higher accuracy on unseen data. Formally, this is obtained by minimizing the following risk functional:

$$\hat{R}(f) = \frac{1}{m} \sum_{i \in M} V(y_i, f(\mathbf{x}_i)) + \lambda \|f\|_K^2, \qquad (1)$$

where the first term, based on the loss function V, is termed *empirical risk,* and represents the empirical error on the training set S_m, whereas the second term is related to the generalization capability of f. Here, $K = K(\cdot, \cdot)$ is a given symmetric positive definite function named *kernel,* $\|f\|_K^2$ denotes the norm of f in the reproducing kernel Hilbert space induced by K, and λ is a parameter that controls the trade-off between the two terms.

This leads to the minimization of the expression:

$$\frac{1}{m}\sum_{i \in M} V(y_i, f(\mathbf{x}_i)) + \lambda \frac{\|\mathbf{w}\|_2^2}{2},$$

where $\quad \|\mathbf{w}\|_2^2 = \sum_{j \in N} w_j^2 \qquad (2)$

The second term in (2) is the reciprocal of the margin of separation, defined as the distance between the pair of parallel canonical supporting hyperplanes $\mathbf{w'x} - b - 1 = 0$ and $\mathbf{w'x} - b + 1 = 0$. The geometrical interpretation of the canonical hyperplanes and the margin is given in Figure 1.

According to the SRM principle, the first term in (1) and (2) expresses the misclassification rate of the examples in the training set. However, for computational reasons, SVM replace this term with a continuous proxy of the sum of the distances of the misclassified examples from the separating hyperplane.

More specifically, for SVM the loss function V takes the form,

$$V(y_i, f(\mathbf{x}_i)) = \left|1 - y_i(\mathbf{w'x}_i - b)\right|_+, \qquad (3)$$

where $|t|_+ = t$ if t is positive and zero otherwise.

Let $d_i, i \in M$, a nonnegative slack variable such that the following linear constraints are satisfied:

$$y_i(\mathbf{w'x}_i - b) \geq 1 - d_i, \quad i \in M. \qquad (4)$$

The optimal separating hyperplane can therefore be determined by solving the following quadratic optimization problem:

$$\min \quad \sum_{i \in M} d_i + \frac{\lambda}{2}\|\mathbf{w}\|_2^2$$

s. to $y_i(\mathbf{w'x}_i - b) \geq 1 - d_i, \quad i \in M$ (QSVM)

$$d_i \geq 0, i \in M; \mathbf{w}, b \text{ free}, \qquad (5)$$

where λ is a parameter available to control the trade-off between the generalization capability of the classifier and the misclassification error.

Figure 1. Margin maximization for linearly nonseparable sets

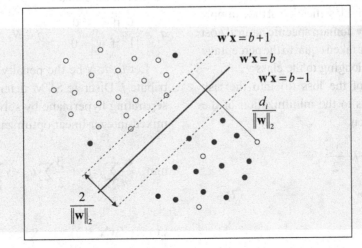

The solution of the quadratic problem (QSVM) is obtained via Lagrangean duality, and also provides the interpretation of the support vectors (Vapnik, 1995). Taking advantage of the dual formulation and of suitable kernel functions (Cristianini & Shawe-Taylor, 2000; Schölkopf & Smola, 2002), SVM proceed by projecting the original examples into a higher dimensional feature space, in which the linear separation is derived, allowing to efficiently obtain nonlinear discriminations in the original space.

A different family of classification models, termed *discrete* SVM, has been introduced in Orsenigo and Vercellis (2003, 2004). and is motivated by an alternative loss function that, according to the SRM principle, counts the number of misclassified examples instead of measuring their distance from the separating hyperplane. The distinctive trait of discrete SVM is the accurate representation of the empirical error, by using the total misclassification error in the objective function, in place of the sum of the slacks considered in (QSVM) by traditional SVM. Hence, the rationale behind discrete SVM is that a precise evaluation of the empirical error could possibly lead to a more accurate classifier.

In this case, the loss function is given by,

$$V(y_i, f(\mathbf{x}_i)) = c_i \theta (1 - y_i(\mathbf{w}'\mathbf{x}_i - b)), \qquad (6)$$

where $\theta(t)=1$ if t is positive and 0 otherwise, and c_i is a penalty for the misclassification of the example $\mathbf{x}_i$. In absence of any domain specific natural cost attribution, c_i can be taken equal to the percentage of examples not belonging to the class y_i.

The inclusion of the loss (6) into the risk functional (2) leads to the minimization of the following expression:

$$\frac{1}{m}\sum_{i \in M}| y_i - f(\mathbf{x}_i)| + \lambda \frac{\|\mathbf{w}\|_2^2}{2},$$

where $\|\mathbf{w}\|_2^2 = \sum_{j \in N} w_j^2.$ \qquad (7)

Notice that the first term in (7) precisely evaluates the empirical risk and expresses the accuracy of the classifier on the training set through the percentage of misclassified examples.

In order to formulate the optimization problem corresponding to the minimization of the risk (7), discrete SVM first replace the 2–norm in the second term of (7) with the 1–norm,

$$\|\mathbf{w}\|_1 = \sum_{j \in N} |w_j|.$$

In order to express the empirical risk, we introduce the following binary variables,

$$p_i = \begin{cases} 0 & \text{if } \mathbf{x}_i \text{ is correctly classified} \\ 1 & \text{if } \mathbf{x}_i \text{ is misclassified} \end{cases} \qquad i \in M$$

used to count the number of classification errors.

The complexity of the discriminating rule is related to the number of attributes that directly contribute, with a nonzero coefficient, to the separating hyperplane. By reducing the complexity of the discriminating rules, it is likely to increase the generalization capability of a model learned on the training set. Furthermore, the induced rules become simpler and more suitable to the interpretation of domain experts. Hence, in order to take into account the complexity of the rule, discrete SVM introduce a second set of binary variables, defined as:

$$q_j = \begin{cases} 0 & \text{if } w_j = 0 \\ 1 & \text{if } w_j \neq 0 \end{cases} \qquad j \in N.$$

Let $h_j, j \in N$ be the penalty cost for using attribute j. Discrete SVM determine the optimal separating hyperplane by solving the following mixed-integer linear optimization problem,

$$\min \quad \frac{\alpha}{m}\sum_{i \in M} c_i p_i + \frac{\beta}{2}\sum_{j \in N} u_j + \gamma \sum_{j \in N} h_j q_j \qquad \text{(DSVM)}$$

$$\text{s.to} \quad y_i(\mathbf{w}'\mathbf{x}_i - b) \geq 1 - S p_i, \quad i \in M \qquad (11)$$

$$-u_j \leq w_j \leq u_j, \quad j \in N \tag{12}$$

$$u_j \leq Rq_j, \quad j \in N \tag{DSVM}$$

$$p_i \in \{0,1\}, i \in M;$$
$$u_j \geq 0, q_j \in \{0,1\}, j \in N; \mathbf{w}, b \text{ free} \tag{13}$$

The objective function of problem (DSVM) is composed of the weighted sum of three terms, expressing a trade-off between accuracy and potential of generalization, regulated by the parameters (α, β, γ). The first term represents the empirical error. The second term expresses the 1−norm computed with respect to a linear kernel, and its role is to restore the well-posedness of the optimization problem and to increase the predictive capability of the classifier. Finally, the third term is aimed at further increasing the generalization capability of the model by minimizing the number of attributes used in the classification rule. Constraints (11) are required in order to correctly evaluate the empirical error through the binary variables $\mathbf{p}$, as each of them forces to the value 1 the binary variable p_i whenever example $\mathbf{x}_i$ is misclassified, that is, when it falls on the wrong side of the corresponding canonical hyperplane. Here S is an appropriate large constant. Constraints (12) ensure that the components of $\mathbf{u}$ bound the absolute value of the elements of the vector $\mathbf{w}$; hence, upper bounding its 1−norm $\|\mathbf{w}\|_1$. Finally, constraints (13) imply that the binary variable q_j takes the value 1 whenever $w_j > 0$, that is, whenever the j-th attribute is actively used in the optimal separating hyperplane. Like S, R is a large constant.

Model (DSVM) is a mixed binary linear optimization problem, notoriously more difficult to solve to optimality than continuous linear optimization. However, it can be solved by means of an efficient heuristic procedure, based on a sequence of linear optimization problems, for obtaining suboptimal solutions. Model (DSVM) can be used as a linear perceptron; alternatively, it can be framed within a recursive procedure for the generation of oblique classification trees, to derive an optimal separating hyperplane at each node of the tree, as in Orsenigo and Vercellis (2003, 2004). In the quoted references, it was shown, by means of extensive testing, that the increase in model complexity is justified by a more accurate discrimination and a higher generalization capability due to the correct estimation of the empirical misclassification error and the minimization of the number of attributes defining the separating hyperplane.

Here is a short description of the heuristic procedure for determining a feasible suboptimal solution to model (DSVM), based on a sequence of linear optimization (LO) problems. The heuristic starts by considering the LO relaxation of problem (DSVM). Each LO problem $(DSVM)_{t+1}$ in the sequence is obtained by fixing to zero the relaxed binary variable with the smallest fractional value in the optimal solution of the predecessor $(DSVM)_t$. Notice that, if problem $(DSVM)_t$ is feasible and its optimal solution is integer feasible, the procedure is stopped, and the solution generated at iteration t is retained as an approximation to the optimal solution of problem (DSVM). Otherwise, if problem $(DSVM)_{t+1}$ is unfeasible, the procedure modifies the previous LO problem $(DSVM)_t$ by fixing to 1 all of its fractional variables. Problem $(DSVM)_{t+1}$, defined in this way, is feasible, and any of its optimal solutions is integer. Thus, the procedure is stopped and the solution found for $(DSVM)_t$ is retained as an approximation to the optimal solution of (DSVM).

DISCRETE SVM FOR MULTICATEGORY CLASSIFICATION

In this section, we consider two techniques for extending discrete SVM to multicategory classification problems. The first is obtained by framing model (DSVM) within a one-against-all scheme, where the output of K binary classifiers is combined to derive the desired multiclass discrimination. The second method is based on a round-robin scheme, where $K(K-1)/2$ binary classification problems have to be solved.

Suppose we have a collection of binary classifiers B_k, $k=1,2,...,K$, trained to discriminate two subsets of classes in $\mathcal{D}$. In what follows, we will assume that each classifier B_k is based on model (DSVM). In order to specify which combination of classes is presented to each classifier, a matrix $\mathbf{V} \in \{-1,0,1\}^{D \times K}$ is assigned, with the following interpretation: if either $v_{dk}=-1$ or $v_{dk}=1$, the examples for which $y_i=d$ are presented to classifier B_k with revised class value -1 or 1, respectively; if instead $v_{dk}=0$ the examples belonging to class d are not entailed into the k-th binary classification. Denote by $g_k(\mathbf{x}_i)$ the score function assigned by classifier B_k to example $\mathbf{x}_i$. Thus, given a new example $\mathbf{x}$, the multicategory classifier applies the K algorithms B_k, $k = 1,2,...,K$, obtaining a vector of outputs $\mathbf{g}(\mathbf{x}) = (g_1(\mathbf{x}), g_2(\mathbf{x}),...,g_K(\mathbf{x}))$, and predicts for $\mathbf{x}$ the class value d for which row $\mathbf{v}_d$ of the matrix $\mathbf{V}$ is "nearest" to $\mathbf{g}(\mathbf{x})$ with respect to a suitably defined metric. More precisely, the multicategory classifier requires a distance $\delta : \Re^K \times \{-1,0,1\}^K \to \Re$ be defined, and assigns to the new example $\mathbf{x}$ the class value $y = \arg\min_d \delta(\mathbf{g}(\mathbf{x}), \mathbf{v}_d)$. A similar multicategory classification scheme is called *error correcting output codes* (ECOC) (Allwein, Schapire, & Singer, 2000).

The first specific method we consider within the ECOC scheme is the popular *one-against-all* classifier, also termed *one-vs.-others*. In this case, $K = D$ and each classifier B_k is required to discriminate between a single class in one subset and all the remaining classes in the other. Therefore, the $D \times D$ matrix $\mathbf{V}$ is defined as: $v_{dd} = 1$ and $v_{dk} = -1$ for $d \neq k$. As each classifier is represented by model (DSVM), it is natural to define the distance δ as,

$$\delta(\mathbf{g}(\mathbf{x}), \mathbf{v}_d) = -\left[\sum_{k=1}^{K} g_k(\mathbf{x}) \max(v_{dk}, 0) \right]. \quad (8)$$

The effect of this metric is to predict, for the new example $\mathbf{x}$, the class value y for which the score is maximum, that is,

$$y = \arg\max_k g_k(\mathbf{x}) = \arg\max_k (\mathbf{w}_k' \mathbf{x} - b_k). \quad (9)$$

Intuitively, this corresponds to assigning to the example $\mathbf{x}$ the class value y, which is more likely when at least one of the components of the vector $\mathbf{g}(\mathbf{x})$ is positive, that is, $g_k(\mathbf{x}) > 0$ for some k. If instead all the components of $\mathbf{g}(\mathbf{x})$ are nonpositive, the choice of y in (9) corresponds to the less unlikely class assignment. Geometrically, this is equivalent to labeling example $\mathbf{x}$ with the class whose separating hyperplane lies furthest among the classes d for which $\mathbf{g}_d(\mathbf{x}) > 0$; if instead $\mathbf{g}_k(\mathbf{x}) \leq 0$ for every k, then the assignment of y picks up the class for which the separating hyperplane is nearest to $\mathbf{x}$. The multicategory classifier derived by embedding model (DSVM) within the one-against-all scheme described will be denoted as DSVM_{OAA}.

A second multicategory classifier that can be defined within the ECOC framework is represented by a *round-robin* or *pairwise decomposition* scheme, in which each class is discriminated against each other in all $K=D(D-1)/2$ possible ways. The resulting matrix $\mathbf{V}$ is formed by all D dimensional columns that include exactly a -1 and a 1, having all the remaining entries equal to 0. In this case, a more appropriate choice for δ is represented by the *Hamming* distance, also known as *voting*, defined as,

$$\delta(\mathbf{g}(\mathbf{x}), \mathbf{v}_d) = 0.5 \sum_{k=1}^{K} (1 - \text{sgn}(g_k(\mathbf{x})) v_{dk}$$

$$= 0.5 \sum_{k=1}^{K} (1 - f_k(\mathbf{x})) v_{dk}. \quad (10)$$

This distance is equivalent to assigning to example $\mathbf{x}$ the class value y that received most votes among all $K=D(D-1)/2$ binary pairwise classifiers applied to **x**. The multicategory classifier obtained by embedding model (DSVM) within the round-robin scheme described will be denoted as $DSVM_{RR}$.

COMPUTATIONAL TESTS

The prediction ability of discrete SVM has been evaluated on two benchmark datasets that have been extensively used in the context of protein folding recognition in order to compare alternative classification methods (Ding & Dubchak, 2001; Dubchak, Muchnik, Holbrook, & Kim, 1995; Dubchak, Muchnik, Mayor, Dralyuk, & Kim, 1999). These datasets do not contain highly homologous protein sequences. In particular, the first dataset, collected by Dubchak et al. (1999) and generally used for the training process, consists of 313 proteins having no more than 35% identity with each other. The second dataset, utilized as an independent test sample, is the PDB-40D set developed by the authors of the SCOP database (Andreeva, Howorth, Brenner, Hubbard, Chothia, & Murzin, 2004; Murzin, Brenner, Hubbard, & Chothia, 1995). It contains 385 proteins possessing less than 40% of the sequence identity. Furthermore, all the proteins in the testing sample have less than 35% identity with the proteins contained in the training dataset. The two datasets refer to 27 folds, each represented by at least seven proteins. These folds are grouped with respect to proteins structural classes, which can be one of four categories: all α, all β, α + β, and α / β. Table 1 shows the folds included into each structural class, with the number of proteins falling into the training and the testing datasets.

Table 1. Structural classes, folds, and cardinalities of the training and the test datasets

Structural class	Fold index	n° train	n° test
α α-helix secondary structure	1	13	6
	3	7	9
	4	12	20
	7	7	8
	9	9	9
	11	7	9
β β-sheet secondary structure	20	30	44
	23	9	12
	26	16	13
	30	7	6
	31	8	8
	32	13	19
	33	8	4
	35	9	4
	39	9	7
α / β mixed or alternating α-helix and β-sheet segments	46	29	48
	47	11	12
	48	11	13
	51	13	27
	54	10	12
	57	9	8
	59	10	14
	62	11	7
	69	11	4
α + β α-helix and β-sheet segments not mixed	72	7	8
	87	13	27
	110	12	27

In order to perform protein folding recognition by means of machine learning methods, proteins must be converted into vectors of numerical or categorical attributes. In Ding and Dubchak (2001), six sets of attributes are extracted from amino acids sequences. These are amino acids composition "C" (20 attributes), predicted secondary structure "S" (21 attributes), hydrophobicity "H" (21 attributes), normalized van der Waals volume "V" (21 attributes), polarity "P" (21 attributes), and polarizability "Z" (21 attributes). In order to train the classification algorithms, these sets of attributes can be used alone or can be mixed in different combinations. For instance, one might train a classifier based on the combination of the attributes in the sets C, S, and H. In the sequel, the notation "CSH" means the *union* of the attributes in the sets C, S and H.

The effectiveness of a classifier is generally evaluated in terms of the overall prediction accuracy it is able to achieve on a test sample. Specifically, if z denotes the number of correctly classified proteins and a is the size of the test sample, the overall accuracy is computed as $Q = z / a$. However, for protein folding classification, it is worth also to evaluate the accuracy for each fold, obtaining 27 values on our test sample, given by the ratios $Q_d = z_d / a_d$, $d \in \mathcal{D}$ where z_d and a_d denote, respectively, the number of correctly classified proteins and the total number of proteins for fold d.

Two main classifiers were considered for the computational testing: the one-against-all and the round-robin methods, denoted respectively as $DSVM_{OAA}$ and $DSVM_{RR}$ and described in the previous section, each embedding discrete SVM

Table 2. Prediction accuracy (%) on the test sample achieved by direct methods

Method	Attributes					
	C	CS	CSH	CSHP	CSHPV	ALL
SVM_{OAA}*	43.5	31.5	45.2	-	-	-
SVM_{UOAA}*	49.4	36.2	51.1	-	-	-
$DSVM_{OAA}$	28.1	37.1	39.0	41.7	41.2	38.3
$DSVM_{RR}$	37.7	47.5	**51.4**	43.3	48.2	42.3

* Results from Ding and Dubchak (2001)

Table 3. Prediction accuracy (%) on the test sample achieved by alternative voting methods

Method	Voting predictions					
	{C}	{C+CS}	{C+CS+CSH}	{C+CS+CSH+ CSHP}	{C+CS+CSH+ CSHP+CSHPV}	{ALL}
$V\text{-}NN_{OAA}$*	20.5	36.8	40.6	41.1	41.2	41.8
$V\text{-}SVM_{OAA}$*	43.5	43.2	45.2	43.2	44.8	44.9
$V\text{-}SVM_{RR}$*	44.9	52.1	56.0	56.5	55.5	53.9
$V\text{-}SVM_{UOAA}$*	49.4	48.6	51.1	49.4	50.9	49.6
$V\text{-}DSVM_{OAA}$	28.1	39.7	42.6	44.7	43.9	42.3
$V\text{-}DSVM_{RR}$	37.7	50.4	57.4	**58.2**	57.1	53.2

* Results from Ding and Dubchak (2001)

model (DSVM) as the base binary classifier. Each algorithm was trained using different combinations of the set of attributes formed by C, CS, CSH, CSHP, CSHPV, and CSHPVZ, where this latter will be denoted as ALL hereafter. By this way, the combination of the six attributes groupings with the two methods $DSVM_{OAA}$ and $DSVM_{RR}$

originated 12 distinct classifiers, indicated as *direct* in the sequel.

Besides these models, 12 further ensemble classifiers were derived by applying the voting scheme proposed in Ding and Dubchak (2001). To explain how these ensemble methods were generated, consider, for example, the three mod-

Table 4. Prediction accuracy (%) for each fold of the test sample using $DSVM_{OAA}$ in the voting framework

Voting – V-DSVM$_{OAA}$						
Fold index	{C}	{C+CS}	{C+CS+CSH}	{C+CS+CSH+CSHP}	{C+CS+CSH+CSHP+CSHPV}	{ALL}
1	83.3	83.3	83.3	83.3	83.3	83.3
3	66.7	77.8	88.9	88.9	88.9	88.9
4	20.0	20.0	20.0	25.0	20.0	20.0
7	25.0	37.5	50.0	50.0	50.0	50.0
9	100.0	100.0	100.0	100.0	100.0	100.0
11	33.3	55.6	55.6	55.6	44.4	44.4
20	25.0	54.5	68.2	79.5	79.5	68.2
23	41.7	50.0	50.0	50.0	50.0	33.3
26	15.4	15.4	15.4	15.4	15.4	15.4
30	33.3	50.0	50.0	50.0	50.0	50.0
31	50.0	62.5	62.5	62.5	62.5	62.5
32	10.5	10.5	10.5	10.5	10.5	10.5
33	25.0	25.0	25.0	25.0	25.0	25.0
35	25.0	25.0	25.0	25.0	25.0	25.0
39	28.6	28.6	28.6	28.6	28.6	28.6
46	22.9	54.2	58.3	58.3	58.3	58.3
47	8.3	8.3	8.3	8.3	8.3	8.3
48	30.8	38.5	38.5	38.5	38.5	38.5
51	7.4	11.1	11.1	18.5	18.5	18.5
54	16.7	16.7	16.7	16.7	16.7	16.7
57	12.5	12.5	12.5	12.5	12.5	12.5
59	21.4	21.4	21.4	21.4	21.4	21.4
62	14.3	28.6	28.6	28.6	28.6	28.6
69	25.0	25.0	25.0	25.0	25.0	25.0
72	12.5	12.5	12.5	12.5	12.5	12.5
87	3.7	22.2	22.2	22.2	22.2	22.2
110	77.8	85.2	88.9	88.9	85.2	88.9
Acc.	28.1	39.7	42.6	**44.7**	43.9	42.3

els generated by training $DSVM_{OAA}$ using the datasets composed by CS, CSH, and CSHP. The predictions obtained on the test sample by each of these three direct classifiers were combined, deriving the predictions for a new classifier based on a majority-voting scheme. If two or more folds classes receive the same number of votes for a protein, the tie is resolved by assigning the class fold corresponding to the largest score among the alternative folds classes predicted for that protein. The same procedure was repeated using the six combinations of direct classifiers {C},

Table 5. Prediction accuracy (%) for each fold of the test sample using $DSVM_{RR}$ in the voting framework

Voting – V-DSVM$_{RR}$						
Fold index	{C}	{C+CS}	{C+CS+CSH}	{C+CS+CSH+CSHP}	{C+CS+CSH+CSHP+CSHPV}	{ALL}
1	83.3	83.3	83.3	83.3	83.3	83.3
3	77.8	88.9	88.9	88.9	88.9	88.9
4	35.0	35.0	35.0	35.0	40.0	40.0
7	25.0	50.0	50.0	50.0	50.0	37.5
9	44.4	66.7	66.7	66.7	66.7	55.6
11	22.2	22.2	22.2	22.2	22.2	11.1
20	54.5	68.2	86.4	90.9	88.6	88.6
23	16.7	16.7	16.7	16.7	16.7	8.3
26	30.8	53.8	53.8	53.8	53.8	46.2
30	33.3	50.0	50.0	50.0	50.0	33.3
31	50.0	50.0	50.0	50.0	37.5	37.5
32	21.1	36.8	36.8	36.8	36.8	36.8
33	25.0	50.0	50.0	50.0	25.0	25.0
35	25.0	25.0	25.0	25.0	25.0	25.0
39	39	28.6	42.9	42.9	42.9	42.9
46	46	64.6	72.9	89.6	91.7	91.7
47	47	66.7	50.0	58.3	58.3	58.3
48	48	23.1	46.2	46.2	46.2	46.2
51	51	18.5	40.7	40.7	40.7	40.7
54	54	33.3	33.3	33.3	33.3	33.3
57	57	0.0	25.0	25.0	25.0	12.5
59	59	21.4	42.9	42.9	42.9	35.7
62	62	14.3	42.9	42.9	42.9	42.9
69	69	0.0	25.0	25.0	25.0	25.0
72	72	12.5	12.5	12.5	12.5	12.5
87	87	18.5	33.3	48.1	48.1	48.1
110	110	48.1	70.4	92.6	92.6	92.6
Acc.	Acc.	37.7	50.4	57.4	**58.2**	57.1

126

{C+CS}, {C+CS+CSH}, {C+CS+CSH+CSHP}, {C+CS+CSH+CSHP+CSHPV}, and {C+CS+CS H+CSHP+CSHPV+CSHPVZ} (this latter denoted by {ALL}) for each of the methods $DSVM_{OAA}$ and $DSVM_{RR}$. The ensemble classifiers, based on the voting scheme, are denoted as $V\text{-}DSVM_{OAA}$ and $V\text{-}DSVM_{RR}$.

In order to choose the best set of parameters for each of the resulting 24 classifiers, 10-fold cross-validation was applied on the training dataset before evaluating the accuracy on the test sample.

Tables 2 and 3 show the overall accuracy on the test sample for the 24 classifiers, as well as for other competing methods considered in Ding and Dubchak (2001). In particular, Table 2 indicates, in rows 3-4, the performance exhibited by the direct classifiers $DSVM_{OAA}$ and $DSVM_{RR,}$ by using different sets of attributes during the training process. Rows 1-2, in Table 2, report the accuracy values obtained in Ding and Dubchak (2001) for SVM, based on the one-against-all (SVM_{OAA}) and the unique-one-against-all (SVM_{UOAA}) schemes trained with the same attributes. Table 3 contains the results provided according to distinct combinations of predictions generated by alternative direct classifiers embodied into the voting mechanism. The first four rows in Table 3 report the accuracy values obtained in Ding and Dubchak (2001) for neural networks ($V\text{-}NNO_{OAA}$) and SVM ($V\text{-}SV\text{-}M_{OAA}$) based on the one-against-all scheme, for SVM framed within the round-robin ($V\text{-}SVM_{RR}$) framework, and for SVM based on a unique-one-against-all procedure ($V\text{-}SVM_{UOAA}$). These four methods were derived from a majority-voting scheme that used a combination of the predictions, as indicated in the columns. The last two rows refer, instead, to the ensemble classifiers $V\text{-}DSVM_{OAA}$ and $V\text{-}DSVM_{RR}$.

The results presented in Tables 2 and 3 suggest some empirical findings. First, notice that discrete SVM achieve the highest accuracy of 51.4%, marked in bold, among the direct meth-

ods considered in Table 2. When discrete SVM are adopted as the base binary classifier, the round-robin scheme constantly dominates the one-against-all framework. This remark can be intuitively explained by the fact that, due to the large number of folds, the one-against-all scheme can hardly distinguish between a single fold, often represented by a few proteins, and a relatively large set of proteins, including heterogeneous folds. Moreover, the voting mechanism notably improves the accuracy for all the classifiers, playing an effective regularization role. Finally, the best overall accuracy of 58.2% on the test sample is achieved by discrete SVM, embodied into the round-robin scheme and subject to the voting mechanism applied to the combination of predictions {C+CS+CSH+CSHP}. Notice that by using a larger number of predictions, derived from the direct classifiers, the overall accuracy decreases, since the two sets of attributes "V" and "Z" introduce some disturbing noise.

Tables 4 and 5 provide detailed accuracy results, at the level of singular folds, for the voting methods $V\text{-}DSVM_{OAA}$ and $V\text{-}DSVM_{RR}$. As one might expect, the accuracy is higher for the most populated folds.

Although the prediction accuracy achieved by the different methods might appear low with respect to other classification tasks, one should bear in mind that there are 27 class folds, so that a random classifier would obtain an average accuracy of 3.7%. Furthermore, the use of machine learning methods leads to great benefits, since the recognition in vitro of the folding of a new protein is a costly and complex activity.

REFERENCES

Allwein, E. L., Schapire, R. E,. & Singer, Y. (2000). Reducing multiclass to binary: A unifying approach for margin classifiers. *Journal of machine learning research, 1,* 113-141.

Andreeva, A., Howorth, D., Brenner, S. E., Hubbard, T. J., Chothia, C., & Murzin, A. G. (2004). SCOP database in 2004: Refinements integrate structure and sequence family data. *Nucleic Acids Res., 32*, D226–D229.

Baldi, P., & Brunak, S. (1998). *Bioinformatics: The machine learning approach*. Cambridge, MA: MIT Press.

Breiman, L., Friedman, J. H., Olshen, R. A., & Stone, C. J. (1984). *Classification and regression trees*. Belmont: Wadsworth International.

Craven, M. W., Mural, R. J., Hauser, L. J., & Uberbacher, E. C. (1995). Predicting protein folding classes without overly relying on homology. *ISMB, 3*, 98–106.

Cristianini, N., & Shawe-Taylor, J. (2000). *An introduction to support vector machines and other kernel-based learning methods*. Cambridge, UK: Cambridge University Press.

Dietterich, T., & Bakiri, G. (1995). Solving multiclass learning problems via error-correcting output codes. *Journal of Artificial Intelligence Research, 2*, 263-286.

Ding, C. H., & Dubchak, I. (2001). Multi-class protein fold recognition using support vector machines and neural networks. *Bioinformatics, 17*, 349-358.

Dubchak, I., Muchnik, I., Holbrook, S. R., & Kim, S. H. (1995). Prediction of protein folding class using global description of amino acid sequence. *Proc. Natl Acad. Sci. USA, 92*, 8700-8704.

Dubchak, I., Muchnik, I., Mayor, C., Dralyuk, I., & Kim, S. H. (1999). Recognition of a protein fold in the context of the Structural Classification of Proteins (SCOP) classification. *Proteins, 35*, 401-407.

Durbin, R., Eddy, S., Krogh, A., & Mitchison, G. (1998). *Biological sequence analysis*. Cambridge University Press.

Guruswami, V., & Sahai, A. (1999). Multiclass learning, boosting, and error-correcting codes. In *12th Annual Conference on Computational Learning Theory* (pp. 145-155).

Holm, L., & Sander, C. (1999). Protein folds and families: Sequence and structure alignments. *Nucleic Acids Res., 27*, 244–247.

Kreßel, U. (1999). Advances in kernel methods—Support vector learning. In B. Scholkopf, C. Burges, & A. J. Smola (Eds.), *Pairwise classification and support vector machines*. Cambridge, MA: MIT Press.

Murthy, S. K. (1998). Automatic construction of decision trees from data: A multi-disciplinary survey. *Data Mining and Knowledge Discovery, 2*, 345-389.

Murzin, A. G., Brenner, S. E., Hubbard, T., & Chothia, C. (1995). SCOP: A structural classification of protein database for the investigation of sequence and structures. *J. Mol. Biol., 247*, 536-540.

Orsenigo, C., & Vercellis, C. (2003). Multivariate classification trees based on minimum features discrete support vector machines. *IMA Journal of Management Mathematics, 14*, 221-234.

Orsenigo, C., & Vercellis, C. (2004). Discrete support vector decision trees via tabu-search. *Journal of Computational Statistics and Data Analysis, 47*, 311-322.

Orsenigo, C., & Vercellis, C. (2007a). Multicategory classification via discrete support vector machines. *Computational Management Science*.

Orsenigo, C., & Vercellis, C. (2007b). Accurately learning from few examples with a polyhedral classifier. *Computational Optimization and Applications*.

Quinlan, J. R. (1993). *C4.5: Programs for machine learning*. San Mateo: Morgan Kaufmann.

Schölkopf, B., & Smola, A. J. (2002). *Learning with kernels. Support vector machines, regularization, optimization and beyond.* Cambridge, MA: MIT Press.

Vapnik, V. (1995). *The nature of statistical learning theory.* New York: Springer.

Vapnik, V. (1998). *Statistical learning theory.* New York: Wiley.

Chapter VIII
Hierarchical Profiling, Scoring, and Applications in Bioinformatics

Li Liao
University of Delaware, USA

ABSTRACT

Recently, clustering and classification methods have seen many applications in bioinformatics. Some are simply straightforward applications of existing techniques, but most have been adapted to cope with peculiar features of the biological data. Many biological data take a form of vectors, whose components correspond to attributes characterizing the biological entities being studied. Comparing these vectors, aka profiles, are a crucial step for most clustering and classification methods. We review the recent developments related to hierarchical profiling where the attributes are not independent, but rather are correlated in a hierarchy. Hierarchical profiling arises in a wide range of bioinformatics problems, including protein homology detection, protein family classification, and metabolic pathway clustering. We discuss in detail several clustering and classification methods where hierarchical correlations are tackled in effective and efficient ways, by incorporation of domain-specific knowledge. Relations to other statistical learning methods and more potential applications are also discussed.

INTRODUCTION

Profiling entities based on a set of attributes and then comparing these entities by their profiles is a common, and often effective, paradigm in machine learning. Given profiles, frequently represented as vectors of binary or real numbers, the comparison amounts to measuring "distance" between a pair of profiles. Effective learning hinges on proper and accurate measure of distances.

In general, given a set A of N attributes, $A = \{a_i | i = 1, \dots, N\}$, profiling an entity x on A gives a mapping $p(x) \rightarrow \Re^N$, namely, $p(x)$ is an N vector of real values. Conveniently, we also use x to denote its profile $p(x)$, and x_i the i-*th* component of $p(x)$. If all attributes in *A* can only have two discrete values 0 and 1, then $p(x) \rightarrow \{0,1\}^N$ yields a binary profile. The distance between a pair of profiles x and y is a function: $D(x, y) \rightarrow \Re$. Hamming distance is a straightforward, and also one of the most commonly used, distance measures for binary profiles; it is a simple summation of difference at each individual component:

$$D(x, y) = \Sigma_i^n d(i) \tag{1}$$

where $d(i) = |x_i - y_i|$. For example, given $x = (0, 1, 1, 1, 1)$ and $y = (1, 1, 1, 1, 1)$, then $D(x, y) = \Sigma_{i=1}^5 d(i) = 1+0+0+0+0 = 1$. A variant definition of $d(i)$, which is also very commonly used, is that $d(i) = 1$ if $x_i = y_i$ and $d(i) = -1$ if otherwise. In this variant definition, $D(x, y) = \Sigma_{i=1}^5 d(i) = -1+1+1+1+1 = 3$.

The Euclidean distance, defined as $D = \sqrt{\Sigma_i^n (x_i - y_i)^2}$, has a geometric representation: a profile is mapped to a point in a vector space where each coordinate corresponds to an attribute. Besides using Euclidean metric, in vector space the distance between two profiles is also often measured as dot product of the two corresponding vectors: $x \cdot y = \Sigma_i^n x_i y_i$. Dot product is a key quantity used in Support Vector Machines (Vapnik, 1997, Cristianini & Shawe-Taylor, 2000, Scholkopf & Smola, 2002). Many clustering methods applicable to vectors in Euclidean space can be applied here, such as K-means.

While Hamming distance and Euclidean distance are the commonly adopted measures of profile similarity, both of them imply an underlying assumption that the attributes are independent and contribute equally in describing the profile. Therefore, the distance between two profiles is simply a sum of distance (i.e., difference) between them at each attribute. These measures become inappropriate when the attributes are not equally contributing, or not independent, but rather correlated to one another. As we will see, this is often the case in the real-world biological problems.

Intuitively, nontrivial relations among attributes complicate the comparisons of profiles. An easy and pragmatic remedy is to introduce scores or weighting factors for individual attributes to adjust their apparently different contribution to the Hamming or Euclidean "distance" between profiles. That is, the value of $d(i)$ in equation (1) now depends not only on the values of x_i and y_i, but also on the index i. Often, scoring schemes of this type are also used for situations where attributes are correlated, sometimes in a highly nonlinear way. Different scoring schemes thereby are invented in order to capture the relationships among attributes. Weighting factors in these scoring schemes are either preset *a priori* based on domain knowledge about the attributes, or fixed from the training examples, or determined by a combination of both. To put into a mathematical framework, those scoring based approaches can be viewed as approximating the correlations among attributes, which, without loss of generality, can be represented as a polynomial function. In general, a formula that can capture correlations among attributes as pairs, triples, quadruples, and so forth, may look like the following:

$$D' = \Sigma_i^n d(i) + \Sigma_{i \neq j}^n d(i)c(i,j)d(j) + \Sigma_{i \neq j \neq k}^n d(i)d(j)d(k)c(i,j,k) + \ldots \tag{2}$$

where the coefficients $c(i,j)$, $c(i,j,k)$, …, are used to represent the correlations. This is much like introducing more neurons and more hidden layers in an artificial neural network approach, or introducing a nonlinear kernel functions in kernel-based methods. Because the exact relations among attributes are not known *a priori*, an open formula like equation (2) is practically useless: as the number of these coefficients grows exponentially with the profile size, solving it would be computationally intractable, and there would not be enough training examples to fit these coefficients.

However, it was found that the situation would become tractable when these correlations could be structured as a hierarchy—a quite loose requirement and readily met in many cases as we shall see later. In general, a hierarchical profile of entity x can be defined as $p(x) \rightarrow \{0,1\}^L$, where L stands for the set of leaves in a rooted tree T. A hierarchical profile is no longer a plain string of zeros and ones. Instead, it may be best represented as a tree with the leaves labeled by zeros and ones for binary profiles, or by real value numbers for real value profiles.

As the main part of this chapter, we will discuss in detail several clustering and classification methods where hierarchical profiles are coped with effectively and efficiently, noticeably due to incorporation of domain specific knowledge. Relations to other statistical learning methods, e.g., as kernel engineering, and more possible applications to other bioinformatics problems are also discussed, towards the end of this chapter.

HIERARCHICAL PROFILINGS IN BIOINFORMATICS

Functional Annotations

Hierarchical profiling arises naturally in many bioinformatics problems. The first example is probably from the phylogenetic profiles of proteins and using them to assign functions to proteins (Pellegrini et al., 1999). To help understand the key concepts and motivations, and also to introduce some terminologies for later discussions, we first briefly review the bioinformatics methodologies for functional annotation.

Determining protein functions, also called as functional annotation, has been and remains a central task in bioinformatics. Over the past 25 years many computational methodologies have been developed toward solving this task. The development of these computational approaches can be generally broken into four stages by both the chronological order and algorithmic sophistication (Liao & Noble, 2002). For our purpose here, we can categorize these methods into three levels according to the amount and type of information used.

The methods in level one compare a pair of proteins for sequence similarity. Among them are the landmark dynamic programming algorithm by Smith and Waterman (1980) and its heuristic variations BLAST (Altschul et al., 1990) and FASTA (Pearson, 1990). The biological reason behind these methods is protein homology; two proteins are homologous if they share a common ancestor. Therefore, homologous proteins are similar to each other in the primary sequence and keep the same function as their ancestor's, until the evolutionary divergence—mutations, deletions, or insertions from the ancestral protein (or gene, to be more precise) is significant enough to cause any change. A typical way to annotate a gene with unknown function is to search against a database of genes whose functions are already known, such as GenBank (http://www.ncbi.nlm.nih.gov/Genbank), and assign to the query gene the function of a homologous gene found in the database.

The next level's methods use multiple sequences from a family of proteins with same or similar functions, in order to collect aggregate statistic for more accurate/reliable annotation. Profiles (Gribskov et al., 1987) and hidden Markov models (Krogh et al., 1994; Durbin et al., 1998) are two popular methods to capture and represent these aggregate statistics from whole sequences for protein families. More refined and sophisticated methods using aggregate statistics are developed, such as PSI-BLAST (Altschul et al., 1997), SVM-Fisher (Jaakkola et al., 1999, 2000), Profile-Profile (Sadreyev & Grishi, 2003; Mittelman et al., 2003), SVM-Pairwise (Liao & Noble, 2002, 2003). The aggregate statistic may also be represented as patterns or motifs. Methods based on patterns and motifs include BLOCKs (Henikoff & Henikoff, 1994), MEME (Bailey &

Elkan, 1995), Meta-MEME (Grundy et al., 1997), and eMotif (Nevill-Manning et al., 1998).

The third level's methods go beyond sequence similarity to utilize information such as DNA Microarray gene expression data, phylogenetic profiles, and genetic networks. Not only can these methods detect distant homologues—homologous proteins with sequence identity below 30%, but they also can identify proteins with related functions, such as those found in a metabolic pathway or a structural complex.

Given a protein, its phylogenetic profile is represented as a vector, where each component corresponds to a genome and takes a value of either one or zero, indicating respectively the presence or absence of a homologous protein in the corresponding genome. Protein phylogenetic profiles were used in Pellegrini, Marcotte, Thompson, Eisenberg, and Yeates (1999) to assign protein functions based on the hypothesis that functionally linked proteins, such as those participating in a metabolic pathway or a structural complex, tend to be preserved or eliminated altogether in a new species. In other words, these functionally linked proteins tend to co-evolve during evolution. In Pellegrini et al. (1999), 16 then-fully-sequenced genomes were used in building the phylogenetic profiles for 4290 proteins in *E. coli* genome. The phylogenetic profiles, expressed as 16-vector, were clustered as following; proteins with identical profiles are grouped and considered to be functionally linked, and two groups are called *neighbors* when their phylogenetic profiles differ by one bit. The results based on these simple clustering rules supported the functional linkage hypothesis. For instance, homologues of ribosome protein RL7 were found in 10 out of 11 eubacterial genomes and in yeast but not in archaeal genomes. They found that more than half of the *E. coli* proteins with the RL7 profile or profiles different from RL7 by one bit have functions associated with the ribosome, although none of these proteins share significant sequence similarity with the RL7 protein. That is, these proteins are unlikely to be annotated as RL7 homologues by using sequence similarity based methods.

There are some fundamental questions regarding the measure of profile similarity that can affect the results from analysis of phylogenetic profiles. Can we generalize the definition of similar profiles? In other words, can we devise a measure so we can calculate similarity for any pair of profiles? What threshold should be adopted when profile similarity is used to infer functional linkage?

A simple measure for profile similarity first brought up was Hamming distance. In Marcotte, Xenarios, van Der Bliek, and Eisenberg (2000), phylogenetic profiles are used to identify and predict subcellular locations for proteins; they found mitochondrial and non-mitochondrial proteins.

The first work that recognizes phylogenetic profiles as a kind of hierarchical profiling is Liberles, Thoren, von Heijne, and Elofsson (2002), a method that utilizes the historical evolution of two proteins to account for their similarity (or dissimilarity). Evolutionary relationships among organisms can be represented as a phylogenetic tree where leaves correspond to the current organisms and internal nodes correspond to hypothetical ancient organisms. So, rather than simply counting the presence and absence of the proteins in the current genomes, a quantity called differential parsimony is calculated that minimize the number of times when changes have to be made at tree branches to reconcile the two profiles.

Comparative Genomics

Another example of hierarchical profiling arises from comparing genomes based on their metabolic pathway profiles. A main goal of comparing genomes is to reveal the evolutionary relationships among organisms, which can be represented as phylogenetic trees.

Originally phylogenetic trees were constructed based on phenotypic—particularly morphological—features of organisms. Nowadays, molecular reconstruction of phylogenetic trees is most commonly based on comparisons of small sub-unit ribosomal RNA (16S rRNA) sequences (Woese, 1987). The small sub-unit rRNAs is orthodoxly used as gold standard for phylogeny study, mainly due to two factors: their ubiquitous presence and relative stability during evolution. However, the significance of phylogenies based on these sequences have been recently questioned with growing evidence for extensive lateral transfer of genetic material, a process which results in blurring the boundaries between species. Phylogenetic trees based on individual protein/gene sequence analysis—thus called gene tree—are not all congruent with species trees based on 16S rRNA (Eisen, 2000). Attempts of building phylogenetic trees based on different information levels, such as pathways (Dandekar et al., 1999; Forst & Schulten, 2001) or some particular molecular features such as folds (Lin & Gerstein 2000), have also led to mixed results of congruence with the 16S rRNA-based trees.

From a comparative genomics perspective, it makes more sense to study evolutionary relationships based on genome-wide information, instead of a piece of the genome, be it an rRNA or a gene. As more and more genomes are fully sequenced, such genome-wide information becomes available. One particularly interesting type of information is the entire repertoire of metabolic pathways in an organism, as the cellular functions of an organism are carried out via these metabolic pathways. A metabolic pathway is a chain of biochemical reactions, together fulfilling certain cellular functions. For example, *Glycolysis* is a pathway existed in most cells, which consists of 10 sequential reactions converting glucose to pyruvate while generating the energy that the cell needs. Because most of these reactions require enzymes as catalyst, therefore in an enzyme centric scheme, pathways are represented as sequences of component enzymes. It is reasonable to set the necessary condition for a pathway to exist in an organism as that all the component enzymes of that pathway are available. Enzymes are denoted by enzyme commission (EC) numbers which specifies the substrate specificity. Most enzymes are proteins. Metabolic pathways in a completely sequenced genome are reconstructed by identifying enzyme proteins that are required for a pathway (Gaasterland & Selkov, 1995; Karp et al., 2002).

The information about presence and absence of metabolic pathways in genomes can be represented as a binary matrix, as shown in Figure 1, where an entry $(i,j) = 1/0$ represents whether pathway j is present/absent in genome i. Therefore, each row serves as a profile of the corresponding genome (termed metabolic pathway profiles), and each column serves as a profile of the corresponding pathway (termed phyletic profiles). It is reasonable to believe that valuable information about evolution and co-evolution is embedded in these profiles, and comparison of profiles would reveal, to some degree, the evolutionary relations among entities (either genomes or pathways) represented by these profiles.

Once again, the attributes used for building these profiles are not independent but correlated to each other. Because different metabolic pathways may be related to one another in terms of physiological functions, e.g., one pathway's absence/presence may be correlated with another pathway's absence/presence in a genome, these relationships among pathways, as attributes of metabolic pathway profiles, should be taken into account when comparing genomes based on their MPPs. The relationships among various pathways are categorized as a hierarchy in the WIT database (Overbeek et al., 2000), which can be found at the following URL (http://compbio.mcs.anl.gov/puma2/cgi-bin/functional_overview.cgi).

COMPARING HIERARCRHICAL PROFILES

In the last section, we showed that the data and information in many bioinformatics problems can be

Figure 1. Binary matrix encoding the presence/absence of pathways in genomes; O_1 to O_m represent m genomes, and P_1 to P_n represent n pathways

	P_1	P_2			P_n
O_1	0	1			0
O_2	1	1			1
.		.		.	
.		.		.	
O_M	0	1			1

represented as hierarchical profiles. Consequently, the clustering and classification of such data and information need to account for the hierarchical correlations among attributes when measuring the profile similarity. While a generic formula like equation (2) posits to account for arbitrary correlations theoretically, its demand of exponentially growing amount of training examples and lacking of an effective learning mechanism render the formula practically useless. In hierarchical profiles, however, the structure of relations among attributes is known, and sometimes the biological interpretation of these relations is also known. As a result, the learning problems will become rather tractable.

P-Tree Approach

Realizing that the hierarchical profiles contain information not only in the vector but also in the hierarchical correlations, it is natural to first attempt at treating them as trees. How to compare trees is itself an interesting topic with wide applications, and has been the subject of numerous studies. In Liao, Kim, and Tomb, (2002), a p-tree based approach was proposed to measure the similarity of metabolic pathway profiles. The relationships among pathways are adopted from

the WIT database and represented as a *Master tree*. About 3300 known metabolic pathways are collected in the WIT database and these pathways are represented as leaves in the Master tree. Then, for each genome, a *p-Tree* is derived from the Master tree by marking off leaves whose corresponding pathways are absent from the organism. In this representation, a profile is no longer a simple string of zeros and ones, where each bit is treated equally and independently. Instead, it is mapped into a p-Tree so that the hierarchical relationship among bits is restored and encoded in the tree.

The comparison of two p-Trees thus evaluates the difference between the two corresponding profiles. To take into account the hierarchy, a scoring scheme ought to weight (mis)matches at bits i and *j* according to their positions in the tree, such as i and *j* are sibling, versus i and *j* are located distantly in the tree. For that, the (mis)matches scores are transmitted bottom-up to the root of the Master tree in four steps: (1) overlay the two p-Trees; (2) score mismatches and matches between two p-Trees and label scores at the corresponding leaves on the master tree; (3) average scores from siblings (weight breadth) and assign the score to the parent node; (4) iterate step 3 until the root is reached. An algorithm implementing these steps is quite straightforward and has a time complexity linear with the size of the Master tree.

To demonstrate how the algorithm works, let us look at an example of two organisms, org_i and org_j, and three pathways p_1, p_2, and p_3. Two hypothetical cases are considered and are demonstrated in Figures 2 and 3 respectively. In case one, org_i contains pathways p_1 and p_3, and org_j contains p_2 and p_3. The metabolic pathway profiles for org_i and org_j are shown in panel B and their corresponding p-Trees are displayed in panel C. In the panel D, two p-Trees are superposed. Matches and mismatches are scored at the leaves and the scores are propagated up to the root. In case two, org_i contains pathways p_1 and p_2, whereas org_j contains all three pathways, and

Figure 2. A comparison of two organisms orgi and orgj, with respect to the three pathways p1, p2, and p3 (weighted scoring scheme, Case 1); Panel A featuring the master tree, where each pathway is represented by a circle and each square represents an internal node; in Panel B, the presence(absence) of a pathway is indicated by 1(0); Panel C features the p-trees, where a filled (empty) circle indicates the presence (absence) of the pathway; Panel D, the bottom-up propagation of the scores is illustrated; a cross (triangle) indicates a mismatch (match)

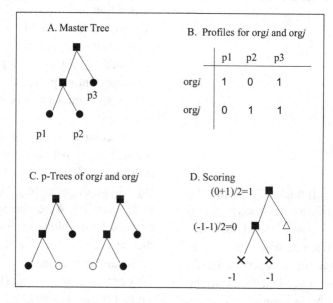

Figure 3. Case 2: A comparison of two organisms org$_i$ and org$_j$, with respect to the three pathways p1, p2, and p3 (weighted scoring scheme, Case 2); the same legends as in Figure 2

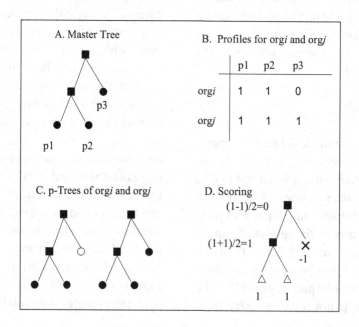

similar calculation is shown in Figure 3. In this example, given the topology of the master tree, the two cases have the same final score 0 in the p-Tree scoring scheme. This is in contrast to the variant Hamming scoring scheme, where the score for case 1 equals to $(-1-1+1) = -1$ and the score for case 2 equals to $(1+1-1) = 1$. Evidently, the p-Tree scoring scheme has taken into account the pathway relationships present in the Master tree: p_1 and p_2, being in the same branch, are likely to have a similar biochemical or physiological role, which is distinct from the role of the p_3 pathway.

Using pathway hierarchical categories in the WIT database and the scoring scheme described above, we compared 31 genomes based on their metabolic pathway profiles (MPP) (see Figure 4). Relations among 31 genomes are represented as a MPP-based tree and are compared with the phylogeny based on 16s rRNA. While the MPP-based tree is congruent with the 16S rRNA-based

tree at several levels, some interesting discrepancies were found. For example, the extremely radiation resistant organism, *D. radiodurans* is positioned in *E. coli* metabolic lineage. The noted deviation from the classical 16s rRNA phylogeny suggests that pathways have undergone evolution transcending the boundaries of species and genera. The MPP-based trees can be used to suggest alternative production platform for metabolic engineering. A different approach to this problem on the same dataset was later proposed in Heymans and Singh (2003).

Parameterized Tree Distance

Evolution of pathways and the interaction with evolution of the host genomes can be further investigated by comparing the columns (phyletic profiles) in the binary matrix in Figure 1. The intuition is that the co-occurrence pattern of

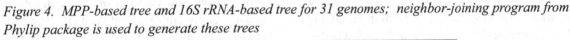

Figure 4. MPP-based tree and 16S rRNA-based tree for 31 genomes; neighbor-joining program from Phylip package is used to generate these trees

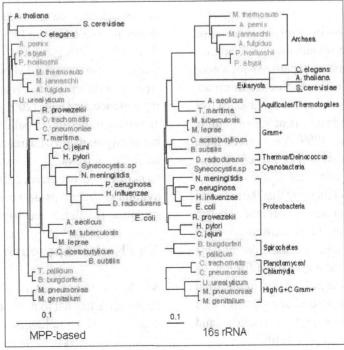

pathways in a group of organisms would coincide with the evolutionary path of these host organisms, for instance, as a result of the emergence of enzymes common to those pathways. Therefore, such patterns would give useful information, e.g., organism-specific adaptations. These co-occurrences can be detected by clustering pathways based on their phyletic profiles. Again, the phyletic profiles here are hierarchical profiles, and the distance measure between them should be calculated using the scheme that we developed above. In Zhang, Liao, Tomb, and Wang (2002), 2719 pathways selected from 31 genomes in the WIT database were clustered into 69 groups of pathways, and pathways in each group co-occur in the organisms.

Further insights were achieved by studying the evolution of enzymes that are components in co-occurred pathways. For completely sequenced genomes, sequences of these enzymes are available and can be used to build individual gene trees (in contrast to species trees). Comparisons of the gene trees of component enzymes in co-occurred pathways would serve as a basis for investigating how component enzymes evolve and whether they evolve in accordance with the pathways. For example, two pathways $p1 = (e1, e2, e3)$ and $p2 = (e4, e5)$ co-occur in organisms o1, o2, and o3. For component enzyme e1 of pathway p1, a gene tree T_{e1} is built with e1's homologues in organisms o1, o2 and o3, by using some standard tree reconstruction method such as neighbor joining algorithm (Felsenstein, 1989). This can be done for other component enzymes e2 to e5 as well. If, as an ideal case, gene trees T_{ei} for i = 1 to 5 are identical, we would then say that the pathway co-occurrence is congruent to speciation. When gene trees for component enzymes in co-occurred pathways are not identical, comparisons of gene trees may reveal how pathways evolve differently, possibly due to gene lateral transfer, duplication or loss. Recent work also shows that analysis of metabolic pathways may explain causes and evolution of enzymes dispensability.

To properly address tree comparisons, several algorithms were developed and tested (Zhang et al., 2002). One similarity measure is based on leaf overlap. Let T1 and T2 be two trees. Let S1 be the set of leaves of T1 and S2 be the set of leaves of T2. The leaf-overlap-based distance is defined as

$$D_x(T1, T2) = |S1 \cap S2| / |S1 \cup S2|, \qquad (3)$$

where $|.|$ denotes the set cardinality. A more elaborated metric, called parameterized distance as an extension from the editing distance between two unordered trees, was proposed to account for the structural difference between ordered and rooted trees. A parameter c is introduced to balance the cost incurred at different cases such as deleting, inserting, and matching subtrees. A dynamic programming algorithm is invoked to calculate the optimal distance. For example, in Figure 5, the distance between D1 and p is 1.875 and the distance between D2 and p is 1.813 when the parameter c is set at value of 0.5. On the other hand, the editing distance between D1 and p and between D2 and p are both 6, representing the cost of deleting the six nodes not touched by the dotted mapping lines in Figure 5. Noting that D1 differs from D2 topologically, this example shows that the parameterized distance better reflects the structural difference between trees than the editing distance. In Zhang (2002), the 523 component enzymes of these 2719 pathways were clustered, based on parameterized tree distance with c being set at 0.6, into exactly the same 69 clusters based on pathway co-occurrence. This suggests that our hypothesis about using co-occurrence to infer evolution is valid, at least approximately.

Tree Kernel Approach

As shown before, use of phylogenetic profiles for proteins has led to methods more sensitive to detect remote homologues. In the following we discuss classifiers using support vector machines

Figure 5. Illustration of parameterized distances between trees. Red dotted lines map tree P to trees D1 and D2, the parameter C = 1 gives the editing distances; the editing distance between D1 and P is 6, because six deletions of nodes other than a, b, and c in D1 will make D1 identical to tree P, although D1 and D2 are topologically different, their editing distances to P are both equal to 6

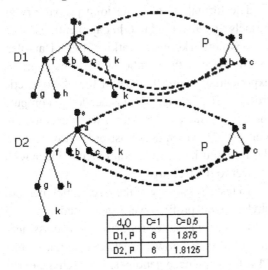

d,0	C=1	C=0.5
D1, P	6	1.875
D2, P	6	1.8125

to explore similarity among phylogenetic profiles as hierarchical profiles.

As a powerful statistical learning method, support vector machines (SVMs) have recently been applied with remarkable success in bioinformatics problems, including remote protein homology detection, microarray gene expression analysis, and protein secondary structure prediction. SVMs have been applied to problems in other domains, such as face detection and text categorization. SVMs possess many nice characteristics a good learning method shall have: it is expressive; it requires fewer training examples; it has an efficient learning algorithm; it has an elegant geometric interpretation; and above all, it generalizes well. The power of SVMs comes partly from the data representation, where an entity, e.g., a protein, is represented by a set of attributes instead of a single score. However, how

those attributes contribute to distinguishing a true positive (filled dot in Figure 6) from a true negative (empty circle) may be quite complex. In other words, the boundary line between the two classes, if depicted in a vector space, can be highly nonlinear (dashed line in the left panel of Figure 6), and nonetheless, it is the goal of a classifier to find this boundary line. The SVMs method will find a nonlinear mapping that transform the data from the original space, called input space, into a higher dimensional space, called feature space, where the data can be linearly separable (right panel of Figure 6). The learning power of SVMs comes mostly from the use of kernel functions, which define how the dot product between two points in the feature space can be calculated as a function of their corresponding vectors in the input space. As the dot product between vectors is the only quantity needed to find a class boundary in the feature space, kernel functions therefore contain sufficient information, and more importantly they avoid explicit mapping to high dimensional feature space; high dimensionality often poses difficult problems for learning such as overfitting, thus termed the curse of dimensionality. The other mechanism adopted by SVMs is to pick the boundary line that has the maximum margin to both classes. A maximum margin boundary line has low Vapnik-Chervonenkis dimension, which ensures good generalization. Because of the central role played by kernel functions, how to engineer kernel functions to incorporate domain specific information for better performance has been an active research activity. It is worth noting that, as compared to other similar learning methods such as artificial neural networks, the SVMs require fewer training examples, which is a great advantage in many bioinformatics applications.

Vert (2002) proposed a tree kernel to compare not just the profiles themselves but also their global patterns of inheritance reflected in the phylogenetic tree. In other words, the tree kernel takes into account phylogenetic histories for two genes—when in evolution they transmitted together or not – rather than by just comparing phylogenetic profiles organism per organism, at the leaves of the phylogenetic tree. A kernel function is thus defined as:

$$K(x, y) = \Sigma_{i=1 \text{ to } D} \ \Phi_i(x) \ \Phi_i(y) \qquad (4)$$

where $\Phi_i(x)$ is an inheritance pattern i for profile x. An inheritance pattern of x gives an explanation of x, that is, the presence (1) or absence (0) of gene x at each current genome is the result of a series of evolutionary events happened at the ancient genomes. If we assume a gene is possessed by the ancestral genome, at each branch of the phylogenetic tree which corresponds to a speciation, the gene may either be retained or get lost. An inheritance pattern corresponds to a series of assignments of retain or loss at all branches such that the results at leaves match the profile of x. Because of the stochastic property of these evolutionary events, we cannot be certain whether a gene is retained or lost. Rather, the best we know may be the probability of either case. Let $F_i(x) = \Phi(x \mid i)$, which is the probability that profile x can be interpreted by the inheritance pattern i, then $K (x, y) = \Sigma_{i=1 \text{ to } D} \ P(x|i) P(y|i)$ is the joint probability that both profiles x and y are

resulted from all possible pattern i. Intuitively, not all possible inheritance patterns occur at the same frequency. Let P(i) be the probability that pattern i actually occurs during the evolution, then the so called tree kernel is refined as:

$$K (x, y) = \Sigma_{i=1 \text{ to } D} \ P(i) \ P(x|i) \ P(y|i) \qquad (5)$$

The formalism of using joint probability as kernels first appeared in other applications, such as convolution kernels (Watkins, 1999; Haussler, 1999). Because the number of patterns D grows exponentially with the size of the phylogenetic tree, an efficient algorithm is needed to compute the kernel. Such an algorithm was developed in Vert (2002), which uses post-order traversals of the tree and has a time complexity linear with respect to the tree size.

To test the validity of the tree kernel method, phylogenetic profiles were generated for 2465 yeast genes, whose accurate functional classification are already known, by BLAST search against 24 fully-sequenced genomes. For each gene, if a BLAST hit with E-value less than 1.0 is found in a genome, the corresponding bit for that genome is then assigned as 1, otherwise is assigned as 0. The resulting profile is a 24-bit string of zeros and ones. Two assumptions were made in calculating the tree kernel for a pair of genes x and y.

Figure 6. Schematic illustration of nonlinear mapping of data from input space to feature space for a SVM

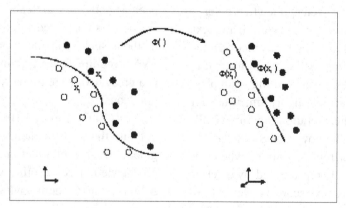

Although their exact probabilities may never be known, it is reasonable to assume that losing a gene or obtaining a new gene is relatively rare as compared to keeping the status quo. In Vert, the probability that an existing gene is retained at a tree branch (i.e., speciation) is set at 0.9, and the probability that a new gene is created at a branch is set at 0.1. It was further assumed that such a distribution remains the same at all branches for all genes. Even with these crude assumptions, in the cross validation experiments on those 2465 yeast genes, the tree kernel's classification accuracy already significantly exceed that of a kernel using just the dot product $x \cdot y = \Sigma_{i=1 \text{ to } 24} x_i y_i$.

Extended Phylogenetic Profiles

The tree-kernel approach's improvement at classification accuracy is mainly due to engineering the kernel functions. In Narra and Liao (2004, 2005) further improvement is attained by both data representations and kernel engineering. As the reader should have been convinced by now, these phylogenetic profiles contain more information than just the string of zeros and ones; the phylogenetic tree provides relationships among the bits in these profiles. In Narra and Liao, a two-step procedure is adopted to extend phylogenetic profiles with extra bits encoding the tree structure: (1) a score is assigned at each internal tree node; (2) the score labeled tree is then flatten into an extended vector. For an internal tree node in a phylogenetic tree, as it is interpreted as ancestor of the nodes underneath it, one way to assign a score for it is to take the average of the scores from its children nodes. This scoring scheme works top-down recursively until the leaves are reached: the score at a leaf is just the value of the corresponding component in the hierarchical profile. The same scoring scheme was also used in p-tree approach. Unlike p-Tree approach that keeps just the score the root node and thus inevitably causes information loss, the scores at all internal nodes are retained and then mapped

into a vector via a post-order tree traversal. This vector is then concatenated to the original profile vector forming an extended vector, which is called tree-encoded profile. The scheme works for both binary and real-valued profiles. In order to retain information, real value profiles for yeast genes are used; the binary profiles for tree-kernel are derived real value profiles by imposing a cutoff at E-values.

Given a pair of tree-encoded profiles x and y, the polynomial kernel is used classification:

$$K(x, y) = [1 + s\, D(x, y)]^d \qquad (6)$$

where s and d are two adjustable parameters. Unlike ordinary polynomial kernels, D(x, y) is not the dot product of vectors x and y, but rather, a generalized Hamming distance for real value vectors:

$$D(x, y) = \Sigma_{i=1 \text{ to } n} (S(|x_i - y_i|)) \qquad (7)$$

where the *ad hoc* function S has value 7 for a match, 5 for a mismatch by a difference less then 0.1, 3 for a mismatch by a difference less than 0.3, and 1 for a mismatch by a difference less than 0.5. The values in the *ad hoc* function S are assigned based on the E-value distribution of the protein dataset. The methods are tested on the same data set by using the same cross-validation protocols as in Vert. The classification accuracy of using the extended phylogenetic profiles with E-values and polynomial kernel generally outperforms the tree-kernel approach at most of the 133 functional classes of 2465 yeast genes in Vert.

MORE APPLICATIONS AND FUTURE TRENDS

We have seen in the last two sections some problems in bioinformatics and computational biology where relationships can be categorized as hierarchy, and how such hierarchical structure can be utilized to facilitate the learning. Because of the central role played by evolution theory in biology, and the fact that phylogeny is natively expressed as a hierarchy, it is no surprise that hierarchical profiling arises in many biological problems.

In Siepel and Haussler (2004), methods are developed that combine phylogenetic and hidden Markov models for biosequence analysis. Hidden Markov models, first studied as a tool for speech recognition, were introduced to bioinformatics field around 1994 (Krogh et al., 1994). Since then, hidden Markov models have been applied to problems in bioinformatics and computational biology including gene identification, protein homology detection, secondary structure prediction, and many more.

In sequence modeling, hidden Markov models essentially simulate processes along the length of the sequence, mostly ignoring the evolutionary process at each position. On the other hand, in phylogenetic analysis, the focus is on the variation across sequences at each position, mostly ignoring correlations from position to position along their length. From a hierarchical profiling viewpoint, each column in a multiple sequence alignment is a profile, not binary but 20ary, and provides a sampling for the current sequences. It is a very attractive idea to combine these two apparently orthogonal models. In Siepel and Haussler, a simple and efficient method was developed to build higher-order states in the HMM, which allows for context-sensitive models of substitution, leading to significant improvements in the fit of a combined phylogenetic and hidden Markov model. Their work promises to be a very useful tool for some important biosequence analysis applications, such as gene finding and secondary structure prediction.

Hierarchical relations also exist in applications where phylogeny is not the subject. For example, in Holme, Huss, and Jeong (2003), biochemical networks are decomposed into sub-networks. Because of the inherent non-local features pos-

sessed by these networks, a hierarchical analysis was proposed to take into account the global structure while doing decomposition. In another work (Gagner et al., 2004), the traditional way of representing metabolic networks as a collection of connected pathways is questioned: such representation suffers the lack of rigorous definition, yielding pathways of disparate content and size. Instead, they proposed a hierarchical representation that emphasizes the gross organization of metabolic networks in largely independent pathways and sub-systems at several levels of independence.

While hierarchical relations are widely existed, there are many applications where the topology of the data relation can not be described as a hierarchy but rather as a network. Large scale gene expression data are often analyzed by clustering genes based on gene expression data alone, though a priori knowledge in the form of biological networks is available. In Hanish, Zien, Zimmer, and Lengauer (2002), a co-clustering method is developed that makes use of this additional information and has demonstrated considerably improvements for exploratory analysis.

CONCLUSION

As we have seen, hierarchical profiling can be applied to many problems in bioinformatics and computational biology, from remote protein homology detection, to genome comparisons, to metabolic pathway clustering, wherever the relations among attribute data possess structure as a hierarchy. These hierarchical relations may be inherent or as an approximation to more complex relationships. To properly deal with such relationships is essential for clustering and classification of biological data. It is advisable to heed the relations that bear artificially hierarchical structure; hierarchical profiling on these relations may yield misleading results.

In many cases, the hierarchy, and the biological insight wherein embodied, can be integrated into the framework of data mining. It consequently facilitates the learning and renders meaningful interpretation of the learning results. We have reviewed the recent developments in this respect. Some methods treat hierarchical profile scoring as a tree comparison problem, some as an encoding problem, and some as a graphical model with a Bayesian interpretation. The latter approach is of particular interest, since most biological data are stochastic by nature.

A trend is seen in bioinformatics that combines different methods and models so the hybrid method can achieve a better performance. In the tree-kernel method, the hierarchical profiling and scoring are incorporated as kernel engineering task of the support vector machines. In sequence analysis, the phylogenetic techniques and hidden Markov models are combined to account for the relationships exhibited in sequences that either method alone can not handle properly. As more and more biological data with complex relationships being generated, it is reasonable to believe the hierarchical profiling will see more applications, either serve by itself as a useful tool for analyzing these data, or serve as a prototype for developing more sophisticated and powerful data mining tools.

REFERENCES

Altschul, S. F., Gish, W., Miller, W., Myers, E., & Lipman, D. J. (1990). Basic local alignment search tool. *Journal of Molecular Biology, 215,* 403-410.

Altschul, S. F., Madden, T. L., Schäffer, A. A., Zhang, J., Zhang, Z., Miller, W., et al. (1997). Gapped BLAST and PSI-BLAST: A new generation of protein database search programs. *Nucleic Acids Res., 25,* 3389-3402.

Bailey, T. L., & Elkan, C. P. (1995). Unsupervised learning of multiple motifs in biopolymers using EM. *Machine Learning, 21*(1-2), 51-80.

Burges, C. J. C. (1998). A tutorial on support vector machines for pattern recognition. *Data Mining and Knowledge Discovery, 2,* 121-167.

Cristianini, N., & Shawe-Taylor, J. (2000). *An introduction to support vector machines and other kernel-based learning methods.* Cambridge, UK: Cambridge University Press.

Dandekar, T., Schuster, S., Snel, B., Huynen, M., & Bork, P. (1999). Pathway alignment: application to the comparative analysis of glycolytic enzymes. *Biochemical Journal, 343,* 115-124.

Durbin, R., Eddy, S., Krogh, A., & Mitchison, G. (1998). *Biological sequence analysis: Probabilistic models of proteins and nucleic acids.* Cambridge, UK: Cambridge University Press.

Eisen, J. A. (2000). Horizontal gene transfer among microbial genomes: New insights from complete genome analysis. *Current Opinion in Genetics & Development, 10,* 606-611.

Felsenstein, J. (1989). PHYLIP—Phylogeny Inference Package (Version 3.2). *Cladistics, 5,* 164-166.

Forst, C. V., & Schulten, K. (2001). Phylogenetic analysis of metabolic pathways. *Journal of Molecular Evolution, 52,* 471-489.

Gaasterland, T., & Selkov, E. (1995). Reconstruction of metabolic networks using incomplete information. In *Proceedings of the Third International Conference on Intelligent Systems for Molecular Biology* (pp. 127-135). Menlo Park, CA: AAAI Press.

Gagneur, J., Jackson, D., & Casar, G. (2003). Hierarchical analysis of dependence in metabolic networks. *Bioinformatics, 19,* 1027-1034.

Gribskov, M., McLachlan, A. D., & Eisenberg, D. (1987). Profile analysis: Detection of distantly related proteins. *Proc. Natl. Acad. Sci. USA, 84* (pp. 4355-4358).

Grundy,W. N., Bailey, T. L., Elkan, C. P., & Baker, M. E. (1997). Meta-MEME: Motif-based hidden Markov Models of biological sequences. *Computer Applications in the Biosciences, 13*(4), 397-406.

Hanisch, D., Zien, A., Zimmer, R., & Lengauer, T. (2002). Co-clustering of biological networks and gene expression data. *Bioinformatics, 18,* S145-S154.

Haussler, D. (1999). *Convolution kernels on discrete structures* (Technical Report UCSC-CRL-99-10). Santa Cruz: University of California.

Henikoff, S., & Henikoff, J.G. (1994). Protein family classification based on search a database of blocks. *Genomics, 19,* 97-107.

Heymans, M., & Singh, A. J. (2003). Deriving phylogenetic trees from the similarity analysis of metabolic pathways. *Bioinformatics, 19,* i138-i146.

Holme, P., Huss, M., & Jeong, H. (2003). Sub-network hierarchies of biochemical pathways. *Bioinformatics, 19,* 532-538.

Jaakkola, T., Diekhans, M., & Haussler, D. (1999). Using the Fisher Kernel Method to detect remote protein homologies. In *Proceedings of the Seventh International Conference on Intelligent Systems for Molecular Biology* (pp. 149-158). Menlo Park, CA: AAAI Press.

Jaakkola, T., Diekhans, M., & Haussler, D. (2000). A discriminative framework for detecting remote protein homologies. *Journal of Computational Biology, 7,* 95-114.

Karp, P.D., Riley, M., Saier, M., Paulsen, I. T., Paley, S. M., & Pellegrini-Toole, A. (2000). The EcoCyc and MetaCyc databases. *Nucleic Acids Research, 28,* 56-59.

Krogh, A., Brown, M., Mian, I. S., Sjolander, K., & Haussler, D. (1994). Hidden Markov models in computational biology: Applications to protein modeling. *Journal of Molecular Biology, 235,* 1501-1531.

Liao, L., Kim, S., & Tomb, J-F. (2002). Genome comparisons based on profiles of metabolic pathways", In *The Proceedings of The Sixth International Conference on Knowledge-Based Intelligent Information & Engineering Systems* (pp. 469-476). Crema, Italy: IOS Press.

Liao, L., & Noble, W. S. (2003). Combining pairwise sequence similarity and support vector machines for detecting remote protein evolutionary and structural relationships. *Journal of Computational Biology, 10,* 857-868.

Liberles, D. A., Thoren, A., von Heijne, G., & Elofsson, A. (2002). The use of phylogenetic profiles for gene predictions. *Current Genomics, 3,* 131-137.

Lin, J., & Gerstein, M. (2002). Whole-genome trees based on the occurrence of folds and ortho-logs: Implications for comparing genomes on different levels. *Genome Research, 10,* 808-818.

Marcotte, E. M., Xenarios, I., van Der Bliek, A. M., & Eisenberg, D. (2000). Localizing proteins in the cell from their phylogenetic profiles. *Proc. Natl. Acad. Sci. USA,* 97, (pp. 12115-12120).

Mittelman, D., Sadreyev, R., & Grishin, N. (2003). Probabilistic scoring measures for profile-profile comparison yield more accurate shore seed alignments. *Bioinformatics, 19,* 1531-1539.

Narra, K., & Liao, L. (2004). Using extended phylogenetic profiles and support vector machines for protein family classification. In *The Proceedings of the Fifth International Conference on Software Engineering, Artificial Intelligence, Networking, and Parallel/Distributed Computing* (pp. 152-157). Beijing, China: ACIS Publication.

Narra, K., & Liao, L. (2005). Use of extended phylogenetic profiles with E-values and support vector machines for protein family classification. *International Journal of Computer and Information Science, 6*(1).

Nevill-Manning, C. G., Wu, T. D., & Brutlag, D. L.(1998). Highly specific protein sequence motifs for genome analysis. *Proc. Natl. Acad. Sci. USA, 95*(11), 5865-5871.

Noble, W. (2004). Support vector machine applications in computational biology. In B. Scholkopf, K. Tsuda, & J-P. Vert. (Eds.), *Kernel methods in computational biology* (pp. 71-92). Cambridge, MA: The MIT Press.

Overbeek, R., Larsen, N., Pusch, G. D., D'Souza, M., Selkov Jr., E., Kyrpides, N., Fonstein, M., Maltsev, N., & Selkov, E. (2000). WIT: Integrated system for high throughput genome sequence analysis and metabolic reconstruction. *Nucleic Acids Res., 28*, 123-125.

Pearson, W. (1990). Rapid and sensitive sequence comparison with FASTP and FASTA. *Meth. Enzymol., 183*, 63-98.

Pellegrini, M., Marcotte, E. M., Thompson, M. J., Eisenberg, D., & Yeates, T. O. (1999). Assigning protein functions by comparative genome analysis: Protein phylogenetic profiles. *Proc. Natl. Acad. Sci. USA, 96*, (pp. 4285-4288).

Rabiner, L. R. (1989). A tutorial on hidden Markov models and selected applications in speech recognition. *Proc. IEEE, 77*, 257-286.

Sadreyev, R., & Grishin, N. (2003). Compass: A tool for comparison of multiple protein alignments with assessment of statistical significance. *Journal of Molecular Biology, 326*, 317-336.

Scholkopf, B., & Smola, A. J. (2001). *Learning with kernels: Support vector machines, learning).* Cambridge, MA: The MIT Press.

Siepel, A., & Haussler, D. (2004). Combining phylogenetic and hidden Markov Models in biosequence analysis. *J. Comput. Biol., 11*(2-3), 413-428.

Smith, T. F., & Waterman, M. S.(1981). Identification of common molecular subsequences. *Journal of Molecular Biology, 147*, 195-197.

Vapnik, V. (1998). *Statistical Learning Theory*: *Adaptive and learning systems for signal processing, communications, and control.* New York: Wiley.

Vert, J-P. (2002). A tree kernel to analyze phylogenetic profiles. *Bioinformatics, 18*, S276-S284.

Watkins, C. (1999). Dynamic alignment kernels. In A. J. Smola, P. Bartlett, B. SchÄolkopf, & C. Schuurmans (Ed.), *Advances in large margin classifiers.* Cambridge, MA: The MIT Press.

Woese, C. (1987). Bacterial evolution. *Microbial Rev., 51*, 221-271.

Zhang, K., Wang, J. T. L., & Shasha, D. (1996). On the editing distance between undirected acyclic graphs. *International Journal of Foundations of Computer Science, 7*, 43-58.

Zhang, S., Liao, L., Tomb, J-F., Wang, J. T. L. (2002). Clustering and classifying enzymes in metabolic pathways: Some preliminary results. In *ACM SIGKDD Workshop on Data Mining in Bioinformatics,* Edmonton, Canada (pp. 19-24).

Chapter IX
Hierarchical Clustering Using Evolutionary Algorithms

Monica Chiş
Avram Iancu University, Romania

ABSTRACT

Clustering is an important technique used in discovering some inherent structure present in data. The purpose of cluster analysis is to partition a given data set into a number of groups such that data in a particular cluster are more similar to each other than objects in different clusters. Hierarchical clustering refers to the formation of a recursive clustering of the data points: a partition into many clusters, each of which is itself hierarchically clustered. Hierarchical structures solve many problems in a large area of interests. In this chapter a new evolutionary algorithm for detecting the hierarchical structure of an input data set is proposed. This problem could be very useful in economy, market segmentation, management, biology taxonomy, and other domains. A new linear representation of the cluster structure within the data set is proposed. An evolutionary algorithm evolves a population of clustering hierarchies. The proposed algorithm uses mutation and crossover as (search) variation operators. The final goal is to present a data clustering representation to find fast a hierarchical clustering structure.

INTRODUCTION

Clustering is an important technique used in the simplification of data sets or in discovering some inherent structure present in data. The purpose of cluster analysis is to partition a given data set into a number of groups such that data in a particular cluster are more similar to each other than objects in different clusters (Jain & Dubes, 1998).

Clustering in data mining is a discovery process that groups a set of data such that the intracluster similarity is maximized and the intercluster similarity is minimized. These discovered clusters can be used to explain the characteristics of the underlying data distribution and thus, serve as the foundation for other data-mining and analysis techniques.

Clustering is an important problem, with applications in areas such as data mining and knowledge discovery, data compression and vector quantization, and pattern recognition and pattern classification.

One-level or "flat" clustering gives no information about the relationship existing between clusters. Hierarchical clustering groups together data into a tree structure; thus, building a multilevel representation capable of revealing intercluster relationships.

Hierarchical clustering refers to the formation of a recursive clustering of the data points: a partition into many clusters, each of which is itself hierarchically clustered. This method is very useful for clustering purposes.

Hierarchical clustering constructs trees of clusters of objects in which any two clusters are disjoint, or one includes the other. The cluster of all objects is the root of the tree (Jain & Dubes, 1998; Johnson, 1967).

Agglomerative algorithms require a definition of dissimilarity between clusters: the most common ones are maximum or complete linkage, in which the dissimilarity between two clusters is the maximum of all pairs of dissimilarities between pairs of points in the different clusters; minimum or single linkage or nearest neighbor, in which the dissimilarity between two clusters is the minimum over all those pairs of dissimilarities; and average linkage, in which the dissimilarity between the two clusters is the average, or suitably weighted average, over all those pairs of dissimilarities.

Agglomerative algorithms begin with an initial set of singleton clusters consisting of all the objects. It proceeds by agglomerating the pair of clusters of minimum dissimilarity to obtain a new cluster, removing the two clusters combined from further consideration. This agglomeration step repeats until a single cluster containing all the observations is obtained. The set of clusters obtained along the way forms a hierarchical clustering (Jain & Dubes, 1998).

The problem of hierarchical clustering assumes a significant role in a variety of research areas such as data mining, pattern recognition, economics, biology, autonomous mobile robots, and so forth.

Standard methods for detecting hierarchical cluster structure of a set of objects are either divisive or agglomerative (Dumitrescu, 1999).

In this chapter, an evolutionary algorithm for detecting the hierarchical structure of a data set is proposed. A sequence describing a binary tree is used for representing the cluster hierarchy. A new representation of cluster hierarchy is proposed. This representation is linear and may describe binary tree structures. An evolutionary algorithm with mutation and crossover as variation operators is used. Binary tournament selection is considered.

The rest of the chapter is organized as follows. Section 2 gives an overview of related hierarchical clustering algorithms. Section 3 presents the hierarchical clustering algorithm. Section 4 gives some experimental results. Section 5 draws conclusions and directions for future work.

RELATED WORK

In this section, a brief description of some existing hierarchical clustering algorithms is made. The basic process of hierarchical clustering is described. Previous hierarchical clustering algorithms (ROCK, CURE, and CHAMELEON) are discussed.

Jonhson (1967) defined the basic process of hierarchical clustering. Given a set of N items to be clustered, and an $N*N$ distance (or similarity) matrix, the basic process of hierarchical clustering is described as follows:

1.   Start by assigning each item to a cluster so that if you have N items, you now have

N clusters, each containing just one item. Let the distances (similarities) between the clusters be the same as the distances (similarities) between the items they contain.

2. Find the closest (most similar) pair of clusters and merge them into a single cluster, so that now you have one cluster less.

3. Compute distances (similarities) between the new cluster and each of the old clusters.

4. Repeat Steps 2 and 3 until all items are clustered into a single cluster of size N. There is no point in having all the N items grouped in a single cluster, but once you have got the complete hierarchical tree, if you want *k* clusters you just have to cut the k–1 longest links.

Step 3 can be done in different ways (Johnson, 1967), which is what distinguishes *single- linkage* from *complete-linkage* and *average-linkage* clustering.

In *single-linkage* clustering, also called the *connectedness* or *minimum* method, the distance between one cluster and another cluster is considered to be equal to the shortest distance from any member of one cluster to any member of the other cluster. If the data consist of similarities, the similarity between one cluster and another cluster is considered to be equal to the greatest similarity from any member of one cluster to any member of the other cluster.

In *complete-linkage* clustering, also called the *diameter* or *maximum* method, the distance between one cluster and another cluster is considered to be equal to the greatest distance from any member of one cluster to any member of the other cluster.

In *average-linkage* clustering, the distance between one cluster and another cluster is considered to be equal to the average distance from any member of one cluster to any member of the other cluster.

A variation on average-link clustering is the UCLUS method of D'Andrade (D'Andrade, 1978),

which uses the median distance, which is much more outlier-proof than the average distance.

This kind of hierarchical clustering is called *agglomerative* because it merges clusters iteratively. There is also a *divisive* hierarchical clustering that does the reverse by starting with all objects in one cluster and subdividing them into smaller pieces. Divisive methods are not generally available, and rarely have been applied.

Agglomerative hierarchical clustering is a bottom-up clustering method where clusters have subclusters, which in turn have subclusters, and so forth. The classic example of this is species taxonomy.

Agglomerative hierarchical clustering (HAC) starts with one datum per cluster (singleton), then recursively merges the two clusters with the smallest distance between them into a larger cluster until only one cluster is left (Jain & Dubes, 1998). The hierarchy within the final cluster has the following properties:

- Clusters generated in early stages are nested in those generated in later stages.
- Clusters with different sizes in the tree can be valuable for discovery.

The advantages of agglomerative clustering are:

- It can produce an ordering of the objects, which may be informative for data display.
- Smaller clusters are generated, which may be helpful for discovery.

The disadvantages are of this method are the fact that no provision can be made for a relocation of objects that may have been "incorrectly" grouped at an early stage. The result should be examined closely to ensure it makes sense.

Use of different distance metrics for measuring distances between clusters may generate different results. Performing multiple experiments and comparing the results is recommended to support the veracity of the original results.

CURE and ROCK are clustering algorithms that belong to the class of agglomerative hierarchical clustering algorithms (Guha, Rastogi, & Shim, 1998).

CURE (**C**lustering **U**sing **RE**presentatives) measures the similarity of two clusters based on the similarity of the closest pair of representative points belonging to different clusters, without considering the internal closeness of two clusters involved (Guha et al., 1998).

ROCK (RObust Clustering using linKs) measures the similarity of two clusters by comparing the aggregate interconnectivity of two clusters against a user-specified static interconnectivity model, and thus ignores the potential variations in the interconnectivity of different clusters within the same dataset. Most of these algorithms breakdown when the data consist of clusters that are of diverse shape, densities, sizes, noise, and artifacts (Guha et al., 2000).

CHAMELEON is another hierarchical clustering algorithm that measures the similarity of two clusters based on a dynamic model. In the clustering process, two clusters are merged only if the interconnectivity and closeness (proximity) between two clusters are comparable to the internal interconnectivity of the clusters and closeness of items within the clusters (Karypis, Han, & Kumar, 1999).

Divisive hierarchical clustering is a top-down clustering method and is less commonly used. It works in a similar way to agglomerative clustering, but in the opposite direction. This method starts with a single cluster containing all objects, and then successively splits resulting clusters until only clusters of individual objects remain.

In Dumitrescu, Lazzerini, and Hui (2000), hierarchical data structure detection using evolutionary algorithms is proposed. A linear chromosome representation is used.

EVOLUTIONARY HIERARCHICAL CLUSTERING

In this section, the evolutionary hierarchical clustering (EvHiCA) algorithm is described. EvHiCA uses linear chromosome for solution representation.

Solution Representation

Let $X=\{x^1, x^2,..., x^p\}$ be a data set and d a distance on X. For applying an evolutionary algorithm for solving hierarchical classification problems, hierarchical clustering scheme is described. Each individual (candidate solution or chromosome) describes a hierarchy. A linear sequence is used for representing the cluster hierarchy. The sequence is translated into a binary classification tree. Each node of the tree represents a potential cluster (a class).

Classes of the classification tree are labeled by parsing each level from left to right and top to bottom. Root node is considered as a class containing the entire data set X. The root node has label 0.

Data points are assigned to the terminal classes only. A class is called *terminal* if it belongs to the front of binary tree. Unless otherwise stated, a class that has no descendent is a terminal class. Classes corresponding to nonterminal nodes are called nonterminal classes.

Solutions are represented by trees having a variable number of levels (variable depth). For a given problem, the maximum *tree depth* is specified by the parameter h_{max}. Hence, the binary tree, which describes the cluster hierarchy, has h_{max} +1 levels labeled from 0 to h_{max}. The level that contains only a class that represents the entire data set X is labeled by 0.

Each level h of the proposed binary classification tree contains a number of nodes, denoted by *levnod* and given by:

$$levnod = 2^h \qquad (3.1)$$

The maximum number of binary tree nodes (classes) (different from the root node) is *Nrnodes* and is given by,

$$Nrnodes = 2^{h_{max}+1} - 2 \qquad (3.2)$$

It is not necessary to occupy every class of this binary tree. Finally, *Nrnodes* not represented the number of classes of proposed hierarchies. The number of classes could be less than *Nrnodes* but not less than 2. The number of classes does not include the root node, which is labeled by 0 and contains the entire classification data set *X*.

The first step when apply an evolutionary algorithm for solving a problem is to find a suitable representation for candidate solutions. For applying an evolutionary algorithm to solve a hierarchical classification problem, a hierarchical representation scheme is built. Each candidate solution (individual) describes a clustering hierarchy. Representation indicates how data points are assigned to classes (tree nodes).

The individual length is equal to cardinality of classification data set, denoted by *p*.

An individual is represented as a vector:

$$c = (c_1, c_2, ..., c_p) \qquad (3.3)$$

where c_j is an integer number,

$$2^{h_{max}} - 1 \le c_j \le 2^{h_{max}+1} - 2 \qquad (3.4)$$

The value c_j of the gene *j* indicates the tree node to which the point x^j is assigned. x^j is assigned only to a terminal node (class). In this case, terminal nodes are all the nodes in the h_{max} level of the binary tree. The other nodes, called *nonterminal* nodes, are formed using the two descendents classes of each other by parsing each level less than h_{max} level from left to right and top to bottom.

Each node in the tree corresponds to a potential cluster. To assure that proposed hierarchical representation of *X* is a binary, well-balanced tree, the genes of the individual are represented in that way:

If *p* is an odd number, then at least $k = (p-1)/2$ genes of chromosome are given by,

$$2^{h_{max}} - 1 \le c_j \le 3 \cdot 2^{h_{max}-1} - 1, \qquad (3.5)$$

and the others are given by,

$$3 \cdot 2^{h_{max}-1} - 1 \le c_j \le 2^{h_{max}+1} - 2, \qquad (3.6)$$

If *p* is an even number, then at least $k = p/2$ genes of the chromosome are given by (3.5), and the others are given by (3.6).

In order to realize a feasible hierarchical representation scheme, a class that has a single descendent that contains points from data set will be a terminal class, and the descendent class will be eliminated from the hierarchical structure.

Example of Proposed Representation

Two examples of proposed representation are listed nest.

Example 1

Consider data set $X = \{x^1, x^2, x^3, x^4, x^5\}$ and $h_{max} = 2$. Thus we may have binary trees with maximum $(2^3 - 2)$ nodes. Consider the individual,

c = (3, 4, 4, 5, 6).

This individual describes the hierarchy depicted in Figure 1.

Example 2

Let us consider data set $X = \{x^1, x^2, ..., x^8, x^9, x^{10},\}$ and $h_{max} = 3$. The maximum binary trees nodes are 14. Consider the individual,

$c = (7, 8, 9, 10, 10, 11, 12, 13, 13, 14)$.

Figure 1. Tree encoding of the individual c = (3, 4, 4, 5, 6)

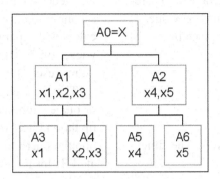

Figure 2. Tree encoding of the individual c = (7,8,9,10,10,11,12,13,13,14)

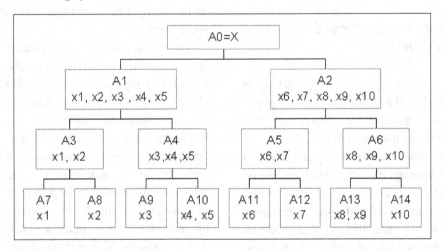

This individual describes the hierarchy depicted in Figure 2.

Fitness Function

The chromosomes have to be evaluated for comparison. The evaluation is done by means of a quality (or fitness) function.

The data set $X=\{x^1,x^2,...,x^p\}$ is considered. The aim is to find the hierarchy that represents the best cluster structure of X. For this purpose, a fitness function to measure the hierarchy quality is needed. Fitness function must have the best value for the hierarchy that is the most representative for input data set.

The fitness function is used as a basis for selecting solutions (individuals) for recombination.

The clustering criterion is based around minimising the sum of squared distances of objects from their cluster (class) centres (prototypes) divided by the cardinality of each cluster (class).

For this purpose, Euclidean distance is used. The optimal clusters hierarchy of data set X corresponds to the minimal fitness function value.

The prototype of a class A_i is given by,

$$l^i = \frac{\sum_{x \in A_i} x}{p_i} \qquad (3.7)$$

where p_i is the cardinality of A_i.

Let n be the number of classes. The fitness function listed next is used.

$$f(c) = \sum_{i=1}^{n} \sum_{x_k \in A_i} \frac{d^2(L^i, x^k)}{p_i} \qquad (3.8)$$

Fitness function is minimised.

Evolutionary Hierarchical Clustering Algorithm (EvHiCA)

Using the proposed solution representation, an evolutionary algorithm is used to evolve a population of clustering hierarchies. Search operators used are crossover and mutation.

One of the issues in evolutionary algorithms is the relative importance of two search operators: mutation and crossover. Genetic algorithms and genetic programming stress the role of crossover, while evolutionary programming and evolution strategies stress the role of mutation. The existence of many different forms of crossover further complicates the issue. Despite theoretical analysis, it appears difficult to decide a priori which form of crossover to use, or even if crossover should be used at all. One possible solution to this difficulty is to have the EA be self-adaptive, that is, to have the evolutionary algorithm dynamically modify which forms of crossover to use and how often to use them, as it solves a problem (Spears, 1995; Spears & De Jong, 1991; Syswerda, 1989).

EvHiCa uses a uniform crossover operator. Uniform crossover does not use predefined crossover points. For each gene of an offspring,

a global parameter indicates the probability that this gene should come from either the first or the second parent. Each position of an offspring is calculated separately (Bäck, Fogel, & Michalewicz, 1997; Dumitrescu, Lazzerini, Jain, & Dumitrescu, 1999).

The mutation probability stands for a parameter of our evolutionary algorithm. Consider p_m the mutation rate. For each gene of the chromosome population, a uniform random number q is generated. If for the i-th gene, the condition $q < p_m$ is fulfilled, that gene is selected for mutation.

Binary tournament selection is considered. Binary tournament selection implies that two individuals directly compete for selection. Tournament selection used is without reinsertion of the competing individuals into the original population.

Evolutionary hierarchical clustering algorithm (EvHiCA) starts with a randomly chosen population of individuals. The number of individuals is chosen first. In the second step, the maximum number of levels for the hierarchy is selected. The following steps are repeated until a termination condition is reached. Two parents are chosen at each step using binary tournament selection. The selected individuals are recombined with a fixed crossover probability p_c. One offspring is obtained by recombining two parents. The offspring is mutated and the best of them replaces the worst individual in population.

The algorithm keeps the best individual obtained up to each generation t. The problem solution is the best individual obtained from the best individual of each generation. The best individual is the individual that has the minimum value for the fitness function (Bäck et al., 1997).

NUMERICAL EXPERIMENTS

In this section, experimental evaluation of EvHiCA is presented.

Data Sets

EvHiCA hierarchical clustering performance is evaluated on two different groups of data. The data set is described as follows:

1. **2-D synthetic data:** This group of two-dimensional synthetic data sets is proposed to exhibit features of hierarchical data structure.
2. **Real data:** Four real data sets from machine learning repository (Blake & Merz, 1998) are used, which are summarized in Table 1. The used data are not necessarily designed

for unsupervised (clustering) methods, but we include these data because they allow us to be more confident about any general conclusions according to our method.

Algorithm Parameters

The parameters of EvHiCa algorithm are given in Tables 2 and 3.

Experimental Results

EvHiCA is a hierarchical clustering algorithm; this means that it gives the number of clusters

Table 1. Data sets

Data set	No of records	No of clusters	Number of features
Iris	150	3	4
Wine	178	3	13
Dermatology	366	6	34
Breast Cancer Wisconsin	699	2	10

Table 2. EvHiCA algorithm parameters for solving hierarchical clustering problems

Parameter	Value
Population Size	According to dataset used
Chromosome length	The cardinality of dataset
Crossover probability	0.9
Crossover type	Uniform crossover
Mutation probability	0.5, 0.1, 0.8

Table 3. EvHiCA algorithm parameters for solving hierarchical clustering problems with a large dataset

Parameter	Value
Population Size	250
Chromosome length	The cardinality of dataset
Crossover probability	0.9
Crossover type	Uniform crossover
Mutation probability	0.9

and the point in each cluster and also produces a dendrogram of possible clustering solutions at different levels of granularity.

For the data sets used in the 2-D space, the accuracy of the solution is 90 %. This means that the clustering is correct in most cases, but sometimes the hierarchy is not good. Generally, the number of misclassified points increases with the number of points and decreases with the number of generations.

For the synthetic dataset presented in Figure 3, EvHiCA gives the following dendrogram.

For this kind of data set, the hierarchy is correctly established and the data points are correctly classified.

The results of the real datasets are discussed as follows.

Figure 3. Dataset 1 in the 2-D space

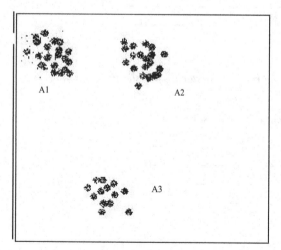

Figure 4. Dendrogram for datasets in figure 3

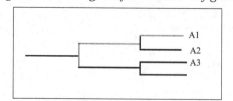

Iris dataset is a classical dataset used in discrimination tasks (Fisher, 1936). For Iris data set, a perfect clustering would show 50 in each of the three clusters and all of same species were in the same cluster. Using our algorithm, the results are, for *Setosa,* all 50 cases are clustered in the same cluster; for *Virginica,* 42 cases were in cluster with this kind of data and 8 was misclassified in cluster. *Versicolor* showed 23 correctly classified, but 27 misclassified as *Virginica.*

It is very important to establish the maximum level, which is a parameter of the proposed algorithm, according to number of points in the classification data set. The algorithm is faster and the results are correctly classified after a great number of generations.

The Wine data represents 13 different chemical constituents of 178 Italian wines derived from three different cultivars. The data of the Wine set is normalized to facilitate the calculation of Euclidian distance. Wine datasets consist of three clusters, 178 data points, and each data point has 13 attributes. The hierarchical clustering result is displayed as a dendrogram. The data that belong to the same cluster will be in the same subtree. Figure 3 presents the clustering results of the application of EvHiCA on the Wine data. The label A, B, C represents three kinds of wine. Figure 5 shows that 2 class "A" Wine data have been misclustered, 14 class of "B" Wine data have been misclustered, and 4 class "C" Wine data have been misclustered in the hierarchical trees result from our clustering algorithm.

Experimental results with Dermatology dataset and Breast Cancer Wisconsin dataset are analyzed (Wolberg & Mangasarian, 1990). The Breast Cancer Wisconsin data has a total number of instances in this dataset of 699. Number of attributes is 10. Records belong to one of two classes. The classes are benign and malignant. Four hundred and fifty eight records belong to class benign and 241 records belong to class malignant.

Figure 5. The EvHiCA hierarchical clustering result of Wine data set

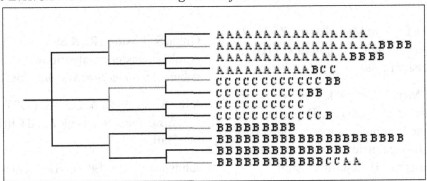

The clustering result is displayed as a dendrogram. The clustering results with EvHiCA show that 106 (23.14 %) of class benign are misclustered and 37 of malignant classes are misclustered. (15.35 %).

The differential diagnosis of erythmato-sqamous is a real problem in dermatology. Types share the clinical features of erythema and scaling, with very few differences (Wolberg & Mangasarian, 1990). Dataset contains 34 attributes, out of which 33 are linear and one is nominal. Dataset consists of 366 records, classified into six classes of diseases. Total records belonging to each class are 112, 61, 72, 49, 52, and 20, respectively. Distribution of datasets among classes is proportional, and thus all the classes are subdivided into four different subsets.

The EvHiCA find after 150 generations these four subsets in a correct way. Only 5-10 % of each class is incorrectly classified.

Solution accuracy generally increases with the number of generations. Usually, it detects structures that represent well-balanced binary hierarchies.

CONCLUSION AND FUTURE RESEARCH

In this chapter, a new evolutionary technique for detecting hierarchical structure in a dataset was presented.

The importance of this chapter is the new representation of the hierarchical structure and the fitness function.

Further research will explore different fitness functions and restrictions in order to provide the ability to handle large data collections, and will try to adapt the genetic operators for increasing the results. A correspondence in the maximum level of the proposed binary classification tree and the cardinality of data set will be finding. The maximum level will be encoding in the chromosome representation.

Another future work is to define special genetic operators that improve the hierarchical structures scheme.

REFERENCES

Bäck, T., Fogel, D. B., & Michalewicz, Z. (1997). *Handbook of evolutionary computation.* Oxford: Oxford University Press.

Blake, C., & Merz, C. (1998). *UCI repository of machine learning databases.* Department of Information and Computer Sciences, University of California, Irvine Retrieved from http://www.ics.uci.edu/~mlearn/MLRepository.html

Chiş, M. (2000). A new evolutionary hierarchical clustering technique. In Babeş-Bolyai University, *Research Seminars, Seminar on Computer Science* (pp. 13-20).

Chiş, M., & Dumitrescu, D. (2002). Evolutionary hierarchical clustering for data mining. In *Proceedings of the Symposium "Zilele Academice Clujene", Computer Science Section, 14-22 June 2002, Seminar on Computer Science* (pp. 12-18).

D'Andrade, R. (1978). U-statistic hierarchical clustering. *Psychometrika, 4,* 58-67.

Dumitrescu, D. (1999). *Mathematical foundations of the classification theory.* Bucharest, Romania: Romanian Academy Press.

Dumitrescu, D., Hui, L., & Lazzerini, D. (1998). Hierarchical data structure detection using evolutionary algorithms, (2), 3-12.

Dumitrescu, D., Lazzerini, B., Jain, L., & Dumitrescu, A. (1999). *Evolutionary computation.* Boca Raton, FL: C.R.C. Press.

Fisher, R. A. (1936). The use of multiple measurements in taxonomic problems. *Annals of Eugenics, 7,* 179-188.

Guha, S., Rastogi, R., & Shim, K. (1998). CURE: An efficient clustering algorithm for large databases. In *Proccedings of ACG SIGMOD International Conference on Management of Data* (pp. 73-82).

Guha, S., Rastogi, R., & Shim, K. (2000). ROCK: A robust clustering algorithm for categorical attributes. *Information Systems, 25*(5), 345-366.

Jain, A. K., & Dubes, R. C. (1998). *Algorithms for clustering data.* Englewood Cliffs, NJ: Prentice Hall.

Johnson, S. C. (1967). Hierarchical clustering schemes. *Psychometrika, 2,* 241-254.

Karypis, G., Han, E., & Kumar, V. (1999). CHAMELEON: A hierarchical clustering algorithm using dynamic modeling. *IEEE Computer, 32*(8), 68-75.

Spears, W. M. (1995). Adapting crossover in evolutionary algorithms. In J. R. McDonnell, R. G. Reynolds, & D. B. Fogel (Eds.), *Proceedings of the Fourth Annual Conference on Evolutionary Programming* (pp.367-384). Cambridge: MIT Press.

Spears, W. M., & De Jong, K. A. (1991). On the virtues of parameterized uniform crossover. In R. Belew & L. Booker (Eds.), *Proceedings of the Fourth International Conference on Genetic Algorithms* (pp. 230-236). San Mateo, CA: Morgan Kaufmann.

Syswerda, G. (1989). Uniform crossover in genetic algorithms. In J. Schaffer (Ed.), *Proceedings of the Third International Conference on Genetic Algorithms* (pp. 2-9), Los Altos: Morgan Kaufmann Publishers.

Wolberg W. H., & Mangasarian, O. L. (1990). Multisurface method of pattern separation for medical diagnosis applied to breast cytology, In *Proceedings of the National Academy of Sciences, 87* (pp. 9193-9196).

Chapter X
Exploratory Time Series Data Mining by Genetic Clustering

T. Warren Liao
Louisiana State University, USA

ABSTRACT

In this chapter, we present genetic-algorithm(GA)-based methods developed for clustering univariate time series with equal or unequal length as an exploratory step of data mining. These methods basically implement the k-medoids algorithm. Each chromosome encodes, in binary, the data objects serving as the k-medoids. To compare their performance, both fixed-parameter and adaptive GAs were used. We first employed the synthetic control-chart data set to investigate the performance of three fitness functions, two distance measures, and other GA parameters such as population size, crossover rate, and mutation rate. Two more sets of time series with or without a known number of clusters were also experimented: one is the cylinder-bell-funnel data and the other is the novel battle simulation data. The clustering results are presented and discussed.

INTRODUCTION

Before prediction models can be built in data mining or knowledge discovery, it is often advisable to first explore the data. Clustering is known to be a good exploratory data-mining tool. The goal of clustering is to create structure for unlabeled data by objectively forming data into homogeneous groups, where the within-group object similarity and the between-group object dissimilarity are

optimized. The bulk of clustering analyses has been performed on data associated with static features, that is, feature values that do not change with time, or the changes are negligible.

Two major classes of clustering methods are partitioning and hierarchical clustering. Well-known partitioning-based clustering methods include k-means (MacQueen, 1967), k-medoids (Kaufman & Rousseeuw, 1990), and the corresponding fuzzy versions: fuzzy c-means

(FCM) (Bezdek, 1987) and fuzzy *c*-medoids (Krishnapuram, Joshi, Nasraoui, & Yi, 2001). Hierarchical clustering methods are either of the agglomerative type or the divisive type. Lately, soft computing technologies, including neural networks and genetic algorithms, have emerged as another class of clustering techniques. Two prominent methods of the neural network approach to clustering are competitive learning and self-organizing feature maps. Most genetic clustering methods implement the spirit of partitioning methods, especially the *k*-means algorithm (Krishna & Murty, 1999; Maulik & Bandyopadhyay, 2000), and the fuzzy *c*-means algorithm (Hall, Özyurt, & Bezdek, 1999).

Just like static feature data, forming groups of similar time series given a set of unlabeled time series is often desirable. These unlabeled time series could be monitoring data collected during different periods from a particular process, or from several different processes. These processes could be natural, engineered, business, economical, or medical related. Unlike static feature values, the time series of a feature consists of dynamic values, that is, values changed with time. This greatly increases the dimensionality of the problem and calls for somewhat different, and often more complicated, clustering methods. This study will focus only on time-series data.

In surveying work related to time-series clustering, Liao (2005) distinguished three different time-series clustering approaches: those working with full data either in the time or frequency domain, those working with extracted features, and model-based approaches with models built from the raw data. An example of the first approach is Golay *et al.* (Golay, Kollias, Stoll, Meier, Valavanis, & Boesiger, 1998). They applied the fuzzy *c*-means algorithm to provide the functional maps of human brain activity on the application of a stimulus. In their study, three different distances (the Euclidean distance and two cross-correlation-based distances) were

alternately used for comparison purposes. Goutte *et al.* (Goutte, Toft, & Rostrup, 1999) and Fu *et al.* (Fu, Chung, Ng, & Luk, 2001) took the feature-based approach. Goutte *et al.* clustered functional magnetic resonance imaging (fMRI) time series in groups of voxels with similar activations using two algorithms: *k*-means and Ward's hierarchical clustering. The cross-correlation function, instead of the raw fMRI time series, was used as the feature space. Fu *et al.* described the use of self-organizing maps for grouping similar temporal patterns dispersed along the time series. Two enhancements were made: consolidating the discovered clusters by a redundancy removal step, and introducing the perceptually important point-identification method to reduce the dimension of the input data sequences.

Three model-based time-series clustering methods are described next. Li and Biswas (1999) described a clustering methodology for temporal data using hidden Markov model representation with a sequence-to-model likelihood distance measure. The temporal data was assumed to have the Markov property. Time series were considered similar when the models characterizing individual series were similar. Policker and Geva (2000) presented a model for nonstationary time series with time varying mixture of stationary sources, comparable to the continuous hidden Markov model. Fuzzy clustering methods were applied to estimate the continuous drift in the time-series distribution, and the resultant membership matrix was given an interpretation as weights in a time varying, mixture probability distribution function. Kalpakis *et al.* (Kalpakis, Gada, & Puttagunta, 2001) studied the clustering of ARIMA time-series by using the Euclidean distance between the Linear Predictive Coding Cepstra of two time-series as their dissimilarity measure and the Partition around Medoids (PAM) method as the clustering algorithm.

To the best of our knowledge, the only study that applied genetic algorithms to cluster time series is Baragona (2001). He evaluated three metaheuristic methods to partition a set of time series into clusters in such a way that (1) the maximum absolute cross-correlation value between each pair of time series that belong to the same cluster is greater than some given threshold, and (2) the k-min cluster criterion is minimized with a specified number of clusters. The cross-correlations were computed from the residuals of the models of the original time series. Among all methods evaluated, Tabu search was found to perform better than single linkage, pure random search, simulation annealing, and genetic algorithms, based on a simulation experiment on 10 sets of artificial time series generated from low-order univariate and vector ARMA models. For each time series, 300 observations were generated.

This study proposes k-medoids-based genetic algorithms for clustering time-series data of equal or unequal length as an exploratory step of data mining. Though few selected time series were tested in this study due to space constraint, the methods proposed can definitely be generalized to time-series data in other domains. We chose to directly process raw data to avoid the need either to extract features or to fit some appropriate models for the data at hand. Our GA differs from that of Baragona in three major aspects: (1) working with the original data rather than the residuals, (2) capable of handling unequal length of time series by using the dynamic time warping distance rather than the cross-correlation function, and (3) implementing different GAs. In the next two sections, the proposed genetic clustering method and the distance measures used are presented. The subsequent three sections present the test results of the synthetic control-chart data, cylinder-bell-funnel data, and battle simulation data, respectively. A discussion then follows and finally, the chapter is concluded.

GENETIC CLUSTERING OF TIME-SERIES DATA

Genetic algorithms have the following elements: population of chromosomes, selection according to fitness, crossover to produce new offspring, and random mutation of new offspring (Mitchell, 1996). In this section, the proposed genetic algorithms for clustering time-series data are detailed element by element.

In summary, four different chromosome representations have been employed by the genetic clustering techniques that implemented either the k-means or fuzzy c-means algorithm. They include integer-coded cluster for each datum with length equal to the number of data points, as in Krishna and Murty (1999); real-coded cluster centers, as in Maulik and Bandyopadhyay (2000) and Bandyopadhyay and Maulik (2002); binary-coded cluster centers (in gray coding), as in Hall *et al.* (1999); and binary-coded representation of p-median problems with length equaling to the number of data points, in which a digit value of "1" denotes a median and "0" not a median (Lorena & Furtado, 2000). These chromosome representation methods can be extended to time-series data, though they were initially designed for static feature data.

We implemented the k-medoids algorithm rather than k-means (or fuzzy c-means) because of the difficulty involved in defining the cluster centers for time series with unequal length. Since the data dimension is relatively large in our application, either binary-coded or real-coded cluster centers are inappropriate. Therefore, we elected to use a binary-coded representation of data objects serving as the cluster medoids. (The other alternative is integer-coded representation, which we will investigate, and hope to share the results once they become available). The prespecified number of cluster medoids and the number of digits used to represent each medoid together determine the chromosome length.

Each chromosome in the population is evaluated in two steps: first distributing each data point to the closest medoid according to some distance measure, and then computing the fitness value. The dynamic time warping distance, d_{DTW}, is chosen because of its ability to handle time series with varying lengths. More details are presented in the next section. The cluster of each datum is determined to be the one medoid closest to it, based on the nearest neighbor concept. Three fitness functions, as given next, are evaluated in this study.

1. 10^6 / (TWCV × (1 + nbr0 × a large integer)), where TWCV and nbr0 denote the total within-cluster distance and number of clusters with zero members, respectively. Krishna and Murty (1999) used the total within-cluster distance for static feature values but not time series. Our implementation differs from theirs also in the distance measure (dynamic time warping instead of Euclidean), the clustering algorithm (*k*-medoids instead of *k*-means), the chromosome representation, and other GA details. Let v_i, i =1, …, c be the cluster medoids representing cluster C_i, i =1, …, c and x_j, j = 1,…, n be the data vectors. If $x_j \in C_i$, then w_{ij} = 1; else, w_{ij} = 0. In this study, the TWCV is computed as:

$$TWCV = \sum_{i=1}^{c} WCV_i = \sum_{i=1}^{c}\sum_{j=1}^{n} w_{ij} d_{DTW}(v_i, x_j).$$

(1)

2. 10^6 / (DB × (1 + nbr0 × a large integer)), where DB is the Davies-Bouldin index that was initially proposed as a measure of the validity of the clusters by Davies and Boulden (1979) and later used by Bandyopadhyay and Maulik (2002) in their study of genetic clustering of satellite images. Let n_i denote

the number of data points in cluster i. For this study, the DB index is modified as:

$$DB = \frac{1}{c}\sum_{i=1}^{c} \max_{j=1, j \neq i}^{c} \left\{ \frac{WCV_i / n_i + WCV_j / n_j}{d_{DTW}(v_i, v_j)} \right\}.$$

(2)

3. 10^6 / (R_m(V) × (1 + nbr0 × a large integer)), where R_m(V) is the reformulated FCM functional (Hall *et al.*, 1999). In the following equation, we replace the original Euclidean distance with the DTW distance and m is the fuzzy weight (m>1).

$$Rm(V) = \sum_{j=1}^{n}\left(\sum_{i=1}^{c} d_{DTW}(v_i, x_j)^{\frac{1}{(1-m)}} \right)^{1-m}.$$

(3)

Note that each fitness function includes a penalty term to discourage the formation of clusters with zero members ("a large integer" was consistently set at 10 in this study). For comparing time-series data with equal length, Euclidean distance, d_E, was also implemented. In this case, d_{DTW}, in the previous equations, is replaced by d_E.

Standard roulette-wheel selection is used to reproduce offspring for the next generation. Each current chromosome in the population has a roulette-wheel slot, sized in proportion to its fitness. Chromosomes with a higher fitness value thus have a higher probability of contributing more offspring. Pairs of chromosomes are randomly chosen to perform the one-point crossover operation according to the specified crossover rate. The mutation is then performed to flip some of the bits in a chromosome from "0" to "1" or "1" to "0," according to the specified rate. If the resultant chromosome contains an invalid cluster medoid, its fitness is then set to zero to prevent it surviving to the next generation (a simple repair procedure to take care of the infeasible solutions).

The GA process has the following steps:

- Set the generation value to zero.
- Initialize the population.
- Evaluate the population.
- While the maximum number of generations is not reached,
- Select chromosomes.
- Perform the crossover operation.
- Perform the mutation operation.
- Evaluate the new pool of chromosomes.
- Increment the generation value.

Several GA parameters need to be set for the GA process to run. They include the population size, s, the crossover rate, p_c, the mutation rate, p_m, and the maximum number of generations, g_{max}. In addition, a different selection strategy could be used. Proper selection of these parameters greatly affects the GA behavior that is strongly determined by the balance between exploration (to explore new and unknown areas in the search space) and exploitation (to make use of knowledge acquired by exploration to reach better positions in the search space). Most GA studies choose these parameters by trial and error, without a systematic investigation.

Attempts to find the optimal and general set of parameters have been made by testing a wide range of problems (Grefenstette, 1986). To determine the relevancy and relative importance of these parameters, Rojas *et al.* (Rojas, González, Pomares, Merelo, Castillo, & Romero, 2002) applied the analysis of the variance (ANOVA) technique. The response variables used to perform the statistical analysis were the maximum fitness in the last generation that measures the capacity to find a local/global optimum, and the average fitness in the last generation that measures the diversity in the population. In terms of the best solution, all variables were found significant, with the first three most significant ones being the selection operator, the population size, and the type of mutation. Regarding the diversity,

the significant variables in descending order were the type of selection operators, the mutation rate, the mutation type, and the number of generations. These results were obtained based on their tests on a 0/1 knapsack problem, the Riolo function, the prisoner's dilemma problem, and three Michalewicz's functions.

Another school of approaches for setting GA parameters is using some mechanism to adapt them depending upon the state of the GA learning process, instead of fixing them from the outset. Srinivas and Patnaik (1994) proposed the adaptive genetic algorithm (AGA) for multimodal function optimization to realize the dual goals of maintaining diversity in the population and sustaining the convergence capacity of the GA. The AGA varies the probabilities of crossover and mutation depending upon the fitness values of the solutions. Let f_{max} and f_{avg} be the maximum fitness and the average fitness of the entire population, f' be the larger of the fitness values of the solutions to be crossed, and f be the fitness values of the solutions to be mutated. The expressions for p_c and p_m are given as,

$$p_c = \begin{cases} k_1(f_{max} - f')/(f_{max} - f_{avg}), & if\ f' \geq f_{avg} \\ k_3, & if\ f' < f_{avg} \end{cases}$$

(4)

and

$$p_m = \begin{cases} k_2(f_{max} - f)/(f_{max} - f_{avg}), & if\ f \geq f_{avg} \\ k_4, & if\ f < f_{avg} \end{cases}$$

(5)

where k_1, k_2, k_3, $k_4 \leq 1.0$. They set k_2 and k_4 to 0.5 to ensure the disruption of those solutions with average or subaverage fitness. To force all solutions with a fitness value less than or equal to the average fitness to undergo crossover, they assigned k_1 and k_3 a value of 1.

Eiben *et al.* (Eiben, Hinterding, & Mizhalewicz, 1999) classified parameter control (or parameter adaptation) studies based on two aspects: how the mechanism works and what component of the evolutionary algorithms (that includes GA) is affected by the mechanism. They classified parameter control mechanisms into three categories: deterministic in the sense that parameter values are altered by some deterministic rule, adaptive by using feedback from the search to determine the direction and/or magnitude of the change, and self-adaptive by encoding the parameters into the chromosomes that undergo mutation and recombination. They identified six components being adapted: representation, evaluation function, mutation operators and their probabilities, crossover operators and their probabilities, parent selection, and replacement operator. Most previous parameter adaptation studies used one mechanism to adapt one or two components. For example, the work of Srinivas and Patnaik (1994) employed an adaptive mechanism to vary two components: mutation rate and crossover rate. To date, there are insufficient research studies and results to conclude how much parameter control is most useful. Herrera and Lozano (2003) reviewed different aspects of fuzzy adaptive genetic algorithms (FAGA) from three points of view: design, taxonomy, and future directions. The steps for designing the fuzzy logic controller used by FAGAs were shown with an example. They categorized FAGAs based on two criteria: how the rule base is obtained, and the level where the adaptation takes place. They also discussed future directions and some challenges for FAGA research.

The parameter adaptation approaches have become more popular than the selection approaches as one gradually realizes the difficulty in coming up with a general rule, and that different problems really require different GA parameters for satisfactory performance. Therefore, this study will employ a parameter adaptation approach. Specifically, we employ the AGA proposed by Srinivas and Patnaik (1994) to adapt both the crossover rate and the mutation rate (or only the mutation rate) in order to evaluate its performance in clustering time-series data. Proposing a new

parameter adaptation method is beyond the scope of this study. Nevertheless, both the results with adapted parameters and those with fixed parameters are obtained and compared.

SIMILARITY/DISTANCE MEASURES

One key issue in clustering is how to measure the similarity between two data objects being compared. For static feature values, Euclidean distance or generalized Mikowski distance is often used. The Euclidean distance has been used to measure the distance between two time series of the same length, for example, by Pham and Chan (1998). The Euclidean distance or generalized Mikowski distance is applicable only when the lengths of time-series data are equal, which is not the case for the battle simulation data to be studied in the sequel. Therefore, we resort to the dynamic-time-warping distance that is known capable of coping with unequal time series.

Dynamic time warping (DTW) is a generalization of classical algorithms for comparing discrete sequences to sequences of continuous values. Given two time series, $Q = q_1, q_2, ..., q_i, ..., q_n$ and $C = c_1, c_2, ..., c_j, ..., c_m$, DTW aligns the two series so that their difference is minimized. To this end, an n by m distance matrix was used in which the (i, j) element contains the distance $d(q_i, c_j)$ between two points q_i and c_j that is often measured by the Euclidean distance. A warping path, $W = w_1, w_2, ..., w_k, ..., w_K$ where $\max(n, m) \leq K \leq m+n-1$, is a set of matrix elements that satisfies three constraints: boundary condition, continuity, and monotonicity. The boundary condition constraint requires the warping path to start and to finish in diagonally opposite corner cells of the matrix. That is $w_1 = (1, 1)$ and $w_K = (n, m)$. The continuity constraint restricts the allowable steps to adjacent cells. The monotonicity constraint forces the points in the warping path to be monotonically spaced in time. Of interest is the warping path that has the minimum distance between the two series.

Mathematically,

$$d_{DTW} = \min \frac{\sum_{k=1}^{K} w_k}{K}. \qquad (6)$$

Dynamic programming can be used to effectively find this path by evaluating the recurrence function given as equation (7), which defines the cumulative distance as the sum of the distance of the current element and the minimum of the cumulative distances of the adjacent elements:

$$d_{cum}(i,j) = d(q_i, c_j)$$
$$+ \min\{d_{cum}(i-1,j-1), d_{cum}(i-1,j), d_{cum}(i,j-1)\} \qquad (7)$$

CLUSTERING RESULTS OF CONTROL-CHART DATA

The genetic clustering methods were first applied to clustering 30 synthetic control-chart data, taken from the UCI Data Mining Archive (http://kdd.ics. uci.edu/), that were initially generated by Pham and Chan (1998). Figure 1 shows the 30 synthetic control-chart data used in this study. There are six known patterns (or clusters): normal, cyclic, increasing trend, decreasing trend, upward shift, and downward shift. For each pattern, there are five time series. It is not difficult to see that there is some overlapping between similar patterns (adjacent clusters), for example, between increasing trend and upward shift. We were able to compute the clustering accuracy rates for the control-chart data set because the ground truth is known. Just like any clustering algorithm, the proposed genetic clustering algorithm arbitrarily labels each cluster. This inconsistent labeling makes the accuracy-checking task somewhat tedious.

Table 1 summarizes the clustering results of using fixed parameter GAs. The fixed GA parameters include population size of 20, maximum generation of 30, crossover rate of 0.8, and three mutation rates: 0.01, 0.05, and 0.1.

For each combination of GA parameters, five repetitions were made with different random seeds. Among all the fixed parameter GAs tested, the highest average clustering accuracy of 0.733 was obtained when using the TWCV-based fitness function, DTW distance measure, and mutation rate of 0.01. This particular combination also produced the highest clustering accuracy of 86.7% among all runs executed for this dataset.

Figure 1. 30 synthetic control-chart data

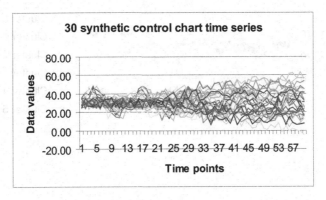

Table 1. Clustering results of control-chart data by fixed parameter GAs

Fitness Function	Distance	Mutation Rate	Run Accuracy	Avg. Accuracy
TWCV	Euclidean	0.01	.667, .667, .8, .567, .667	.673
TWCV	Euclidean	0.05	.633, .667, .7, .7, .633	.667
TWCV	Euclidean	0.1	.667, .733, .633, .7, .7	.687
TWCV	DTW	0.01	.667, **.867**, .767, .7, .667	**.733**
TWCV	DTW	0.05	.667, .7, .6, .7, .667	.667
TWCV	DTW	0.1	.6, .733, .633, .8, .667	.687
DB	Euclidean	0.01	.4, .6, .467, .533, .467	.493
DB	Euclidean	0.05	.433, .6, .433, .467, .433	.473
DB	Euclidean	0.1	.433, .4, .467, .5, .467	.453
DB	DTW	0.01	.567, .633, .5, .7, .667	.601
DB	DTW	0.05	.6, .6, .467, .633, .567	.573
DB	DTW	0.1	.533, .567, .667, .567, .6	.587
Rm	Euclidean	0.01	.467, .533, .533, .567, .7	.560
Rm	Euclidean	0.05	.8, .633, .733, .8, .533	.700
Rm	Euclidean	0.1	.5, .667, .533, .633, .633	.593
Rm	DTW	0.01	.533, .6, .567, .6, .567	.573
Rm	DTW	0.05	.533, .667, .567, .567, .5	.567
Rm	DTW	0.1	.567, .533, .533, .533, .633	.560

Figure 2. Cluster means of a selected fixed parameter GA run

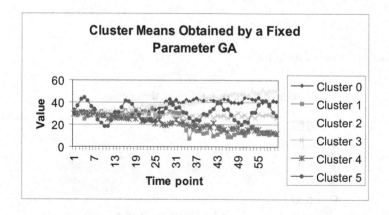

Figure 2 shows the cluster medoids of one of the five runs for this particular fixed parameter GA. Note that the six patterns can be clearly seen in the figure.

Table 2 summarizes the clustering results of using AGAs that adapt both crossover and mutation rates. For the AGAs, the GA parameter fixed is maximum generation of 30. Factors varied are fitness function, distance measure, and population size. For each combination of GA parameters, five repetitions were made with different random seeds. Among all the AGAs tested, the highest average clustering accuracy of 0.7 was attained when using the TWCV-based fitness function, Euclidean distance, and population size of 60. Note that the highest clustering accuracy produced by AGA is 83.3%, which was produced by using the *Rm*-based fitness function, DTW distance, and population size of 20. Figure 3 shows the cluster

medoids of one of the five runs for this particular AGA. Note that the six patterns can also be clearly seen as in Figure 2.

To determine the effect of fitness function, distance measure, and GA parameter, we performed ANOVA tests on the clustering results given in Tables 1 and 2. The results indicate that for both fixed parameter GAs and AGAs, the fitness function, and the interaction between the fitness function and the distance measure are highly significant (with p value < 0.005).

To evaluate the performance of adapting only one parameter, we also tried the adaptive GA that adapts only the mutation rate. Three levels of crossover rates were experimented: 0.7, 0.8, and 0.9. The two fixed parameters are the population size at 20 and the maximum number of generation at 30. Table 3 summarizes the clustering accuracies. Among all the AGAs tested, the

Table 2. Clustering results of control-chart data by AGA

Fitness Function	Distance	Population Size	Run Accuracy	Avg. Accuracy
TWCV	Euclidean	20	.633, .7, .667, .733, .633	.673
TWCV	Euclidean	40	.667, .767, .633, .767, .633	.693
TWCV	Euclidean	60	.667, .733, .7, .7, .7	**.700**
TWCV	DTW	20	.633, .7, .7, .6, .6	.647
TWCV	DTW	40	.7, .6, .667, .6, .667	.647
TWCV	DTW	60	.667, .633, .6, .667, .733	.660
DB	Euclidean	20	.433, .633, .367, .467, .567	.493
DB	Euclidean	40	.5, .4, .433, .5, .467	.460
DB	Euclidean	60	.433, .4, .5, .4, .467	.440
DB	DTW	20	.6, .667, .667, .533, .5	.593
DB	DTW	40	.533, .667, .633, .633, .6	.601
DB	DTW	60	.5, .633, .633, .633, .6	.600
Rm	Euclidean	20	.533, .633, .767, .667, .533	.627
Rm	Euclidean	40	.533, .633, .767, .667, .533	.627
Rm	Euclidean	60	.533, .533, .533, .533, .633	.553
Rm	DTW	20	.7, .7, .6, .533, **.833**	.673
Rm	DTW	40	.533, .633, .633, .7, .467	.593
Rm	DTW	60	.567, .567, .5, .567, .5	.540

Figure 3. Cluster means of a selected AGA run

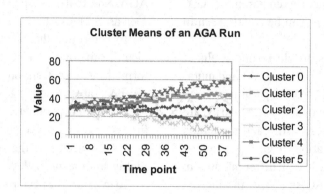

Table 3. Clustering results of control-chart data by AGA that adapts mutation rate

Fitness Function	Distance	Crossover Rate	Run Accuracy	Avg. Accuracy
TWCV	Euclidean	0.7	.6, .7, .633, .633, .733	.660
TWCV	Euclidean	0.8	.7, .7, .7, .633, .633	.673
TWCV	Euclidean	0.9	.7, .733, .7, .733, .677	**.707**
TWCV	DTW	0.7	.7, .633, .7, .633, .7	.673
TWCV	DTW	0.8	.8, .567, .633, .8, .677	.693
TWCV	DTW	0.9	.7, .733, **.833**, .6, .677	**.707**
DB	Euclidean	0.7	.6, .467, .467, .433, .467	.487
DB	Euclidean	0.8	.467, .467, .367, .467, .467	.447
DB	Euclidean	0.9	.6, .467, .567, .433, .467	.507
DB	DTW	0.7	.633, .567, .533, .733, .667	.627
DB	DTW	0.8	.5, .6, .633, .567, .633	.587
DB	DTW	0.9	.6, .667, .667, .6, .633	.633
Rm	Euclidean	0.7	.633, .667, .533, .767, .6	.640
Rm	Euclidean	0.8	.567, .567, .533, .633, .8	.620
Rm	Euclidean	0.9	.633, .567, .533, .567, .567	.573
Rm	DTW	0.7	.733, .7, .6, .533, .5	.613
Rm	DTW	0.8	.667, .567, .6, .567, .5	.580
Rm	DTW	0.9	.733, .7, .6, .533, .533	.620

highest average clustering accuracy was attained at 70.7% when using the TWCV-based fitness function and crossover rate of 0.9, regardless the distance measure. The highest clustering accuracy produced by AGA is 83.3%, which was produced by using the TWCV-based fitness function, DTW distance, and crossover rate of 0.9. The ANOVA test indicates that the fitness function, the distance measure, and their interaction are significant in affecting the clustering accuracy (with p value < 0.005). From Tables 2 and 3, it can be observed that adapting both crossover rate and mutation rates does not have any advantage over adapting only the mutation rate in terms of finding the highest clustering accuracy.

Comparing Table 2 (only those results based on population size of 20) and Table 3 (only those results based on crossover rate of 0.8) with Table 1, one can easily see that for the control-chart data,

AGAs do not always perform better than fixed parameter GAs. Depending upon the combination of fitness-function and distance measure, AGA could be better than all, none, or some fixed parameter GAs tested. Therefore, how to devise an AGA that always performs better than all fixed-parameter GAs does require further investigation.

CLUSTERING RESULTS OF CYLINDER BELL FUNNEL DATA

This section presents results obtained from one relatively larger data set of univariate time series with known number of clusters. The data set contains 300 series generated by implementing the cylinder, bell, and funnel equations given in the UCR Time Series Data Mining Archive (http://www.cs.ucr.edu/~eamonn/TSDMA/). One hundred series were generated for each pattern, with each series having a length of 80 data points.

Table 4. Clustering results of cylinder-bell-funnel data by AGA that adapts mutation rate

Fitness Function	Distance	Crossover Rate	Run Accuracy	Avg. Accuracy
TWCV	Euclidean	0.7	.823, .843, .823, .873, .853	.843
TWCV	Euclidean	0.8	.877, .820, .847, **.883**, .807	.847
TWCV	Euclidean	0.9	.867, .857, .840, .847, .857	.854
TWCV	DTW	0.7	.607, .537, .647, .460, .710	.592
TWCV	DTW	0.8	.643, .570, .543, .647, .760	.633
TWCV	DTW	0.9	.757, .763, .640, .517, .660	.667
DB	Euclidean	0.7	.800, .817, .827, .817, .847	.822
DB	Euclidean	0.8	.830, .820, .713, .817, .693	.775
DB	Euclidean	0.9	.863, .880, .823, .833, .880	**.856**
DB	DTW	0.7	.517, .607, .527, .563, .540	.551
DB	DTW	0.8	.533, .617, .553, .557, .567	.565
DB	DTW	0.9	.593, .503, .487, .547, .543	.535
Rm	Euclidean	0.7	.623, .610, .653, .557, .527	.594
Rm	Euclidean	0.8	.777, .610, .710, .697, .773	.713
Rm	Euclidean	0.9	.703, .617, .670, .760, .627	.675
Rm	DTW	0.7	.520, .550, .520, .603, .650	.569
Rm	DTW	0.8	.777, .570, .543, .613, .493	.599
Rm	DTW	0.9	.443, .607, .483, .560, .493	.515

We run the adaptive GA (that adapts only the mutation rate) 18 times by varying the fitness function, the distance measure, and the crossover rate. In each run, the GA was repeated five times. Three levels of crossover rates were experimented: 0.7, 0.8, and 0.9. The two fixed parameters are the population size at 60 and the maximum number of generation at 30. Table 4 summarizes the clustering results. The ANOVA test on these results reveals that the fitness function, the distance measure, and their interaction are significant in affecting the clustering accuracy (with p value < 0.005). Obviously, the Euclidean distance outperforms the dynamic time warping for this dataset. As far as the fitness function is concerned, the TWCV-based fitness function is the best and the R_m-based index is the worst. The best average clustering accuracy is 85.6%, obtained by the combination using the DB-based fitness, Euclidean distance, and crossover rate of 0.9. The highest clustering accuracy among all runs is 88.3%, which was produced in one of the five replicates by the combination using the TWCV-based fitness, Euclidean distance, and crossover rate of 0.8.

CLUSTERING RESULTS OF BATTLESIMULATION DATA

The OneSAF combat simulation software was used to create a battle scenario for our experiments (Heilman *et al.*, 2002). Time-series data were collected using a modified version of the Killer-Victim Scoreboard (KVS) method originally developed by O'May and Heilman (2002) to collect static-feature data. The KVS method modifies OneSAF to provide critical battlespace data. A set of three files was generated during each simulation execution. These files include entity identification data, firer-target interaction data, and logistics and appearance data. Each run of the raw experimental data collected was then processed into five time series (by arranging them in the order of timestamps) reflecting the state of ongoing battle from the viewpoint of the "blue" force, which includes:

- Relative territory ownership (denoted by *g* in figures)
- Relative firepower strength (*s*)
- Relative ammunition support (*a*)
- Relative fuel support (*f*)
- Relative firing intensity (*i*)

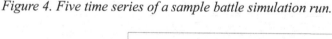

Figure 4. Five time series of a sample battle simulation run.

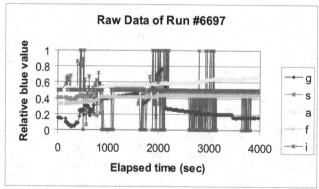

Time-series data of 15 battle simulation runs were prepared for this study according to the outlined procedure. Note that these data were nonuniformly sampled as a result of the event-triggered data-collection mechanism. The result of a sample run is shown in Figure 4.

We intentionally use only a few runs in this study because one often can afford only a limited number of simulation runs in actual applications, due to time constraint. To enable real-time response, which is desirable in order to provide fast decision support, our analyses also attempt to use as few attributes and data points as possible. First, we applied the linear interpolation method to convert the original nonuniform series into uniform ones. Figure 5 shows the interpolated results of a battle simulation run. The uniform interval is consistently set at 100 seconds, which is relatively large.

Comparing Figure 5 with Figure 4, it is observed that the overall trends are retained for all series, except that some high-frequency activities of the intensity series are lost. The empirical results indicate that this relatively large down sampled size is sufficient for the clustering task. Nevertheless, it might be desirable to determine the optimal down sampled size in the future by

investigating the tradeoff between improved clustering (an unknown) and increased computational cost. Naturally, a better interpolation method might also exist for our data. To limit the scope of this study, we elect to address these issues in the future.

Correlation analyses were performed on the five interpolated time series shown in Figure 5. It was discovered that series g and s are highly correlated (0.858); the same is true for f and a (0.771). These correlations generally apply to all simulation runs. Therefore, in the following, we will only discuss three indicators: g, a, and i.

Figures 6, 7, and 8 show the 15 interpolated time series for each one of the three indicators: g, a, and i, respectively. In these figures, the series numbered 0, 1, ..., and 14 come from simulation run x0521, x4672, x0996, x5703, x7553, x3017, x0620, x8646, x4757, x0250, x2739, x6687, x5414, x7554, and x2514, respectively. Based on the clustering results of control-chart data and the cylinder-bell-funnel data, we chose the TWCV-based fitness function for both the fixed parameter GAs and AGAs. The DTW distance measure was chosen to handle these time series of unequal length. The maximum number of generations was kept at 30 throughout. For better results, the population size

Figure 5. Interpolated results of time series shown in Figure 4

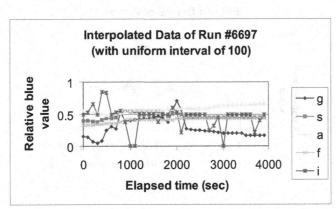

of 20 and mutation rate of 0.01 were selected for the fixed parameter GAs, whereas the population size of 60 was chosen for the AGAs. For each GA, five repetitions were made.

Tables 5-7 summarize the clustering results of territory ownership series, ammunition series, and firing intensity series, respectively, generated by both fixed parameter GAs and AGAs. In all tables, the medoids used by GA to form clusters are underlined. The best fitness value is shown in boldface for each indicator series. Based on these tables, the following observations can be made:

Figure 6. Fifteen series of relative territory ownership of the blue forces

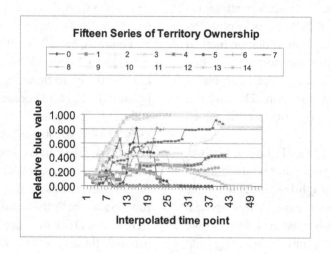

Figure 7. Fifteen series of relative ammunition values of the blue forces

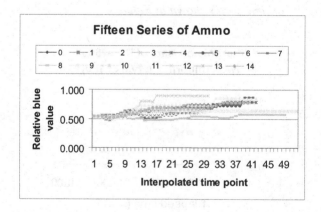

Figure 8. Fifteen series of relative fire intensity values of the blue forces

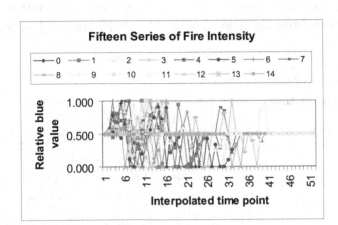

1. The AGAs, in general, generate more consistent results (with the only exception in generating four clusters of territory ownership series) and are able to find solutions with equivalent or higher best fitness than the fixed parameter GAs. Therefore, the results generated by AGAs are more trustworthy.

2. As the number of clusters increases, the consistency decreases for both AGAs and fixed parameter GAs.

Table 5. Clustering results of territory ownership series

Method	Number of clusters	Repl.	Best fitness	Clusters
FPGA	2	1-4	3308602	{0, 1, 4, 5, 8, 11, 12, 14}{2, 3, 6, 7, 9, 10, 13}
		5	3224082	{0, 1, 4, 5, 8, 11, 12, 14}{2, 3, 6, 7, 9, 10, 13}
	3	1-2	3907545	{1, 5, 12}{0, 4, 6, 8, 9, 11, 14}{2, 3, 7, 10, 13}
		3	3948684	{1, 5, 12}{0, 4, 8, 11, 14}{2, 3, 6, 7, 9, 10, 13}
		4	3842933	{1, 5, 12}{0, 4, 8, 11, 14}{2, 3, 6, 7, 9, 10, 13}
		5	3884901	{0, 1, 5, 8, 12}{2, 4, 6, 9, 11, 14}{3, 7, 10, 13}
	4	1-3	4365496	{1, 5, 12}{0, 8, 11, 14}{2, 4, 6, 9}{3, 7, 10, 13}
		4	4424450	{1, 5, 12}{0, 8, 11, 14}{2, 4, 6, 9}{3, 7, 10, 13}
		5	**4462072**	{1, 5, 12}{0, 8, 11, 14}{2, 4, 6, 9}{3, 7, 10, 13}
AGA	2	1-4	3308602	{0, 1, 4, 5, 8, 11, 12, 14}{2, 3, 6, 7, 9, 10, 13}
		5	3229174	{0, 1, 4, 5, 8, 11, 12, 14}{2, 3, 6, 7, 9, 10, 13}
	3	1-3	3978621	{1, 5, 12}{0, 4, 8, 11, 14}{2, 3, 6, 7, 9, 10, 13}
		4-5	3948684	{1, 5, 12}{0, 4, 8, 11, 14}{2, 3, 6, 7, 9, 10, 13}
	4	1	4386211	{1, 5, 12}{0, 8, 11, 14}{2, 4, 6, 9}{3, 7, 10, 13}
		2	4259005	{0}{1, 5, 8, 12}{2, 4, 6, 9, 11, 14}{3, 7, 10, 13}
		3	4294529	{1, 5, 8, 12}{0, 11, 14}{2, 4, 6, 9}{3, 7, 10, 13}
		4	4248009	{0}{1, 5, 12}{4, 8, 11, 14}{2, 3, 6, 7, 9, 10, 13}
		5	**4462072**	{1, 5, 12}{0, 8, 11, 14}{2, 4, 6, 9}{3, 7, 10, 13}

3. A different set of medoids could lead to the same clustering results. For instance, the fixed-parameter GA forms the same set of four clusters of territory ownership series based on three different sets of medoids (for replications 1-3, 4, and 5, respectively).

Table 6. Clustering results of ammunition series

Method	Number of clusters	Repl.	Best fitness	Clusters
FPGA	2	1-5	**4505133**	{8} {0, 1, 2, 3, 4, 5, 6, 7, 9, 10, 11, 12, 13, 14}
	3	1-3	4302312	{8} {2} {0, 1, 3, 4, 5, 6, 7, 9, 10, 11, 12, 13, 14}
		4	4297057	{8} {2, 3, 4} {0, 1, 5, 6, 7, 9, 10, 11, 12, 13, 14}
		5	4178824	{5} {11} {0, 1, 2, 3, 4, 6, 7, 8, 9, 10, 12, 13, 14}
	4	1	4086145	{8} {2} {14} {0, 1, 3, 4, 5, 6, 7, 9, 10, 11, 12, 13}
		2	4081404	{8} {2, 3, 4} {14} {0, 1, 5, 6, 7, 9, 10, 11, 12, 13}
		3	4070924	{8} {11} {5} {0, 1, 2, 3, 4, 6, 7, 9, 10, 12, 13, 14}
		4	4029205	{2} {3, 4} {11} {0, 1, 5, 6, 7, 8, 9, 10, 12, 13, 14}
		5	4060230	{8} {2, 3, 4} {7} {0, 1, 5, 6, 9, 10, 11, 12, 13, 14}
AGA	2	1-5	**4505133**	{8} {0, 1, 2, 3, 4, 5, 6, 7, 9, 10, 11, 12, 13, 14}
	3	1-4	4302312	{8} {2} {0, 1, 3, 4, 5, 6, 7, 9, 10, 11, 12, 13, 14}
		5	4297057	{8} {2, 3, 4} {0, 1, 5, 6, 7, 9, 10, 11, 12, 13, 14}
	4	1-3	4107672	{8} {2} {3, 4} {0, 1, 5, 6, 7, 9, 10, 11, 12, 13, 14}
		4	4101704	{8} {2, 3, 4} {11} {0, 1, 5, 6, 7, 9, 10, 12, 13, 14}
		5	4086145	{8} {2} {14} {0, 1, 2, 3, 5, 6, 7, 9, 10, 11, 12, 13}

Table 7. Clustering results of fire intensity series

Method	Number of clusters	Repl.	Best fitness	Clusters
FPGA	2	1	3794554	{0, 8, 11} {1, 2, 3, 4, 5, 7, 9, 10, 12, 13, 14}
		2-5	3820013	{0, 11, 14} {1, 2, 3, 4, 5, 6, 7, 8, 9, 10, 12, 13}
	3	1	3919665	{6} {0, 11, 14} {1, 2, 3, 4, 5, 7, 8, 9, 10, 12, 13}
		2	3892864	{6} {0, 8, 11} {1, 2, 3, 4, 5, 7, 9, 10, 12, 13, 14}
		3	3811803	{6} {4, 14} {0, 1, 2, 3, 5, 7, 8, 9, 10, 11, 12, 13}
		4	3870329	{8} {0, 11, 14} {1, 2, 3, 4, 5, 6, 7, 9, 10, 12, 13}
		5	3864623	{6} {11, 14} {0, 1, 2, 3, 4, 5, 7, 8, 9, 10, 12, 13}
	4	1	3956146	{6} {11, 14} {0} {1, 2, 3, 4, 5, 7, 8, 9, 10, 12, 13}
		2-3	3969378	{0, 8} {6} {11, 14} {1, 2, 3, 4, 5, 7, 9, 10, 12, 13}
		4	3900812	{6} {4, 14} {0} {1, 2, 3, 5, 7, 8, 9, 10, 11, 12, 13}
		5	3938876	{2} {0, 11, 14} {6} {1, 3, 4, 5, 7, 8, 9, 10, 12, 13}
AGA	2	1-5	3820013	{0, 11, 14} {1, 2, 3, 4, 5, 6, 7, 8, 9, 10, 12, 13}
	3	1-5	3919665	{6} {0, 11, 14} {1, 2, 3, 4, 5, 7, 8, 9, 10, 12, 13}
	4	1-4	**3972659**	{8} {6} {0, 11, 14} {1, 2, 3, 4, 5, 7, 9, 10, 12, 13}
		5	3956145	{0} {6} {11, 14} {1, 2, 3, 4, 5, 7, 8, 9, 10, 12, 13}

Since there is no ground truth for this data set, it is not easy to conclude which clustering result is better because the clustering accuracy cannot be computed for this particular data set. As a standard practice in this situation, we resort to visual inspection for qualitative evaluation. Figures 9-11 show the four-cluster results generated by the AGA for each indicator series, respectively.

Note that in the previous three figures, the values along the vertical axis were changed from the original (for the series in three clusters) in order to separate one cluster from the others. Also note

Figure 9. Four-cluster of territory ownership series generated by AGA

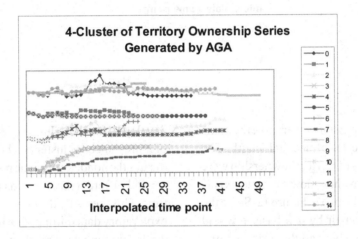

Figure 10. Four-cluster of ammunition series generated by AGA

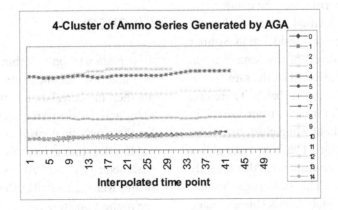

Figure 11. Four-cluster of fire intensity series generated by AGA

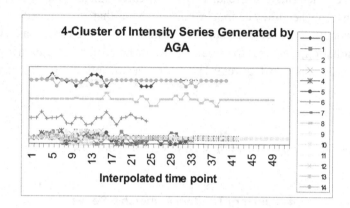

that the time-warping distance warps the time axis in comparing two series of unequal length. The No. 2 and No. 6 series of territory ownership were taken as examples to show how the time axis is warped to compare the two in Figure 12. Since the warping is done on a pair-by-pair basis, it would take a lot of space to show all the warped pairs. To better visualize the similarity between any pair of two series, one must also exercise this warping ability, as demonstrated in Figure 12.

Based on these results, we argue that the proposed genetic clustering method is effective for grouping time-series data produced in battle-simulation experiments. The notable feature of the proposed method is that it takes the entire battle sequence into consideration. Similar battle sequences, grouped in the same cluster, indicate that they are affected by the same set of mechanisms in play during the battle. Therefore, a model should be built for each cluster, to explain the mechanisms. Without using clustering as an exploratory step, one might build just one global model for all runs by assuming that there is only one mechanism in play. The predictions thus derived from this single global model are expected to be less accurate than from a set of local models

developed for each cluster of data. The existence of more than one cluster for the same scenario indicates the nondeterministic nature of battles with several different mechanisms in play at the same time. This is exactly the knowledge discovered by exploratory data mining. The intriguing question is what tips one set of mechanisms from the other. Further investigation is necessary to answer this question. The study by Bodt *et al.* (Bodt, Forester, Hansen, Heilman, E., Kaste, & O'May, 2002) is a step in this direction, except that they did not consider the entire battle sequence.

DISCUSSION

To explain why one combination of fitness function and distance measure performs better than another, the correlation between the best fitness value and the clustering accuracy was computed for the synthetic control-chart data, as given in Table 8. It is interesting to see that for both fixed-parameter GAs and AGAs, the best clustering results were generated by the combination having the highest positive correlation between the best fitness value and the clustering accuracy. The

Figure 12. An example of dynamic time warping

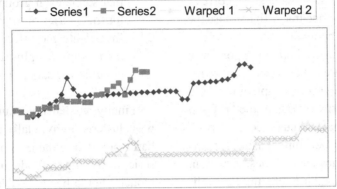

negative correlations are undesirable because such fitness functions are counterproductive in the sense that maximizing the fitness by the GA actually leads to lower clustering accuracy. The low positive correlation indicates that the best fitness function used in this study is marginal, and a better one should be developed in order to achieve even higher clustering accuracy. This will be a topic for future studies.

For comparison, the clustering results of *k*-means were also obtained for both the synthetic control-chart data and the cylinder-bell-funnel data. For each data set, five runs were made. The clustering accuracies obtained are 63.3%, 70%, 60%, 56.7%, and 60% for the synthetic control-chart data, and are 87%, 87.3%, 87%, 87.3%, and 87% for the cylinder-bell-funnel data. Therefore,

genetic clustering methods could obtain higher accuracy than *k*-means if an appropriate combination of fitness function and distance measure is used.

CONCLUSION

This chapter presented an exploratory data-mining study of time-series data. To this end, genetic algorithms were developed for clustering the data. First, the effects of three fitness functions, two distance measures, and some selected GA parameters were investigated using the synthetic control-chart data that is available in the public domain. The results indicate that the two most significant factors are the fitness function and

Table 8. Correlation between best fitness and accuracy

Fitness function	Distance measure	Fixed parameter GAs	AGAs
TWCA	Euclidean	0.327	0.264
TWCA	DTW	-0.100	0.267
DB	Euclidean	-0.783	-0.605
DB	DTW	-0.458	0.182
Rm	Euclidean	-0.129	-0.556
Rm	DTW	-0.676	-0.241

the interaction between the fitness function and the distance measure. It was also discovered that the best fitness value has a low positive correlation with the clustering accuracy. To increase the clustering accuracy, a better fitness function must be developed, which poses both an opportunity and a challenge for researchers interested in this area of research. In addition, adaptive GAs did not always outperform fixed-parameter GAs. Better adaptive schemes are thus needed. Further testing on the cylinder-bell-funnel data confirms once again the significance of fitness function, distance measure, and their interaction. In addition, it was found that dynamic time warping did not work as well as Euclidean for this dataset. The GA parameters such as population size, crossover rate, and mutation rate are relatively insignificant compared with fitness function and distance measure in the context of clustering.

The proposed genetic clustering methods were also applied to time-series data acquired in battle simulation experiments. The potential use of the proposed methods in grouping similar battle profiles and separating dissimilar battle profiles are shown and discussed. As far as future study is concerned, this study can be further improved by investigating the following general topics:

1. Develop a better fitness function.
2. Develop a better GA parameter adaptation method.
3. Try other chromosome coding schemes, such as integer coding rather than binary coding.
4. Develop a viable validation index, either application dependent or application independent, to determine how many clusters are optimal.
5. Extend the current method, which is currently applicable only to univariate time-series data, to multivariate time-series data.

Specific topics for the battle simulation study in the future include:

1. Complement this study with an in-depth analysis of the driving mechanisms/events that shape each group of battle sequences (or time-series profiles).
2. Investigate the effect of the interpolation method and sampling interval in converting nonuniform data into uniform data.

Finally, we shall point out that each time series was clustered individually in this study. For the battle simulation data, some multivariate time-series clustering methods such as hidden Markov models should be applicable. Alternatively, one can make use of a two-stage clustering procedure that was recently developed by Liao (2007), which involves converting multivariate time series into a univariate discrete time series in the first step and then applying the proposed genetic clustering algorithm in the second step. The results of the first stage have been presented in Liao *et al.* (Liao, Bodt, Forester, Hansen, Heilman, Kaste, & O'May, 2002). The complete two-stage procedure and results will be presented in our future work.

ACKNOWLEDGMENT

Dr. Liao acknowledges the support of ASEE-ARL Postdoctoral Fellowship (contract number DAAL01-96-C-0038) that helped make this study possible.

REFERENCES

Bandyopadhyay, S., & Maulik, U. (2002). Genetic clustering for automatic evolution of clusters and application to image classification. *Pattern Recognition, 35,* 1197-1208.

Baragona, R. (2001). A simulation study on clustering time series with metaheuristic methods. *Quaderni di Statistica, 3,* 1-26.

Bezdek, J. C. (1987). *Pattern recognition with fuzzy objective function algorithms.* New York: Plenum Press.

Bodt, B., Forester, J., Hansen, C., Heilman, E., Kaste, R., & O'May, J. (2002). Data mining combat simulations: a new approach for battlefield parameterization. *Army Research Conference,* Orlando FL, December 2002.

Davies, D. L., & Bouldin, D. W. (1979). A cluster separation measure. *IEEE Trans. Pattern Analysis and Machine Intelligence, 1,* 224-227.

Eiben, A. E., Hinterding, R., & Mizhalewicz, Z. (1999). Parameter control in evolutionary algorithms. *IEEE Transactions on Evolutionary Computation, 3*(2), 124-141.

Fu, T.-C., Chung F.-L., Ng, V., & Luk, R. (2001). Pattern discovery from stock time series using self-organizing maps. Paper presented at the *KDD 2001 Workshop on Temporal Data Mining,* August 26-29, San Francisco.

Golay, X., Kollias, S., Stoll, G., Meier, D., Valavanis, A., & Boesiger, P. (1998). A new correlation-based fuzzy logic clustering algorithm for fMRI. *Magnetic Resonance in Medicine, 40,* 249-260.

Goutte, C., Toft, P., & Rostrup, E. (1999). On clustering fMRI time series. *Neuroimage, 9*(3), 298-310.

Grefenstette, J. J. (1986). Optimization of control parameters for genetic algorithms. *IEEE Transactions on Systems, Man, and Cybernetics,* SMC-16, 122-128.

Hall, L. O., Özyurt, B., & Bezdek, J. C. (1999). Clustering with a genetically optimized approach. *IEEE Transactions on Evolutionary Computation, 3*(2), 103-112.

Heilman, E., *et al.* (2002) Identifying battlefield metrics through experimentation. *Proceedings of the 7th International Command & Control Research & Technology Symposium.*

Herrera, F., & Lozano, M. (2003). Fuzzy adaptive genetic algorithms: design, taxonomy, and future directions. *Soft Computing, 7,* 545-562.

Kalpakis, K., Gada, D., & Puttagunta, V. (2001). Distance measures for effective clustering of ARIMA time-series. *Proceedings of the 2001 IEEE International Conference on Data Mining* (pp. 273-280)., San Jose, CA, Nov. 29 – Dec. 2, 2001.

Kaufman, L., & Rousseeuw, P. J. (1990). *Finding groups in data: An introduction to cluster analysis.* New York: John Wiley & Sons.

Krishna, K., & Murty, M. N. (1999). Genetic k-means algorithms. *IEEE Transactions on Systems Man and Cybernetics - Part B: Cybernetics, 29*(3), 433-439.

Krishnapuram, R., Joshi, A., Nasraoui, O., & Yi, L. (2001). Low-complexity fuzzy relational clustering algorithms for web mining. *IEEE Transactions on Fuzzy Systems, 9*(4), 595-607.

Li, C., & Biswas, G. (1999). Temporal pattern generation using hidden Markov model based unsupervised classification. In D. J. Hand, J. N. Kok, M. R. Berthold (Eds.), IDA '99, *LNCS 164,* Springer-Verlag: Berlin, 245-256.

Liao, T. W. (2005). Clustering of time series data - a survey. *Pattern Recognition, 38*(11), 1857-1874.

Liao, T. W. (2007). A clustering procedure for exploratory mining of vector time series. *Pattern Recognition,* doi:10.1016/j.patcog.2007.01.005.

Liao, T. W., Bodt, B., Forester, J., Hansen, C., Heilman, E., Kaste, R., & O'May, J. (2002). *Army Research Conference,* Orlando FL, December 2002.

Lorena, L. A. N., & Furtado, J. C. (2001). Constructive genetic algorithms for clustering problems. *Evolutionary Computation, 9*(3), 309-327.

MacQueen, J. (1967). Some methods for classification and analysis of multivariate observations. In L. M. LeCam & J. Neyman (Eds.), *Proceedings of the 5th Berkeley Symposium on Mathematical Statistics and Probability*, Vol. 1 (pp. 281-297), University of California Press, Berkeley.

Maulik, U., & Bandyopadhyay, S. (2000). Genetic algorithm-based clustering technique. *Pattern Recognition, 33*, 1455-1465.

Mitchell, M. (1996). *An introduction to genetic algorithms*. Cambridge, MA: The MIT Press.

O'May, J., & Hailman, E. (2002). OneSAF killer/victim scoreboard capability. ARL-TR-2829.

Pham, D. T., & Chan, A. B. (1998). Control chart pattern recognition using a new type of self-organizing neural network. *Proc. Instn. Mech. Engrs., 212*(1), 115-127.

Policker, S., & Geva, A. B. (2000). Nonstationary time series analysis by temporal clustering. *IEEE Transactions on Systems, Man, and Cybernetics—Part B: Cybernetics, 30*(2), 339-343.

Rojas, Ignacio, González, J., Pomares, H., Merelo, J. J., Castillo, P. A., & Romero, G. (2002). Statistical analysis of the main parameters involved in the design of a genetic algorithm. *IEEE Transactions on Systems, Man, and Cybernetics-Part C: Applications and Reviews, 32*(1), 31-37.

Srinivas, M., & Patnaik, L. M. (1994). Adaptive probabilities of crossover and mutation in genetic algorithms. *IEEE Transactions on Systems, Man, and Cybernetics, 24*(4), 656-666.

Chapter XI
Development of Control Signatures with a Hybrid Data Mining and Genetic Algorithm

Alex Burns
The University of Iowa, USA

Shital Shah
Rush University Medical Center, Health Systems Management, USA

Andrew Kusiak
The University of Iowa, USA

ABSTRACT

This chapter presents a hybrid approach that integrates a genetic algorithm (GA) and data mining to produce control signatures. The control signatures define the best parameter intervals leading to a desired outcome. This hybrid method integrates multiple rule sets generated by a data-mining algorithm with the fitness function of a GA. The solutions of the GA represent intersections among rules providing tight parameter bounds. The integration of intuitive rules provides an explanation for each generated control setting and it provides insights into the decision-making process. The ability to analyze parameter trends and the feasible solutions generated by the GA with respect to the outcomes is another benefit of the proposed hybrid method. The presented approach for deriving control signatures is applicable to various domains, such as energy, medical protocols, manufacturing, airline operations, customer service, and so on. Control signatures were developed and tested for control of a power plant boiler. These signatures discovered insightful relationships among parameters. The results and benefits of the proposed method for the power plant boiler are discussed in the chapter.

INTRODUCTION

Optimizing process controls is imperative in the energy, medical, and service applications. Due to the increase in the volume of data, decision making in real time is becoming more difficult, and may potentially lead to inefficiencies and hazardous situations. Intelligent control systems have proven to be effective in optimizing complex processes. Merging computational concepts, such as neural networks, genetic algorithms, data mining, and fuzzy logic, can lead to robust controls (Krishnakumar & Goldberg, 1992; Lee, Perakis, Sevcik, Santoso, Lausterer, & Samad, 2000). Though current intelligent approaches may improve operations, they provide limited insights into the decision-making process.

This chapter describes a hybrid method that integrates data mining (Cios, Pedrycz, & Swiniarski, 1998; Fayyad, Piatetsky-Shapiro, Smyth, Uthurusamy, 1995) and genetic algorithm (GA) (Goldberg, 1989; Holland, 1975; Lawrence, 1987; Michalewicz, 1992) concepts to define robust and explicit parameter set points. The hybrid approach consists of partitioning data, developing classifiers for each data set, and combining and analyzing the classifiers (Mitra, Pal, & Mitra, 2002). These steps lead to the development of control signatures (Kusiak, 2002) that define ranges of parameter settings, producing a desired outcome (decision). Control signatures are helpful in learning the interactions and relationships between the parameters.

Data mining is the process of discovering interesting and previously unknown patterns in data sets (Cios, Pedrycz, & Swiniarski, 1998; Fayyad, Piatetsky-Smapiro, Smyth, & Uthurusamy, 1995). A typical data-mining algorithm, applied to partitioned data sets, generates multiple rule sets that describe parameter relationships. The GA provides a global search mechanism to discover the intersections among the decision rules. The intersections among decision rules could be analyzed through visualization (Kusiak, 2001). As the number of rules increases, the graphical presentation and analysis becomes tedious. Furthermore, this kind of analysis does not provide the additional information leading the tighter parameter bounds, control signatures, and increased prediction accuracy.

In this research, the process data is transformed into multiple knowledge bases that are used as the foundation of a GA fitness function. The GA mechanism strongly promotes the solutions that are in the feasible region formed by the rule intersections. It not only allows for exclusion of the less reliable (feasible) regions, but also defines the complex commonality among multiple rules. This in turn provides tighter bounds on various parameters. Complex nonlinear applications can be analyzed as the parameter relationships are preserved by the rule sets that are incorporated into the GA fitness function.

The analysis of the control signatures across different outcomes provides information regarding the general parameter trends. Analyzing these trends and the feasible solutions generated by the GA is a way of visualizing the complex relationships and identifying key parameters. The proposed hybrid approach provides increased level of insight and was applied and tested on the data from a power plant boiler. The signatures defined ideal parameter ranges for boiler operations.

METHOD

This section describes a hybrid approach that integrates genetic algorithm (GA) and data mining to define control signatures (Figure 1). The use of GA is novel, due to the incorporation of data-mining output in the fitness function. Data mining defines relationships among parameters, while a standard GA provides the global search mechanism to identify ranges for parameter settings. The basic concepts of data mining and GA that are relevant to the development of control signatures are discussed next.

The hybrid approach requires a data set with known outcomes, but it can handle continuous, discrete, and categorical data. The GA is facilitated by data preprocessing through the normalization of continuous parameters. Discrete and categorical parameters are assigned a value based on their probability of occurrence. All parameter values are transformed back to their respective ranges at the termination of the GA. The data set with various parameters and known outcomes is used as a starting point in the analysis outlined in Figure 1.

In order to increase the number of rule sets, the initial data set is partitioned into subsets of parameters and data records to form multiple data subsets (Figure 1) (Kusiak, Sham, & Dixon, 2003). This allows the investigation of various regions of the solution space. The partial data sets can be formed by utilizing domain expert's knowledge, or relevant knowledge from the literature or at random. A data-mining algorithm is applied independently to each data subset, thus, producing several rule sets (classifiers). High cross-validation accuracy of the extracted rules increases the confidence in the derived control signatures. The significant rules extracted from these data sets form the GA fitness function. The diversity and redundancy of the rule sets is reflected in the fitness function as explained in detail in Section 2.3.

The analysis of the GA feasible population can be conducted by various measures such as min, max, average, frequency distributions, best solution, and so forth.

Figure 1. Computation of GA fitness function.

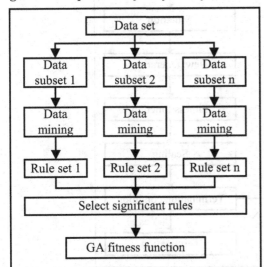

Data-Mining Concepts

Discovering hidden patterns in the data may represent valuable knowledge expressed in the form of control signatures. There are various data-mining algorithms available in the literature. Learning (classification) systems of potential interest to this research fall into 10 categories (Kusiak, 2001a):

a.   Classical statistical methods (e.g., linear discriminant, quadratic discriminant, and logistic discriminant analyses).

b.   Modern statistical techniques (e.g., projection pursuit classification, density estimation, k-nearest neighbor, casual networks, Bayes theorem).

c.   Neural networks (e.g., backpropagation, linear vector quantifiers, and radial function networks).

d.   Support vector machines.

e.   Decision tree methods (e.g., C4.5, ID3, CN2).

f.   Decision rule algorithms (e.g., AQ15, LERS and numerous other algorithms based on the rough set theory).

g.   Association rule algorithms (e.g., DB2IntelligentMiner).

h.   Learning classifier systems (e.g., GOFFER, MonaLysa, and XCS).

i.   Inductive learning algorithms.

j.   Text learning algorithms.

In this research, the learning algorithms of category (e) will be explored for two major reasons:

•   Generation of explicit knowledge in the form acceptable by a user. The user is able to understand the extracted knowledge, assess its usefulness, and learn new and interesting concepts.

- Controlled prediction accuracy. This characteristic is due to the nature of the decision-making approach applied in this research.

The C4.5 decision tree algorithm (Quinlan, 1986) used in this research produces rules in the following format:

IF Boiler Master <= -0.53 AND Air Master > -1.5 AND Air Fuel Ratio > 0.13 AND Avg Mid Temp > -0.42 AND Biomass Feed Rate > 0.4 THEN Interval = 88_90 [Rule strength = 0.857]

Each rule includes a premise and conclusion, and is assigned strength. The rule strength describes the percentage of observations in a given class (outcome) that match both the rule condition and the action.

Multiple applications of the data-mining algorithm to various data subsets produce rules representing different solutions. Each of the solutions potentially describes a setting that results in the desired outcome.

Genetic Algorithm

A GA is a search procedure based on the concepts of natural genetics (Goldberg, 1989; Holland, 1975; Lawrence, 1987; Michalewicz, 1992; Obitko, 2004). It is initiated with a set of solutions (represented by chromosomes) called the population. Each solution in the population is evaluated in terms of its fitness. Solutions chosen to form new chromosomes (offspring) are selected according to their fitness, that is, the more suitable they are, the higher likelihood they will reproduce. This is repeated until some stopping criterion (for example, the number of populations or improvement of the best solution) is satisfied. GA operators and solution encoding scheme need to be carefully chosen to appropriately explore and exploit the solution space (Pham & Karaboga, 2000).

The steps of the GA algorithm are outlined in Figure 2 (Goldberg, 1989). First, the solution-encoding scheme is designed and the problem specific fitness function is formulated. An initial (random) population of n chromosomes (solutions)

Figure 2. Genetic algorithm

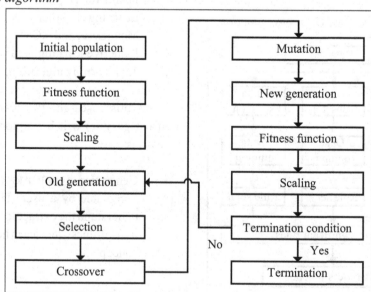

is generated, and these solutions are evaluated based on the value of their scaled fitness. Depending upon the fitness function value some solutions from the old population are chosen for mating using a selection scheme, for example, roulette wheel or tournament selection. The crossover operator is applied to the selected parents. Generally, a single point crossover is recommend if the chromosome size is small (Goldberg, 1989; Pham & Karaboga, 2000). Crossover is performed until *m* new offspring are generated, where *m* is equal to *n* of the old population.

There is a possibility that the initial population was created in such a way that the GA cannot generate some solutions of interest. To avoid this and make the GA more robust, a mutation operator is used. Mutation operator alters the value at some genes using the mutation probability (Goldberg, 1989; Pham & Karaboga, 2000). This allows the exploration of the previously unattainable search space. After crossover and mutation are carried out, a new population is generated. The fitness of the new solutions is evaluated to measure their goodness. If the GA termination condition is not satisfied, then the new population becomes the old population and all the steps are repeated.

The hybrid approach presented in this research employs a single-point standard crossover operator, varying mutation operator, and standard tournament selection operator. These operators ensure a balance between exploration and ex-ploitation for the GA search mechanism. The encoding scheme consists of a gene representing each parameter. Thus, the length of the chromosome is equal to the number of parameters. The phenotype representations of the chromosome are the actual values. The genotype representation is a z-transform of the original values. The encoding scheme for categorical parameters is shown in Appendix A.

Fitness Function

The fitness function is computed by matching the rule sets generated from mining data subsets (Figure 1). The rules define a set of feasible regions for each parameter, which provides the foundation of the control signature. The fitness function locates the rule set intersection, thus, optimizing the multirule set fitness function. For example, consider the following three rules:

- **Rule 1:** IF A < 1.2 AND B = 4 AND C = S THEN Decision = 1
- **Rule 2:** IF A > 0.7 AND D = 6 THEN Decision = 1
- **Rule 3:** IF A < 1.5 AND E = 8 AND C = S THEN Decision = 1

The range of values of parameter A satisfying all the three rules is graphically represented in Figure 3. The rule set intersection for parameter

Figure 3. Minimum rule set intersection for parameter A

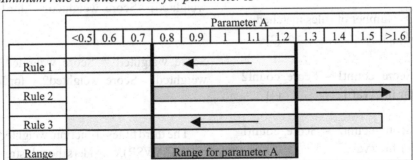

A is between 0.8 and 1.2. The fitness function is evaluated for the rule set intersections of all parameters involved in forming the control signature. The fitness functions discussed next involve three decision intervals. These functions could be extended to include as many intervals as required.

A fitness function depends on the nature of the problem at hand. The first type of fitness function is the rule match (GARM). The goal is to match as many rules as possible with the desired GA interval (outcome) (see equations 1 - 6). Thus, if the GA solution matches a rule in a given interval, then it is awarded score = 1, else 0. The GA solution can match various rules from different intervals, and the scores for each interval are computed separately. Depending upon the GA optimization interval, the scores of other intervals are subtracted from the score of the desired GA interval. This guarantees that there is higher incentive for matching the GA desired interval and penalizing the matching rules of other intervals. GARM also tries to promote GA solutions with minimum contradiction within intervals. A high penalty is assigned for a GA solution not satisfying a single rule from any interval. The ultimate objective is to maximize the number of rules matched.

$$\text{Score\_count1} = \sum_i \text{Number of rules matched}$$
i: Rule set – Interval 1 $\quad\quad$ (1)

$$\text{Score\_count2} = \sum_i \text{Number of rules matched}$$
i: Rule set – Interval 2 $\quad\quad$ (2)

$$\text{Score\_count3} = \sum_i \text{Number of rules matched}$$
i: Rule set – Interval 3 $\quad\quad$ (3)

Score_count = Score_count1 - Score_count2 - Score_count3 for Interval 1 $\quad$ (4)

Score_count = Score_count2 - Score_count1 - Score_count3 for Interval 2 $\quad$ (5)

Score_count = Score_count3 - Score_count1 - Score_count2 for Interval 3 $\quad$ (6)

The GARM fitness function maximizes the number of rules matched without due consideration to the strength of the rules. The weighted score fitness function (GAWS) weights the matched rules (see equations 7 - 12). Thus, if the GA solution matches a rule in a given interval, then it is awarded score = one*(scaled rule strength), else 0. The individualized scores are calculated for each interval and, depending upon the desired GA optimizing interval, a similar approach to the GARM is incorporated.

$$\text{Score\_weighted1} = \sum_i \sum_j \text{Rules matched} * \text{Rule strength}$$
i: Rule set – Interval 1 $\quad$ j: Rule number in a set – Interval 1 $\quad\quad$ (7)

$$\text{Score\_weighted2} = \sum_i \sum_j \text{Rules matched} * \text{Rule strength}$$
i: Rule set – Interval 2 $\quad$ j: Rule number in a set – Interval 2 $\quad\quad$ (8)

$$\text{Score\_weighted3} = \sum_i \sum_j \text{Rules matched} * \text{Rule strength}$$
i: Rule set – Interval 3 $\quad$ j: Rule number in a set – Interval 3 $\quad\quad$ (9)

Score_weighted = Score_weighted1 - Score_weighted2 - Score_weighted3 for Interval 1 $\quad$ (10)

Score_weighted = Score_weighted2 - Score_weighted1 - Score_weighted3 for Interval 2 $\quad$ (11)

Score_weighted = Score_weighted3 - Score_weighted1 - Score_weighted2 for Interval 3 $\quad$ (12)

The third fitness function, weighted scores and rules (GAWSR), considers the rule strength and the number of rules matched by the GA solution (see equations 13 - 18). This ensures that the strongest rules do not dominate the GA population. Here, the same procedure as the GAWS is adopted, except the number of matched rules is multiplied by the GAWS score for each respective solution. Thus, an effort is made by the GA search to satisfy as many rules as possible for each rule set.

$$Score1 = \sum_i Number\ of\ rules\ matched\ (\sum_j Rules\ matched\ {}_*\ Rule\ strength)$$

(i: Rule set number and j: Rule number in a set)
– Interval 1   (13)

$$Score2 = \sum_i Number\ of\ rule\ matched\ (\sum_j Rules\ matched\ {}_*\ Rule\ strength)$$

(i: Rule set number and j: Rule number in a set)
– Interval 2   (14)

$$Score3 = \sum_i Number\ of\ rules\ matched\ (\sum_j Rules\ matched\ {}_*\ Rule\ strength)$$

(i: Rule set number and j: Rule number in a set)
– Interval 3   (15)

Score = Score1 - Score2 - Score3   for Interval 1   (16)
Score = Score2 - Score1 - Score3   for Interval 2   (17)
Score = Score3 - Score1 - Score2   for Interval 3   (18)

The domain knowledge is incorporated in the fitness function as a penalty for violating the domain rules. These domain rules can be in the form of illegal parameter values, prohibited combinations of parameters, and so forth. Improved knowledge-based fitness functions enhance both the ability of GA search mechanism and control intervals. The performance of the GA depends on the complexity of the problem and the type of fitness function. Thus, the complexity of GAWSR fitness function may lead to lengthy GA runs.

Control Signature Development

The generation of control signatures requires running the GA several times due to maximization of each of the three fitness functions at least once for each outcome. The maximization of each fitness function ensures that a large solution space is explored. Repeating the GA n times for each fitness function further enhances the exploration and exploitation of the solution space. The next step in developing control signatures is running the GA for each interval. This allows for the development of ideal (desirable) parameter set points across all intervals of interest. Operating at the ideal parameter settings will improve the confidence of achieving the desired outcomes.

The analysis of the trends, graphically and statistically between various outcomes, increases insight in the process expressed in the data. Each feasible setting generated by the GA is stored in a database. The volume of data is sufficiently large to develop histograms as well as to perform other statistical analyses. Plotting the histogram for each parameter provides more insights into the behavior of the parameters (i.e., critical process control parameters). Critical parameters may be indicated by unique histograms for each outcome. These histograms may display narrow operating conditions related to the desired outcome. Noncritical parameters may have similar histograms for each outcome. This indicates that regardless of outcome, the possible settings obtained through the GA are the same for that parameter.

Table 1 shows the parameters values that maximize each of the three fitness functions. In this example, each fitness function was maximized twice at every interval. The final step in deriving control signatures is determining the minimum, maximum, and average for each parameter over all intervals. Confidence intervals and other statistics could also be used. The application will dictate the appropriate tool. Example control signatures were computed using the data in Table 1 for both parameters, as shown in Table 2. The minimum and maximum values define the range of acceptable values for each parameter.

The next section presents the development of control signatures for a power plant application. These control signatures are used to describe ideal parameter settings for several boiler efficiency intervals.

INDUSTRIAL APPLICATION

Reduction of production costs and waste is critical to operations of power plants. Currently, operators are required to make decisions regarding the operation of the boiler system based on multiple pieces of information. This information is complex and changes in real time. Control signatures will increase efficiency, lower operating costs, reduce fuel consumption, and decrease emission of air pollutants. Some of the model-based controllers published in the literature are briefly discussed next.

The topic of intelligent control in power system applications is an active area of research.

Table 1. Example output from the genetic algorithm

A. Interval 1			
Fitness function	**Run**	**Parameter 1**	**Parameter 2**
Matched rules (GARM)	GARM 1	2.6	10.1
	GARM 2	3.1	11.8
Weighted score (GAWS)	GAWS1	2.7	9.3
	GAWS2	2.8	9.8
Rules and weighted score (GAWSR)	GAWSR1	3.2	11.6
	GAWSR2	2.6	11.5
B. Interval 2			
Fitness function	**Run**	**Parameter 1**	**Parameter 2**
Matched rules (GARM)	GARM 1	4.9	22.6
	GARM 2	5.1	23
Weighted score (GAWS)	GAWS1	4.8	24.1
	GAWS2	4.6	24
Rules and weighted score (GAWSR)	GAWSR1	6.2	22.8
	GAWSR2	6.1	22.9

Table 2. Control signatures for parameters 1 and 2

A. Control signature for parameter 1				B. Control signature for parameter 2		
Range	**Interval 1**	**Interval 2**		**Range**	**Interval 1**	**Interval 2**
Min	2.6	4.6		Min	9.3	22.6
Average	2.8	5.3		Average	10.7	23.2
Max	3.2	6.2		Max	11.8	24.1

Chong *et al.* (Chong, Wilcox, & Ward, 2000) applied a neural network to represent the formation of pollutant emissions resulting from the combustion of coal. The resultant "black-box" models of the pollutant emissions, namely the nitrogen oxides and carbon monoxide emissions, represented the dynamics of the process and delivered reasonably accurate estimates over a wide range of test data. The authors pointed out that the neural network model, although lacking in model transparency produce estimates of the derivatives of combustion with acceptable accuracy, relative to the simplicity of the model design.

Ghezelayagh and Lee (2002) proposed an intelligent predictive controller for a fossil fuel power unit. This controller was based on a self-organized neurofuzzy identifier to predict the response of the plant. The control inputs were optimized by an evolutionary programming algorithm to minimize the error of the identifier outputs and reference set points.

Stephan *et al.* (Stephan, Debes, Gross, Wintrich, & Wintrich, 2001) presented a control scheme based on reinforcement learning for an industrial hard-coal combustion process in a power plant. They minimized the nitrogen oxides emission to comply with the tightening environmental protection requirements, while keeping other process parameters within specified limits. They demonstrated that the proposed multiagent system significantly reduced the overall air consumption of the combustion process at the power plant.

Booth and Roland (1991) applied a neural network-based system to several types of combustion facilities. The neural network-based system optimized the boiler operation by accommodating equipment performance changes due to wear and maintenance activities, adjusting to fluctuations in fuel quality, and improving operating flexibility. The system dynamically adjusted combustion set points and bias settings in closed-loop supervisory control to reduce NOx emissions and improve heat rate simultaneously.

Li, Thompson, and Peng, (2002) presented a method for building an artificial neural network (ANN) based on a GA. The proposed ANN was used to predict NO_2 emissions in a coal-burning power plant. This hybrid approach was effective in building an ANN model that was highly accurate in predicting NO_2 emissions.

Genetic algorithms have also been applied to optimize power flow problems (Bakirtzis, Biskas, Zoumas, & Petridis, 2002), design boiler system controls (Dimeo & Lee, 1995) and generation expansion (Park, Park, Won, & Lee, 2000) with varying degrees of success.

The approaches published in the literature may provide improved control settings, however, the decision-making process is indiscernible or a "black box." The hybrid approach presented in this research is novel due to the utilization of explicit rules.

To demonstrate the methods outlined in the chapter, control signatures were developed for a circulating fluidized boiler (CFB) at the University of Iowa Power Plant. These control signatures define ideal parameter values for three boiler efficiency intervals (84-86, 86-88, and 88-90) measured in percentages (Table 4). These three intervals were selected due to the fact they comprised a significant portion of the observations. There was not sufficient data to define control signatures for the higher efficiency zones (i.e., efficiencies greater than 90%). This could be attributed to the quality of the coal used for combustion, the environmental conditions, the skill and consistency of the operators, and so on. As the efficiency of the process increases, the higher efficiency data will be collected and used to develop future control signatures.

For this research, efficiency was defined as the theoretical energy output divided by the input (equation (19)).

$$\text{Efficiency (\%)} = 100 * (H_{steam} * n_{steam}) / (B_{coal} * n_{coal})$$
$$(19)$$

where:

B_{coal} = Average btu/lb of fuel
n_{coal} = Total number of lbs of fuel fed into boiler
H_{steam} = Enthalpy of steam
n_{steam} = Total lbs of steam produced by boiler

The average btu/lb of fuel was obtained from historical data and was used as a constant for the computation of the efficiency.

Data on 13 parameters selected by a domain expert was collected in 1-minute intervals. In this research, a dataset of over 10,000 observations was considered. The list of the parameters is shown in Table 3.

During the preprocessing phase, all observations collected during the daily calibration of the instruments and gauges were removed. Based on computational experimentation, the original data was transformed using a 20-minute moving average. This reduced the noise of the parameters and improved prediction accuracy of knowledge produced by the data-mining algorithm. Experiments have also led to the discretization of the efficiency into to 2% intervals (see Table 4). Discretizing efficiency into crisp intervals will likely increase the error rate of the classifier and result in some reduction of information. That is to say, there is a small difference between 85.95% and 86.01%, but these efficiencies will be labeled into two distinct intervals (i.e., 84-86 and 86-88). Utilizing fuzzy logic could overcome this concern. This research

Table 3. List of parameters

Parameter List	
Boiler Master	Average Middle Bed
Air Master	Average Lower Bed
Biomass Feed Rate	Bed Pressure Median
Furnace Draft	Air Fuel Ratio
Average O$_2$	Ratio SA / PA
SA Fan Flow	DTemp
PA Fan Flow	

Table 4. Discretized values of efficiency

Category	Efficiency	Category	Efficiency
L_80	Efficiency $\leq$ 80%	90_92	90% < Efficiency $\leq$ 92%
80_82	80% < Efficiency $\leq$ 82%	92_94	92% < Efficiency $\leq$ 94%
82_84	82% < Efficiency $\leq$ 84%	94_96	94% < Efficiency $\leq$ 96%
84_86	84% < Efficiency $\leq$ 86%	96_98	96% < Efficiency $\leq$ 98%
86_88	86% < Efficiency $\leq$ 88%	G_98	98% > Efficiency
88_90	88% < Efficiency $\leq$ 90%		

focuses on defining control settings leading to a relative improvement of boiler efficiency, making the potential loss of information acceptable. The proposed interval approach is also supported by a measurement error of some parameters that could be significant and varying in time.

Data Mining

To generate multiple rule sets the initial data set was partitioned into two data subsets (Figure 1). Each data subset consisted of at least 2,500 observations. To further increase the number of rule sets,

different partial-parameter subsets were created for each data subset. Each data subset contained a unique combination of parameters. This created seven knowledge bases for the data-mining algorithm (C4.5, Quinlan, 1986) to explore. At least eight parameters were used in every application of the algorithm and all applications resulted in a 10-fold cross validation (Stone, 1974) with classification accuracy greater than 90%.

The 10 rules with the highest strength (Mitra, Pal, & Mitra, 2002) that described the three efficiency intervals were selected from each rule set. The selected rules were then incorporated

Figure 4. Coding scheme

Bit position	B1	B2	B3	....	B11	B12	B13
Features	Boiler Master	Air Master	Air Fuel Ratio	....	AverageO_2	SA Fan Flow	Biomass Feed Rate
Phenotype	109.80	102.97	10.28	....	5.39	57.01	13.27
Genotype	2.74	2.00	-1.12	....	2.66	0.21	2.93

Figure 5. GA input screen

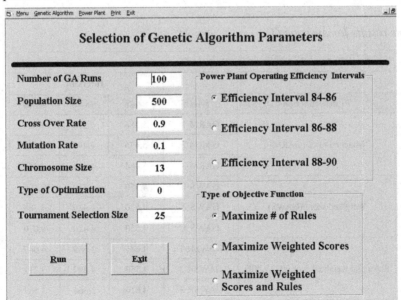

into the fitness functions discussed in Section 2.7. This amounted to 70 rules that described each of the three intervals. In this application, score1 is equivalent to the efficiency interval 84-86, score2 is equal to interval 86-88, and score3 coincides with the interval 88-90.

All the parameters consisted of continuous real values and were standardized using the z-transformation. A sample of the 13-bit chromosome is shown in Figure 4.

Control Signature Generation

Standard settings were applied at each iteration of the GA. The GA used a single-point crossover with a probability of 0.9 and an initial mutation rate of 0.1. The mutation rate was set to incrementally increase up to 0.4 after 10 generations of obtaining the same solution. The population size was set to 500, and a tournament selection of 25 observations was employed to select chromosomes for breeding. The user interface is shown in Figure 5.

To identify control signatures, the GA was run nine times for each efficiency interval. The nine trials were comprised of three replications for

Table 6. Control signature for Average O_2

Average O_2			
	Interval		
	84-86	**86-88**	**88-90**
Min	4.530	3.639	3.567
Average	4.860	4.192	3.825
Max	5.484	4.420	4.084

each of the three unique fitness functions. Table 5 displays the results from the nine GA trials for the parameter, Average O_2.

The control signature was determined by computing the minimum, maximum, and average for each efficiency interval. Table 6 presents the control signature obtained for the Average O_2. Control signatures were obtained for all of the 13 parameters.

Computational Results

The development control signatures identify trends in parameter settings that contribute to

Table 5. GA runs results for average O_2

Average O_2				
		Interval		
Fitness function	**Trial**	**84-86**	**86-88**	**88-90**
Matched rules (GARM)	GARM 1	5.386	4.290	4.084
	GARM 2	5.436	4.295	3.797
	GARM 3	5.484	4.404	3.751
Weighted score (GAWS)	GAWS 1	4.550	4.310	3.606
	GAWS 2	4.530	4.420	3.788
	GAWS 3	4.550	4.420	4.070
Rules and weighted score (GAWSR)	GAWSR 1	4.641	3.639	4.047
	GAWSR 2	4.550	4.290	3.711
	GAWSR 3	4.610	3.664	3.567

Figure 6. DTemp control signature

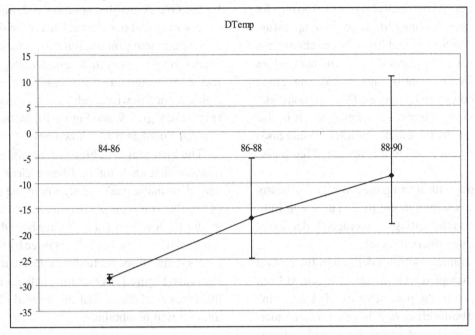

Figure 7. Average O$_2$ control signature

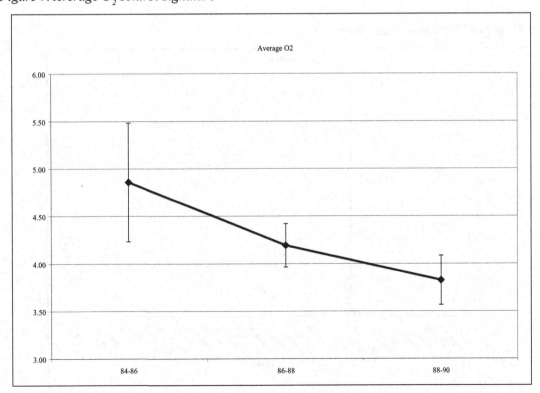

improved understanding of the boiler process. Figure 6 depicts the control signature settings for the parameter DTemp (difference in temperature between the middle and lower boiler chambers). The error bars represent the min and max values defined by the control signatures. The graph clearly shows a trend that the DTemp should decrease to obtain increased efficiencies. It is also evident that as the efficiency interval increases the range of acceptable settings for DTemp also increases.

The graph illustrating the Average O_2 control signature is shown in Figure 7. The graph indicates that as the setting of Average O_2 decreases the efficiency interval increases.

The parameter settings defined by the control signature coupled with the insights gained from the analysis of the parameter trends lead to the increased boiler efficiency. To gain further understanding into the parameter settings, histograms

were developed for all feasible solutions obtained by the GA. A feasible solution is defined as an observation that matches at least one rule for the desired efficiency interval and does not match any rules related to other intervals. A histogram was then constructed on all the feasible solutions for a given efficiency interval. The three histograms (Figure 8, Figure 9, and Figure 10) for the parameter Air Fuel Ratio are shown next.

The histograms in Figure 8 through 10 demonstrate that each interval has a clear, unique, and identifiable peak value, where the majority of feasible solutions occur. The identified peaks led to the development of tighter bounds for the Air Fuel Ratio. The bounds obtained by the histograms define three mutually exclusive operating intervals (Figure 11). Operating in these intervals increases confidence that the desired efficiency interval will be obtained.

Figure 8. Air Fuel Ratio histogram: Interval 84-86

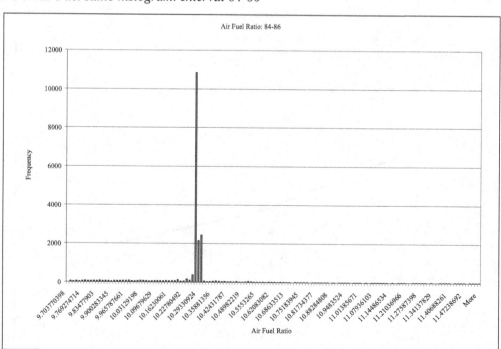

Figure 9. Air Fuel Ratio histogram: Interval 86-88

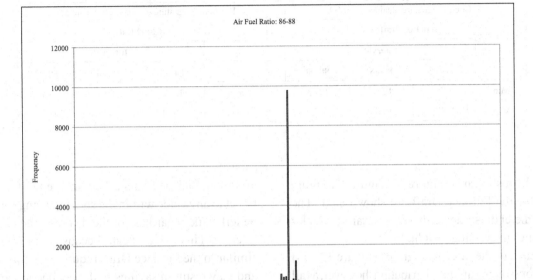

Figure 10. Air Fuel Ratio histogram: Interval 88-90

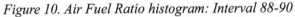

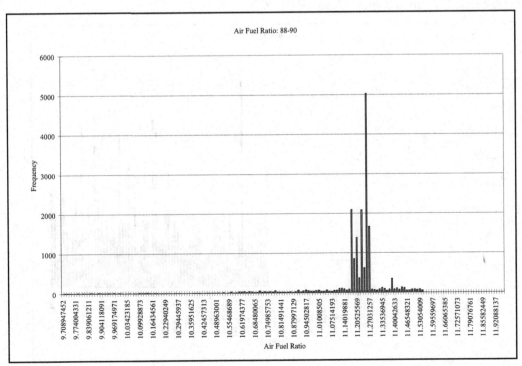

Figure 11. Control signatures comparison for Air Fuel Ratio

	A. Control signature obtained from GA				B. Control signature obtained from histograms		
	Air Fuel Ratio				**Air Fuel Ratio**		
	Interval				Interval		
	84-86	**86-88**	**88-90**		**84-86**	**86-88**	**88-90**
Min	10.28	10.33	10.78	Min	10.28	10.54	11.17
Max	10.73	10.65	11.25	Max	10.31	10.58	11.73

The histograms (Figure 12, Figure 13, Figure 14) for the Furnace Draft are shown next. The parameter illustrated in these histograms provides another interesting insight.

Each of the three histograms (Figure 12 - 14) demonstrates a unique distribution, however, as all share interesting similarities. Each histogram has approximately the same lower and upper bound, -0.92 to -0.12, respectively. This fact indicates that this particular parameter is robust to changes in all three efficiency intervals. Furnace Draft can be set to any value within the specified range and be left static regardless of the desired efficiency interval. This is significant because parameters similar to the Furnace Draft require less control and may result in savings in data collection and storage.

The three histograms for the parameter Bed Pressure Median are illustrated in Figure 15, Figure 16, and Figure 17. These figures represent a

Figure 12. Furnace Draft histogram: Interval 84-86

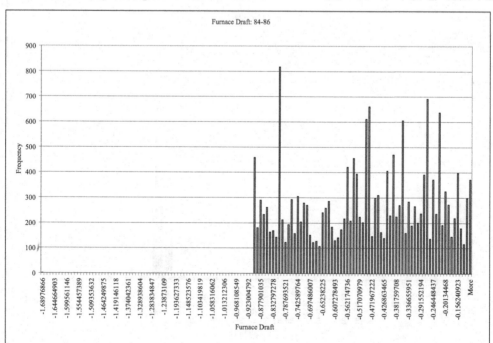

Figure 13. Furnace Draft histogram: Interval 86-88

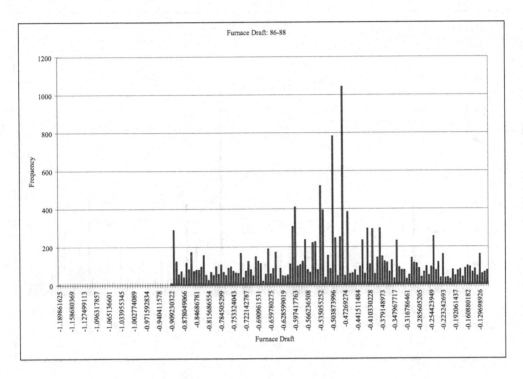

Figure 14. Furnace Draft histogram: Interval 88-90

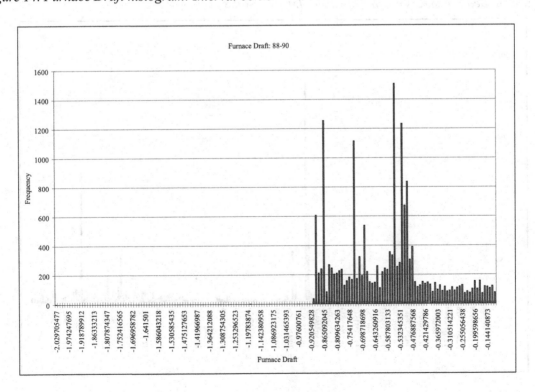

Figure 15. Bed Pressure Median histogram: Interval 84-86

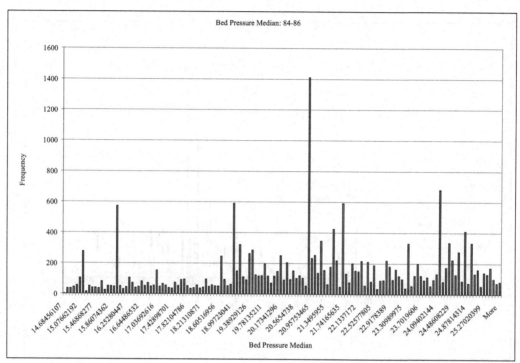

Figure 16. Bed Pressure Median histogram: Interval 86-88

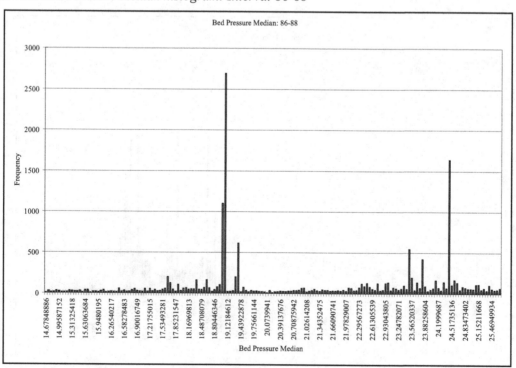

Figure 17. Bed Pressure Median histogram: Interval 84-86

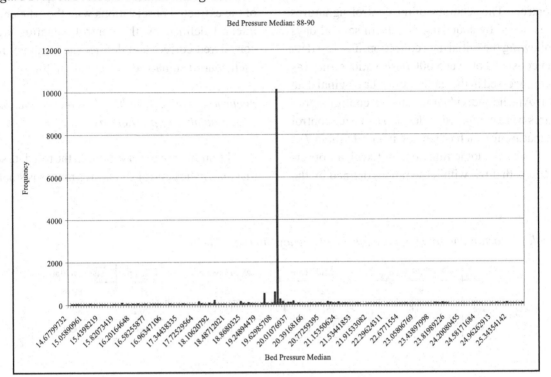

third type of parameter action within the feasible solution space.

The histograms (Figure 15 - 17) for Bed Pressure Median show a parameter that reacts differently depending on the efficiency interval. In Figure 15, the space of feasible solutions is robust, and any setting greater than 14.5 results in the 84-86 efficiency interval. The interval 86-88 (Figure 16) shows a bimodal distribution with peak values at 19 and 24.5. These two peaks will more likely yield efficiency in the 86-88 interval. The final histogram (Figure 17) demonstrates a single definitive peak similar to the histograms for the parameter Air Fuel Ratio (Figure 8 through Figure 10).

The three parameters (Air Fuel Ratio, Furnace Draft, and Bed Pressure Median) demonstrate the three main pattern types discovered in this research. Mutually exclusive operating intervals were seen in the histograms for the parameter Air

Fuel Ratio. These types of operating conditions are desirable due to the fact that at each interval, there are distinct and tight operating ranges. The second pattern was demonstrated with the parameter Furnace Draft. The histograms developed from that parameter display large robust operating ranges that remained constant between the efficiency intervals. This can be advantageous because it allows for selection of any setting within the operating range. The last pattern can be seen in the Bed Pressure Median histograms. The histogram for each efficiency interval has a different pattern. These types of parameters may require further investigation.

Testing

The testing accuracies are different then cross-validation/data-mining accuracies. They are based on different test data sets that are obtained without the application of control signatures for process controls. The accuracy of the control signatures was tested by acquiring data from several days following the initial data collection. The test data set consisted of over 5,000 observations and was preprocessed in the same way as the original data set. All the individual parameter control signatures were combined to develop a master control signature for each efficiency interval (Table 7).

Using the logic functions in Excel, all observations that fell within the ranges defined by the control signature were extracted. The percentage of the extracted observations with the efficiency interval defined by the control signature was considered correct and defines the accuracy for each control signature (see equation 20).

$$Accuracy = 100\% * Number\ of\ correct\ predictions / Total\ number\ of\ predictions \qquad (20)$$

The number of observations in the test data set that match the control signatures ranges is small;

Table 7. Partial control signature for the efficiency interval: 84-86

	Air Fuel Ratio	Avg Mid Temp	Avg Lower Bed	DTemp	PA Fan Flow
Min	9.6	1521.3	1537.3	-31.5	69.6
Average	10.6	1565.1	1585.3	-20.2	88.3
Max	12.1	1606.4	1626.8	16.0	106.7

Figure 18. Comparison of control signature accuracy

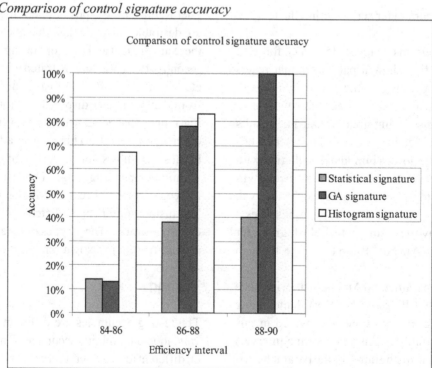

however, if the observations are within the ranges, then the accuracy for identifying the appropriate outcome is high. Due to the tight control signature ranges defined by the GA, the constraints were relaxed in order to increase the number of predictions. The signatures were relaxed by reducing the number of parameters used in making the predictions. The relaxed signatures used between 4 and 8 of the 13 parameters. The accuracy was computed for the three efficiency intervals (84-86, 86-88, and 88-90) of interest.

The accuracy was also computed for control signatures derived from a classical statistical method. The statistical control signature was obtained by calculating the min, max, and average for each parameter for the three intervals. These values were computed using the original boiler data set. The accuracy was also calculated from the histograms analysis. The results are shown in Figure 18.

The results demonstrate that the histogram signature has the highest accuracy regardless of the efficiency interval. The statistical control signature has the worst performance except for the 84-86 interval. The relaxing of the GA control signatures may account for a reduction of accuracy. The true GA control signature accuracy should be higher than the relaxed accuracy.

The results demonstrate that the tighter bounds defined through the GA and histogram analysis are, on average, more accurate (see Table 8) than the statistical control signature.

Table 8. Average accuracy of control signatures

Control signature	Average accuracy
Statistical signature	31%
GA signature	64%
Histogram signature	83%

CONCLUSION

A hybrid method combining genetic algorithm and data mining for deriving control signatures was proposed and successfully applied to improve the operations of a power plant boiler. The genetic algorithm evaluated a multiobjective fitness function comprised of rules obtained from several partial data sets. A control signature defines the feasible region marked by the rule intersections, thus providing tighter bounds for operating parameters. Operating at these ideal parameter settings will improve the confidence of achieving the desired outcomes. The hybrid method has the capability of visualizing the complex relationships through the analysis of trends and the feasible solutions generated by the genetic algorithm.

The application of the control signatures to the power plant data has led to the development of robust and at times mutually exclusive parameter settings. The hybrid method provided valuable insight regarding the Air Fuel Ratio, Furnace Draft, and so on, with respect to the different efficiency intervals. The control signatures reduced the parameter ranges by 53% (i.e., tightening the bounds) while increasing the testing accuracy by 52% over the standard statistical method. The control signatures increased boiler efficiency, resulting in decreased operating costs and the reduction in fuel consumption. The method outlined can be applied to diverse domains with similar results. The research reported in this chapter contributes to the development of real-time intelligent control systems.

REFERENCES

Bakirtzis, A., Biskas, P., Zoumas, C., & Petridis, V. (2002). Optimal power flow by enhanced genetic algorithm. *IEEE Transactions on Power Systems, 15*(1), 229-236.

Booth, R. C., & Roland, W. B. (1991). Neural network-based combustion optimization reduces NOx emissions while improving performance. In *Proceedings of the 1991 IEEE Workshop on Dynamic Modeling Control Applications for Industry Applications* (pp. 1-6).

Chong, A. Z. S., Wilcox, S. J., & Ward, J. (2000). Neural network models of the combustion derivatives emanating from a chain grate stoker fired boiler plant. In *Proceedings of the. 2000 IEEE Seminar Advanced Sensors and Instrumentation Systems for Combustion Processes* (pp. 61-64).

Cios, K. J., Pedrycz, W., & Swiniarski, R. (1998). *Data mining methods for knowledge discovery.* The Netherlands: Kluwer, Dordrecht.

Dimeo, R., & Lee, K.Y. (1995). Boiler-turbine control system design using a genetic algorithm. *IEEE Transactions on Energy Conversion, 10*(4), 752-759.

Fayyad, U., Piatetsky-Shapiro, G., Smyth, P., & Uthurusamy, R. (1995). *Advances in knowledge discovery and data mining.* Cambridge, MA: AAAI/MIT Press.

Ghezelayagh, H., & Lee, K. Y. (2002). Intelligent predictive control of a power plant with evolutionary programming optimizer and neuro-fuzzy identifier. In *Proceedings of the 2002 IEEE World Congress on Computational Intelligence* (pp. 1-6).

Goldberg, D. E. (1989). *Genetic algorithms.* New York: Addison Wesley Longman, Inc,.

Holland, J. H. (1975) *Adaptation in natural and artificial systems.* Cambridge, MA: MIT Press.

Krishnakumar, K., & Goldberg, D. E. (1992). Control system optimization using genetic algorithms. *Journal of Guidance, Control, and Dynamics, 15*(3), 735-740.

Kusiak, A. (2001). Rough set theory: A data mining tool for semiconductor manufacturing.

IEEE Transactions on Electronics Packaging Manufacturing, 24(1), 44-50.

Kusiak, A. (2001a). Feature transformation methods in data mining. *IEEE Transactions on Electronics Packaging Manufacturing, 24*(3), 214-221.

Kusiak, A. (2002). A data mining approach for generation of control signatures. *ASME Transactions: Journal of Manufacturing Science and Engineering, 124*(4), 923-926.

Kusiak, A., Shah, S., & Dixon, B. (2003). Data mining enhanced decision-making approach for kidney dialysis survival time. In *Proceedings of the 5th IFAC Symposium on Modeling and Control in Biomedical Systems*, Melbourne, Australia, August 2003 (pp. 35-39).

Lawrence, D. (1987). Genetic algorithms in search, optimization, and machine learning. Addison-Wesley.

Li, K., Thompson, S., & Peng, J. (2002). GA based neural network modeling of NOx emission in a coal-fired power generation plant. In *Proceedings of the 15th IFAC World Congress on Automatic Control.*

Lee, K. Y., Perakis, M., Sevcik, D., Santoso, N. I., Lausterer, G., & Samad, T. (2000). Intelligent distributed simulation and control of power plants. *IEEE Transactions on Energy Conversion, 15*(1), 116-123.

Michalewicz, Z. (1992). Genetic algorithms + data structures = evolution programs. Berlin: Springer-Verlag.

Mitra, S., Pal, S. K., & Mitra, P. (2002). Data mining in soft computing framework: A survey. *IEEE Transactions on Neural Networks, 13*(1), 3-14.

Obitko, M. (2004). *Introduction to genetic algorithms*, 1998. Retrieved from http://cs.felk.cvut.cz/~xobitko/ga/, Accessed on 02/03/2004

Park, J. B., Park, Y. M., Won, J. R., & Lee, K. (2000). An improved genetic algorithm for generation expansion planning. *IEEE Transactions on Energy Conversion, 15*(3), 913-922.

Pham, D., & Karaboga, D. (2000). *Intelligent optimization techniques.* London: Springer-Verlag.

Quinlan, J. R. (1986). Induction of decision trees. *Machine Learning, 1*(1), 81-106.

Stephan, V., Debes, K., Gross, H. M., Wintrich, F., & Wintrich, H. (2001). A new control scheme for combustion process using reinforcement learning based on neural networks. *International Journal of Computational Intelligence and Applications, 1*(2), 121-136.

Stone, M. (1974). Cross-validatory choice and assessment of statistical predictions. *Journal of the Royal Statistical Society, 36*, 111-147.

APPENDIX A

Solution-Encoding Scheme

The GA solution-encoding scheme used in this method arranges all the numeric parameters at the beginning of a chromosome followed by the categorical parameters (Figure A1). Thus, each chromosome is split into two parts, namely numeric section (N-section) and categorical section (C-section). Each gene position represents a parameter making the length of the chromosome equal to the number of parameters. The phenotype representation of the N-section and C-section chromosome are the actual values for respective parameters. The genotype representation for N-section is a z-transform of the original value, while for C-section, it is based on a probability values. The genotype representation is used in the GA.

Proposed Crossover and Mutation Operators

For numerical parameters, Z-transform was used to standardize the parameters. The standardization procedure eliminated the need to develop special GA operators, as crossover (single/multiple point) and mutation operation can be performed without worrying about maintaining the feasibility of the parameter values. For categorical parameters, they are placed at the end of the coding scheme, and two phase crossover and individualized mutation operators were used.

In the first phase, standard crossover and mutation with respect to the numerical parameters is performed (Table A1, Figure A2, and Table A2). In the second phase, individualized mutation operators, with respect to each categorical parameter, are carried out (Table A1, Figure A3, and Table A2). This scheme requires one crossover probability for N-section (0.9), and different mutation probabilities for N-section (0.1) and the C-section (0.6), respectively. The C-section mutation operator ensures the feasibility of each parameter and is set at a higher rate to facilitate the faster exploration process.

Figure A1. Coding scheme

Position	Numeric parameter				Categorical parameter		
	N1	N2	N3	N4	C1	C2	C3
Parameters	Boiler Master	Air Master	Air Fuel Ratio	Avg Mid Temp	Test 1	Test 2	Test 3
Phenotype	109.803	102.971	10.28	1524.029	A	C	T
Genotype	2.737	1.996	-1.116	-2.212	A	C	T

Table A1. Parent chromosomes

	N1	N2	N3	N4	N5	N6	C1	C2	C3	C4
Chromosome 1	0.23	0.58	1.25	-2.57	0.89	1.25	A	C	T	C
Chromosome 2	0.58	0.92	1.75	0.01	-0.75	2.25	T	T	A	C

Figure A2. N-section crossover and mutation

	Phase I						
	Parameter	N1	N2	N3	N4	N5	N6
	Chromosome 1	0.23	0.58	1.25	-2.57	0.89	1.25
Crossover	Chromosome 2	0.58	0.92	1.75	0.01	-0.75	2.25
	Post crossover						
	Chromosome 1	0.23	0.58	1.25	0.01	-0.75	2.25
	Chromosome 2	0.58	0.92	1.75	-2.57	0.89	1.25
	Parameter	N1	N2	N3	N4	N5	N6
	Chromosome 1	0.23	0.58	1.25	**0.01**	-0.75	2.25
Mutation	Chromosome 2	0.58	0.92	1.75	-2.57	0.89	1.25
	Post mutation						
	Chromosome 1	0.23	0.58	1.25	1.45	-0.75	2.25
	Chromosome 2	0.58	0.92	1.75	-2.57	0.89	1.25

Figure A3. C-section mutation

	Phase II				
	Parameter	C1	C2	C3	C4
	Chromosome 1	A	**C**	**T**	C
	Chromosome 2	T	**T**	A	C
Mutation	Parameter	C1	C2	C3	C4
	Chromosome 1	A	T	A	C
	Chromosome 2	T	G	A	A

Table A2. Children chromosomes

	N1	N2	N3	N4	N5	N6	C1	C2	C3	C4
Chromosome 1	0.23	0.58	1.25	1.45	-0.75	2.25	A	T	A	C
Chromosome 2	0.58	0.92	1.75	-2.57	0.89	1.25	T	G	A	A

Chapter XII
Bayesian Belief Networks for Data Cleaning

Enrico Fagiuoli
Università degli Studi di Milano-Bicocca, Italy

Sara Omerino
ETNOTEAM S.p.A., Italy

Fabio Stella
Università degli Studi di Milano-Bicocca, Italy

ABSTRACT

The importance of data cleaning and data quality is becoming increasingly clear, as evidenced by the surge in software, tools, consulting companies, and seminars addressing data quality issues. In this contribution, the authors present and describe how Bayesian computational techniques can be exploited for data-cleaning purposes to the extent of reducing the time to clean and understand the data. The proposed approach relies on the computational device named Bayesian belief network, which is a general statistical model that allows the efficient description and treatment of joint probability distributions. This work describes the conceptual framework that maps the Bayesian belief network computational device to some of the most difficult tasks in data cleaning, namely imputing missing values, completing truncated datasets, and outliers detection. The proposed framework is described and supported by a set of numerical experiments performed by exploiting the Bayesian belief network programming suite named HUGIN.

INTRODUCTION

Every data analysis task starts by gathering, characterizing, and cleaning a new, unfamiliar dataset (Dasu & Johnson, 2003). After this process, the data can be analyzed and the results delivered. It is well established that the first step is far more difficult and time consuming than the second.

Indeed, data gathering is complicated by sociological (turf sensitivity) and technological problems (different software and hardware platforms make transferring and sharing data very difficult). Once the data are in place, acquiring the metadata (data description and business rules) is another challenge. Indeed, very often the metadata are poorly documented, and when we are ready to analyze the data, its quality is suspect. Fortunately, automated techniques can be applied to understand the data through *exploratory data mining,* and to ensure *data quality* through *data cleaning.*

Data cleaning and quality monitoring is an incessant and continuous activity starting right from data gathering stage to the ultimate choice of analysis and interpretation of the results. It is needed to update the static conventional definitions and metrics of data quality to reflect the continuous and flexible nature of data quality process and metrics required to effectively measure and monitor data quality (Scannapieco, Missier, & Batini, 2005).

According to a study conducted by The Data Warehouse Institute, commissioned by DataFlux (The Data Warehouse Institute, 2003), current data quality problems cost U.S. business more than 600 billion dollars a year. Furthermore, a survey from conversation to practitioners of *data mining* leads to assert that between 30% to 80% of the data-analysis task is spent in cleaning and understanding the data. Therefore, the importance of data cleaning and data quality is becoming increasingly clear as evidenced by the surge in software, tools, consulting companies, and seminars addressing data quality issues.

A taxonomy of data-quality problems, addressed by data cleaning, together with an overview of the main solution approaches has been proposed by Rahm and Hai Do (2000).

Several contributions, devoted to cleaning databases containing corrupted data, have been proposed in the specialized literature.

Guyon, Matic, and Vapnik (1996) emphasize the link between informative patterns and data cleaning, and describe how machine learning approaches can be exploited to remove noise from a database.

Schwarm and Wolfman (2000) pointed out that many techniques have attempted to use learners to predict problems with class values in datasets, with the same approach being extended to correct errors in data. However, these techniques suffer several problems: they can only actually correct noise in the class attribute, do not fully leverage dependencies among attributes, and are inappropriate for datasets with no distinguished class attribute.

As far as the authors know, Schwarm and Wolfman were the first to propose the use of Bayesian belief networks for data cleaning. However, the approach described through this work significantly differs from their approach in the sense that it is unsupervised, and therefore it does not require the availability of any subset of precleaned instances from the database to be cleaned, as required by Schwarm and Wolfman (2000).

Arning, Agrawal, and Raghavan (1996) described a linear time method for detecting deviations in a database. The authors assume that all records should be similar, which may not be true in unsupervised learning tasks, and that an entire record is either noisy or clean.

The assumption that entire records are noisy or clean is also common in outlier and novelty detection (Hampel, Rousseeuw, Ronchetti, & Stahel, 1986; Huber, 1981).

However, as pointed out in Kubika and Moore (2003), a significant downside to looking at noise on the scale of records is that entire records are thrown out, and useful, uncorrupted data may be lost. In datasets where almost all records have at least a few corrupted cells, this may prove disastrous. The approach described in Kubika and Moore (2003) is particularly interesting, and uses the data to learn a probabilistic model containing three components: a generative model of the clean records, a generative model of the noise values, and a probabilistic model of the corruption process.

In this work, the authors introduce and describe a conceptual framework that maps the *Bayesian belief network* computational device to some of the most difficult tasks of data cleaning, namely *imputing missing values, completing truncated datasets,* and *outliers detection.* Furthermore, the conceptual framework is supported by a set of numerical experiments performed by means of three sample databases. Numerical experiments have been performed through the implementation of a software prototype that relies on the Bayesian belief network programming suite offered by the HUGIN software package (Andersen, Olesen, & Jensen, 1990).

The rest of the work is organized as follows. The section titled "Bayesian Belief Networks" is devoted to present the main characteristics of the Bayesian belief network computational device. The "BBNs for Data Cleaning" section describes, through a sample database, the conceptual framework for the data-cleaning task by using the Bayesian belief network computational device. Furthermore, this section provides a quantitative comparison between the performance of the Bayesian belief network device and a commercial tool for outliers detection, namely the GritBot software package (Rulequest Research). Finally, conclusions and directions for further research are reported in the last section.

BAYESIAN BELIEF NETWORKS

Bayesian belief networks (BBNs) (Jensen, 1996; Neapolitan, 1990; Pearl, 1988), which emerged within a more general framework of Bayesian statistics, are general statistical models that allow the efficient description and treatment of joint probability distributions. BBNs proved to be a very useful tool for combining informal expert knowledge with statistical techniques for distribution evaluation. BBNs are specifically designed for cases when the vector of random parameters can have considerable dimension and/or it is difficult to come up with traditional parametric models of the joint distribution of random parameters. BBNs have been utilized in application domains such as, for example, image processing (Geman & Geman, 1984), medical diagnosis (Spiegelhalter, Dawid, Lauritzen, & Cowell, 1993), modeling the Internet and the Web (Baldi, Frasconi, & Smyth, 2003), and reliability analysis of integrated circuits manufacturing (Gaivoronski & Stella, 1998).

A Bayesian belief network (BBN) is a graphical model used to describe dependencies in a probability distribution function defined over a set of variables. Namely, dependencies among variables are represented in a graphical fashion, and exploited to decompose (factor) the joint distribution in terms of conditional independence relations defined over subsets of variables. In other words, a BBN is a graphical way of representing a particular joint distribution factorization. Formally, a BBN model M consists of a set of n discrete random variables $X_1,..,X_n$, and an underlying directed acyclic graph (DAG) $G = (V, E)$, such that each random variable is uniquely associated with a vertex of the DAG. The BBN model M is completely specified by means of the DAG G together with a set of conditional probability tables $P(X_i \mid pa[X_i])$, $i = 1,...,n$, where $pa[X_i]$ denotes the parents of node X_i, that is, the set of variables that directly influence the random variable X_i.

The main characteristic of the BBN model M is that its joint probability distribution, that is, the joint probability distribution for the random vector $(X_1,...,X_n)$, can be represented through the factorization of the conditional probabilities for each random variable X_i according to:

$$P(X_1,...,X_n) = \prod_{i=1}^{n} P(X_i \mid pa[X_i]). \qquad (1)$$

In particular, in the case when the random variable X_i has no parents (no directed links oriented towards the node associated with X_i), that is, the set $pa[X_i]$ is empty, the conditional probability $P(X_i|pa[X_i])$ is simply its marginal probability $P(X_i)$. As already mentioned, BBNs are often represented in a graphical form that helps in emphasizing the causal nature of the directed links between the random variables of the DAG. A classic example of BBN (*Figure 1*) is provided by the "Hypothetical Medical Belief Networks" (Cooper, 1984; Pearl, 1988). This model consists of five binary (yes, no) random variables (nodes) namely,

- X_1, Metastatic Cancer
- X_2, Serum Calcium
- X_3, Brain Tumor
- X_4, Coma
- X_5, Severe Headaches

The directed links represent causal relationships between random variables. According to *Figure 1*, the "Metastatic Cancer" is a direct cause for "Serum Calcium" as well as for "Brain Tumor,"

"Serum Calcium" and "Brain Tumor" are direct causes for "Coma," and finally, "Brain Tumor" is direct cause for "Severe Headaches."

The BBN in *Figure 1* describes the joint probability distribution $P(X_1, X_2, X_3, X_4, X_5)$ according to equation (1) as follows:

$$P(X_1, X_2, X_3, X_4, X_5)$$
$$= P(X_1)P(X_2 \mid X_1)P(X_3 \mid X_1)$$
$$P(X_4 \mid X_2, X_3)P(X_5 \mid X_3). \qquad (2)$$

As shown through the classic example of BBN in *Figure 1*, the power of BBNs is that one can infer conditional variable dependencies by visually inspecting the DAG and exploiting the concept of *conditional independence* (Pearl, 1988). The conditional independence property of BBNs is very important because it has a direct impact on the inference task that consists of deducing what a distribution over a particular subset of random variables is, given that one knows the states of some other variables in the network. More precisely, one needs to efficiently calculate a particular conditional or marginal probability distribution function from the one defined by the net.

Figure 1. Medical Bayesian belief network

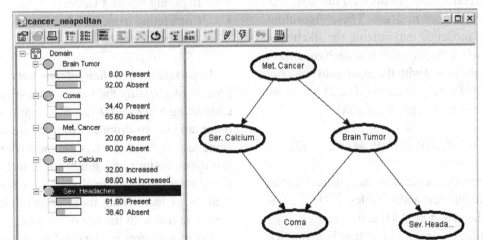

Another important feature of BBNs is that many algorithms for learning are available. There are several levels of learning in BBNs, ranging from learning the entire graph structure (the edges in the model) to learning the conditional distributions when the structure is known (Baldi et al., 2003). As a first approximation, four different situations can be considered, depending on whether the structure of the network is known or not, and whether the network contains unobserved data, such as hidden variables, which are completely unobserved in the model. In the case when the structure is known and no hidden variables are suspected, the problem is relatively simple and consists of estimating conditional probabilities $P(X_i|pa[X_i])$ from observed frequencies. On the contrary, learning both the structure, that is, the DAG, and the parameters, that is, the conditional probabilities $P(X_i|pa[X_i])$ of a BBN where hidden variables are suspected, can be a prohibitive task. Reasonable computational approaches are available in the intermediate cases: in particular, in the case when the structure is known but hidden variables are suspected, the Expectation Maximization (EM) algorithm (Dempster, Laird, & Rubin, 1977). When the structure is unknown, but no hidden variables are assumed, a variety of search algorithms can be formulated to search for the structure (and parameters) that optimize some performance measure. These algorithms select the model M maximizing the likelihood $P(D|M)$, while the Bayesian approach consists of selecting the model with the maximum posterior probability $P(M|D)$ given the data D where we average over parameter uncertainty,

$$P(M\,|\,D) \propto \int_{\Theta} P(D\,|\,\theta, M)P(\theta\,|\,M)d\theta, \quad (3)$$

where θ represents the BBN parameters, that is, the conditional probability tables, $P(D\,|\,\theta, M)$ is the likelihood and $P(\theta\,|\,M)$ is the prior.

According to equation (3), there is an implicit penalty in effect; indeed, by averaging over the parameter space, instead of picking the most likely parameter: more complex models will in effect be penalized for having a higher-dimensional parameter space, and will only "win" if their predictive power on the training data can overcome the inherent penalty that arises from having to integrate over a higher-dimensional space.

Heckerman (1999) provides a full and detailed discussion of how the Bayesian estimation approach can be exploited to automatically construct BBNs from data.

BBNS FOR DATA CLEANING

The data-cleaning task deals with several aspects of data quality improvement, as described by Dasu and Johnson (2003).

There is no panacea, no single tool that can solve a majority of data quality problems. Indeed, data quality problems are highly complex and context dependent, requiring extensive domain knowledge and involving solutions that often need to be chosen case by case. Therefore, according to the data cleaning and data quality taxonomy introduced and described in Dasu and Johnson (2003), we decided to focus the attention to the following specific aspects of data cleaning:

- Imputing missing values
- Completing truncated datasets
- Outliers detection

In particular, *imputing missing values* is the process of guessing the values of missing data; *completing truncated datasets* is required when observations are dropped from the dataset for some reason, for example, customers who spend less than a dollar a year might not be included in a customer database; finally, *outliers detection* consists of identifying those observations that are not in line with the rest of the data.

Let us now present the database that will be used in the sequel of this work to illustrate how the

BBN computational device can be appropriately exploited for solving the three data-cleaning tasks listed previously.

The available database consists of 1,000 records, concerning the *Italian School-University system*, where the following fields are recorded:

- **Age**; ranges from 16 to 27
- **Gender**; binary variable (Male, Female)
- **HighSchool**; binary variable (yes, no) indicating whether or not the subject has the High School degree
- **1st Level Degree**; binary variable (yes, no) indicating whether or not the subject has the 1st Level Degree
- **2nd Level Degree**; binary variable (yes, no) indicating whether or not the subject has the 2nd Level Degree
- **PhD**; binary variable (yes, no) indicating whether or not the subject has the PhD

The first task that has been performed is the structural learning of the BBN model *M* from the available database *D*. We exploited the *Learning Wizard* offered by the HUGIN Researcher 6.3 software package that maximizes the posterior probability $P(M|D)$ given the database *D* according to equation (3). The *Level of Significance* parameter was set to *0.05* and the structural learning algorithm was set to *PC*. The resulting BBN model is depicted in *Figure 2*, where the right pane shows the learned BBN model, while the left pane visualizes, for each field, its marginal distribution over the corresponding support.

Once, the BBN model, DAG, and conditional probability tables has been recovered, it is possible to proceed further and implement the three data-cleaning tasks considered through this work.

The *imputing missing values* task can be accomplished by achieving optimality. Indeed, it is well known that the Bayesian decision rule is optimal and that the *Bayes Risk* is the best performance that can be achieved (Duda, Hart, & Stork, 2001). In order to clarify how the imputing missing values task is accomplished, by using the BBN model depicted in *Figure 2*, let us consider the following record:

Figure 2. BBN model for the School-University database

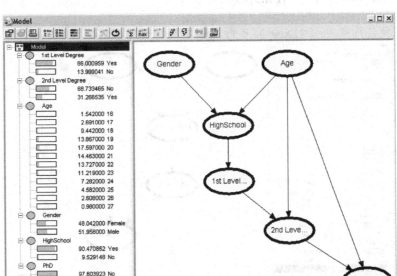

Field	Value
Age	*23*
Gender	*Female*
HighSchool	*yes*
1st Level Degree	*yes*
2nd Level Degree	*yes*
PhD	*?*

where the question mark (?) means that, for the given record, the **PhD** field value is missing.

The imputing missing values task is accomplished by first entering into the BBN model in *Figure 2*, left panel, the available information (evidence), values associated with the available fields for the given record, and then finding the state, for the **PhD** field, that maximizes the posterior probability given the available information. Formally, the imputing missing values task is accomplished by solving the following optimization problem,

$$\max_{\{yes,no\}} P(PhD \mid E), \qquad (4)$$

where in abbreviated form,

E=(Age=23,Gen=Female,HS=yes,1stLD=yes,2nd LD=yes),

is the available information that is usually named *evidence* within the Bayesian framework.

The optimization problem in equation (4) when solved, using the HUGIN Researcher 6.3 software package, suggests to replace the missing value with "**PhD**=*no*". Indeed, the posterior probability for the state "no" is $P(PhD=no \mid E)$=0.9986 while the posterior probability for the state "yes" is $P(PhD=yes \mid E)$=0.0014 as shown in *Figure 3*. Notice that the missing replacement "**PhD**=*no*" would be the most probable replacement, irrespectively of the available data. Indeed, by ignoring the data for the given record, we obtain $P(PhD=no)$=0.9760 and $P(PhD=yes)$=0.0240 as shown in *Figure 4*. Therefore, apparently the BBN model gives no advantage for solving the imputing missing values task. The advantage and power of BBNs becomes clear in the case when considering the following record,

Figure 3. Posterior probability P(PhD|E)

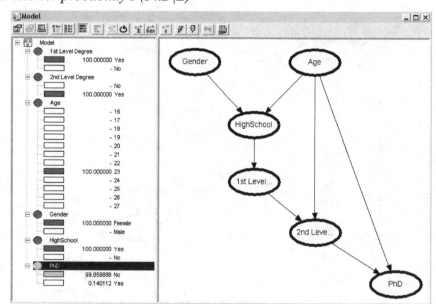

Field	Value
Age	*27*
Gender	*Female*
HighSchool	*yes*
1st Level Degree	*yes*
2nd Level Degree	*yes*
PhD	*?*

Indeed, the solution of the optimization problem in equation (4), where,

E=(Age=27,Gen=Female,HS=yes,1stLD=yes,2nd LD=yes),

suggests the missing replacement "**PhD**=*yes*" where $P(PhD=yes)=0.8421$ and $P(PhD=no)=0.1579$ (*Figure 5*), while the most probable replacement irrespectively of the available data is obviously the same as before, that is, "**PhD**=*no*".

A different methodology for the treatment of missing data was introduced by Zaffalon (2002).

In this approach, it is not tried to determine the most probable value to replace the missing observation, but the observation is allowed to assume the whole set of possible values of the variable. It follows that the probability of an event concerning a variable with missing observations is no longer a single, precise value (point probability), but assumes a set of values,

$$p \in (\underline{p}, \overline{p}),$$

according to the specific value hypothesized for the missing value (the probability is then called an *imprecise probability*, in the one dimensional case), while, in a higher dimensional case, the probability vector ranges in a convex polytope (called a *credal set*). Algorithms for the computation over credal sets of probabilities exist in literature, for some specific models (Fagiuoli & Zaffalon, 1998; Zaffalon & Fagiuoli, 2003) they are, in general, more complex than the equivalent algorithms for point probabilities; their treatment, however, goes beyond the scope of this work.

Figure 4. Prior probability P(PhD)

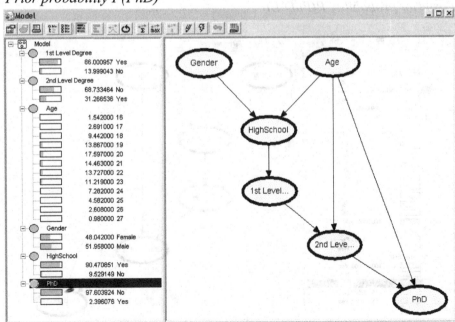

The second data-cleaning task, namely the *completing truncated datasets* task, can be accomplished by exploiting the same principle as the one exploited for the imputing missing values task. In this case, however, the optimization problem to be solved involves more than one decision variable. In order to clarify how the completing truncated datasets task can be accomplished, let us consider the following record,

Field	Value
Age	26
Gender	*Male*
HighSchool	*yes*
1st Level Degree	*yes*
2nd Level Degree	?
PhD	?

In this case, the optimization problem to solve is the following,

$$\max_{\{(yes,yes),(yes,no),(no,yes),(no,no)\}} P(2ndLD, PhD \mid E),$$

(5)

where,

$$E = (Age = 26, Gen = Male, HS = yes, 1st\ LD = yes)$$

The solution of the optimization problem in equation (5) suggests that the available record has to be completed with the following assignment,

2nd Level Degree=*yes* and **PhD**=*yes*,

whose posterior probability is $P(2ndLD=yes, PhD=yes|E)=0.6556$ while the posterior probabilities for the other configurations are $P(2ndLD=no, PhD=no|E)=0.0015$, $P(2ndLD=yes, PhD=no|E)=0.0029$, $P(2ndLD=no, PhD=yes|E)=0.3400$.

Notice that, when the available information $E=(Age=26, gen=Male, HS=yes, 1st\ LD=yes)$, is not exploited, we obtain the probabilities, for the pair of database fields (**2nd Level Degree, PhD**), reported in Table 1. Thus the considered record will be completed, according to the maximum probability criteria, as follows : **2nd Level Degree**=*no* and **PhD**=*no*.

Figure 5. Posterior probability P(PhD|E)

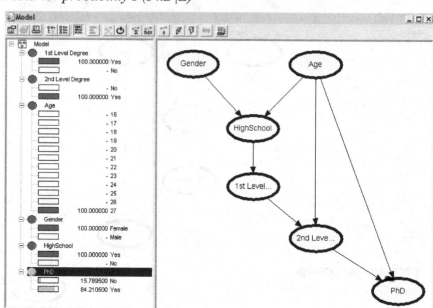

Therefore, it is evident how the record, completed by assuming the independence between the database fields and using their prior probabilities, is nonoptimal, leading to the wrong record completion.

An alternative for trying to recover the right probability distribution for the considered pair of fields (**2nd Level Degree, PhD**) would be to isolate, from the entire database, those cases where the remaining fields assume the joint assignment, (*Age*=26,*Gen=Male,HS,yes*,1*stLD=yes*), and then to compute an estimate of the joint probability for the pair of fields (**2nd Level Degree, PhD**), similar to the one reported in *Table 1*. However, in the case when either the number of records to be completed and/or the number of fields in the database is great, this procedure will lead to a computationally intensive set of queries and subsequent computations. Furthermore, in the case where the number of fields to be completed is greater than two, as shown in our example, it will be very costly to compute an approximation for the joint probability similar to the one depicted in *Table 1*. Indeed, it is evident how the computational complexity of the described procedure is exponential in both the number of fields to be completed and in the cardinality of their supports.

All these motivations support the interest for the proposed BBNs approach to accomplish the completing truncated datasets task. Indeed, BBNs are capable to efficiently memorize the database content, and to efficiently scan it to recover local probability distributions.

Finally, let us describe how the BBN computational device can be properly exploited to accomplish the *outliers detection* task. Starting from the definition of outlier, that is, an observation that is not in line with the rest of the data, it is straightforward to conclude that BBNs are well suited for accomplishing such a complex task. Indeed, given a database, and once the structural learning of the BBN model has been performed, it is possible to efficiently evaluate high dimensional probability distributions. Therefore, given the BBN model *M* for the considered database *D*, the outlier detection task can be accomplished through the solution of the following optimization problem,

$$\min P(X_1,...,X_n). \tag{6}$$

The optimization problem in equation (6) can be solved efficiently by exploiting the BBN model, that is, by exploiting the factorization of the joint probability $P(X_1,...,X_n)$ that, according to equation (1), can be written as follows,

$$P(X_1,...,X_n) = \prod_{i=1}^{n} P(X_i \mid pa[X_i]).$$

Thus, the optimization problem in equation (6) can be decomposed into a set of easier optimization problems, that is, a set of optimization problems where few decision variables (database fields) are involved. A detailed discussion about computational algorithms for solving the optimization problem on BBNs in equation (6) can be found in Gaivoronski and Stella (1998).

The outliers detection task, that is, the solution of problem in equation (6) related to records belonging to the given database and concerning the BBN depicted in *Figure 2*, allows to obtain a set of candidate outliers, some of which are

Table 1. Joint probability for the random pair (2nd Level Degree, PhD)

	2nd Level Degree=no	2nd Level Degree=yes
PhD=no	0.6708	0.3052
PhD=yes	0.0165	0.0075

reported together with their probability, that is, $P(X_1,...,X_n)$, in *Table 2*.

For matter of brevity, the candidate outliers reported in *Table 2* are only a fraction of those identified by means of the BBN model.

The analysis of the candidate outliers in *Table 2* allows the following comments. The candidate *outlier 1* is clearly such due to the fact that it is not possible to obtain the **2nd Level Degree** without the **1st Level Degree**. Candidate *outliers 2 and 3* are explained by the fact that it is not possible to have a **PhD** without the **2nd Level Degree**. Furthermore, for candidate *outlier 2* also the **Age** of *19* is suspect, due to the fact that it is unlikely that one can achieve a **PhD** at the **Age** of *19*. The candidate *outlier 4* is suspect due to the fact that it is unlikely that one can achieve a **2nd Level Degree** at the **Age** of *20*, even though this can

occur. The candidate *outlier 5* is suspect for a reason similar to the one just described. Indeed, it is unlikely that one can have a **PhD** at the **Age** of *23*. Other candidate outliers can be commented on a similar basis.

To provide a performance comparison between the BBN approach and other techniques, for the outliers detection task, we decided to scan the available database by using a trial license of the GritBot software package. This software package allows different filtering levels to detect outliers, but we decided to exploit its default values by obtaining 10 possible outliers reported in *Table 3*, together with their GritBot relevance measure, which measures the reliability of the outlier.

All the outliers identified by GritBot appear to be such. Indeed, the candidate *outliers 1* and *2* are explained by the fact that it is unlikely to

Table 2. BBN candidate outliers

N°	Age	Gender	HighSchool	1st Lev Deg	2nd Lev Deg	PhD	Prob.
1	25	Male	Yes	No	Yes	No	0.00002206
2	19	Male	Yes	Yes	No	Yes	0.00004351
3	21	Female	Yes	Yes	No	Yes	0.00004543
4	20	Male	Yes	Yes	Yes	No	0.00004982
5	23	Female	Yes	Yes	Yes	Yes	0.00004992
6	24	Female	Yes	Yes	No	No	0.00005035
7	19	Female	Yes	Yes	Yes	No	0.00005120
8	17	Male	No	No	Yes	No	0.00005179
9	26	Male	No	No	No	Yes	0.00005588
10	18	Male	Yes	Yes	No	Yes	0.00008853
11	18	Female	Yes	Yes	No	Yes	0.00009105
12	24	Female	Yes	Yes	No	No	0.00010012
13	24	Female	Yes	Yes	Yes	Yes	0.00010012
14	24	Male	Yes	Yes	Yes	Yes	0.00010022
15	24	Male	Yes	Yes	No	No	0.00010022
16	22	Male	Yes	Yes	Yes	Yes	0.00014861
17	27	Male	Yes	No	No	No	0.00019471
18	27	Female	Yes	No	No	No	0.00019984
19	25	Female	No	No	No	No	0.00029797
20	19	Female	No	No	No	No	0.00030428

have a **2ⁿᵈ Level Degree** at the **Age** of *20* and *19*. The candidate *outlier 3* is explained by the fact that it is impossible to have the **2ⁿᵈ Level Degree** without the **1ˢᵗ Level Degree** and the **HighSchool**. Furthermore, the **Age** of *17* is also suspect. Candidate *outliers 4* and *5* are motivated by the fact that within the considered database, it is unlikely that at the **Age** of 24 one does not have the **2ⁿᵈ Level Degree**. The candidate *outlier 6* is explained by the fact that it is not possible to have the **2ⁿᵈ Level Degree** without the **1ˢᵗ Level Degree**. Candidate *outliers 7, 8,* and *9* are explained by the fact that it is unlikely that, within the considered database, at the **Age** of 23 one does not have the **2ⁿᵈ Level Degree**, even though this could happen. Finally, the candidate *outlier 10* is such due to the fact that it is not possible to have a **PhD** without having the **2ⁿᵈ Level Degree**. Furthermore, also the values for the fields **HighSchool** and **1ˢᵗ Level Degree** are suspect.

To evaluate the capability of the BBN in *Figure 2*, for accomplishing the outlier detection task, we introduce the following quantities:

- *Agreement w.r.t. GritBot*; given the candidate outliers, detected by means of the GritBot software package, it represents the ratio of the BBN correctly identified outliers to the number of the GritBot candidate outliers. This measure ranges from zero, that is, none of the candidate outliers from GritBot have been identified by the BBN, to one, that is, all the candidate outliers from GritBot have been identified by the BBN,

- *Percentage of database cases classified as outliers;* is the ratio of the number of candidate outliers identified by means of the BBN to the number of cases in the considered database. It ranges from 0% to 100%, that is, all the database cases are identified by means of the BBN as candidate outliers.

The results obtained for the School-University database are reported through *Table 4* and depicted in *Figure 6*.

From *Table 4* it is possible to conclude that the outliers detection task, accomplished by solving the optimization problem (equation (6)), in the case when the BBN model depicted in *Figure 2* is considered, allows to identify all the GritBot candidate outliers when classifying the 0.7% of the analyzed database cases as outliers.

However, the GritBot software package does not identify all the candidate outliers correctly identified by means of the BBN model depicted in *Figure 2*. In particular, the GritBot software

Table 3. GritBot candidate outliers (default parameters)

N°	Age	Gender	HighSchool	1st Lev Deg	2nd Lev Deg	PhD	Rel.
1	20	Male	Yes	Yes	Yes	No	0.002
2	19	Female	Yes	Yes	Yes	No	0.002
3	17	Male	No	No	Yes	No	0.003
4	24	Female	Yes	Yes	No	No	0.004
5	24	Male	Yes	Yes	No	No	0.004
6	25	Male	Yes	No	Yes	No	0.005
7	23	Female	Yes	Yes	No	No	0.008
8	23	Female	Yes	Yes	No	No	0.008
9	23	Female	Yes	Yes	No	No	0.008
10	26	Male	No	No	No	Yes	0.009

Figure 6. BBN-GritBot agreement: School-University database

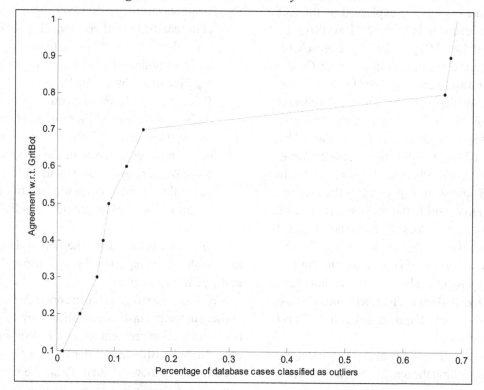

Table 4. BBN-GritBot agreement

Agreement w.r.t. GritBot	Percentage of database cases classified as outliers
0.1	0.01
0.2	0.04
0.3	0.07
0.4	0.08
0.5	0.09
0.6	0.12
0.7	0.15
0.8	0.67
0.9	0.68
1.0	0.69

package does not detect the *outliers* number *2, 3, 5, 10, 11, 13, 14, 16, 17, 18, 19,* and *20* that are listed in *Table 2*.

The comparison between the BBNs approach and the GritBot software package has been extended also to two more databases described in the specialized literature, that is, the "*churn*" and the "*hypothyroid*" databases, which are made available together with the evaluation copy of the GritBot software package.

In particular, the "*churn*" database consists of 5,000 records, 20 attributes of which 16 continuous, 3 binary, and 1 that can assume 51 possible values. The GritBot package applied to the "*churn*" database, when using a *filtering level* equal to *25%* together with the parameter *maximum conditions* equals to *4*, identifies 8 possible outliers. The corresponding BBN-GritBot agreement graph is reported in *Figure 7*.

Figure 7 shows that 7 out of 8 possible outliers, identified by means of the GritBot software package, are correctly identified by means of the BBN approach in the case when classifying the 6.2% of

Figure 7. BBN-GritBot agreement: Churn database

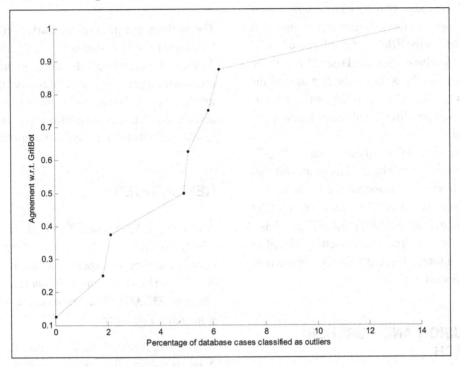

Figure 8. BBN-GritBot agreement: Hypothyroid database

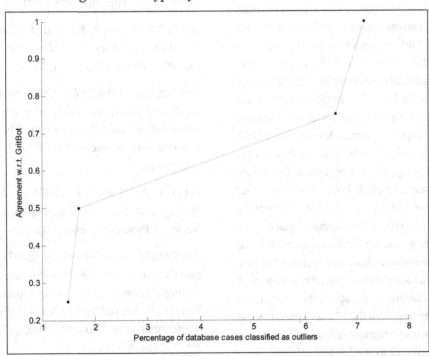

the "*churn*" database cases as outliers. However, 1 out of the 8 candidate outliers identified by means of the GritBot software package is not evaluated as such by means of the BBN model unless the 13% of the "*churn*" database cases are classified as outliers. This problem maybe is due to the fact that all the continuous attributes (16 out of 20) of the "*churn*" database have been discretized before learning the BBN model.

The "*hypothyroid*" database consists of 3,772 records or cases; 23 attributes, of which 6 are continuous; 15 are binary; 1 can assume 6 possible values; and 1 can assume 4 possible values. The GritBot package applied to the "*hypothyroid*" database, when using the default parameters setting, identifies 4 possible outliers. The BBN-GritBot agreement graph is reported in *Figure 8*.

CONCLUSION AND FURTHER RESEARCH

In this work, the authors described how the BBN computational device can be efficiently exploited for data cleaning and data quality improvement. The imputing missing values, completing truncated datasets, and outliers detection tasks have been addressed and discussed within the framework of BBNs. The examples described through the section named "BBNs for Data Cleaning," as well as the results of the performed numerical experiments, emphasize the importance and relevance of the BBN computational device to efficiently deal with data cleaning and data quality improvement. Directions for further research include both enlarging the set of data cleaning and data quality tasks that can be dealt with by using the BBN computational device. Futhermore, it would be desirable to exploit the BBNs expressiveness to allow the efficient treatment of databases including continuous attributes as well as textual data enlarging the application domain of BBNs by trying to exploit the BBNs expressiveness to allow the efficient treatment of databases including continuous attributes as well as textual data.

ACKNOWLEDGMENT

The authors are grateful to Carlo Vercellis and Giovanni Felici for organizing the "*Mathematical Methods for Learning - Advances in data mining and knowledge discovery*" conference and for editing this volume. The authors would like to thank the anonymous referees for useful comments, which resulted in substantial improvement of the work.

REFERENCES

Andersen, S. K., Olesen, K. G., & Jensen, F. V. (1990). HUGIN—A shell for building Bayesian belief universes for expert systems. In G. Shafer & J. Pearl (Eds.), *Readings in uncertain reasoning* (pp. 332-337). San Francisco, CA: Morgan Kaufmann Publishers.

Arning, A., Agrawal, R., & Raghavan, P. (1996). A linear method for deviation detection in large databases. In *Proceeding of the Second International Conference on Knowledge Discovery and Data Mining* (pp. 164–169, Portland, Oregon.

Baldi, P., Frasconi, P., & Smyth, P. (2003). *Modeling the internet and the WEB: Probabilistic methods and algorithms*. Chichester: Wiley.

Cooper, G. F. (1984). *NESTOR: A computer-based medical diagnostic aid that integrates causal and probabilistic knowledge*. Ph.D. Thesis, Medical Information Sciences, Stanford, CA: Stanford University.

Dasu, T., & Johnson, T. (2003). *Exploratory data mining and data cleaning*. Hoboken, NJ: Wiley Series in Probability and Statistics.

The Data Warehouse Institute, (2003). Data quality and the bottom line: Achieving business success through a commitment to high quality data, ZD Net UK, Retrieved the 13th April 2003, from http://whitepapers.zdnet.co.uk/0,39025945,60096183p-39000581q,00.htm.

Dempster, A. P., Laird, N. M., & Rubin, D. B. (1977). Maximum likelihood from incomplete data via the EM algorithm. *Journal of the Royal Statistical Society, Series B, 39*, 1-38.

Duda, R. O., Hart, P. E., & Stork, D. G. (2001). *Pattern classification*. New York: Wiley Inter-science.

Fagiuoli, E., & Zaffalon, M. (1998). 2U: An exact interval propagation algorithm for polytrees with binary variables. *Artificial Intelligence, 106*(1), 77-107.

Gaivoronski, A., & Stella, F. (1998). Stochastic optimization with structured distributions: The case of Bayesian nets. *Annals of Operations Research, 81*, 189-211.

Geman, S., & Geman, D. (1984). Stochastic relaxation, gibbs distribution, and the Bayesian restoration of images. *IEEE Transactions on Pattern Analysis and Machine Intelligence, 6*(6), 721-741.

Guyon, I., Matic, N., & Vapnik, V. (1996). Discovering informative patterns and data cleaning. In U. M. Fayyad, G. Piatetsky-Shapiro, P. Smyth, & R. Uthurusamy (Eds.), *Advances in Knowledge Discovery and Data Mining* (pp. 181-203). Menlo Park, CA: AAAI Press.

Hampel, F. R., Rousseeuw, P. J., Ronchetti, E. M. & Stahel, W. A. (1986). *Robust statistics: The approach based on influence functions*. New York: Wiley.

Heckerman, D. (1999). A tutorial on learning with Bayesian networks. In M. Jordan (Ed.), *Learning in Graphical Models* (pp. 301-354). Cambridge, MA: The MIT Press.

Huber, P. J. (1981). *Robust statistics*. New York: John Wiley & Sons.

Jensen, F.V. (1996). *An introduction to Bayesian networks*. London: UCL Press.

Kubika, J., & Moore, A. (2003). Probabilistic noise identification and data cleaning. In X. Wu, A. Tuzhilin, & J. Shavlik (Eds.), *The Third IEEE International Conference on Data Mining* (pp. 131-138). IEEE Computer Society.

Neapolitan, R.E. (1990). *Probabilistic reasoning in expert systems: Theory and algorithms*. New York: John Wiley & Sons.

Pearl, J. (1988). *Probabilistic reasoning in intelligent systems: Networks of plausible inference*. San Mateo, CA: Morgan Kaufmann Publishers.

Rahm, E., & Hai Do, H. (2000). Data cleaning: Problems and current approaches. *IEEE Data Engineering Bulletin, 23*(4), 3-13.

Rulequest Research. Gritbot. Retrieved from http://www.rulequest.com

Scannapieco, M., Missier, P., & Batini, C. (2005). Data quality at a glance. *Datenbank-Spektrum, 14*, 6-14.

Schwarm, S., & Wolfman S. (2000). Cleaning data with Bayesian methods. Final project report for CSE574, University of Washington, Retrieved the 23rd April 2002, from http://www.cs.washington.edu/homes/wolfwork/

Spiegelhalter, D. J., Dawid, A. P., Lauritzen, S. L., & Cowell, R.G. (1993). Bayesian analysis in expert systems. *Statistical Science, 8*(3), 219-283.

Zaffalon, M. (2002). Exact credal treatment of missing data. *Journal of Statistical Planning and Inference, 105*(1), 105-122.

Zaffalon, M., & Fagiuoli, E. (2003). Tree-based credal networks for classification. *Reliable computing, 9*(6), 487-509.

Chapter XIII
A Comparison of Revision Schemes for Cleaning Labeling Noise

Chuck P. Lam
Lama Solutions LLC., USA

David G. Stork
Ricoh Innovations, Inc., USA

ABSTRACT

Data quality is an important factor in building effective classifiers. One way to improve data quality is by cleaning labeling noise. Label cleaning can be divided into two stages. The first stage identifies samples with suspicious labels. The second stage processes the suspicious samples using some revision scheme. This chapter examines three such revision schemes: (1) removal of the suspicious samples, (2) automatic replacement of the suspicious labels to what the machine believes to be correct, and (3) escalation of the suspicious samples to a human supervisor for relabeling. Experimental and theoretical analyses show that only escalation is effective when the original labeling noise is very large or very small. Furthermore, for a wide range of situations, removal is better than automatic replacement.

INTRODUCTION

Most pattern recognition systems are built by fitting a model to a training dataset. The accuracy of the resulting classifier is dependent on both the choice of model and the quality of the dataset.

However, based on practical experience, most of the standard models (Duda, Hart, & Stork, 2001) tend to have comparable accuracy. In fact, simple models, such as Nearest Neighbor and Decision Trees, have remained very popular among practitioners because they are easy to work with while being almost as accurate as the much more elaborate models. Some researchers have concluded that the choice of a model is insignificant when one has a large, high-quality training dataset (Ho & Baird, 1997).

Many researchers have therefore examined ways to improve the quality of datasets by cleaning up noise in the data. Generally, there are two types of noise: attribute noise and labeling noise. Attribute noise refers to errors in the attribute values of data samples; labeling noise refers to errors in the class labels of data samples. In this chapter, we will focus on the issue of cleaning labeling noise.

An important factor in building good pattern classifiers is the quality of training data. A leading source of degradation of training data is labeling noise. Cleaning such noisy data can be divided into two stages. The first stage is the identification of possibly mislabeled samples; such samples are called *suspects*. The second stage is the processing of those suspicious samples. There are three major schemes to this revision process: (1) *removal* simply removes the suspects from the dataset (Brodley & Friedl, 1996; Brodley & Friedl, 1999; Zhu, Wu, & Chen, 2003), (2) *replacement* changes the labels of the suspects to what the machine believes are the correct classes (Castelli, Hutchins, Li, & Turek, 2001; Gimlin & Ferrell, 1974; Gowda & Krishna, 1979; Shanmugam & Breipohl, 1971; Teng, 1999), and (3) *escalation* asks a human supervisor (labeler) to look at the samples and provide the correct labels (Guyon, Matić, & Vapnik, 1996).

Choosing among the three revision schemes would be a moot issue if the first stage of identification is perfect. For such an ideal situation, all three schemes would result in noise-free datasets[1], and one would certainly choose to use automatic replacement since removal would have created a smaller dataset, while escalation has the added expense of getting a human supervisor involved.

In practice, one cannot perfectly identify mislabeled samples automatically. Thus, the choice of the proper revision scheme is complicated. Mistakes in the identification stage can propagate to the revision stage and may even be magnified there. For example, removal can erroneously remove a correctly-labeled sample, and replacement can erroneously change a correct label. Relabeling by a human expert, on the other hand, is much more robust, yet it is also more costly.

Most literature on cleaning mislabeled data has focused on the identification stage, while the choice of revision scheme was arbitrary or based on unique factors. Removal, for example, may be chosen if there is a desire to shrink the training dataset for reasons of storage or computational efficiency. Escalation, for example, may be chosen if high accuracy is strongly desired and labelers are essentially free (Guyon et al., 1996). There has been little systematic comparison of the improvement in accuracy from each of the revision schemes. This chapter attempts to bring some insights on how to choose the right revision scheme.

RELATED WORK

While labeling noise can occur in almost all application areas, it is especially common in some domains such as remote sensing (Brodley & Friedl, 1999; Smyth, Fayyad, Burl, & Perona, 1996), medical diagnosis (Dawid & Skene, 1979; Gamberger, Lavrač, & Grošelj, 1999), and natural language processing (Blaheta, 2002; Eskin, 2000). For remote sensing and medical diagnosis, in principle, one can obtain completely accurate labels, but in practice it is usually not the case, due to the exorbitant cost involved. One has to settle for approximate judgments from indirect information.

For natural language processing, several researchers have looked into domain specific ways to mitigate the labeling noise problem. Eskin (2000) has examined probabilistic methods for detecting anomalous tags in the Penn Treebank corpus.

These anomalies are considered mislabels. Blaheta (2002) has examined deterministic methods for cleaning common types of tagging errors.

While labeling noise presents an obvious problem for training, it also complicates the evaluation. Any automatic cleaning of the testing data would skew its distribution and bias the evaluation. Some recent works have addressed issues around labeling noise in the test dataset (Lam & Stork, 2003; Ng, 1997). For example, Lam and Stork (2003) have demonstrated that under realistic assumptions, the true error rate of a classifier is bounded by the apparent error rate, plus or minus the mislabeling rate of the test dataset.

Identifying Mislabeled Data

There has been extensive literature on identifying mislabeled data. We review some of them in this section.

The k-nearest-neighbor algorithm and its variants (Instance-Based Learning, Lazy Learning, Case-Based Learning, etc.) have received considerable attention on the issue of label noise cleaning, especially by the removal scheme. However, the main motivation behind much of that work has been to reduce the training set size for storage and computational reasons, rather than improving accuracy. Removing mislabeled samples is one of the logical approaches to shrink the dataset, and it is often combined with other approaches, such as removing inconsequential samples far from the decision boundaries.

One of the earliest cleaning algorithms is Wilson Editing (Wilson, 1972). Under Wilson Editing, a sample is removed (i.e., "edited") if its label is different from the labels of the majority of its neighbors.

Aha et al. (Aha, 1991) introduce the instance-based learning algorithms IB1, IB2, and IB3. IB3 in particular removes noisy samples. It sequentially examines each sample in a training set, and a sample is kept if it has been misclassified by previous samples but contributes significantly to the correct classification of later samples (using the nearest-neighbor classifier). The idea is to remove unimportant samples that do not border the decision boundaries, as well as noisy samples that do not contribute to correct classification.

Wilson and Martinez (1997) present three reduction techniques: RT1, RT2, and RT3. Their techniques involve the concept of *associate*. A sample is considered an associate of **x** if **x** is one of the sample's k nearest neighbors, where k is generally a small odd integer. Their heuristics remove **x** if its removal does not increase the number of its associates being misclassified. Again the concept is to remove samples that do not make important contributions to the overall classification accuracy; some of those samples are probably mislabeled.

Brighton and Mellish (2002) has a review of removal methods for kNN, including their own iterative case filtering algorithm (ICF). The ICF algorithm first uses Wilson Editing for removing suspicious samples. Afterward ICF further reduces training set size by eliminating redundant samples in the interior of decision regions. Brighton and Mellish have made the theoretical observation that, for nearest neighbor classification, it is always possible to reduce the training set to the size of the testing set without affecting the classifier's accuracy *on the testing set*.

Brodley and Friedl (1996, 1999) describe a general identification scheme that does not require the samples to be in a metric space. The scheme uses m learning algorithms and n-fold cross-validation to identify mislabeled training data. That is, the training data is first divided into n parts. For each part, the m algorithms are trained on the other n-1 parts, and a sample is considered mislabeled if m' or more of the m algorithms misclassify that sample. Brodley and Friedl call the case $m'=m$ as *consensus filtering* and the case $m' = (m+1)/2$ as *majority filtering*.

Gamberger et al. (1999) developed the *saturation filter*. Theoretically, if a training dataset D is noiseless and saturated (containing enough data points to find a correct target hypothesis), and D_n is the union of D and $\{(\mathbf{x}; \tilde{y})\}$, where $(\mathbf{x}; \tilde{y})$ is a noisy data point not correctly classified by the target hypothesis, then the CLCH (Complexity of the Least Complex Hypothesis) value $g(D)$ is less than $g(D_n)$. The saturation filter thus tries to remove data points that would reduce the CLCH value of the dataset. Gamberger et al. applied the saturation filter to a medical domain and obtained favorable results.

John (1995) developed ROBUST-C4.5, an extension of the C4.5 decision tree induction with built-in removal of mislabeled data. ROBUST-C4.5 first trains a decision tree, using C4.5 with pruning. It then removes all samples in the training set that the decision tree misclassifies. After that the remaining data is used to train a new decision tree, and the process is repeated until all samples are classified correctly.

Teng (1999) developed a procedure, called polishing, that attempts to correct both attribute and labeling noise. It is an example of labeling cleaning using the replacement scheme (Castelli et al., 2001; Gimlin & Ferrell, 1974; Gowda & Krishna, 1979; Shanmugam & Breipohl, 1971). Polishing exploits interdependencies among attributes as well as interdependencies between attributes and target class. These interdependencies are used to predict and correct attribute and label noise.

Guyon et al. (1996) is the only published work we know of that escalates suspicious samples for relabeling by a human expert labeler. They propose a cleaning method where a human supervisor checks those samples that have the largest information criterion and are therefore most "surprising." While most automatic cleaning methods assume surprising patterns to be garbage, Guyon et al. argue that surprising patterns can also be informative and human judgment should be exercised to discriminate between the two.

Many of the cleaning programs discussed have trouble scaling up to large datasets. Recently, Zhu et al. (2003) developed a scheme called partition filter to address the problem of label cleaning in large, distributed datasets.

Unfortunately, with the exception of Aha et al. (1991) and Gowda and Krishna (1979), none of the filtering experiments we found in the literature has tried to characterize the effectiveness of filtering under different severity of mislabeling. In fact, most experiments have only compared a classifier trained on a dataset with the same classifier trained on the filtered dataset. It is assumed that labeling noise exists in the original dataset, and the improved accuracy from training on the filtered dataset is mainly due to the removal of the mislabeled samples. However, another legitimate line of argument is that data filtering would have improved accuracy even if there was no mislabeling. For example, filtering may just be a data driven form of regularization that prevents overfitting.

k-k' NEAREST-NEIGHBOR IDENTIFICATION

Many identification schemes exist, including the instance-based (IB) algorithms of Aha et al. (1991) and the reduction techniques (RT) of Wilson and Martinez (1997). The general approach is to find unusual samples whose removal does not decrease classification accuracy significantly.

The identification scheme we examine is the *k-k'* nearest-neighbor procedure (Gimlin & Ferrell, 1974). Consider a sample $\mathbf{x}$ with label $y \in \{\omega_1, \omega_2\}$. Let k be an odd integer and k' be an integer greater than $k/2$. Of the k nearest neighbors to $\mathbf{x}$, if k' or more of the patterns do not belong to class y, then sample $\mathbf{x}$ is considered a suspect. Note that Wilson's editing algorithm (Wilson, 1972) is a specialization of this procedure, with $k' = (k + 1)/2$, and filtering as the default revision scheme.

Note that the original *k-k'* nearest-neighbor procedure studied by Gimlin and Ferrell (1974) is an online version of our batch algorithm. That is, the procedure receives samples one at a time, and it categorizes them as either suspects or not, based only on the samples it had seen previously.

In practice, the choice of online vs. batch cleaning will generally depend on the learning algorithm, which in turn depends on the application domain. An online learning algorithm should probably use an online cleaning procedure, while a batch learning algorithm can use either batch cleaning or online cleaning. Batch cleaning will clean better, but if storage capacity is a concern and one can keep only a subset of training data, then online filtering makes sense. This chapter considers only the batch procedure, so that the samples' ordering effect will not create unnecessary variance in the experiments.

The parameters k and k' together specify the level of confidence necessary before one can categorize a sample as suspicious. That is, it determines the level of aggressiveness in cleaning the data. For example, the pairs ($k = 5$; $k' = 3$) will clean aggressively, whereas the pairs ($k = 5$; $k' = 5$) will process a sample only if the algorithm is very confident that the sample has been mislabeled.

THEORETICAL ANALYSIS

In the following theoretical analysis, we assume a nearest-neighbor classifier and a training dataset of infinite size, so that a test sample would have a nearest neighbor at exactly the same point in the feature space. We denote this neighbor $\mathbf{x}$, and it has a true but unknown label y. The dataset has a noisy version $\tilde{y}$, which is modeled as a random flipping (with probability ε) of y. Cleaning algorithms try to substitute $\tilde{y}$ with $\tilde{y}_\dagger$, where the subscript (†) denotes a cleaned version of $\tilde{y}$. The hope is that $\tilde{y}_\dagger$ is more likely than $\tilde{y}$ to equal to y. Note that in the removal scheme, we technically would have removed the neighbor and its label. However, in the limiting case of infinite data samples, the next closest neighbor will be at the same location. Thus we still denote the next closest neighbor as $\mathbf{x}$, but its label is now $\tilde{y}_\dagger$.

To simplify notation, we denote $\Pr[y = \omega_i \mid \mathbf{x}]$ as just $p(\omega_i \mid \mathbf{x})$. The Bayes' error rate at point $\mathbf{x}$ is therefore $r^*(\mathbf{x}) = \min[p(\omega_1 \mid \mathbf{x}), p(\omega_2 \mid \mathbf{x})]$. The error rate of the nearest-neighbor classifier using true labels y is,

$$r_y(\mathbf{x}) = p(\varpi_1 \mid \mathbf{x})p(\varpi_2 \mid \mathbf{x}) + p(\varpi_2 \mid \mathbf{x})p(\varpi_1 \mid \mathbf{x})$$
$$= 2r^*(\mathbf{x})(1 - r^*(\mathbf{x})). \tag{1}$$

and this error rate is well known to be less than $2r^*(\mathbf{x})$ (Cover & Hart, 1967). The noisy labels are assumed to be distributed according to $\Pr[\tilde{y} = \omega_i \mid \mathbf{x}] = (1-\varepsilon)\,p(\omega_i \mid \mathbf{x}) + \varepsilon(1 - p(\omega_i \mid \mathbf{x}))$, where mislabeling rate $\varepsilon < 0:5$. The error rate using data with noisy labels $\tilde{y}$ is,

$$r_{\tilde{y}}(\mathbf{x}) = (1-\varepsilon)r_y(\mathbf{x}) + \varepsilon(1 - r_y(\mathbf{x}))$$
$$= r_y(\mathbf{x}) + \varepsilon(1 - 2r_y(\mathbf{x}))$$
$$= \varepsilon + (1 - 2\varepsilon)r_y(\mathbf{x})$$

It can be seen from the last two equations that the error rate from the noisy data is higher than both $r_y(\mathbf{x})$, the error rate given true labels, and the mislabeling rate ε. In addition, the error rate increases linearly with ε.

The error rate using data with cleaned labels $\tilde{y}_\dagger$ is derived analogously,

$$r_{\tilde{y}_\dagger}(\mathbf{x}) = \Pr[\tilde{y}_\dagger = y \mid \mathbf{x}]r_y(\mathbf{x}) + \Pr[\tilde{y}_\dagger \neq y \mid \mathbf{x}](1 - r_y(\mathbf{x}))$$
$$= r_y(\mathbf{x}) + \Pr[\tilde{y}_\dagger \neq y \mid \mathbf{x}](1 - 2r_y(\mathbf{x}))$$

and it also increases linearly with $\Pr[\tilde{y}_\dagger \neq y \mid \mathbf{x}]$.

We proceed to find the mislabeling rate $\Pr[\tilde{y}_\dagger \neq y \mid \mathbf{x}]$ of the cleaned data under the three different revision schemes.

Replacement

We denote the binomial distribution as $\text{Bin}(k, i, p) = \binom{k}{i} p^i (1-p)^{k-i}$. The mislabeling rate with replacement, as derived in Gimlin and Ferrell (1974), is,

$$\Pr[\tilde{y}_\dagger \neq y \mid \mathbf{x}] = .p(\varpi_1 \mid \mathbf{x}) \sum_{i=0}^{k-k'} \mathrm{Bin}(k,i,\Pr[\tilde{y}=\varpi_1 \mid \mathbf{x}])$$
$$+ p(\varpi_2 \mid \mathbf{x}) \sum_{i=0}^{k-k'} \mathrm{Bin}(k,i,\Pr[\tilde{y}=\varpi_2 \mid \mathbf{x}])$$
$$+ \varepsilon \cdot \sum_{i=k-k'+1}^{k'-1} \mathrm{Bin}(k,i,\Pr[\tilde{y}=\varpi_1 \mid \mathbf{x}])$$

If a sample has k' or more of its k nearest neighbors belong to one class, then it is forced to belong to that class also, regardless of the original value of its label. The first two terms in the equation represent the mistake of forcing a label to the wrong value. The last term is due to not revising an incorrect label.

Let $f_\varepsilon(\mathbf{x}) = (1 - \varepsilon)r^*(\mathbf{x}) + \varepsilon(1 - r^*(\mathbf{x}))$. Note that $f_\varepsilon(\mathbf{x}) = \Pr[\tilde{y} = \omega_1 \mid \mathbf{x}]$ if $r^*(\mathbf{x})= p(\omega_1|\mathbf{x})$ and $f_\varepsilon(\mathbf{x}) = \Pr[\tilde{y}=\omega_2|\mathbf{x}]$ if $r^*(\mathbf{x})= p(\omega_2|\mathbf{x})$. We can leverage the symmetry of the mislabeling rate $\Pr[\tilde{y}_\dagger \neq y|\mathbf{x}]$ to rewrite it in terms of ε and $r^*(\mathbf{x})$,

$$\Pr[\tilde{y}_\dagger \neq y \mid \mathbf{x}] = .\left[r^*(\mathbf{x}) \cdot \sum_{i=0}^{k-k'} \mathrm{Bin}(k,i,f_\varepsilon(\mathbf{x})) \right]$$
$$+ \left[(1-r^*(\mathbf{x})) \cdot \sum_{i=0}^{k-k'} \mathrm{Bin}(k,i,1-f_\varepsilon(\mathbf{x})) \right]$$
$$+ \left[\varepsilon \cdot \sum_{i=k-k'+1}^{k'-1} \mathrm{Bin}(k,i,f_\varepsilon(\mathbf{x})) \right]$$

Escalation

Under escalation, a suspicious label is escalated to a human supervisor for correction. Assuming the mislabeling rate of this human labeler is $\varepsilon_h(\mathbf{x})$, the mislabeling rate of the cleaned data is, (see Box 1).

The first two terms represent the probability the suspicious labels will be relabeled incorrectly. The next two terms represent the probability that a label agrees with k' or more of its k neighbors, but is in fact wrong. The last term represents the probability of mislabeling when fewer than k' neighbors can agree on a class.

Note that,

$$\Pr[y = \varpi_1, \tilde{y} = \varpi_2 \mid \mathbf{x}]. = \Pr[\tilde{y}=\varpi_2 \mid \mathbf{x}, y = \varpi_1]p(\varpi_1 \mid \mathbf{x})$$
$$= \Pr[\tilde{y} \neq y \mid \mathbf{x}, y = \varpi_1]p(\varpi_1 \mid \mathbf{x})$$
$$= \varepsilon \cdot p(\varpi_1 \mid \mathbf{x})$$

and analogously $\Pr[y=\varpi_2, \tilde{y}=\varpi_1|\mathbf{x}] = \varepsilon \cdot p(\varpi_2|\mathbf{x})$. Again, recognizing the formal symmetry in this equation, we can rewrite the mislabeling rate $\Pr[\tilde{y}_\dagger \neq y|\mathbf{x}]$ as, (see Box 2).

Removal

We think of removal in the infinite data scenario as a form of relabeling. That is, removal is equivalent

Box 1.

$$\Pr[\tilde{y}_\dagger \neq y \mid \mathbf{x}] = .\varepsilon_h(\mathbf{x}) \cdot \Pr[\tilde{y}=\varpi_1 \mid \mathbf{x}] \cdot \sum_{i=0}^{k-k'} \mathrm{Bin}(k,i,\Pr[\tilde{y}=\varpi_1 \mid \mathbf{x}])$$
$$+ \varepsilon_h(\mathbf{x}) \cdot \Pr[\tilde{y}=\varpi_2 \mid \mathbf{x}] \cdot \sum_{i=0}^{k-k'} \mathrm{Bin}(k,i,\Pr[\tilde{y}=\varpi_2 \mid \mathbf{x}])$$
$$+ \Pr[y = \varpi_1, \tilde{y} = \varpi_2 \mid \mathbf{x}] \cdot \sum_{i=0}^{k-k'} \mathrm{Bin}(k,i,\Pr[\tilde{y}=\varpi_1 \mid \mathbf{x}])$$
$$+ \Pr[y = \varpi_2, \tilde{y} = \varpi_1 \mid \mathbf{x}] \cdot \sum_{i=0}^{k-k'} \mathrm{Bin}(k,i,\Pr[\tilde{y}=\varpi_2 \mid \mathbf{x}])$$
$$+ \varepsilon \cdot \sum_{i=k-k'+1}^{k'-1} \mathrm{Bin}(k,i,\Pr[\tilde{y}=\varpi_1 \mid \mathbf{x}])$$

Box 2.

$$
\begin{aligned}
\Pr[\tilde{y}_\dagger \neq y \,|\, \mathbf{x}] = & \,.\left[\varepsilon_h(\mathbf{x}) \cdot f_\varepsilon(\mathbf{x}) \cdot \sum_{i=0}^{k-k'} \mathrm{Bin}(k,i,f_\varepsilon(\mathbf{x})) \right] \\
& + \left[\varepsilon_h(\mathbf{x}) \cdot (1 - f_\varepsilon(\mathbf{x})) \cdot \sum_{i=0}^{k-k'} \mathrm{Bin}(k,i,(1 - f_\varepsilon(\mathbf{x}))) \right] \\
& + \left[\varepsilon \cdot r^*(x) \cdot \sum_{i=0}^{k-k'} \mathrm{Bin}(k,i,f_\varepsilon(\mathbf{x})) \right] \\
& + \left[\varepsilon \cdot (1 - r^*(x)) \cdot \sum_{i=0}^{k-k'} \mathrm{Bin}(k,i,(1 - f_\varepsilon(\mathbf{x}))) \right] \\
& + \left[\varepsilon \cdot \sum_{i=k-k'+1}^{k'-1} \mathrm{Bin}(k,i,f_\varepsilon(\mathbf{x})) \right]
\end{aligned}
$$

to letting the label of the nearest neighbor (still at $\mathbf{x}$) take the original suspected label's place. If the nearest neighbor's label is suspicious also, then the *next* nearest neighbor (still at $\mathbf{x}$) is used, and so on.

The repetition of the process can lead to complicated dependencies in mislabeling rates. To reduce that dependency, we will assume that cleaning is batch rather than sequential. That is, there is no ordering effect. We are not removing one sample at a time and then clean the other samples with that one sample missing. For analysis, we then make the approximation that a sample's neighbor, being suspicious, is independent of the fact that the original sample itself is suspicious. This allows us to say that the probability of error when removing a suspicious label is just $r_{\tilde{y}_\dagger}(\mathbf{x})$, or $r_y(\mathbf{x})$ + $\Pr[\tilde{y}_\dagger \neq y|\mathbf{x}](1 - 2r_y(\mathbf{x}))$. The mislabeling rate can then be written as, (see Box 3).

Rewriting in terms of $r^*(\mathbf{x})$ and $f_\varepsilon(\mathbf{x})$, and then solving for $\Pr[\tilde{y}_\dagger \neq y|\mathbf{x}]$, the resulting mislabeling rate is shown next, where $r_y(\mathbf{x})$ is defined in equation 1 as $2\,r^*(\mathbf{x})\,(1 - r^*(\mathbf{x}))$. (See Box 4.)

Box 3.

$$
\begin{aligned}
\Pr[\tilde{y}_\dagger \neq y \,|\, \mathbf{x}] = & \,.r_{\tilde{y}_\dagger}(\mathbf{x}) \cdot \Pr[\tilde{y} = \varpi_1 \,|\, \mathbf{x}] \cdot \sum_{i=0}^{k-k'} \mathrm{Bin}(k,i,\Pr[\tilde{y} = \varpi_1 \,|\, \mathbf{x}]) \\
& + r_{\tilde{y}_\dagger}(\mathbf{x}) \cdot \Pr[\tilde{y} = \varpi_2 \,|\, \mathbf{x}] \cdot \sum_{i=0}^{k-k'} \mathrm{Bin}(k,i,\Pr[\tilde{y} = \varpi_2 \,|\, \mathbf{x}]) \\
& + \Pr[y = \varpi_1, \tilde{y} = \varpi_2 \,|\, \mathbf{x}] \cdot \sum_{i=0}^{k-k'} \mathrm{Bin}(k,i,\Pr[\tilde{y} = \varpi_1 \,|\, \mathbf{x}]) \\
& + \Pr[y = \varpi_2, \tilde{y} = \varpi_1 \,|\, \mathbf{x}] \cdot \sum_{i=0}^{k-k'} \mathrm{Bin}(k,i,\Pr[\tilde{y} = \varpi_2 \,|\, \mathbf{x}]) \\
& + \varepsilon \cdot \sum_{i=k-k'+1}^{k'-1} \mathrm{Bin}(k,i,\Pr[\tilde{y} = \varpi_1 \,|\, \mathbf{x}])
\end{aligned}
$$

Box 4.

$$\left[1-(1-2r_y(\mathbf{x}))\cdot\left(f_\varepsilon(\mathbf{x})\cdot\sum_{i=0}^{k-k'}\text{Bin}(k,i,f_\varepsilon(\mathbf{x}))+(1-f_\varepsilon(\mathbf{x}))\cdot\sum_{i=0}^{k-k'}\text{Bin}(k,i,(1-f_\varepsilon(\mathbf{x})))\right)\right]^{-1}*$$

$$\left[r_y(\mathbf{x})\cdot\left(f_\varepsilon(\mathbf{x})\cdot\sum_{i=0}^{k-k'}\text{Bin}(k,i,f_\varepsilon(\mathbf{x}))+(1-f_\varepsilon(\mathbf{x}))\cdot\sum_{i=0}^{k-k'}\text{Bin}(k,i,(1-f_\varepsilon(\mathbf{x})))\right)\right.$$

$$+\varepsilon\cdot r^*(\mathbf{x})\cdot\sum_{i=0}^{k-k'}\text{Bin}(k,i,f_\varepsilon(\mathbf{x}))+\varepsilon\cdot(1-r^*(\mathbf{x}))\cdot\sum_{i=0}^{k-k'}\text{Bin}(k,i,(1-f_\varepsilon(\mathbf{x})))$$

$$\left.+\varepsilon\cdot\sum_{i=k-k'+1}^{k'-1}\text{Bin}(k,i,f_\varepsilon(\mathbf{x}))\right]$$

Discussion of Theoretical Analysis

Figure 2 plots the mislabeling rate $\Pr[\tilde{y}_{\dagger}\neq y\mid\mathbf{x}]$ for the three revision schemes. For the escalation case, we have assumed $\varepsilon_h(\mathbf{x})=0$. Diagonal lines where $\Pr[\tilde{y}_{\dagger}\neq y\mid\mathbf{x}]=\varepsilon$ are drawn to represent the mislabeling rate of noisy data. Note that if the mislabeling rate is above the diagonal lines, then the cleaning procedure has actually *degraded* data quality.

It is not surprising that escalation never degrades data quality, since we have assumed that escalation is done by a (human) oracle with $\varepsilon_h(\mathbf{x})=0$. On the other hand, both automatic replacement and removal can degrade data quality, and the degradation is a function of the Bayes' error rate. The higher the Bayes' error rate, the less effective replacement and removal are.

EXPERIMENTS

To corroborate the theoretical findings, we performed various experiments on the UCI machine learning datasets. We show the results for the OptDigit (digits "8" and "9"), Ionosphere, and Sonar datasets here. We vary the experiments on four dimensions. The first dimension is train-ing set size. We randomly sampled 80%, 40%, and 10% of the original dataset for training. The second dimension is the amount of labeling noise ε injected. With probability ε, a label in the training data is flipped to its opposite state. The third dimension of the experiments is the aggressiveness of label cleaning, expressed by various combinations of (k, k') values: (5,5), (5,4), and (5,3). The final dimension is the three revision schemes described previously. For the escalation scheme, we assume the human labeler is perfect. Escalation is thus simulated by revealing the true labels of all suspected data samples.

Results using clean training data (before any noise is injected) and results for the noisy training data (before any cleaning) are also shown for comparison. For each combination of training set size, labeling noise level, (k, k') tuple, and revision scheme, 30 iterations of the experiment were run, and the average is reported. In all experiments, the nearest-neighbor classifier is used for testing.

For our analysis, the main difference between the OptDigit, Ionosphere, and Sonar datasets is that OptDigit is the easiest dataset to learn while Sonar is the most difficult. Even when only 10% of the training data is used, OptDigit has an accuracy of 97%. We will note later that this ease-of-learning does have an effect on label cleaning.

Results

It is not surprising that clean, noiseless data always provide the best classification accuracy; *in general,* noisy training data without any cleaning provide the worst accuracy. (As we will see, this is not always the case.) Using the nearest-neighbor classifier without any data cleaning, random labeling noise has a linear relationship with classification accuracy. While larger training datasets increase classifier accuracy, this gain can easily be taken away if just a few percent of labeling noise is added and not properly handled.

Of the three revision schemes, escalation to a perfect oracle consistently performs better than the other two schemes, especially for datasets with high labeling noise. This is explained by the fact that, as the noise level ε approaches 0.5, there is less and less information in the dataset for any identification method to find mislabeled samples. The identifier's assumption that anomalous samples are mislabeled breaks down. When $\varepsilon = 0.5$, all the automated methods do is distribute the noise to neighboring regions. For the escalation scheme, at least some (random) samples will be corrected by a human labeler, and some accuracy will therefore be gained.

The experiments show that the two automated cleaning schemes provide the most benefit under a medium amount of labeling noise; that is, the labeling noise is much less than 50% but far from nonexistent. This observation is hinted by and consistent with Shanmugam and Breipohl (1971) and Gimlin and Ferrell (1974). The intuitive explanation for the case of high labeling noise has already been stated. The case of low labeling noise takes a bit more explaining.

In the Ionosphere and Sonar dataset, at low levels of labeling noise, the removal and replacement schemes actually do *worse* than using the uncleaned data. To the authors' knowledge, the fact that cleaning low levels of labeling noise may decrease classifier accuracy has not been discussed in the literature before.[2]

However, the intuition about this phenomenon is not difficult to develop. If the labeling noise level is lower than the error rate of the identification algorithm, then the identification stage misidentifies more suspects than truly mislabeled samples. In turn, automatic propagation of identification errors under the removal and replacement schemes hurts the overall cleaning procedure. Furthermore, the accuracy in identifying mislabeled samples is theoretically bounded by the Bayes' error rate of the feature domain. (No automated identification algorithm can identify mislabeled samples at a higher accuracy; otherwise, one can use the identification algorithm as a classifier and violate the accuracy bound.) In practice the accuracy is further lowered by the finite training set size. Therefore, there is generally a lower limit to the labeling noise level beyond which *automated* cleaning schemes tend to inject noise rather than clean. Fortunately, escalation transcends this limit by involving a human labeler who has a higher accuracy than the Bayes' error rate, because she can examine the data from a higher dimensional space (i.e., the original data sample rather than an abstract feature vector).

One of our initial concerns was that automatic removal of suspicious samples may work well in large datasets, but for small datasets, it may be discarding too much information. This effect was not observed in the OptDigits experiments, where there is enough redundancy such that a large reduction in training set size has a negligible effect on overall accuracy, but the degradation is definitely seen for the Ionosphere and Sonar domains.

However, while replacement does not discard attribute information as removal does, replacement does equal or worse than removal in all our experiments, although this difference shrinks for more conservative (e.g., ($k = 5$; $k' = 5$)) identification schemes. This suggests that removal is more robust to identification mistakes. That is, accidentally replacing a good sample with the wrong label does far more harm than accidentally removing that sample. On the other hand, correcting a mislabeled sample is only marginally better than removing that sample, and lets its neighbors "smooth" over that part of the feature space.

The aggressiveness of identification (and cleaning) only magnifies the effect of the revision scheme. Given that now one knows what situation to apply cleaning schemes, it only makes sense for one to either not clean at all (because doing so may actually degrade data quality) or clean aggressively.

CONCLUSION

Our analysis shows that automatic replacement is always a poorer revision scheme than removal. However, while removal works fine under a medium mislabeling rate, it performs poorly when the mislabeling rate is either low or high. Removal can degrade data quality when mislabeling rate is low relative to the Bayes' error rate. For any given situation, it is usually best to either clean aggressively or not clean at all. Only in some rare cases does cleaning conservatively improve data quality, but cleaning aggressively degrades it.

Figure 1. Legend for interpreting the plot lines in Figures 2, 3, 4, and 5. The lines represent results from using the clean training set (no noise injected), using the noisy training set (no cleaning), and using the three revision schemes. Both "clean" and "noisy" results are shown as solid lines, although it should be obvious from context how to differentiate them.

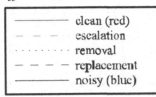

Figure 2. Theoretical mislabeling rates for various revision schemes given infinite data. Figure 1 shows the legend for matching the plot lines with revision schemes. The mislabeling rate for clean data is zero and is not shown. The diagonal line represents the mislabeling rate of the noisy data at $Pr[\tilde{y}_+ \neq y | \mathbf{x}] = \varepsilon$. That is, any cleaning method that changes the mislabeling rate to below this line has improved data quality. Analogously, being above this line means that data quality has degraded.

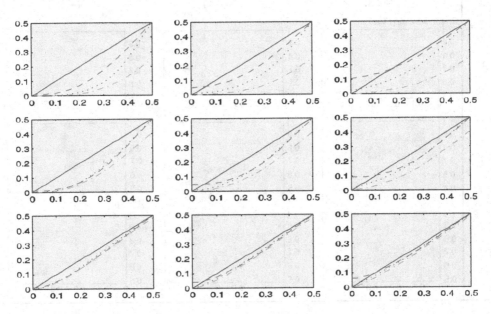

Figure 3. Accuracy of different revision schemes for the digits "8" and "9" in the UCI OptDigits dataset. Each column represents a different training set size. Each row represents different parameter value pairs for the k-k' identification scheme. Figure1 shows the legend for matching lines in the plots with revision schemes.

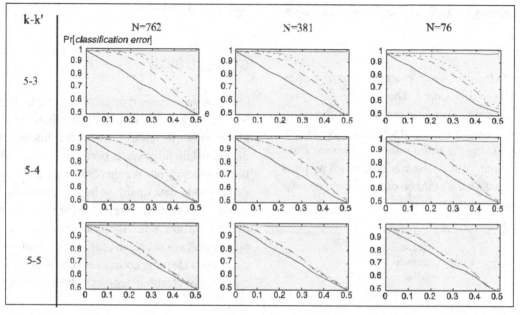

Figure 4. Accuracy of different revision schemes for the UCI Ionosphere dataset. Each column represents a different training set size. Each row represents different parameter value pairs for the k-k' identification scheme. Figure1 shows the legend for matching lines in the plots with revision schemes.

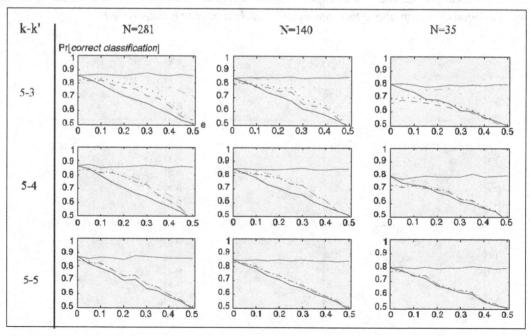

Figure 5. Accuracy of different revision schemes for the UCI Sonar dataset. Each column represents a different training set size. Each row represents different parameter value pairs for the k-k' identification scheme. Figure1 shows the legend for matching lines in the plots with revision schemes.

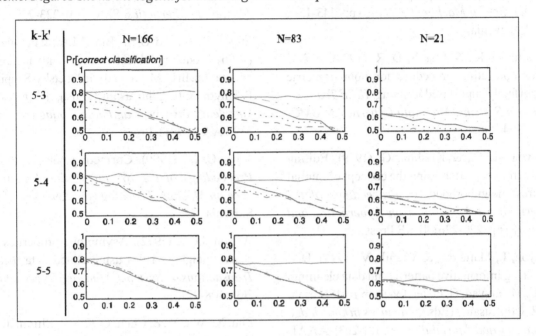

REFERENCES

Blaheta, D. (2002). Handling noisy training and testing data. In *Proceedings of the 7th Conference on Empirical Methods in Natural Language Processing* (pp. 111-116). Philadelphia, PA.

Brighton and Mellish (2002).

Brodley, C. E., & Friedl, M. A. (1996). Identifying and eliminating mislabeled training instances. In *Proceedings of the 13th National Conference on Artificial Intelligence* (pp. 799-805). Portland, OR: AAAI Press.

Brodley, C. E., & Friedl, M. A. (1999). Identifying mislabeled training data. *Journal of Artificial Intelligence Research, 11*, 131-167.

Castelli, V., Hutchins, S. T., Li, C.-S., & Turek, J. J. E. (2001). *Modifying an unreliable training set for supervised classification.* United States Patent 6,298,351.

Cover, T. M., & Hart, P. E. (1967). Nearest neighbor pattern classification. *IEEE Transactions on Information Theory*, IT-13, 21-27.

Dawid, A. P., & Skene, A. M. (1979). Maximum likelihood estimation of observer error-rates using the EM algorithm. *Applied Statistics, 28*(1), 20-28.

Duda, R. O., Hart, P. E., & Stork, D. G. (2001). *Pattern classification* (2nd ed.). John Wiley & Sons.

Eskin, E. (2000). Detecting errors within a corpus using anomaly detection. In *Proceedings of the 1st Conference of the North American Association for Computational Linguistics*. Seattle, WA.

Gamberger, D., Lavrač, N., & Grošelj, C. (1999). Experiments with noise filtering in a medical domain. In *Proceedings of the 16ᵗʰ International Conference on Machine Learning* (pp. 143-151). Bled, Slovenia.

Gimlin, D. R., & Ferrell, D. R. (1974). A *k-k'* error correcting procedure for nonparametric imperfectly supervised learning. *IEEE Transactions on Systems, Man, and Cybernetics, SMC-4*(3), 304-306.

Gowda, K. C., & Krishna, G. (1979). Editing and error correction using the concept of mutual nearest neighborhood. In *Proceedings of the International Conference on Cybernetics and Society* (pp. 222-226). IEEE Press.

Guyon, I., Matić, N., & Vapnik, V. (1996). Discovering informative patterns and data cleaning. In U. M. Fayyad, G. Piatetsky-Shapiro, P. Smyth, & R. Uthurusamy (Eds.), *Advances in knowledge discovery and data mining* (pp. 181-203). AAAI/ MIT Press.

Ho, T. K., & Baird, H. S. (1997). Large-scale simulation studies in image pattern recognition. *IEEE Transactions on Pattern Analysis and Machine Intelligence, 19*(10), 1067-1079.

Lam, C. P., & Stork, D. G. (2003). Evaluating classifiers by means of test data with noisy labels. In *Proceedings of the 18ᵗʰ International Joint Conference on Artificial Intelligence* (pp. 513-518). Acapulco, Mexico.

Ng, A. Y. (1997). Preventing "overfitting" of cross-validation data. In *Proceedings of the 14ᵗʰ International Conference on Machine Learning* (pp. 245-253). Nashville, TN.

Shanmugam, K., & Breipohl, A. M. (1971). An error correcting procedure for learning with an imperfect teacher. *IEEE Transactions on Systems, Man, and Cybernetics, SMC-1*(3), 223-229.

Smyth, P., Fayyad, U. M., Burl, M. C., & Perona, P. (1996). Modeling subjective uncertainty in image annotation. In U. M. Fayyad, G. Piatetsky-Shapiro, P. Smyth, & R. Uthurusamy (Eds.), *Advances in knowledge discovery and data mining* (pp. 517-539). AAAI/MIT Press.

Teng, C. M. (1999). Correcting noisy data. In *Proceedings of the 16ᵗʰ International Conference on Machine Learning* (pp. 239-248). Bled, Slovenia.

Wilson, D. L. (1972). Asymptotic properties of nearest neighbor rules using edited data. *IEEE Transactions on Systems, Man, and Cybernetics, 2*(3), 408-421.

Zhu, X., Wu, X., & Chen, Q. (2003). Eliminating class noise in large datasets. In *Proceedings of the 20ᵗʰ International Conference on Machine Learning* (pp. 920-927). Washington, DC.

ENDNOTES

[1] Here we loosely say that the first stage is perfect if it correctly identifies all mislabeled data and correctly determines their true labels. For two-class problems, the two tasks are equivalent.

[2] One of the experiments in Aha et al. (1991) has shown that the IB3 algorithm can decrease accuracy given a noiseless training dataset. The effect was simply noted and not fully explained.

Chapter XIV
Improving Web
Clickstream Analysis:
Markov Chains Models and
Genmax Algorithms

Paolo Baldini
University of Pavia, Italy

Paolo Giudici
University of Pavia, Italy

ABSTRACT

Every time a user links up to a Web site, the server keeps track of all the transactions accomplished in a log file. What is captured is the "click flow" (clickstream) of the mouse and the keys used by the user during the navigation inside the site. Usually every click of the mouse corresponds to the viewing of a Web page. The objective of this chapter is to show how Web clickstream data can be used to understand the most likely paths of navigation in a Web site, with the aim of predicting, possibly online, which pages will be seen, having seen a specific path of other pages before. Such analysis can be very useful to understand, for instance, what is the probability of seeing a page of interest (such as the buying page in an e-commerce site) coming from another page. Or, what is the probability of entering (or exiting) the Web site from any particular page. From a methodological viewpoint, we present two main research contributions. On one hand, we show how to improve the efficiency of the Apriori algorithm; on the other hand, we show how Markov chain models can be usefully developed and implemented for Web usage mining. In both cases, we compare the results obtained with classical association rules algorithms and models.

INTRODUCTION

In the last few years, the number of people that have used the Internet has enormously increased. Com-panies promote and sell their products on the Web, institutions provide information about their services, and single individuals exploit personal Web pages to be introduced to the whole Internet community.

We will show how the information, concerning the order in which the pages of a Web site are visited, can be profitably used to predict the visit behaviour at the site itself.

Every time a user links up to a Web site, the server keeps track of all the actions accomplished in a *log file*. What is captured is the "click flow" (clickstream) of the mouse and the keys used by the user during the navigation inside the site. Usually every click of the mouse corresponds to the viewing of a Web page. Therefore, we can define the click-stream as the sequence of the Web pages requested. The succession of the pages shown by a single user during navigation inside the Web identifies a user session. Typically, the analysis only concentrates on the part of each user session concerning the access at a specific site. The set of the pages seen, inside a user session, coming from a determinate site, is known with the term server session.

All this information can be profitably used to efficiently design a Web site. A Web page is well designed if it is able to attract users and address them easily to other pages within the site. A very important area in Web mining is the application of data mining techniques to discover usage patterns from Web data in order to optimally design a Web site, and to better satisfy needs of different visitors. This problem is known as Web usage mining, in contrast to Web context mining (analysis of the content of Web sites) and Web structure mining (analysis of Internet links): for more details on this see, for instance, Baldi *et al* (Baldi, Frasconi, & Smyth, 2003) or Chakrabarti (2003).

The objective of our analysis is to use Web clickstream data to understand the most likely paths of navigation in a Web site, with the aim of predicting, possibly online, which pages will be seen, having seen a specific path of other pages before. Such analysis can be very useful to understand, for instance, what is the probability of seeing a page of interest (such as the buying page in an e-commerce site) coming from another page. Or, what is the probability of entering (or exiting) the Web site from any particular page.

The most frequent type of statistical analysis of Web clickstream data is the search of the most interesting association and sequence rules (see, for an introduction, Han & Kamber, 2000 or Hand, Heikki, & Smyth, 2001); this search is accomplished by means of the well known Apriori algorithm (Agrawal, Mannila, Srikant, Toivonen, & Verkamo, 1995).Our research proposal is two-fold: it improves both the statistical analysis, by considering different Markov chain models; and the computational search algorithm, considering a Genmax type proposal.

According to what typically is done in the data-mining literature and practice, we shall compare our proposal with standard approaches by means of a real case study. In the description of the analysis, we shall follow the steps of the data mining process as described, for instance, in Berry and Linoff (1997), Giudici (2003) or Hastie et al. (Hastie, Tibshirani, & Friedman, 2001).

The database from which we start to illustrate our methodology is the result of the elaboration

Table 1. Extract of the considered dataset

c_value	C_time	c_caller
70ee683a6df…	14OCT97:11:09:01	Home
70ee683a6df…	14OCT97:11:09:08	Catalog
70ee683a6df…	14OCT97:11:09:14	Program
70ee683a6df…	14OCT97:11:09:23	Product
70ee683a6df…	14OCT97:11:09:24	Program

of a log file concerning a site of e-commerce, described, for instance, in Giudici (2003). The whole data set contains 250,711 observations, each corresponding to a click, that describe the navigation paths of 22,527 visitors among the 36 pages that compose the site of the Webshop. For illustrative purposes, Table 1 reports a very small extract of the available dataset.

Table 1 describes, the user session of one Web visitor, indexed by the *c_value* 70ee683a6df... Specifically, the column *c_caller* describes the clicks done by the visitor, at the times described by *c_time*.

In order to model the previous data, we have considered two main classes of statistical models: sequence rules and Markov Chains. In order to compare fairly the two approaches, we have used the same statistical tool, the SAS software. In the case of Markov chains, we have programmed part of the code using the IML language of SAS.

Furthermore, in the case of sequence rules, we have compared the results obtained with the Apriori algorithm implemented in SAS with a recent proposal, the GenMax algorithm (Zaki & Hsiao, 2002), that we have implemented with the IML language.

The structure of the chapter is as follows: in Section 2 we briefly recall what sequence rules are, in the context of association rules; in Section 3 we present algorithms to efficiently find such rules, and, specifically, we compare the Apriori algorithm with a GenMax algorithm; in Section 4 we introduce Markov chains for Web mining and, finally, in the last section, we present the experimental results concerning the comparison between classical sequence rules and Markov chains.

SEQUENCE RULES

We now recall what a sequence rule is. For more details, the reader can consult a recent text on data mining, such as Han and Kamber (2001) or, from a more statistical viewpoint, Hand et al (2001), Hastie et al. (2001) and Giudici (2003).

An *association rule* is a statement between two sets of binary variables (itemsets), A and B, that can be written in the form A→B, to be interpreted as a logical statement: *if A, then B*. If the rule is ordered in time, we have a *sequence rule* and, in this case, A preceeds B.

In Web clickstream analysis, a sequence rule is typically *indirect*: namely, between the visit of page A and the visit of page B other pages can be seen. On the other hand, in a *direct* sequence rule, A and B are seen consecutively.

A sequence rule model is, essentially, an algorithm that searches for the most interesting rules in a database. In order to find a set of rules, statistical measures of "interestingness" have to be specified. The measures more commonly used in Web mining to evaluate the importance of a sequence rule are the indexes of support, confidence, and the lift.

In this chapter, we shall consider mainly the confidence index. The confidence for the rule A → B is obtained dividing the number of server sessions that satisfy the rule (the so called "support" of the rule) by the number of sessions containing the page A (the support of the page).

In other words, the confidence approximates the conditional probability that, in a server session in which page A has been seen, page B is subsequently requested. (see e.g., Giudici, 2003).

ALGORITHMS TO FIND SEQUENCE RULES

In this section, we shall compare algorithms to extract association and, therefore, sequence rules, from a transactional database. The algorithms that we shall compare will be the Apriori, the backtracking, and the genmax algorithms.

Let us define a k-itemset a subset of order k of the variables being analysed; in our case, the pages of the Web site under consideration. The previous algorithms aim to find the most frequent itemsets, possibly of any order. Operationally, most frequent means all itemsets whose support passes a fixed threshold.

The Apriori algorithm is the simplest of the three algorithms. It is level-wise: it obtains all frequent itemset of a level (order) before moving on to the following level. It consists of a series of join operations; at each step, it joins the most frequent k-itemsets found in the previous step; thus, generating k+1-itemsets. It then eliminates the k+1 itemsets that contain subsets that are not frequent, according to the following property:

Apriori property: *All subsets of a frequent k-itemset must be frequent.*

To better illustrate the algorithm, consider the example described in Table 2.

Based on the data in Table 2, the search strategy of the Apriori algorithm is described in Figure 1, for a threshold support level equal to three out of six (50%).

From Figure 1 it is clear that the algorithm is level-wise. The dashed blue lines separate the different levels; the itemsets in green are those that do not pass the threshold level, and the itemsets in red are those eliminated by the Apriori property.

The weakness of the Apriori algorithm is that, as the number of frequent itemsets increases (e.g., when the number of Web pages is large), the algorithm gets worse both in terms of occupied storage and in computational capability required to apply the Apriori property.

A first solution to this problem is the backtracking algorithm. The backtracking algorithm employs a depth-first technique, where the search space does not proceed by levels, as in the Apriori

Table 2. Illustrative transactional database

Transaction ID	Items
1	A C T W
2	C D W
3	A C T W
4	A C D W
5	A C D T W
6	C D T

Figure 1. The search space of the Apriori algorithm

algorithm, but rather by branches, generating a tree-like structure.

Operationally, to each k-itemset found is associated with a combine set, that is the set of pages that, added to the k-itemset, generate k+1-itemsets whose support is above the established threshold (for brevity, in the following we shall say that they are frequent).

Figure 2 illustrates the search strategy of the backtracking algorithm, for the example illustrated in Table 2. The combine sets are shown in squared brackets.

In Figure 2, note arrows show the direction of the search, which is clearly branch-wise rather than level-wise as in Figure 1. Notice that the coloured itemsets will not be generated by the search strategy, as no combine set will lead to them. We have left them in the figure to emphasize the difference between the Apriori and the backtracking algorithm.

In other words, the backtracking algorithm exploit the Apriori property without actually computing it. This alleviates the computational burden of the Apriori algorithm, especially when the number of Web pages becomes large.

However, for low values of the support threshold, the search space of the backtracking algorithm keeps a considerable dimension. To improve the efficiency of the algorithm in this case, a number of research papers suggest using maximal frequent itemsets (MFI). Maximal frequent itemsets are frequent k-itemset for which no superset is frequent; in other words, the combine set of an MFI is empty.

Once all MFIs are extracted, all k-itemsets can be obtained as their subsets, as, according to the Apriori property, all subsets of MFIs are all frequent. The GenMax algorithm that we propose in this chapter employs the properties of certain sets, named tidsets, to extract MFIs.

In our problem, the tidset is the set of all users that visit a specific k-itemset. For the data described in Table 2, and k=1, the tidsets are those in Table 3.

Figure 2. The search space of the backtracking algorithm

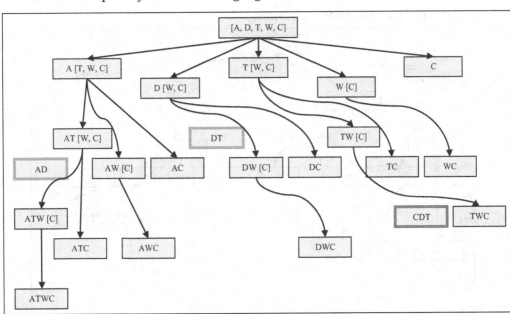

Table 3. Tidsets of all 1-itemsets

Itemset	Tidset
A	1, 3, 4, 5
C	1, 2, 3, 4, 5, 6
T	1, 3, 5, 6
W	1, 2, 3, 4, 5
D	2, 4, 5, 6

1. *if $t(X_i)=t(X_j)$ then $c(X_i)=c(X_j)=c(X_i\cup X_j)$*
2. *if $t(X_i)\subset t(X_j)$ then $c(X_i)\neq c(X_j)$, but $c(X_j)=c(X_i\cup X_j)$*
3. *if $t(X_i)\supset t(X_j)$ then $c(X_i)\neq c(X_j)$ but $c(X_j)=c(X_i\cup X_j)$*
4. *if $t(X_i)\neq t(X_j)$ then $c(X_i)\neq c(X_j)\neq c(X_i\cup X_j)$*

The previous properties can be used to generate an efficient search space. Figure 3 describes the search space of the GenMax algorithm.

Figure 3 contains different colours to emphasize differences with respect to the Apriori and the backtracking algorithm. While the branches of the backtrack algorithm are in black, the branches of the GenMax algorithm are green, if search branches, or red, if obtained as updated frequent itemsets from the previous theorem. All MFIs are in yellow.

By eliminating redundances in the application of the comparisons requested by the theorem, as well as the branches that generate false MFIs, that is, subsets of others, the search space can be further simplified as in Figure 4.

The Genmax algorithm is built on a theorem introduced by Zaki and Hsiao (2002), which follows. Let $I_{k+1}\cup c_i$ indicate the union between a k+1-itemset and the i-th element of its combine set and let $c(.)$ indicate the *closure operator*, that is, the largest itemset with tidset equal to that under consideration.

Let $X_i=I_{l+1}$ and $X_j=I_l\cup\{c_i\}$ *two members of a set of itemsets of level l+1, and let $[X_i, t(X_i)]$ and $[X_j, t(X_j)]$* the corresponding pairs of itemset and associated tidset. The following hold:

Figure 3. The search space of the GenMax algorithm

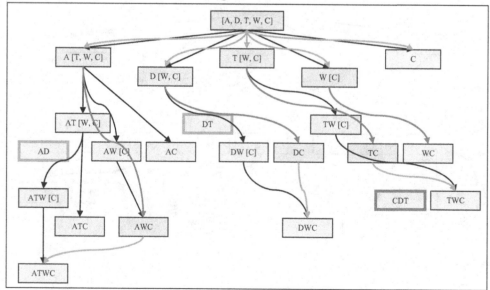

Figure 4. The search space of the modified GenMax algorithm

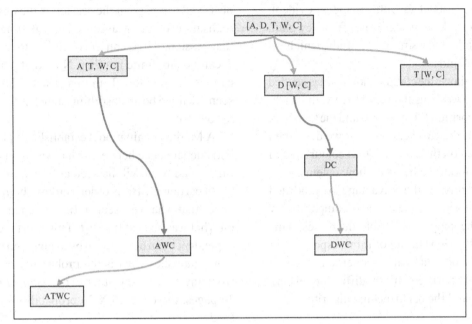

Figure 5. Comparison of computational efficiencies of the backtracking and GenMax algorithms

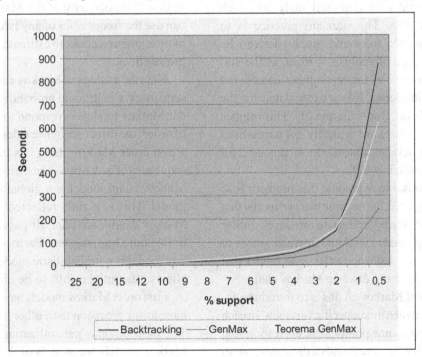

Comparing Figure 4 with Figure 3 and Figure 2, the computational advantages appear evident. In order to check empirically the efficiency gain, we have conducted a simulation study, which we describe in Figure 4.

Figure 4 shows the computational times needed to obtain the MFIs using, respectively, the back-tracking algorithm (blue colour) and the GenMax algorithm in the two versions that we described before, based on the Theorem (red colour) and in the version modified by us (yellow colour).

From Figure 5, the advantage of modified GenMax algorithm is clear, and is comparatively higher as the support threshold decreases. This shows that the GenMax algorithms, especially in the version modified by us, is an efficient search algorithm that can overcome the difficulties of both the Apriori and the backtracking algorithms.

MARKOV CHAINS FOR WEB USAGE MINING

We now consider the statistical analysis of Web clickstream data. The standard practice is to employ sequence rules and graphs derived by them. This type of analysis is local, as the statistical measures of interestingness considered (support, confidence, lift) are calculated for the itemsets at hand, that is marginally. This implies that such measures are typically not normalised (for a more detailed discussion of this see, for example, Giudici, 2003).

Our proposal to overcome this problem is to introduce a global, that is, multivariate model that can also lead, as a byproduct, to measures similar to those employed to evaluate interestingness of association and sequence rules. The model that we propose here is a discrete Markov chain.

The idea of Markov chains is to introduce dependence between time-specific variables. In each session, to each time point i, here corresponding to the i-th click, it corresponds a discrete random variable, with as many categories as the number of pages (these are named states of the chain). The observed i-th page in the session is the observed realisation of the Markov chain, at time i, for that session. Time can go from i=1 to i=T, and T can be any finite number. Note that a session can stop well before T: in this case the last page seen is said to be an absorbing state (end_session for our data).

A Markov chain model establishes a probabilistic dependence between what was seen before time i, and what will be seen at time i.

In particular, a first-order Markov chain establishes that what was seen at time i depends only on what was seen at time $i-1$. This short memory dependence can be assessed by a transition matrix that establishes what is the probability of going from any page to any other in one step, only. For 36 pages, there are 36 X 36 probabilities of this kind.

The conditional probabilities in the transition matrix can be estimated on the basis of the available conditional frequencies. If we add the assumption that the transition matrix is constant in time (homogeneity of the Markov chain), we can use the frequencies of any two adjacent pairs of time-ordered clicks to estimate the conditional probabilities.

Note the analogy of Markov chains with direct sequences. Conditional probabilities in a first-order Markov model correspond to the confidence of order two direct sequence rules and, therefore, a first order Markov chain is a model for direct sequences of order two. The difference is that the Markov chain model is a global and not a local model. This is mainly reflected in the fact that Markov chains consider all pages and not only those with a high support. Furthermore, the Markov model is a probabilistic model and, as such, allows inferential results to be obtained.

First order Markov models have been shown to have lower precision than other Markov models. The most obvious generalizations of first order Markov models are models of the second and of the third order. It can be shown that a second-order

Markov model is a model for direct sequences of order three, a third-order Markov model is a model for sequences of order four, and so on. In general, a k-th order model is described by the following property:

$$P(Pag_{n+1}|Pag_n,Pag_{n-1},...,Pag_0)=P(Pag_{n+1}|Pag_n,...,Pag_{n-k})$$

Such a model demands, therefore, the calculation of all M^k possible combinations of the states, increasing the dimension of the transition matrix space. This because the states of these models are not just the pages (as in first order models), but all their possible combinations. This increase in

the number of states can limit the use of Markov models for applications in which the speed is essential, or storage memory is limited.

A further problem is that a higher order Markov chain requires more data points for an efficient estimation; it becomes likely, for a high-order chain, that no data is present for a specific state. A simple way to overcome this problem is to use several Markov models at a time. For each state, if the greatest Markov model contains the request, it is used for prediction; otherwise, it is passed to the previous model and so on. This method is called *All-K^{th}-Order Markov model* (Pitkow & Pirolli, 1999).

EXPERIMENTAL RESULTS AND COMPARISONS

We shall now compare the predictive accuracy of sequence rules with that of Markov chain models of different orders, using those described in section 1. We have split the database in a training and validation set, as is done in the cross-validation comparison approach.

Furthermore, in order to better evaluate possible effects of different data structures on

Table 4. Number of MFI associated with each cluster

	Nr. MFI
Cluster 1	156
Cluster 2	363
Cluster 3	38
Cluster 4	826
Cluster 5	381
Cluster 6	65

Table 5. Comparison of predictive performances between different models

	Markov First Order	Markov Second Order	Markov Third Order	Sequenze Order two	Sequence Order three
Cluster 1	58,9 % (0,00 %)	63,8 % (0,72 %)	67,9 % (2,52 %)	51,9 % (0,00 %)	54,1 % (1,59 %)
Cluster 2	73,0 % (0,00 %)	70,7 % (0,10 %)	73,7 % (0,34 %)	43,8 % (0,00 %)	39,8 % (0,84 %)
Cluster 3	75,8 % (0,00 %)	86,2 % (0,22 %)	88,7 % (0,95 %)	71,5 % (0,00 %)	88,8 % (4,72 %)
Cluster 4	66,7 % (0,00 %)	63,9 % (0,17 %)	67,6 % (0,53 %)	42,9% (0,00 %)	39,1 % (0,36 %)
Cluster 5	61,2 % (0,00 %)	67,9 % (0,34 %)	69,2 % (1,00 %)	37,4 % (0,00 %)	41,1 % (0,99 %)
Cluster 6	70,9 % (0,00 %)	73,5 % (0,02 %)	73,4 % (0,42%)	52,4 % (0,00 %)	50,9 % (1,46 %)

prediction accuracy, we have run a preliminary cluster analysis of the data, obtaining six different clusters of navigation behaviours. The number of MFIs associated with each cluster is shown in Table 4.

From Table 4, note the rather different structure of the clusters. For example, while cluster 4 has 826 MFIs, cluster 3 has only 38.

The predictive performance of Markov chains and sequence models is compared, for each cluster, in Table 5.

In Table 5, we report two figures for each combination of models and clusters. The first figure is the prediction accuracy; the second figure (in parentheses) is the percentage of predictions that cannot be made because of absence of data on which to base the estimates. The considered models are first, second, and third order Markov models as well as sequence rules models of order two and three.

We can see that Markov chains perform generally better. The difference between the performance of Markov models and sequence rules is, however, not constant across clusters.

It can be shown that this relative difference is proportional to the number of maximal frequent itemsets: the lower such number, the greater the gain in accuracy, as can be deduced comparing Tables 4 and 5.

We finally compare the statistical efficiency (that is, predictive performance) of the All k-th order Markov model. Table 6 contains the results of such comparison.

From Table 6, it appears that the gain of the all k-th order Markov model is negligible, especially with respect to the third-order model. We believe that the obtained gain does not compensate for the increased complexity and extra computational cost that the model bears.

ACKNOWLEDGMENT

The chapter has benefited from funding from MIUR (Italian Research Ministry), within the project PRIN 2004: "Data mining methodologies for e-business." We thank the referees for useful suggestions that help improve the chapter.

Table 6. Comparison of predictive performances between All k-th order Markov model and low order models

	Markov 1° Ord.	Markov 2° Ord.	Markov 3° Ord.	All-K^th-Order Markov model
Cluster 1	58,9 % (0,00 %)	63,8 % (0,72 %)	67,9 % (2,52 %)	68,3 % (1,50 %)
Cluster 2	73,0 % (0,00 %)	70,7 % (0,10 %)	73,7 % (0,34 %)	73,5 % (0,21 %)
Cluster 3	75,8 % (0,00 %)	86,2 % (0,22 %)	88,7 % (0,95 %)	88.5 % (0,51 %)
Cluster 4	66,7 % (0,00 %)	63,9 % (0,17 %)	67,6 % (0,53 %)	67,6 % (0,34 %)
Cluster 5	61,2 % (0,00 %)	67,9 % (0,34 %)	69,2 % (1,00 %)	69,2 % (0,62 %)
Cluster 6	70,9 % (0,00 %)	73,5 % (0,02 %)	73,4 % (0,42%)	73,5 % (0,03 %)

REFERENCES

Agrawal, R., Mannila, H., Srikant, R., Toivonen, H., & Verkamo, A. I. (1995). Fast discovery of association rules. In *Advances in knowledge discovery and data mining*. Cambridge: AAAI/MIT Press.

Baldi, P., Frasconi, P., & Smyth, P. (2003). *Modelling the Internet and the Web—Probabilistic Methods and Algorithms*. John Wiley & Sons.

Berry, M., & Linoff, G. (1997). *Data mining techniques for marketing, sales, and customer support*. New York: Wiley.

Chakrabarti, S. (2003). *Mining the web: Discovering knowledge from hypertext data*. New York: Morgan Kaufmann.

Deshpande, M., & Karypis, G. (2000). *Selective Markov models for predicting Web-page accesses*. Retrieved from http://www.cs.umn.edu/~karypis

Giudici, P. (2003). *Applied data mining, statistical methods for business and industry*. London: Wiley.

Han J., & Kamber M. (2000). *Data mining: Concepts and techniques*. New York: Morgan Kaufmann.

Hand, D. J., Heikki, M., & Smyth, P. (2001). *Principles of data mining*. Cambridge: MIT Press.

Hastie, T., Tibshirani, R., & Friedman, J. (2001). *The elements of statistical learning: Data mining, inference and prediction*. New York: Springer-Verlag.

Pitkow, J., & Pirolli, P. (1999). Mining longest repeating subsequence to predict world wide web surfing. In *Second USENIX Symposium on Internet Technologies and Systems*, Boulder, CO.

Zaki, M. J., & Hsiao C.-J. (2002). CHARM: An efficient algorithm for closed Itemset mining. In *2nd SIAM International Conference on Data Mining*.

Chapter XV
Advanced Data Mining and Visualization Techniques with Probabilistic Principal Surfaces:
Applications to Astronomy and Genetics

Antonino Staiano
University of Napoli, "Parthenope", Italy

Lara De Vinco
Nexera S.c.p.A., Italy

Giuseppe Longo
University "Federico II" of Napoli Polo delle Scienze e della Tecnologia, Italy

Roberto Tagliaferri
University of Salerno, Italy

ABSTRACT

Probabilistic principal surfaces (PPS) is a nonlinear latent variable model with very powerful visualization and classification capabilities that seem to be able to overcome most of the shortcomings of other neural tools. PPS builds a probability density function of a given set of patterns lying in a high-dimensional space that can be expressed in terms of a fixed number of latent variables lying in a latent Q-dimensional space. Usually, the Q-space is either two- or three-dimensional and thus, the density function can be used to visualize the data within it. The case in which Q = 3 allows to project the patterns on a spherical manifold that turns out to be optimal when dealing with sparse data. PPS may also be arranged in ensembles to tackle complex classification tasks. As template cases, we discuss the application of PPS to two real- world data sets from astronomy and genetics.

INTRODUCTION

The explosive growth in the quantity, quality, and accessibility of data that is currently experienced in all fields of science and human endeavor, has triggered the search for a new generation of computational theories and tools, collectively constituting the field of data mining, capable to assist humans in extracting useful information (knowledge) from huge amounts of distributed and heterogeneous data. This revolution has two main aspects: on the one hand in astronomy, as well as in high energy physics, genetics, social sciences, and in many other fields, traditional interactive data analysis and data visualization methods, have proved to be far inadequate to cope with data sets that are characterized by huge volumes and/or complexity (ten or hundreds of parameter or features per record, cf. Abello, Pardalos, & Resende, 2002, and references therein). In second place, the simultaneous analysis of hundreds of parameters may unveil previously unknown patterns that will lead to a deeper understanding of the underlying phenomena and trends.

Knowledge discovery in databases or KDD is therefore becoming of paramount importance not only in its traditional arena, but also as an auxiliary tool for almost all fields of research. In this chapter, after a short introduction on the latent variable models, we shall first focus on the visualization and classification capabilities of the spherical probabilistic principal surfaces and then on the possibility to build PPS ensembles. Finally, we shall discuss two applications in the fields of astronomy and genetics. All results have been obtained in the framework of the Astroneural collaboration: a joint project between the Department of Mathematics and Informatics of the University of Salerno and the Department of Physical Sciences of the University Federico II of Napoli. The main goal of the collaboration is to implement a user-friendly data-mining tool capable to deal with heterogeneous, high-dimensionality data sets. All software is implemented under the Matlab computing environment exploiting the LANS Pattern Recognition Matlab Toolbox (http://www.lans.ece.utexas.edu/~lans/lans/) and the Netlab Toolbox (Nabney, 2002).

LATENT VARIABLE MODELS

The goal of a latent variable model is to express the distribution $p(\mathbf{t})$ of the variable $\mathbf{t}=(t_1, \ldots, t_D)$ in terms of a smaller number of latent variables $\mathbf{x} = (x_1, \ldots, x_Q)$, where $Q < D$. To achieve this, the joint distribution $p(\mathbf{t},\mathbf{x})$ is decomposed into the product of the marginal distribution $p(\mathbf{x})$ of the latent variables and the conditional distribution $p(\mathbf{t}|\mathbf{x})$ of the data variables, given the latent variables (Bishop, 1999). Expressing the conditional distribution as a factorization over the data variables the joint distribution becomes:

$$p(\mathbf{t},\mathbf{x}) = p(\mathbf{x})p(\mathbf{t} \mid \mathbf{x}) = p(\mathbf{x})\prod_{d=1}^{D} p(t_d \mid \mathbf{x}). \quad (1)$$

The conditional distribution $p(\mathbf{t}|\mathbf{x})$ is then written in terms of a mapping from latent variables to data variables, so that $\mathbf{t}=\mathbf{y}(\mathbf{x};\mathbf{w})+\mathbf{u}$. $\mathbf{y}(\mathbf{x};\mathbf{w})$ is a function of the latent variable $\mathbf{x}$ with parameters $\mathbf{w}$, and $\mathbf{u}$ is an $\mathbf{x}$-independent noise process. If the components of $\mathbf{u}$ are uncorrelated, the conditional distribution for $\mathbf{t}$ will factorize as in (1). Geometrically, the function $\mathbf{y}(\mathbf{x};\mathbf{w})$ defines a manifold in data space given by the image of the latent space. The definition of the latent variable model is completed by specifying the distribution $p(\mathbf{u})$, the mapping $\mathbf{y}(\mathbf{x};\mathbf{w})$, and the marginal distribution $p(\mathbf{x})$. The type of mapping $\mathbf{y}(\mathbf{x};\mathbf{w})$ determines the specific latent variable model. The desired model for the distribution $p(\mathbf{t})$ of the data is then obtained by marginalizing over the latent variables:

$$p(\mathbf{t}) = \int p(\mathbf{t} \mid \mathbf{x})p(\mathbf{x})d\mathbf{x}.$$

Although this integration will, in general, be analytically intractable, there exist specific forms of the distributions $p(\mathbf{t} \mid \mathbf{x})$ and $p(\mathbf{x})$ that lead to an analytic solution.

Probabilistic Principal Surfaces

Probabilistic principal surfaces or PPS (Chang, 2000; Chang & Ghosh, 2001) is a nonlinear extension of principal components in that each node on the PPS is the average of all data points that projects near/onto it. From a theoretical point of view, the PPS may be seen as a generalization of the generative topographic mapping (GTM) (Bishop, Svensen, & Williams, 1998), which, on the other hand, can be seen as a parametric alternative to self-organizing maps (SOM) (Kohonen, 1995).

Some advantages of PPS include the parametric and flexible formulation for any geometry/topology in any dimension, and the guaranteed convergence (indeed the PPS training is accomplished through the expectation-maximization (EM) algorithm (Dempster, Laird, & Rubin,1977)).

It has to be pointed out also that a PPS is governed by its latent topology and, owing to their intrinsic flexibility, a large variety of PPS topologies can be created. Among these, that of a 3-D sphere is particularly appealing since a sphere is finite and unbounded, and all nodes are distributed at the edge of the sphere; thus, making

it ideal for emulating the sparseness and peripheral property of high-D data. The *PPS* generalizes the GTM model by building a unified model, and shares the same formulation as the GTM, except for an oriented covariance structure for nodes in R^D. This means that data points projecting near a principal surface node have higher influences on that node than points projecting far away from it (Figure.1). Finally, the sphere topology (with no edges such as, for instance, is the case for SOM) can be easily comprehended by humans, and thereby be extremely effective for the visualization of high-D data.

Each node $\mathbf{y}(\mathbf{x};\mathbf{w})$, $\mathbf{x} \in \{\mathbf{x}_m\}_{m=1}^{M}$, has covariance,

$$\Sigma(\mathbf{x}) = \frac{\alpha}{\beta} \sum_{q=1}^{Q} \mathbf{e}_q(\mathbf{x})\mathbf{e}_q^T(\mathbf{x}) + \frac{(D - \alpha Q)}{\beta(D - Q)} \sum_{d=Q+1}^{D} \mathbf{e}_d(\mathbf{x})\mathbf{e}_d^T(\mathbf{x}),$$

$$0 < \alpha < \frac{D}{Q}$$

where,

- $\{\mathbf{e}_q(\mathbf{x})\}_{q=1}^{Q}$ is the set of orthonormal vectors tangential to the manifold at $\mathbf{y}(\mathbf{x};\mathbf{w})$.
- $\{\mathbf{e}_d(\mathbf{x})\}_{d=Q+1}^{D}$ is the set of orthonormal vectors orthogonal to the manifold in $\mathbf{y}(\mathbf{x};\mathbf{w})$.

Figure 1. Under a spherical Gaussian model of the GTM, points 1 and 2 have equal influences on the centre node y(x) (a) PPS have an oriented covariance matrix so point 1 is probabilistically closer to the centre node y(x) than point 2 (b) (Figure taken from Chang, 2000).

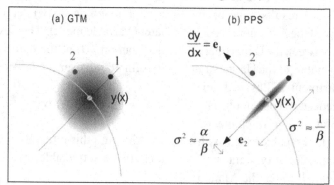

The complete set of orthonormal vectors $\{e_d(\mathbf{x})\}_{d=1}^{D}$ spans R^D. The parameter α is a clamping factor and determines the orientation of the covariance matrix. The unified *PPS* model reduces to GTM for $\alpha = 1$ and to the manifold-aligned GTM for $\alpha > 1$.

$$\Sigma(\mathbf{x}) = \begin{cases} 0 < \alpha < 1 & \perp \text{ to the manifold} \\ \alpha = 1 & I_D \text{ or spherical} \\ 1 < \alpha < D/Q & \| \text{ to the manifold.} \end{cases}$$

The EM algorithm can be used to estimate the PPS parameters $\mathbf{W}$ and β, while the clamping factor is fixed by the user and is assumed to be constant during the EM iterations. If we choose a 3-D latent space, a spherical manifold can be constructed using a PPS with nodes $\{\mathbf{x}_m\}_{m=1}^{M}$ arranged regularly on the surface of a sphere in R^3 latent space, with the latent basis functions evenly distributed on the sphere at a lower density. After a PPS model is fitted to data, the data themselves are projected for visualization purposes into the latent space as points on a sphere (Figure 2).

The latent manifold coordinates $\hat{\mathbf{x}}_n$ of each data point $\mathbf{t}_n$ are computed as,

$$\hat{\mathbf{x}}_n \equiv \langle \mathbf{x} | \mathbf{t}_n \rangle = \int \mathbf{x} p(\mathbf{x} | \mathbf{t}) d\mathbf{x} = \sum_{m=1}^{M} r_{mn} \mathbf{x}_m,$$

where r_{mn} are the latent variable responsibilities defined as:

$$r_{mn} = p(\mathbf{x}_m | \mathbf{t}_n)$$

$$= \frac{p(\mathbf{t}_n | \mathbf{x}_m) P(\mathbf{x}_m)}{\sum_{m'=1}^{M} p(\mathbf{t}_n | \mathbf{x}_{m'}) P(\mathbf{x}_{m'})} = \frac{p(\mathbf{t}_n | \mathbf{x}_m)}{\sum_{m'=1}^{M} p(\mathbf{t}_n | \mathbf{x}_{m'})}.$$

Since $\| x_m \| = 1$ and $\sum_m r_{mn} = 1$, for $n = 1, ..., N$, these coordinates lie within a unit sphere, that is $\| \hat{\mathbf{x}} \| \leq 1$.

SPHERICAL PPS AS DATA VISUALIZATION TOOLS

From the visualization point of view, the software implemented within Astroneural allows to:

a. Interact with data into the latent space in several ways.

b. Visualize the data probability density function in the latent space in order to derive a first understanding about the clusters existing in the data.

c. Select a number of clusters and visualize the individual data points therein.

Figure 2. (a) The spherical manifold in R^3 latent space; (b) The spherical manifold in R^3 data space; (c) Projection of data points t onto the latent spherical manifold (Figure taken from Chang, 2000).

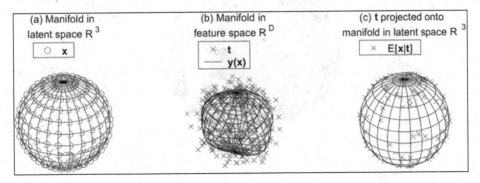

If needed, one can still interact with data by selecting data points in a given cluster and accomplish a number of comparisons and tests.

Interactively Selecting Points on the Sphere

Having projected the data on the latent sphere, a typical task performed by most data analyzers is the localization of the most interesting data points; for instance, the ones lying far away from denser areas (outliers), or those lying in the overlapping regions between clusters, and to investigate their characteristics by linking the data points on the sphere with their position in the original data set. For instance, in the astronomical application described later on, if the images corresponding to the data were available, the user might want to visualize the object on the astronomical image corresponding to the data point selected on the sphere. The user is also allowed to select a latent variable and color all the points for which that specific latent variable is responsible (Figure 3).

Visualizing the Latent Variable Responsibilities on the Sphere

The simple projections of the data points onto the sphere provide only partial information about the clusters inherently present in the data. For instance, if the points are strongly overlapping the user cannot derive any information at all. A first insight on the number of agglomerates localized into the spherical latent manifold is provided by the mean of the responsibility for each latent variable. Furthermore, if we build a spherical manifold that is composed by a set of faces, each one delimited by four vertices, then we can color each face with colors varying in intensity on the basis of the value of the responsibility associated with that given vertex (and hence, to each latent variable). The overall result is that the sphere will contain regions denser than other regions, and this information is easily visible and understandable. Obviously, denser areas of the spherical manifold might contain more than one cluster, and this calls for further investigations.

Figure 3. Data points selection phase. The bold black circles represent the latent variables; the blue points represent the projected input data points. When a latent variable is selected, each projected point for which the variable is responsible is colored.

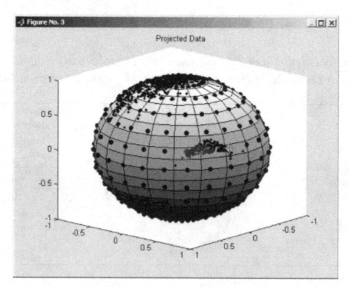

A Method to Visualize Clusters on the Sphere

Once the user has an overall idea of the number of clusters on the sphere, he can exploit this information through the use of classical clustering techniques (such as hard or fuzzy *k*-means (Bezdek, Keller, Krisnapuram, & Pal, 1999)) to find out the prototypes of the clusters and the data therein contained. This task is accomplished by running the clustering algorithm on the projected data. Afterwards, one may proceed by coloring each cluster with a given color (see Figure 4).

The visualization options so far described have been integrated in a user-friendly graphical user interface that provides a unified tool for the training of the PPS model, and next, after the completion of the training phase, also to accomplish all the functions for the visualization, characterization, and further analysis of the data (Staiano, 2004).

PPS AS A CLASSIFICATION TECHNIQUE

The spherical PPS may also be used as a "reference manifold" for classifying high-D data. A reference spherical manifold is computed for each class during the training phase. In the test phase, a data point previously unseen by the model is classified to the class of its nearest spherical manifold. Obviously, the concept of "nearest" implies a distance computation between a data

Figure 4. Clusters computed in the latent space by k-means

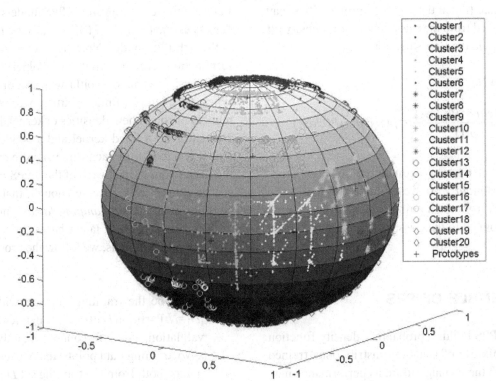

point **t** and the nodes of the manifold. Before doing this computation, the data point **t** must be linearly projected onto the manifold. Since a spherical manifold consists of square and triangular patches, each one defined by three or four manifold nodes, only an approximation of the distance is computed. The PPS framework provides three approximation methods:

- **Nearest Neighbor:** Finds the minimal square distance to all manifold nodes.
- **Grid Projections:** Finds the shortest projection distance to a manifold grid.
- **Nearest Triangulation:** Finds the nearest projection distance to the possible triangulation.

Another way to use PPS as classifier consists in choosing the class C with the maximum posterior class probability for a given new input **t**. Formally speaking, let us suppose to have N labeled data points $\{\mathbf{t}_1, ..., \mathbf{t}_n\}$, with $\mathbf{t}_i \in R^D$ and class labels in the set $\{1, ..., C\}$, then the posterior probabilities may be derived from the class-conditional density $p(\mathbf{t} \mid class)$ via the Bayes theorem:

$$P(class \mid \mathbf{t})$$

$$= \frac{p(\mathbf{t} \mid class)P(class)}{p(\mathbf{t})} \propto p(\mathbf{t} \mid class)P(class)$$

In order to approximate the posterior probabilities $p(class \mid \mathbf{t})$, we estimate $p(\mathbf{t} \mid class)$ and $p(class)$ from the training data. Finally, an input **t** is assigned to the class with maximum $p(class \mid \mathbf{t})$.

ENSEMBLE OF PPS

Since PPS builds a probability density function as a mixture of Gaussian distributions trained through the EM algorithm, its performance may degrade with increasing data dimensionality due to singularities and local maxima in the log-likelihood function. Therefore, we propose two schemes for designing a committee of spherical PPS to gain improved probability density functions and hence, classification rates. The area of ensemble of learning machines is now a well-defined field and has been successfully applied to neural networks, especially in the case of supervised learning algorithms. Fewer cases can be found for unsupervised learning methodologies and for density estimation as well: among these, the works introduced in Ormoneit and Tresp (1998) and Smyth and Wolpert (1999); both exploits consolidated techniques in supervised contexts as stacking (Wolpert, 1992) and bagging (Breiman, 1996) to density estimation, and represent the basis of our proposed schemes.

Stacking probabilistic principal surfaces for density estimation: *StPPS.* The ensemble, herein described, may be seen as an instantiation of the method proposed in Smyth and Wolpert (1999). Let us suppose we are given S PPS models (i.e., S density estimators) $\{PPS_s(\mathbf{t})\}_{s=1,...,S}$, where $PPS_s(\mathbf{t})$ is the s-th PPS model. Note that in the original formulation given in Smyth and Wolpert (1999), the S density estimators could also be of different types, for example, finite mixtures with a fixed number of component densities or kernel density estimate with a fixed kernel and a single fixed global bandwidth in each dimension. Now, going back to our model, each of the S PPS models can be chosen to be diverse enough, that is, by considering different *clamping factors*, number of latent variables, and latent bases. In order to stack the S PPS models, we follow the procedure described next:

1. Let D be the training data set, with size $|D|=N$. Partition D v times, as in v-fold cross validation. The v-th contains exactly $(v-1)$ x N/v training data points and N/v test data points, both from the training set D. For each fold:

a. fit each of the S PPS models to the training subset of D.
b. evaluate the likelihood of each data point in the test partition of D, for each of the S fitted models.

2. At the end of these preliminary steps, we obtain S density estimators for each of the N data points that are organized in a matrix A of size $N \times S$, where each entry a_{is} is $PPS_s(\mathbf{t}_i)$.

3. Use the matrix A to estimate the combination coefficients $\{\pi_s\}_{s=1,...,S}$ that maximize the log-likelihood at the points $\mathbf{t}_i$ of a stacked density model of the form:

$$StPPS(\mathbf{t}) = \sum_{s=1}^{S} \pi_s PPS_s(\mathbf{t}), \text{ which}$$

corresponds to maximize

$$\sum_{i=1}^{N} \ln\left(\sum_{s=1}^{S} \pi_s PPS_s(\mathbf{t}_i)\right),$$

as a function of the weight vector $(\pi_1, \ldots, \pi_S)$. Direct maximization of this function is a nonlinear optimization problem. We can apply the EM algorithm directly by observing that the stacked mixture is a finite mixture density with weights $(\pi_1, \ldots, \pi_S)$. Thus, we can use the standard EM algorithm for mixtures, except that the parameters of the component densities $PPS_s(\mathbf{t})$ are fixed, and the only parameters allowed to vary are the mixture weights.

4. The concluding phase consists in the parameters reestimation of each of the S component PPS models using all of the training data D. The stacked density model is then the linear combination of the so obtained component PPS models, with combining coefficients $\{\pi_s\}_{s=1,...,S}$.

Committee of PPS via Bagging: *BgPPS*. The second ensemble proposed employees bagging as mean to average single density estimators, in our case the PPS, in a way similar to the model

proposed in Ormoneit and Tresp (1998). All we have to do is to train a number S of PPS with S bootstrap replicates of the original learning data set. At the end of this training process, we obtain S different density estimates that are then averaged to form the overall density estimate model. Formally speaking, let D be the original training set of size N and $\{PPSs\}_{s=1,...,S}$ a set of PPS models:

1. Create S bootstrap replicates (with replacement) of D, $\{D_{Boot}(s)\}_{s=1,...,S}$ with size N.
2. Train each of the S PPS models with a bootstrap replicate D_{Boot}.
3. At the end of the training we obtain S density estimates $\{PPS_s\}_{s=1,...,S}$.
4. Average the S density estimates $\{PPS_s\}_{s=1,...,S}$ as $BgPPS(\mathbf{t}) = \frac{1}{S}\sum_{s=1}^{S} PPS_s(\mathbf{t})$.

APPLICATION OF PPS TO GOODS DATA

The Great Observatories Origins Deep Survey (GOODS) is an international project that joins together NASA, ESA (European Space Agency), and some of the most powerful ground-based facilities to survey the distant universe to the faintest flux limits across the broadest range of wavelengths. At the end of the project, GOODS will survey a total of roughly 320 square arc minutes in two fields centered on the Hubble Deep Field North and the Chandra Deep Field South, respectively (Dickinson, 2002). The currently available GOODS catalogue is a catalogue composed by 28,405 objects. Each object has been measured in seven optical bands, namely U,B,V,R,I,J,K bands. For each band, three different parameters, astrometric (positions), geometric (i.e., *Kron* radius, ellipticity, etc.), and photometric (*Flux* and *Magnitudes*) were measured, adding up to several dozens of parameters. Objects are also classified as angularly resolved (or *galaxies*, in the astronomical jargon) and nonresolved (or *stars*).

Moreover, GOODS (and more, in general, astronomical surveys) data present a further peculiarity: the majority of the objects are "drop outs," that is, they are detected only in some bands and not detected in the others, due to either instrumental (different detection limits) or intrinsic (different spectral properties) reasons. Without entering into details, we must stress that the characterization of an object as a "drop out" (i.e., as an object with a strong relative flux difference between two or more spectral regions) is very important from the astronomical point of view, since it allows discrimination among different classes of celestial objects. From our statistical point of view, therefore, the data set contains four classes of objects, namely stars, galaxies, stars that are drop outs, and galaxies that are drop outs (at this stage, we do not take into account the number of bands for which an object is a drop out). In order to evaluate the performance of our tools, we also processed a reference synthetic catalogue kindly provided us by Maurilio Pannella (MPI-Garching, Germany), matching the characteristics of the GOODS data. This catalogue contains 20,000 objects equally divided into two classes formed by 10,000 stars and galaxies, respectively. Each object is described by eight features (parameters); namely, the magnitudes in the corresponding eight optical filters.

Data Visualization for GOODS Catalogue

In Figure 5, we visualize the actual GOODS catalogue. As it may be seen, it exhibits four strongly overlapping classes. As it is apparent, the PCA vi-

Figure 5. From top left to bottom right clockwise: GOODS 3-D PCA projections, PPS projections on the sphere, galaxy density on the sphere, and star density on the sphere.

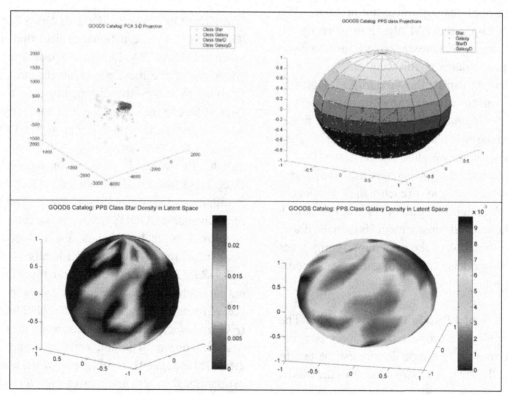

sualization gives no interesting information at all, since it displays only a single condensed group of data. In PCA, the class of dropped galaxies (whose objects are yellow colored), which contains the majority of objects (about 24,000), is near totally hidden. The PPS projection onto the spherical latent manifold appears much more readable than the PCA and contains much more information. The figure also depicts the corresponding latent variable probability density function. By rotating the sphere with density, two high-density regions are highlighted together with a few other regions of lower density.

Classification of GOODS Data

In the case of StPSS, we built a model in which a group of six different PPS models, each one with a fixed α value, are put together in an ensemble via stacking. An important parameter for stacking is the number v of folds in the cross validation procedure. In our experiments, we tried 5-fold and 10-fold cross validation.

In the case of BgPPS we used, instead, a single PPS model with its own parameters setting, and to bag it in order to improve its performance. In our experiments, we bag 10 PPS models (for $\alpha = 0.2, 0.4, ...2.0$) in order to assess the best α value. The PPS models are trained on 20 bootstrap replicates of the training data set (hence, we have a committee of 20 PPS models whose responses are averaged).

In all experiments, the classifiers run 25 times, and each time, new training and test data partitions (60% for training and 40% for testing) are generated. Moreover, for comparison purposes, we accomplished classification by using single PPS models as well, to:

1. Compute the reference manifolds for each class (we denote this classifier as PPSRM).
2. Compute the posterior class probability (hereinafter denoted as PPSPR).

Application to Synthetic Catalogue: StPPS

Parameter settings are listed in Table 1. The results are depicted in Figure 6, and clearly show that 5-fold cross validation works better than 10-fold cross validation in both the mean classification error (1.34 against 1.84, respectively) and standard deviation (0.2606 against 0.4071, respectively).

The minimum error reached is 1.05, as it is shown in Table 2, where it is also shown the corresponding confusion matrix. The difference between 5-fold and 10-fold cross validation could be explained by the fact that the size of the training set is quite high, so a 10-fold cross validation may lead to overfitting problems (recall that in our PPS models we do not employ any regularization method).

Table 1. Synthetic catalogue: StPPS parameter settings

Parameters	PPS_1	PPS_2	PPS_3	PPS_4	PPS_5	PPS_6
α	1	0.5	3	0.2	0.3	0.8
M	266	266	266	266	266	266
L	18	51	51	51	6	51
$Lfac$	2.2	2	2	2	2.5	2
$iter$	100	100	100	100	100	100
ε	0.01	0.01	0.01	0.01	0.01	0.01

Figure 6. Synthetic catalogue: StPPS classification errors over 25 iterations

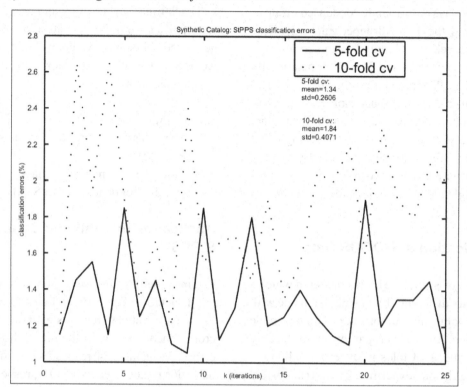

Table 2. Synthetic catalogue: Confusion matrix by StPPS best result

Classifier	Confusion Matrix		
		Star	Galaxy
St PPS (1.05)	Star	3943	27
	Galaxy	57	3973

Table 3. Synthetic catalogue: BgPPS parameter settings

Parameter	Value	Description
M	266	number of latent variables
L	60	number of basis functions
Lfac	1	basis functions width
iter	100	maximum number of
ε	0.01	iteration
		early stopping threshold

Application to Synthetic Catalogue: BgPPS

The parameter settings are shown in Table 3. On synthetic catalogue, bagging performs very well for values of clamping factor α between [1.0, 2.0], where the best mean classification error and standard deviation results are obtained. In particular, for $\alpha = 2.0$

BgPPS reaches its minimum mean classification error (0.24) (see Figure 7).

Application to GOODS Catalogue: StPPS

In GOODS catalogue, the behavior of the stacked model, for which the parameters are set as in

Figure 7. Synthetic catalogue: BgPPS error bars over 25 iterations for each fixed α

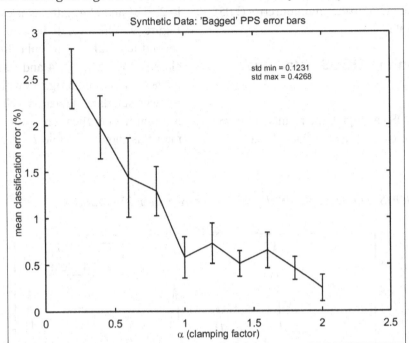

Table 5, is inverted in terms of 5-fold and 10-fold cross validation.

In fact, here we have better results for 10-fold cross validation (mean classification error 2.87 and standard deviation 0.1344) with respect to 5-fold cross validation (mean classification error 3.44 and standard deviation 0.4720), as can be seen from Figure 8. This is reasonable as the number of training data for the first three classes (S, G, and SD) are much less than the number of training data for class GD, so a higher number

Table 4. Synthetic catalogue: Confusion matrix by BgPPS best model

Classifier	Confusion Matrix			α
		Star	Galaxy	
BgPPS(0.05)	Star	3996	0	2.0
	Galaxy	4	4000	

Table 5. GOODS catalogue: StPPS parameter settings

Parameters	PPS_1	PPS_2	PPS_3	PPS_4	PPS_5	PPS_6
α	1.4	1.2	0.8	0.6	1.6	2.0
M	266	266	266	266	615	615
L	18	83	83	83	83	83
Lfac	1	2	1.5	1.1	1.3	2
iter	100	100	100	100	100	100
ε	0.01	0.01	0.01	0.01	0.01	0.01

of folds leads to a better fit to data. Confusion matrix corresponding to the minimum error (1.05) is shown in Table 6.

Application to GOODS catalogue: BgPPS

For the GOODS catalogue, the results are more fluctuating for each of the α values. In fact, the best results are obtained between the interval [0.2, 0.6] and [1.4, 2.0], as it can be seen from Figure 9. The overall best result falls in the second interval, in particular for $\alpha = 1.8$ (mean classification error 2.74 and standard deviation 0.3987) even though BgPSS with $\alpha = 0.6$ obtains a lower standard deviation value (0.1725). The minimum classification error with confusion matrix is shown in Table 7.

Figure 8. GOODS catalogue: StPPS classification errors over 25 iterations

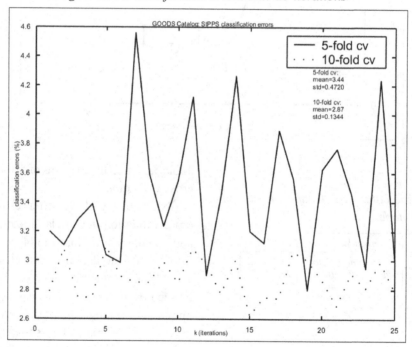

Table 6. GOODS catalogue: Confusion matrices by StPPS best model

Classifier	Confusion Matrix				
		S	*G*	*SD*	*GD*
	S	*92*	*4*	*2*	*0*
	G	*76*	*1234*	*2*	*36*
StP P S(2.62)	*SD*	*0*	*0*	*52*	*36*
	GD	*0*	*8*	*134*	*9688*

Figure 9. GOODS catalogue: BgPPS error bars over 25 iterations for each fixed α

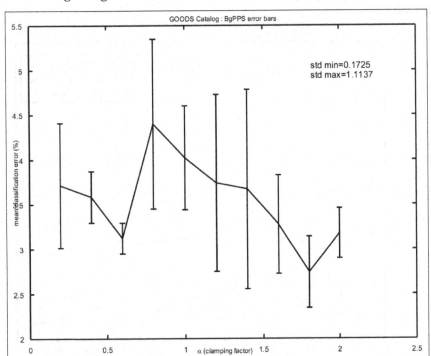

Table 7. GOODS catalogue: Confusion matrix by BgPPS best model

Classifier	Confusion Matrix					α
		S	**G**	**SD**	**GD**	
	S		155	35	12 5	
BgPPS(2.15)	**G**					1.8
		8	1160	6	8	
	SD					
		0		0	64 7	
	GD					
		5		51 108	9740	

Synthetic Catalogue: PPSRM, PPSPR, StPPS and BgPPS Comparison

As can be seen from Figure 10, BgPPS outperforms both PPSRM and PPSPR for near all α values. Moreover, from Figure 11, it is clear that BgPPS outperforms StPPS. This latter model performs best of the single model classifiers on average, even though PPSPR for just one α value reaches a better result.

GOODS Catalogue: PPSRM, PPSPR, StPPS, and BgPPS Comparison

GOODS catalogue classification task is more complex. The four classes are heavily overlapping and even in the best cases, there are classes (i.e.,

Figure 10. Synthetic catalogue: PPSRM, PPSPR, and BgPPS mean classification errors over 25 iterations for each fixed α

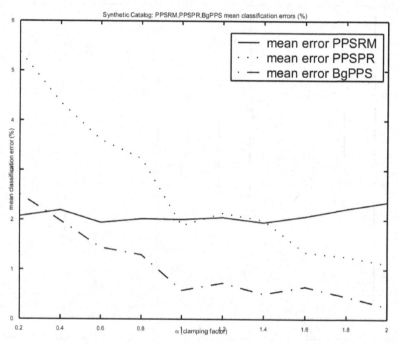

Figure 11. Synthetic catalogue: PPSRM, PPSPR, StPPS, and BgPPS best model statistics

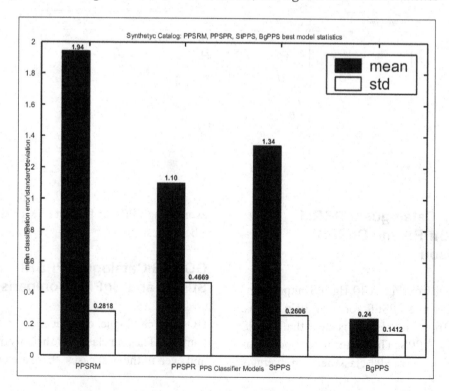

Figure 12. GOODS catalogue: PPSRM, PPSPR, and BgPPS mean classification errors over 25 iterations for each fixed α

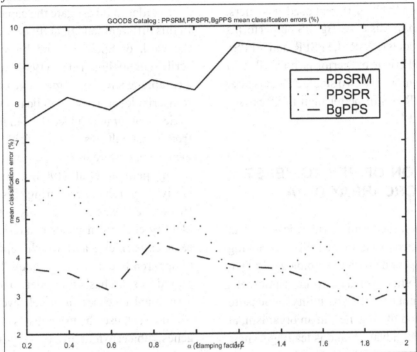

Figure 13. GOODS catalogue: PPSRM, PPSPR, StPPS, and BgPPS best model statistics

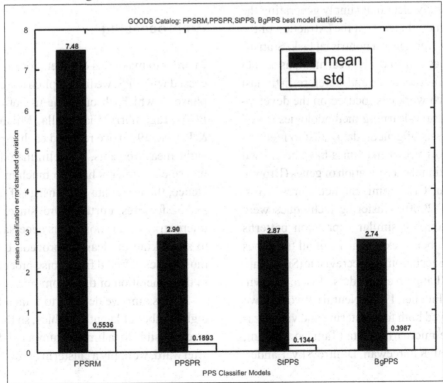

S and SD) whose objects are classified with an error rate of about 60%. This is evident from the results obtained by the different used classifiers. However, even in this case, BgPPS outperforms all the other models (PPSRM, PPSPR, and StPPS). Moreover, StPPS here outperforms both PPSRM and PPSPR. Among the two single PPS classifier models, PPSPR is still better than PPSRM (see Figures 12 and 13).

APPLICATION OF PPS TO YEAST GENES MICROARRAY DATA

Gene-expression microarrays, whose development started in the second half of the 1990s, are having a powerful impact on molecular biology. In fact, although the ability to measure transcription of a single gene is not new, the possibility to measure the transcription of all genes, in an organism, at once, is a recent advance, and is leading to new methods of diagnosis and of treatment for a large number of diseases. However, it is also becoming increasingly clear that simply generating the data is not enough, and that the extraction of the relevant information is a nontrivial task. Statistical techniques and other classical methods of data analysis, are not adequate and therefore, in the last decade, much work has focused on the development of machine-learning methodologies suited for the analysis of genetic data. Just to mention a few, support vector machines have been used for the functional classification of genes (Brown, Grundy, Lin, Cristianini, Sugnet, Furey, Ares, & Haussler, 2000); clustering techniques were used for grouping similar expression patterns across a number of experiments of all the genes of the yeast Saccharomyces cerevisiae (Spellman, Sherlock, Zhang, Iyer, Anders, Eisen, Brown, Botstein, & Futcher, 1998); Neural networks have been employed both for clustering and visualization of gene microarray data (Tamayo, Slonim, Mesirov, Zhu, Kitareewan, Dmitrovsky, Lander,

& Golub 1999; Toronen, Kolehmainen, Wong, & Castren, 1999).

In order to investigate the capabilities of PPS in this different field of activity, we started from the work of Spellman and his colleagues described in Spellman et al. (1998), which provides a comprehensive catalogue of yeast genes whose transcript levels vary periodically within the cell cycle. In order to produce the catalogue, samples from yeast cultures synchronized with different experiments were used.

In Spellman et al. (1998) a type of agglomerative hierarchical clustering (Eisen, Spellman, Brown, & Botstein, 1998) was used in order to identify clusters of genes behaving similarly in each experiment, and which represent groups of apparently coregulated genes. These clusters provide a solid basis for understanding the transcriptional mechanism of cell cycle regulation. The data set, used by us, consists of a set of 6,125 genes, subject to four different experiments. Each experiment consists of measurements at different epochs, for a total of 73 parameters.

Preprocessing

In order to make this data set more apt to be processed with PPS, we first applied a preprocessing phase in which, through the use of a nonlinear PCA (Tagliaferri, Ciaramella, Milano, Barone, & Longo, 1999), we reduced each experiment to eight measurements, and eliminated the genes whose experiments had too much missing data. Hence, the used data set consists of 5,425 genes and 32 features. Furthermore, since, in general, microarray data is noisy, it is necessary to resort to some kind of cleaning procedure to identify those genes affected from noise process involved in the generation of data from microarrays.

At this aim, we decided to train a PPS with a high number of latent variables, so that each one is responsible for a limited number of data points; afterward, we apply a clustering procedure on the

Figure 14. Yeast gene data set: (a) 3-D PCA projection; (b) Data point projections in the latent space; (c) Data probability density in the latent space

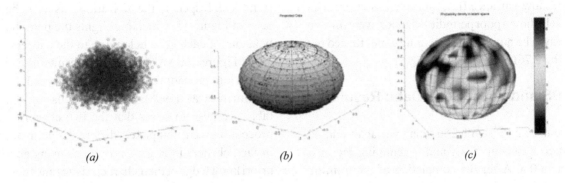

Figure 15. Cluster prototype periodic behaviors and error bars (3σ) showing the standard deviations of genes from the prototypes for a fixed cluster. On top of each subplot, the cluster number and the number of genes within each cluster is reported.

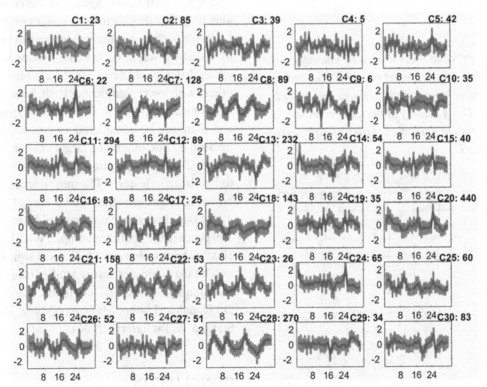

nodes of the manifold in the data space. So doing, a number of identified clusters containing genes with low variance (i.e., genes whose transcript levels show a poor periodic behavior) were thrown away. The number of remaining genes turned out to be 2,761.

PPS and Yeast Gene Data: Results

We used a PPS with 266 latent variables and 40 latent basis functions and a clamping factor α set to 0.5. After the completion of the training phase, we projected the data into the latent space and computed the responsibility for each latent variable, as shown in Figure 14.

On the basis of probability density function visualized in Figure 14, we decided to identify 30

*Figure 16. PPS and Spellman cluster comparisons. On each row are reported the 30 PPS clusters, while on the columns are the clusters computer by Spellman. The A_{ij-th} entry of the table corresponds to the fraction of Spellman cluster **j** falling in the PPS cluster **i**.*

	1	2	3	4	5	6	7	8
1	0	0	0	0	0	0	0	0
2	0	0	0	10	0	0	0	0
3	3.4483	0	0	0	0	0	0	0
4	0	0	0	0	0	0	0	0
5	0	0	0	20	0	0	0	0
6	0	0	0	0	0	0	0	0
7	0	0	0	0	10.345	0	0	3.7037
8	89.655	0	0	0	20.69	0	4.5455	0
9	0	0	0	0	0	0	0	0
10	0	0	0	0	0	0	0	0
11	0	0	0	0	0	0	0	0
12	0	0	0	0	0	0	0	0
13	0	0	0	0	0	0	0	0
14	0	0	0	0	0	0	0	0
15	0	0	0	0	0	0	0	0
16	0	0	0	50	0	0	0	0
17	0	0	0	10	31.034	0	0	0
18	0	0	0	10	3.4483	0	0	0
19	0	0	0	0	0	0	0	0
20	0	0	0	0	13.793	7.6923	0	0
21	6.8966	2.5	100	0	0	23.077	0	0
22	0	0	0	0	0	69.231	0	0
23	0	0	0	0	3.4483	0	0	0
24	0	0	0	0	0	0	0	0
25	0	0	0	0	10.345	0	90.909	0
26	0	0	0	0	0	0	4.5455	0
27	0	0	0	0	0	0	0	0
28	0	97.5	0	0	0	0	0	96.296
29	0	0	0	0	0	0	0	0
30	0	0	0	0	6.8966	0	0	0

clusters through a hierarchical clustering procedure. For each cluster, we plotted the prototype trend with respect to the 32 features, as it can be seen in Figure 15, which highlights the periodic behavior of each gene belonging to the clusters.

In Figure 16, we compare the results of our clustering procedure with those obtained by Spellman et al. (1998). Before discussing the table, we wish to stress that the two clustering procedures were completely different: Spellman, in fact, clustered the gene properties using an a priori knowledge of their characteristics and thus, he worked with only 209 genes, while our algorithm made use only of the statistical properties of the data with no a priori knowledge. In spite of this, some remarkable patterns may be detected: Spellman's cluster number 1 falls near entirely in our cluster 8; Spellman's clusters number 2 and 8 are, statistically speaking, indistinguishable (together they form our cluster number 28); Spellman's cluster number 5 appears to be a sort of statistical waste basket that groups together rather different clusters (7, 8, 17, 20, plus several others with lower significance) that, however, are topological neighbors in the PPS latent space, and can therefore be considered as "substructures" (missed by Spellman) of a larger cluster. Finally, cluster 21 contains, entirely, the genes belonging to Spellman's cluster 3. The most relevant result, however, seems to be the fact that many (13 out of 30) of our clusters are not mapped by any of the 209 genes in the Spellman sample. Whether these clusters have or have not biological significance will be the subject of future studies (see Amato, Ciaramella, Deniskina, Del Mondo, di Bernardo, D., Donalek, Longo, Mangano, Miele, Raiconi, Staiano, & Tagliaferri, 2006).

CONCLUSION

The aim of this work is to propose a new tool to the data mining community. The tool is based upon probabilistic principal surfaces, and we discussed its potentiality and highlighted the flexibility it exhibits in a number of activities fruitful for data mining applications. For this purpose, the discussion focused, in particular, on PPS classification and data visualization capabilities, two important activities for data mining, by showing some results gained with data coming from astronomy (star/galaxy data) and genetics (yeast gene microarray data). Although the obtained experimental results are, for one case (i.e., yeast gene data), in a preliminary phase and must be validated by further experiments, it is undoubtedly from the results themselves what advantages and benefits PPS provide.

Classification Tasks

Even though the basic PPS model provides good classification performance (Chang, 2000), we showed that these abilities can be further improved by employing PPS in an ensemble, proposing, specifically, two combining schemes based on stacked generalization and bagging. We applied the PPS ensemble to astronomical data, and from the experimental results, it can be stated that the committee of PPS perform better than single PPS, even though this is clear for the ensemble of PPS built via bagging. Stacked PPS, instead, has less stable results, but it seems a promising combining schema after all, since we did few experiments by varying the PPS component complexity. We rather focused on the impact of cross validation, which appears as of primary importance.

Data Visualization Tasks

The spherical PPS, which consists of a spherical latent manifold lying in a three-dimensional latent space, is better suitable to visualize high-D data, since the sphere is able to capture the sparsity and periphery of data in large input spaces, which are due to the curse of dimensionality. We proposed a number of visualization possibilities integrated in a user-friendly graphical user interface:

- Interactive selection of regions of sample points projected into the sphere for further analysis. This is particularly useful to profile groups of data.
- Visualization of the latent variable responsibilities onto the sphere as a colored surface plot. It is useful to localize more and less dense areas to find out a first number of clusters existing in the data, and to highlight the regions where lies outliers.
- A method to exploit the information gathered with the previous visualization options through a clustering algorithm to find out the clusters with the corresponding prototypes and data points.

The data visualization tasks have been proved effective for data mining in complex application domains: astronomical data and yeast gene microarray data analysis. Although the study of the methods addressed in this chapter is devoted to the astronomical and genetic applications, the system is general enough to be used in whatever data-rich field to extract meaningful information.

REFERENCES

Abello, J., Pardalos, P. M., & Resende, M. G. C. (Eds). (2002). *Handbook of massive data sets.* Kluwer Academic Publishers.

Amato, R., Ciaramella, A., Deniskina, N., Del Mondo, C., di Bernardo, D., Donalek, C., Longo, G., Mangano, G., Miele, G., Raiconi, G., Staiano, A., & Tagliaferri, R. (2006). A multi-step approach to time series analysis and gene expression clustering. *Bioinformatics, 22*(5), 589-596.

Bezdek, J. C., Keller, J., Krisnapuram, R., & Pal, N. R. (1999). *Fuzzy models and algorithms for pattern recognition and image processing.* Kluwer Academic Publisher.

Bishop, C. M. (1999). Latent variable models. In M. I. Jordan (Ed.), *Learning in graphical models* (pp. 371-403). MIT Press.

Bishop, C. M., Svensen, M., & Williams, C. K. I. (1998). GTM: The generative topographic mapping. *Neural Computation, 10*(1), 215-235.

Breiman, L. (1996). Bagging predictors. *Machine Learning, 26*(2), 123-140.

Brown, M., Grundy, W., Lin, D., Cristianini, N., Sugnet, C., Furey, T., Ares, M. Jr. & Haussler, D. (2000). Knowledge-based analysis of microarray gene expression data by using support vector machines. *Proceedings of the National Academy of Science USA, 97*(1), 262-267.

Chang, K. (2000). *Nonlinear dimensionality reduction using probabilistic principal surfaces*. Unpublished PhD Dissertation. The University of Texas at Austin, USA.

Chang, K.,& Ghosh, J. (2001). A unified model for probabilistic principal surfaces. *IEEE Transactions on Pattern Analysis and Machine Intelligence, 23*(1), 22-41.

Dempster, A. P., Laird, N. M., & Rubin, D. B. (1977). Maximum-likelihood from incomplete data via the EM algorithm. *J. Royal Statistical Society, 39*(1), 1-38.

Dickinson, M. (2002). *The Great Observatories Origins Deep Survey: An overview*. Retrieved from http://www.stsci.edu/science/goods/abstract.html

Eisen, M. B., Spellman, P. T., Brown, P. O., & Botstein, D. (1998). Cluster analysis and display of genome-wide expression patterns. *PNAS, 95,*14863-14868.

Kohonen, T. (1995). *Self-organizing maps*. Berlin: Springer-Verlag.

Nabney, I. T. (2002). *Netlab: Algorithms for pattern recognition*. Springer-Verlag.

Ormoneit, D., & Tresp, V. (1998). Averaging, maximum likelihood and Bayesian estimation for improving gaussian mixture probability density estimates. *IEEE Transaction on Neural Networks, 9*(4), 639-650.

Smyth, P., & Wolpert, D. H. (1999). An evaluation of linearly combining density estimators via stacking. *Machine Learning Journal, 36*, 59-83.

Spellman, P. T., Sherlock, G., Zhang, M. Q., Iyer, V. R., Anders, K., Eisen, M. B., Brown, P. O., Botstein, D., & Futcher, B.(1998). Comprehensive identification of cell cycle-regulated genes of the yeast saccharomyces cerevisiae by microarray hybridization. *Molecular Biology of the Cell, 9*, 3273-3297.

Staiano, A. (2004). *Unsupervised neural networks for the extraction of scientific information from astronomical data*. Unpublished PhD Dissertation, University of Salerno, Italy.

Tagliaferri, R., Ciaramella, A., Milano, L., Barone, F., & Longo, G. (1999). Spectral analysis of stellar light curves by means of neural networks. *Astronomy and Astrophysics Supplement Series, 137*, 391-405.

Tamayo, P., Slonim, D., Mesirov, J., Zhu, Q., Kitareewan, S., Dmitrovsky, E., Lander, E. S., & Golub, T. R. (1999). Interpreting patterns of gene expression with self-organizing maps: Methods and application to hematopoietic differentiation. *PNAS, 96*, 2907-2912.

Toronen, P., Kolehmainen, M., Wong, G., & Castren, E.(1999). Analysis of gene expression data using self-organizing maps. *FEBS Letters, 451*, 142-146.

Wolpert, D. H. (1992). Stacked generalization. *Neural Networks, 5*(2), 241-259.

Chapter XVI
Spatial Navigation Assistance System for Large Virtual Environments:
The Data Mining Approach

Mehmed Kantardzic
University of Louisville, USA

Pedram Sadeghian
University of Louisville, USA

Walaa M. Sheta
Mubarak City for Scientific Research, Egypt

ABSTRACT

Advances in computing techniques, as well as the reduction in the cost of technology, have made possible the viability and spread of large virtual environments. However, efficient navigation within these environments remains problematic for novice users. Novice users often report being lost, disorientated, and lacking the spatial knowledge to make appropriate decisions concerning navigation tasks. In this chapter, we propose the frequent wayfinding-sequence (FWS) methodology to mine the sequences representing the routes taken by experienced users of a virtual environment, in order to derive informative navigation models. The models are used to build a navigation assistance interface. We conducted several experiments using our methodology in simulated virtual environments. The results indicate that our approach is efficient in extracting and formalizing recommended routes of travel from the navigation data of previous users of large virtual environments.

INTRODUCTION

The design and implementation of virtual environments (VEs) has improved significantly in recent times due to advances in both hardware and software technology. Research from various disciplines, such as computer graphics, human-computer interaction, urban design, and psychology, have also contributed to the advancement of the field of VEs. In general, VEs provide a computer-synthesized world in which users can interact with objects, perform various activities, and navigate the environment as if they were in the real world. Applications for VEs are found in various domains including medicine, engineering, oil exploration, and the military (Burdea & Coiffet, 2003; Grady, 2003; Sherman & Craig, 2002).

Navigation is a fundamental activity in VEs (see Figure 1). Therefore, successful use of a VE requires that the user be able to easily and efficiently navigate from one location to another (Darken & Peterson, 2002). However, previous research has shown that novice users of VEs are often disoriented, feel "lost in hyperspace," and lack the spatial knowledge needed to pick an appropriate route due to the deficiency of experience with the VE (Conroy, 2001; Darken & Sibert, 1996; Kantardizc, Rashad, & Sadeghian, 2004; Sade-

ghian, Kantardzic, Lozitskiy, & Sheta, 2005; Sadeghian, Kantardzic, Lozitskiy, & Sheta, 2006a; Sadeghian, Kantardzic, & Rashad, 2006b; van Dijk, op Den, Rieks, & Zwiers, 2003). Traditional tools, such as maps, have demonstrated some success in helping users navigate (Darken & Sibert, 1996; Statalich, 1995); however, researchers are continuously looking for alternative intelligent navigation tools (Chen & Stanney, 1999).

In this chapter, we are introducing the frequent wayfinding-sequence (FWS) methodology to derive a model of the experienced users' navigation behaviors. This process is conducted by transforming the previously recorded navigation data of experienced users into sequences, applying a modified sequence mining algorithm to find frequent sequences corresponding to preferred routes of travel, and forming a final model of routing rules. This model is the basis of a spatial navigation tool that can be used by

Figure 1. Navigation is a fundamental activity in large VEs

novice users to get recommended routes of travel within the VE.

RELATED RESEARCH

One approach of structuring spatial knowledge about a VE is based on three distinct components: landmark knowledge, procedural knowledge, and survey knowledge (Darken & Sibert, 1996; Elvins, Nadeau, & Kirsh, 2001). Landmark knowledge represents information about the shape, size, color, and contextual information of landmarks, or memorable and distinctive objects in a VE. Procedural knowledge is encoded as a series of steps required in following a particular route. Landmarks play a role in procedural knowledge by marking decision points along a route, and helping a traveler recall the procedures required to get to a destination. Survey knowledge provides a bird's eye view of a region in which object locations and interobject distances are encoded in terms of a global frame of reference. Landmarks play a role in survey knowledge by providing regional anchors with which to calibrate distances and directions (Chen & Stanney, 1999; Darken & Sibert, 1996; Elvins et al., 2001).

Through experience and repeated exposure to a VE, a user gains spatial knowledge and becomes efficient at wayfinding. Wayfinding is "the act of traveling to a destination by a continuous, recursive process of making route-choices whilst evaluating previous spatial decisions against constant cognition of the environment," (Conroy, 2001, pp. 26). O'Neill (1992) found that wayfinding performance decreases as an environment's complexity increases. On the other hand, an increase in the familiarity with a VE leads to an increase in wayfinding ability (Ruddle, Payne, & Jones, 1998). The use of different types of technology to navigate (Peterson, Wells, Furness, & Hunt, 1998; Ruddle, Randall, & Jones, 1996), the goal associated with a wayfinding task (Magliano, Cohen, Allen, & Rodrigue, 1995), and the travel techniques used

by a traveler (Bowman, 1999) are also factors that effect wayfinding performance.

Since wayfinding in VEs is not a trivial task, much effort has been put in developing navigation tools. Darken and Sibert (1996) found that the navigation performance of users in a VE equipped with a map was superior to the performance of users in a similar VE equipped with a radial grid. However, users of a VE without any navigation tools performed the worst and had extreme difficulty completing navigation tasks. Other studies have also shown the benefit of map use in improving wayfinding (Statalich, 1995). On the other hand, some studies have found that people are better at finding a target location when using signs or narrative directions than with using maps (Moser, 1988; Streeter, Vitello, & Wonsiewicz, 1985). Concentrating on the importance of landmarks in wayfinding, Elvins, Nadeau, and Kirsh (2001) proposed the use of worldlets, which are 3-D thumbnails of landmarks that can be interactively viewed. Their research showed that VEs equipped with worldlets allowed for more efficient navigation of the environment, compared with VEs equipped with textual or 2-D image representation of landmarks. A series of recent papers have proposed the use of agents to assist in wayfinding (Nijholt, Zwiers, & van Dijk, 2001; van Dijk et al., 2003; van Luin, den Akker, & Nijholt, 2001). In general, agents provide assistance to users by offering navigation advice and by answering questions about navigation within a VE. Other innovative techniques based on the use of sound or natural language processing have also been introduced to assist users with navigation (Gunther, Kazman, & Macgregor, 2004; McNeill, Sayers, Wilson, & McKevitt, 2002).

The emphasis of the previous research has been simply on helping a user of a VE get to a desired destination. However, there is often more than one possible route available to reach a desired destination in large VEs (Kantardzic et al., 2004; Sadeghian et al., 2005, 2006a, 2006b). Navigation tools are needed that not only help the user reach the destination, but also help the user select an appropriate route when more than one route is possible. Our work addresses this concern.

USING DATA MINING TO IMPROVE NAVIGATION TOOL DESIGN

Previous experience with an environment generally leads to an improvement in wayfinding ability (Chen & Stanney, 1999) and thus, much insight can be gained from the knowledge of experienced users (Peterson, Stine, & Darken, 2000). However, few navigation tools have been proposed that log the movements of users of VEs and do any analysis on the data (Chen & Stanney, 1999; Sadeghian et al., 2006b).

Our objective in designing a navigation tool is not to simply get the user of the VE to a desired location (i.e., the classical wayfinding problem), but to lead the user via the most "preferred" route when more than one route is possible. Of course, the most "preferred" route is defined differently for different environments. For example, in the real world the "preferred" route is often the shortest or fastest route (Liu, 1996). However, VEs are used for a variety of purposes, and other criteria may be the driving force in selecting a route. For example, the recommended way to get from one location to another in a large VE may not be the shortest possible route. A longer route may be preferred because it is "more scenic," "has more educational value," or "has the least number of obstacles associated with it," (Sadeghian et al., 2006b; Smyth & McGinty, 2002).

The problem is that these "preferred" routes are not *a priori* knowledge, especially to the novice user. Even the designers of a VE cannot know for sure which route out of several possible routes will be preferred by most of the users, or all the criteria that users will consider in the route selection process. Since these "preferred" routes are often not explicitly obvious, a conscious ef-

fort must be made to find them (Sadeghian et al., 2005, 2006a).

In general, data-mining techniques are used to discover previously unknown patterns, rules, and relationships (Han and Kamber, 2001; Kantardzic, 2002). Thus, methodologies based on data-mining techniques can be used to find these "preferred" routes. By mining the navigation data of previous experienced users, frequently used routes can be discovered. A model of these frequent routes, which are equated as the "preferred" routes of travel, can be developed to form as the basis for navigation tool design.

W-SEQUENCES

According to Agrawal and Srikant (1995), a sequence is defined as follows: Let $I = \{i_1, i_2, i_3, ...\}$ be a set of distinct attributes called items. An itemset is a nonempty set of unordered items. Items within the same itemset are assumed to occur at the same time. A sequence is an ordered list of itemsets. The goal of sequence mining is to find those frequent sequences that occur more than a specified threshold (Soliman, 2004). Applications of sequence mining have been found in a great number of fields including the health care industry, financial industry, telecommunication industry, and bioinformatics (Chan, Fan, Prodromidis, & Stolfo, 1999; Soliman, 2004; Stolfo, Lee, Chan, Fan, & Eskin, 2001). The FWS methodology extends the application of sequence mining to the domain of VEs.

The components that make up a wayfinding-sequence (W-sequence) are derived from the urban design elements that have been popular in the design of VEs (Charitos, 1997; Darken & Sibert, 1996; Ingram & Benford, 1995; Ingram & Benford, 1996; Ingram, Benford, & Bowers, 1996; Lynch, 1960; Modjeska & Waterworth, 2000; Steck & Mallot, 2000; Vinson, 1999). Specifically, the components of a W-sequence are derived from these elements:

1. **Landmarks:** A distinguishable object in the VE. In the FWS model, the symbol representing a specific landmark is denoted uniquely by L_i, where i is an integer from 1 to nl, and nl is the number of landmarks within the VE. For example, L_1 and L_2 refer to the two landmarks (i.e., buildings) in the map pictured in Figure 2.

2. **Paths:** This is a channel of movement, such as a walkway or a street. In the FWS model, the symbol representing a specific path is denoted uniquely by P_i, where i is an integer from 1 to np, and np is the number of paths within the VE. For example, P_1, P_2, P_3, and P_4 refer to the four paths (i.e., streets) in the map pictured in Figure 2.

3. **Nodes:** A node is an intersection of two or more paths. In the FWS model, the symbol representing a specific node is denoted uniquely by N_i, where i is an integer from 1 to nn, and nn is the number of nodes within the VE. For example, N_1, N_2, N_3, and N_4 refer

to the four intersections in the map pictured in Figure 2.

Edges, which are defined as boundaries (e.g., walls), are an important element in designing VEs; however they are not a component of a W-sequence. They primarily serve to restrict the area of movement of the user within the VE. **Districts** are sections of an environment containing the other 4 components. The use of districts will be discussed later to show how they improve the discovery process of the model of frequent W-sequences.

Formally, we define a W-sequence as a serial sequence of n symbols ($n \geq 3$) representing landmarks, paths, and nodes occurring at consecutive time intervals and specifying a route in a VE as follow:

$$WS = \langle X_1, X_2, ..., X_n \rangle \qquad (1)$$

where:

X_1 and X_n are symbols representing landmarks, and

X_2 to X_{n-1} are symbols representing landmarks, paths, or nodes following the rules of a valid W-sequence.

The following three rules apply for forming a *valid* W-sequence (see Table 1 for examples):

1. The beginning symbol and the ending symbol of a W-sequence must be a landmark.
2. A W-sequence cannot have two consecutive symbols in it representing the same type of element.
3. The symbol before and after a landmark or a node is always representing a path.

A W-sequence may contain other W-sequences within it. For example, the W-sequence, $WS_1 = \langle$

Figure 2. The three components of a W-sequence: landmarks, paths, and nodes

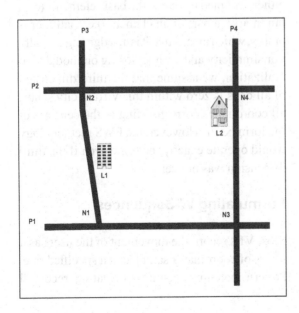

Table 1. Examples of valid and invalid W-sequences for the map in Figure 2

W-sequence	Valid	Invalid	Explanation
$<P_3\,L_1\,P_3\,N_2\,P_2\,L_2>$		X	Violates rule 1, P_3 begins the sequence, but it is not representing a landmark
$<L_1\,P_3\,P_2\,L_2>$		X	Violates rule 2, $P_3\,P_2$ are consecutive symbols representing the same type of element.
$<L_1\,N_2\,P_2\,L_2>$		X	Violates rule 3, N_2 following L_1 is not allowed
$<L_1\,P_3\,N_2\,P_2\,L_2>$	X		Valid W-sequence

$L_1\,P_1\,N_1\,P_2\,L_2\,P_3\,L_3>$, contains the following sub W-sequences:

1. $WS_1 = <L_1\,P_1\,N_1\,P_2\,L_2\,P_3\,L_3>$,
2. $WS_2 = <L_1\,P_1\,N_1\,P_2\,L_2>$, and
3. $WS_3 = <L_2\,P_3\,L_3>$

We define the relation *contained- in*, denoted by the symbol, $\subseteq$, to relate a W-sequence and its subsequences. From this example, $WS_1 \subseteq WS_1$, $WS_2 \subseteq WS_1$, and $WS_3 \subseteq WS_1$. Also, we define the operation of concatenation, denoted by the symbol, +, to join two W-sequences together (eliminating only the repeating landmark at the connection point) to form another W-sequence. From the example above, $WS_1 = WS_2 + WS_3$.

Finally, it is important to note that a valid W-sequence corresponds to a *route* from one landmark to another landmark. For example, $WS_1 = <L_1\,P_1\,N_1\,P_2\,L_2\,P_3\,L_3>$, corresponds to the route "Start at landmark L_1, travel on path P_1, at node N_1 switch to path P_2, travel on path P_2, reach landmark L_2, travel on path P_3, end at landmark L_3." The rules of a valid W-sequence have been specifically formulated in order to guarantee unique translation of a W-sequence to valid route directions.

FREQUENT W-SEQUENCE METHODOLOGY

We focus on designing a tool to help users navigate from one landmark to another via the most recommended route. To accomplish this task, we propose the frequent wayfinding-sequence (FWS) methodology. In this methodology, a modified frequent sequence mining algorithm is used to mine the sequences representing the navigation of previous experienced users, which we refer to as W-sequences. The result is a model of frequent W-sequences that can be used to derive routing rules from one landmark to another. Finally, an interface is provided for the new users in order to access the discovered routing rules.

In order to clarify our discussion and provide examples, we introduce a simple virtual city. A 2-dimensional map of this city can be found in Figure 3. As can be seen, the city occupies a 12 X 12 grid with 4 landmarks, 11 paths, and 8 nodes (i.e., intersections). Each one of the 144 cells is either occupied by one of the basic elements (e.g., landmark), a part of an element (i.e., path), or is empty, which means that it is an edge (e.g., a wall). For simplicity and without losing on model generalization, we assume that the third dimension of all cells is zero within this virtual city. Thus, all coordinates corresponding to the cells are of the form (x, y). However, the FWS methodology would operate exactly the same even if the third dimension was not zero.

Formulating W-Sequences

Most VEs record the movement of the users as a series of coordinates sampled at a specified rate. Pattern-matching techniques relating recorded

Figure 3. A 2-D map of a simple virtual city

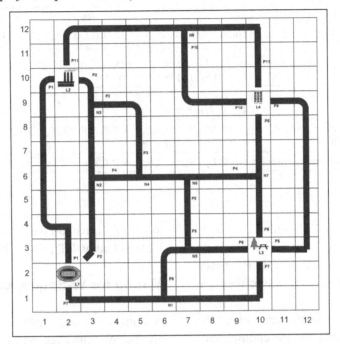

coordinates to W-sequence components need to be used in order to formulate W-sequences. The VEs that we are developing to support the FWS methodology are designed with an underlying 3-dimensional grid made up of a series of cells (Sadeghian et al., 2005, 2006a). Each cell is uniquely identified by a 3-dimensional coordinate and is associated with one of the basic design elements. A database is available relating each element to its cell coordinate(s). When a user is navigating within one of the VEs, at any one point in time he/she is located in one of the cells. The fundamental move that can be made is to move to another neighboring cell. Such a move can be made if the neighboring cell is associated with any element except an edge. Thus, the movements of a user can be recorded as a set of coordinates.

The first step in translating the set of coordinates into a W-sequence is to simply match each recorded coordinate with the symbol for the corresponding element using the coordinate database. Next, some preprocessing is necessary

to transform the resulting string of symbols into a valid W-sequence. For example, any symbol (e.g., P) preceding the first and following the last landmark needs to be eliminated. Also, consecutive symbols that are identical need to be condensed into just one occurrence of the symbol. Table 2 shows the process of translating a set of cell coordinates representing the navigation of a user in the virtual city in Figure 3 into a valid W-sequence.

Pruning the W-Sequences

The next step of the FWS methodology involves pruning the W-sequences to eliminate "noisy" data. One form of pruning that takes place is the elimination of redundant symbols. This means condensing a string of symbols of the form $P_i N_j P_i$ to simply P_i. No "information" is lost by this modification; rather, the W-sequence that goes through this type of pruning is more conducive for direction generation. For example, $P_1 N_1 P_1$

Table 2. Translation of a set of coordinates into a valid W-sequence

Action	Result
Recording of coordinates	(10,5), (10,4), (10,3), (11,3), (12,3), (12,4), (12,5), (12,6), (12,7), (12,8), (12,9), (11,9), (10,9), (10,8), (10,7), (10,6), (10,5), (10,4), (10,3), (9,3), (8,3), (7,3), (6,3), (6,2), (6,1), (5,1), (4,1), (3,1), (2,1), (2,2), (2,3), (2,4), (1,4), (1,5), (1,6), (1,7), (1,8), (1,9), (1,10), (2,10), (3,10)
Translation of coordinates into string of symbols	$< P_8 P_8 L_3 P_9 P_9 P_9 P_9 P_9 P_9 P_9 P_9 P_9 L_4 P_8 P_8 N_7 P_8 P_8 L_3 P_6 P_6 N_5 P_6 P_6 N_1 P_7 P_7 P_7 P_7 L_1 P_1 P_1 P_1 P_1 P_1 P_1 P_1 P_1 P_1 L_2 P_2 >$
Elimination of symbols before first and last landmark symbol	$< L_3 P_9 P_9 P_9 P_9 P_9 P_9 P_9 P_9 P_9 L_4 P_8 P_8 N_7 P_8 P_8 L_3 P_6 P_6 N_5 P_6 P_6 N_1 P_7 P_7 P_7 P_7 L_1 P_1 P_1 P_1 P_1 P_1 P_1 P_1 P_1 P_1 L_2 >$
Elimination of identical consecutive symbols	$< L_3 P_9 L_4 P_8 N_7 P_8 L_3 P_6 N_5 P_6 N_1 P_7 L_1 P_1 L_2 >$

would translate to "travel on path P_1, at node N_1, continue to travel on path P_1." This would be more simply stated as "travel on path P_1." Since landmarks are crucial to wayfinding, they are never eliminated.

A loop is defined as a W-sequence where the starting landmark symbol and the ending landmark symbol are the same. In terms of navigation, this corresponds to starting at a specific landmark and, through a series of movements, ending back up at the same landmark. In a VE, this often translates to backtracking and disorientation in the space (Conroy, 2001; Darken & Sibert, 1996). The FWS methodology finds these loops, registers them in a database, and eliminates them from the W-sequences before the mining process. Finding and registering loops is, in itself, a form of knowledge extraction (Sadeghian et al., 2006b). Information about frequent loops discovered can later be made available to users of the system to help them avoid the mistakes of previous users. Information on loops is also beneficial to the designers of the system, allowing them to improve the design of their system.

Table 3 shows the result of pruning the W-sequence in Table 2. First, the redundant nodes and paths (i.e., P_8, N_7, P_6, N_5) are eliminated. Then the loop (i.e., $< L_3 P_9 L_4 P_8 L_3 >$) is registered and eliminated from the W-sequence. The resulting W-sequence corresponds to the route "Start at landmark L_3, travel on path P_6, at node N_1 switch to path P_7, travel on path P_7, arrive at landmark L_1, travel on path P_1, end at landmark L_2".

Mining for W-Sequences

A sequence-mining algorithm is executed to find all the W-sequences within the preprocessed W-sequence database (Sadeghian et al., 2006a). While examining a W-sequence, *all W-sequences contained within it (i.e., sub-W-sequences) are noted and counted.* We assume that the paths within a VE allow for two-way travel. Therefore, a W-sequence and its reverse are representing the same unique W-sequence. For example, $< L_1 P_1 L_2 >$ is the same as $< L_2 P_1 L_1 >$.

The end result of the mining process is the formation of a data structure with entries for landmark pairs of the form L_i - L_j where $i \neq j$, $1 \leq i \leq nl\text{-}1$ and $i < j \leq nl$. Each L_i - L_j entry will contain all the W-sequences starting with the symbol L_i and ending with the symbol L_j mined from the W-sequence database, their correspond-

Table 3. Pruning the W-sequence in Table 2

Action	Result
Elimination of redundant symbols	$< L_3 P_9 L_4 P_8 L_3 P_6 N_1 P_7 L_1 P_1 L_2 >$
Elimination of the loop	$< L_3 P_6 N_1 P_7 L_1 P_1 L_2 >$

ing count of occurrence, and support. The support for a W-sequence starting with L_i and ending with L_j is defined as the percentage equal to the value of its count of occurrence over the total number of occurrence of all W-sequences for the L_i - L_j pair. Table 4 provides an example of such a data structure found after mining a randomly generated W-sequence database for the VE in Figure 3. Note that for some L_i - L_j pairs, more than one unique W-sequence is found. The W-sequences are followed by the corresponding count of occurrence, support, and the corresponding lower limit of the binomial confidence interval with $\alpha = .05$

Discovering the Model of Frequent W-Sequences

A frequent W-sequence for the pair L_i - L_j meets the following two conditions:

1. It has a statistical confidence above a predefined threshold value.
2. It has the highest statistical confidence among all other W-sequences for the pair L_i - L_j.

In our experiments, we used the lower limit of the binomial confidence as our measure of statistical confidence. Since it is possible that for some L_i - L_j pairs there will not be any W-sequences above the predefined threshold, *it is not guaranteed* that for each possible L_i - L_j pair, there will be a frequent W-sequence.

If a frequent W-sequence is not found for a pair L_i - L_j, it may be possible to combine two other frequent W-sequences to derive one. This can be done by searching for a landmark L_k, where k is an integer and $1 \le k \le nl$, such that there exists frequent W-sequences for the pairs L_i - L_k and L_k - L_j. The frequent W-sequence for the pair L_i - L_j is derived from the following concatenation process:

$$(L_i - L_j)_{FWS} = (L_i - L_k)_{FWS} + (L_k - L_j)_{F.WS} \quad (2)$$

If there is more than one such L_k available, then the L_k in which the smaller confidence of $(L_i - L_k)_{FWS}$ and $(L_k - L_j)_{FWS}$ is higher than the smaller confidence of any other $(L_i - L_k)_{FWS}$ and $(L_k - L_j)_{FWS}$ should be selected. The proposed logic could be extended to consider combing more than two frequent W-sequences at a time using dynamic programming techniques, but the problem quickly becomes computationally expensive (Goodrich & Tamassia, 2001). Figure 4 summarizes the process described for finding the frequent W-sequence for a pair L_i - L_j.

Table 4. W-sequences mined from a hypothetical database for the VE in Figure 3

$L_i - L_j$	W-sequences	Count	Support	Confidence
$L_1 - L_2$	$< L_1 P_1 L_2 >$	11	18.33 %	.0952
	$< L_1 P_2 L_2 >$	20	33.33%	.2169
	$< L_1 P_2 N_2 P_4 N_4 P_3 N_3 P_2 L_2 >$	29	48.33%	.3523
$L_1 - L_3$	$< L_1 P_7 L_3 >$	10	25.00%	.1269
	$< L_1 P_7 N_1 P_6 L_3 >$	30	75.00%	.5880
$L_1 - L_4$	$< \phi >$	0	0%	0
$L_2 - L_3$	$< L_2 P_2 N_3 P_3 N_4 P_4 N_6 P_5 N_5 P_6 L_3 >$	2	100%	.1581
$L_2 - L_4$	$< L_2 P_{11} L_4 >$	5	100%	.4782
$L_3 - L_4$	$< L_3 P_8 L_4 >$	11	29.73%	.1587
	$< L_3 P_9 L_4 >$	26	70.27%	.5302

Figure 4. Deriving the frequent W-sequence for a pair L_i - L_j.

```
Given: A set S of n unique W-sequences for the pair Lᵢ - Lⱼ
such that S = {(Lᵢ - Lⱼ)₁, (Lᵢ - Lⱼ)₂,... (Lᵢ - Lⱼ)ₙ}
Find: (Lᵢ - Lⱼ)ₜ ⊆ S such that
Confidence ((Lᵢ - Lⱼ)ₜ) ≥ threshold value &
Confidence ((Lᵢ - Lⱼ)ₜ) > Confidence ((Lᵢ - Lⱼ)ₘ)
∀ t, m ∈ [1,…n]
if (Lᵢ- Lⱼ)ₜ ≠ φ {
   (Lᵢ- Lⱼ)_FWS = (Lᵢ- Lⱼ)ₜ
   return ((Lᵢ- Lⱼ)_FWS)}
else {
   Find: (Lᵢ - Lₖ)_FWS ≠ φ and (Lₖ - Lⱼ)_FWS ≠ φ such that
   Min (Confidence ((Lᵢ - Lₖ)_FWS), Confidence ((Lₖ - Lⱼ)_FWS)) >
   Min (Confidence ((Lᵢ - Lₘ)_FWS), Confidence ((Lₘ - Lⱼ)_FWS))
   ∀ k, m ∈ [1,…,nl]
   try{
      (Lᵢ- Lⱼ)_FWS = (Lᵢ - Lₖ)_FWS + (Lₖ - Lⱼ)_FWS
      return ((Lᵢ- Lⱼ)_FWS)}
   catch{
      return (φ)}}
```

Table 5. The FWS model

$L_i - L_j$	FWS	Confidence
$L_1 - L_2$	$< L_1 P_2 N_2 P_4 N_4 P_3 N_3 P_2 L_2 >$	.3523
$L_1 - L_3$	$< L_1 P_7 N_1 P_6 L_3 >$	.5880
$L_1 - L_4$	$< L_1 P_7 N_1 P_6 L_3 P_9 L_4 >$	.5302
$L_2 - L_3$	$< L_2 P_{11} L_4 P_9 L_3 >$	.4782
$L_2 - L_4$	$< L_2 P_{11} L_4 >$	.4782
$L_3 - L_4$	$< L_3 P_9 L_4 >$	.5302

Table 5 shows the FWS model when the threshold value is set to 0.25 for the W-sequences found in Table 4. For L_i - L_j pairs (i.e., $L_1 - L_4$ and $L_2 - L_3$) where no frequent W-sequence was found, other frequent W-sequences are concatenated to derive a frequent W-sequence for the particular L_i - L_j pair. The assigned confidence is the lower confidence of the two frequent W-sequences that were concatenated. The FWS model represents the model of the navigation behaviors of the experienced VE users. The recommended routes of travel are developed by translating the frequent W-sequences into informative directions.

Districting

The concept of districts can be used to improve the discovery process of the model of frequent W-sequences (Sadeghian et al., 2006a). Given a VE, the *nl* landmarks are divided into *nd* districts, where *nd* is an integer such that $nd \geq 1$. A specific district is denoted by the symbol D_i, such that $1 \leq i$

$\leq nd$. Each district D_i contains nl_{Di} landmarks such that $2 \leq nl_{Di} \leq nl$, and $D_i \cap D_j = \phi \; \forall \; i, j \in [1,\dots nd]$. The process of assigning a certain landmark to a certain district involves consideration of the spatial distribution of the landmarks and experimentation with the VE. Our experiments showed that proximity of the landmarks, and the number of paths and nodes between landmarks, are important factors of consideration.

After deriving the districts, the FWS methodology operates by finding the FWS model *for each* district. In addition, special frequent W-sequences, termed as C-sequences (connection-sequences), are introduced. A C-sequence for the pair D_i - D_j, $(D_i \neq D_j)$, is defined as a frequent W-sequence starting with L_u, $L_u \in D_i$ and ending with L_v, $L_v \in D_j$, such that its confidence is higher than any other such frequent W-sequence between the selected districts. C-sequences allow for the derivation of frequent W-sequences, where the landmarks represented by the beginning and ending symbols are in different districts. For example, given a C-sequence for the pair of districts D_i - D_j, say $(L_u - L_v)_{C\text{-Sequence}}$, and a FWS model for both districts D_i and D_j, a frequent W-sequence for the pair L_s - L_t where $L_s \in D_i$ and $L_t \in D_j$ is derived by the following concatenation process:

$$(L_s - L_t)_{FWS} = (L_s - L_u)_{FWS} + (L_u - L_v)_{C\text{-Sequence}} + (L_v - L_t)_{FWS} \tag{3}$$

Districting has the advantage of requiring the discovery of a smaller number of frequent W-sequences to build a complete FWS model for the entire VE. The minimum number of C-sequences in a VE divided up into *nd* linearly connected

Figure 5. Interface of a navigation tool designed using the FWS methodology

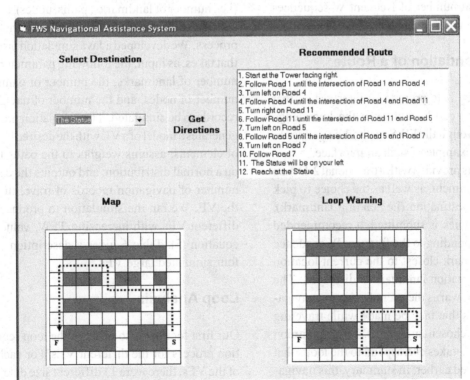

districts that need to be found in order to allow for travel between districts is defined as follows:

Minimum number of C-sequences = nd - 1

$$(4)$$

Equation (5) shows the total number of frequent W-sequences (TFW) needed to be discovered for the entire VE, assuming that the minimum number of C-sequences for its nd districts has already been found. Note, nd = 1 means that no districting is involved. Equation 6 defines the percentage of the frequent W-sequence (PFW) model discovered for an environment with nd districts.

$$\text{TFW}_{nd} = \sum_{i=1}^{i=nd} nl_{Di} C_2 \qquad (5)$$

$$\text{PFW}_{nd} = (nf / \text{TFW}_{nd}) * 100\% \qquad (6)$$

where:

$$nl_{Di} C_2 = (nl_{Di})! / ((nl_{Di} - 2)! * (2))$$

nf is the total number of frequent W-sequences registered in the models

Recommendation of a Route

The final step is to provide an interface so that the VE users can benefit from the knowledge extracted through the FWS methodology. Figure 5 shows an example of such an interface.

The user is provided with a traditional 2-D map of the environment, as well as the choice to pick the desired destination (i.e., ending landmark). When the choice is submitted, a recommended route corresponding to the frequent W-sequence for the landmark closest to the current location and the destination landmark is displayed. The interface also warns the user of common navigation mistakes that may be made while traveling between the chosen landmarks. These common navigation mistakes correspond to the loops that were registered earlier. In summary, this naviga-

tion tool helps the user choose a route of travel from the many available possibilities.

EXPERIMENTS AND DISCUSSION

In our previous papers, we discussed a real-world experiment designed to explore the usefulness and the applicability of the FWS methodology (Sadeghian et al., 2005, 2006a). Our results showed that users of a VE equipped with an FWS-based tool performed better at navigation tasks than users of a similar VE equipped with a traditional 2-D map. Furthermore, the use of the FWS-based tool also led to the improvement in the quality of the human-computer interaction with the VE (Sadeghian et al., 2005, 2006a).

In this chapter, we will discuss several simulation experiments of large VEs. The purpose of our simulation experiments was to study the scalability of the FWS methodology. In particular, we wanted to study how the size of the navigation data available, the complexity of the environment (i.e., number of landmarks, paths, nodes), and the concept of districting would influence the FWS process. We developed a Java simulation program that takes, as input, the following parameters: the number of landmarks, the number of paths, the number of nodes, and the number of navigation records to be simulated. The simulation program generates a model of a VE with the desired number of elements, assigns weights to the paths based on a normal distribution, and outputs the desired number of navigation records of travel through the VE. We ran the simulation to produce four different VEs with increasing TFW_1 value (See equation (5). Table 6 gives a description of the four simulated large VEs.

Loop Analysis

Our first focus was to observe the loop registration process for the different VEs. For each one of the VEs, there were 13 different size databases

Table 6. Description of the four simulated large VEs

ID	nl	np	nn	TFW$_1$
Environment A	10	30	10	45
Environment B	50	150	50	1225
Environment C	100	300	100	4950
Environment D	500	1500	500	124750

of navigation records produced. Each navigation record corresponded to the navigation of a "user" who is visiting five landmarks in the given VE. The "user" is free to start at any landmark and move in any direction, although movements through certain paths are preferred over others due to the weights randomly assigned to the paths. In this experiment, we defined a loop as significant if the count of occurrence is five or more. A nonsignificant loop occurs at least once, but less than five times in the database. Figure 6 shows the number of significant loops discovered in each database for the four VEs, while Figure 7 shows the number of nonsignificant loops discovered in each database for the four VEs.

The number of both nonsignificant and significant loops increases as the sizes of the database increases for a given VE. This is because as there is more data available, there will be higher probability of extracting loops. An interesting

Figure 6. Significant loops for the VEs

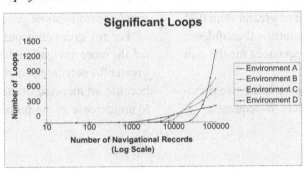

Figure 7. Nonsignificant loops for the VEs

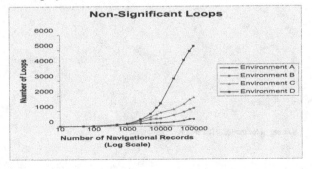

observation is that the more complex the environment, the more records required before the initial discovery of significant loops. This is because the bigger space, the more opportunity for making different types of loops and therefore, less likely that the same loop will get repeated enough times to become significant. However, as a sufficient number of records become available, the number of significant loops extracted from a complex environment becomes much greater than for a less complex environment given the same size database.

FWS Model Discovery

In our next set of experiments, the navigation records contained in different size databases for the four VEs were mined for W-sequences. In the experiments, we set the threshold value of the confidence measure to 0.25 (lower limit of the binomial confidence interval) with $\alpha = .05$ (95% confidence). A frequent W-sequence for a pair L_i - L_j is only registered in the FWS model if,

1. Its confidence is equal or greater than 0.25 **and** it has the highest statistical confidence among all other W-sequences for the pair L_i - L_j, or if
2. It is derived from concatenating two previously registered frequent W-sequences.

Our focus with this set of simulation experiments is in determining the percentage of the model of frequent W-sequences that could be discovered and how districting could be used to improve the discovery process.

Each navigation record for *Environment A* consisted of up to five landmarks before the loop elimination process. Figure 8 shows that approximately 100 navigation records are needed to discover all 45 frequent W-sequences corresponding to 100 percent of the FWS model. Therefore, it is not necessary to have more than one district. However, this is not the case for more complex environments.

Each navigation record for the remaining three VEs consisted of up to 20 landmarks before the loop elimination process. For each environment, the FWS process was applied for several different size W-sequence databases at different values of nd (number of districts). A value of $nd = 1$ means that no districting is involved. Figure 9, Figure 10, and Figure 11 show the differences in model generation for the different W-sequence databases and nd values for the three VEs. Analysis of these graphs leads to some general observations.

For any given environment with any value of nd, the more navigation data that is available, the greater the percentage of the model found. This is because an increase in the amount of data leads to an increase in the probability of the same W-

Figure 8. Discovering the FWS model for Environment A

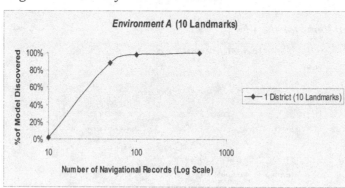

Figure 9. Discovering the FWS model for Environment B

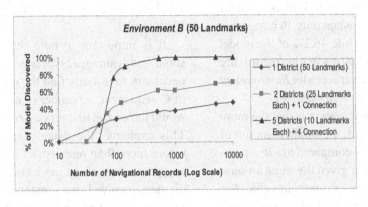

Figure 10. Discovering the FWS model for Environment C

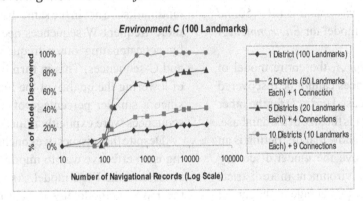

Figure 11. Discovering the FWS model for Environment D

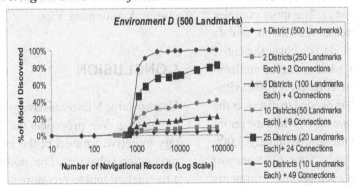

sequence frequently occurring in the database. For example, only 0.7% of the model is discovered for *Environment B* ($nd = 1$) when only 10 navigation records are available, while 46.2% of the model is discovered when 10,000 navigation records are available. Similar results are seen for *Environment A, Environment C,* and *Environment D.*

Another general observation is that the more complex an environment, the smaller amount of the model discovered as compared to a less complex environment, when given the same amount of navigation data and a constant *nd* value. For example, 10,000 navigation records leads to the discovery of only 3.7% of the model for *Environment D* ($nl = 500$, $nd = 1$), but to 21.9% of the model for *Environment C* ($nl = 100$, $nd = 1$), 46.2 % of the model for *Environment B* ($nl = 50$, $nd = 1$), and 100% of the model for *Environment A* ($nl = 10$, $nd = 1$).

Unlike *Environment A*, the entire model of the frequent W-sequences cannot be discovered when districting is not used ($nd = 1$) for the other environments whatever is the size of the database. The simulation results show that districting is an effective way to improve the model discovery process. For any given environment, more districts correspond to a greater percentage of the model discovered, given the same size database. For example, given approximately 1,000 navigation records, 100% of the model is discovered when $nd = 5$ for *Environment B* as opposed to the 37% when $nd = 1$. Similar results are seen for *Environment C* and *Environment D*. The more complex the environment, the greater the amount of data and districts that is needed to discover the entire model, as compared to a less complex environment. For example, approximately 5,000 navigation records and $nd = 10$ is needed to discover the entire model for *Environment C,* while the entire model is discovered with approximately 1,000 navigation records and $nd = 5$ for *Environment B*. However, even for very complex environments

(e.g., *Environment D*), the use of districts eventually leads to the discovery of 100 percent of the FWS model.

It is important to note that districting has some disadvantages associated with it. Since it is necessary to initially find the minimum number of C-sequences, a number of navigation records would have to be first processed for this purpose. This explains "the delay" seen in the graphs where more than one district is involved before any percentage of the model is found. The more districts involved, the greater number of initial records needed to find C-sequences, and thus, the longer "the delay." The other disadvantage of districting is that as the number of districts increases, so does the probability that any two landmarks are not in the same district. Therefore, more frequent W-sequences need to be derived by concatenating other frequent W-sequences and C-sequences. This in turn has the potential of lowering the quality of the final FWS model, since a smaller percentage of the frequent W-sequences were explicitly mined.

Despite these issues, we conclude that districting is an effective way to improve the discovery process of the FWS model. As we have shown, the FWS methodology is able to discover the necessary frequent W-sequences for even complex environments through the use of districts. Therefore, the FWS methodology is an effective approach to extract the spatial knowledge needed to design an intelligent navigation assistance system for complex VEs.

CONCLUSION

Users of large VEs often have problems efficiently navigating. We proposed the FWS methodology to derive a model of the experienced users' navigation behaviors. The model is used to build a navigation tool to recommend routes to novice users of a VE. Our experiments showed the scalability of the approach, and its ability to derive a complete navigation model even for very large and complex VEs.

In the next stage of our research, we plan to improve the route recommendation process. Specifically, we plan to enhance the interface of the FWS-based tool with an interactive image preview of the most salient characteristics of a recommended route. Furthermore, we plan to personalize the route recommendation process. Personalization will be realized by the introduction of informative user profiles that capture the route selection factors deemed as important by the user. Routes matching the user's profile will be recommended. The addition of these techniques will enhance the FWS methodology's ability to further improve the quality of the human-computer interaction within large VEs.

ACKNOWLEDGMENT

This research has been funded by the National Science Foundation (NSF) under grant #0318128.

REFERENCES

Agrawal, R., & Srikant, R. (1995). Mining sequential patterns. In *Proceedings of the International Conference on Data Engineering*, (pp. 3-14). Taipei, Taiwan.

Bowman, D. (1999). Maintaining spatial orientation during travel in an immersive virtual environment. *Presence, 8*(6), 618-631.

Burdea, G., & Coiffet, P. (2003). *Virtual reality technology*. NJ: John Wiley and Sons.

Chan, P., Fan, W., Prodromidis, A., & Stolfo, S. (1999). Distributed data mining in credit card fraud detection. *IEEE Intelligent Systems, 14*(6), 67-74.

Charitos, D. (1997). Designing space in VE's for aiding wayfinding behaviour. In *Proceedings of the 4th UK Virtual Reality Special Interest Group Conference, Brunel University.*

Chen, J., & Stanney, K. (1999). A theoretical model of wayfinding in virtual environments: Proposed strategies for navigational aiding. *Presence, 8*(6), 671-685.

Conroy, R. (2001). *Spatial navigation in immersive virtual environments*. PhD Thesis, University College London.

Darken, R., & Peterson, B. (2002). Spatial orientation, wayfinding, and representation. In K. Stanney (Ed.), *Handbook of virtual environments: Design, implementation, and applications* (pp. 493-518). Mahwah, NJ: Lawrence Erlbaum Associates.

Darken, R., & Sibert, J. (1996). Navigating large virtual spaces. *The International Journal of Human-Computer Interaction, 8*(1), 49-72.

Elvins, T., Nadeau, D., & Kirsh, D. (2001). Worldlets: 3D thumbnails for wayfinding in large virtual worlds. *Presence, 10*(6), 565-582.

Goodrich, M., & Tamassia, R. (2001). *Algorithm design: Foundations, analysis, and Internet examples*. New York: Wiley.

Grady, S. (2003). *Virtual reality: Simulating and enhancing the world with computers*. Facts on File Inc.

Gunther, R., Kazman, R., & Macgregor, C. (2004). Using 3D sound as a navigational aid in virtual environments. *Behaviour and Information Technology, 23*(6), 435-446.

Han, J., & Kamber, M. (2001). *Data mining: Concepts and techniques*. San Francisco: Morgan Kaufmann Publishers.

Ingram, R., & Benford, S. (1995). Legibility enhancement for information visualization. In

Proceedings of IEEE Conference on Visualization (IEEE VIZ'95), Atlanta, GA.

Ingram, R., & Benford, S. (1996). The application of legibility techniques to enhance 3-D information visualizations. *The Computer Journal*, *39*(10), 819-836.

Ingram, R., Benford, S., & Bowers, J. (1996). Building virtual cities: Applying urban planning principles to the design of virtual environments. In *Proceedings of the ACM Symposium on Virtual Reality Software and Technology (VRST96)*, (pp. 83-91). Hong Kong, China.

Kantardzic, M. (2002). *Data mining: Concepts, models, methods, and algorithms*. Piscataway, NJ: IEEE Press.

Kantardzic, M., Rashad, S. & Sadeghian, P. (2004). Spatial navigation assistance system for large virtual environments: The data mining approach. In *Proceedings of the Mathematical Methods for Learning*, (pp. 44-46). Villa Geno, Como, Italy.

Liu, B. (1996). Intelligent route finding: Combining knowledge, cases and an efficient search algorithm. In *Proceedings of the 12th European Conference on Artificial Intelligence (ECAI-96)*, (pp. 380-384). Budapest, Hungary.

Lynch, K. (1960). *The image of the city*. Cambridge, MA: MIT Press.

Magliano, J., Cohen, R., Allen, G., & Rodrigue, J. (1995). The impact of a wayfinder's goal on learning a new environment: Different types of spatial knowledge as goals. *Journal of Environmental Psychology*, *15*, 65-75.

McNeill, M., Sayers, H., Wilson, S., & McKevitt, P. (2002). A spoken dialogue system for navigation in non-immersive virtual environments. *Computer Graphics*, *21*(4), 713-722.

Modjeska, D., & Waterworth, J. (2000). Effects of desktop 3D world design on user navigation and search performance. In *Proceedings of IEEE Information Visualization*, (pp. 215-220). Salt Lake City, Utah.

Moeser, S. (1988). Cognitive mapping in a complex building. *Environment and Behavior*, *20*, 21-49.

Nijholt, A., Zwiers, J., & van Dijk, B. (2001). Maps, agents and dialogue for exploring a virtual world. In *Proceedings of the 5th World Multiconference on Systemics, Cybernetics and Informatics (SCI 2001)*, (pp. 94-99). Orlando, Florida.

O'Neill, M. (1992). Effects of familiarity and plan complexity on wayfinding in simulated buildings. *Journal of Environmental Psychology*, *12*, 319-327.

Peterson, B., Stine, J., & Darken, R. (2000). A process and representation for modeling expert navigators. In *Proceedings of the 9th Conference on Computer Generated Forces and Behavioral Representation*, (pp. 459-470). Orlando, Florida.

Peterson, B., Wells, M., Furness, T., & Hunt, E. (1998). The effects of the interface on navigation in virtual environments. In *Proceedings of Human Factors and Ergonomics Society 1998 Annual Meeting*, (pp. 1496-1505). Chicago, IL.

Ruddle, R., Payne, S., & Jones, D. (1998). Navigating large-scale "desk-top" virtual buildings: Effects of orientation aids and familiarity. *Presence: Teleoperators and Virtual Environments*, *7*(2), 179-192.

Ruddle, R., Randall, S., & Jones, D. (1996). Navigation and spatial knowledge acquisition in large-scale virtual buildings: An experimental comparison of immersive and "desk-top" displays. In *Proceedings of the 2nd International FIVE Conference*, (pp. 125-136).

Sadeghian, P., Kantardzic, M., Lozitskiy, O., & Sheta, W. (2005). Route recommendations: Navigation assistance in complex virtual environments. In *Proceedings of the 20th International*

Conference on Computers and their Applications, (pp. 220-225). New Orleans, LA.

Sadeghian, P., Kantardzic, M., Lozitskiy, O., & Sheta, W. (2006a). The frequent wayfinding-sequence (FWS) methodology: Finding preferred routes in complex VEs. *The International Journal of Human-Computer Studies, 64*(4), 356-374.

Sadeghian, P., Kantardzic, M., & Rashad, S. (2006b). Knowledgeable navigation in virtual environments. In C. Ghaoui (Ed.), *Encyclopedia of human computer interaction* (pp. 389-395). Hershey, PA: Idea Group Inc.

Satalich, G. (1995). *Navigation and wayfinding in virtual reality: Finding proper tools and cues to enhance navigation awareness.* Master Thesis, University of Washington, Seattle, WA.

Sherman, W., & Craig, A. (2002). *Understanding virtual reality.* San Francisco: Morgan Kaufmann.

Smyth, B., & McGinty, L. (2002). The route to personalization: A case-based reasoning perspective. *Expert Update, 5*(2), 27-36.

Soliman, M. (2004). *A model for mining distributed frequent sequences.* Ph.D. Dissertation, University of Louisville, Louisville, KY.

Steck, S., & Mallot, H. (2000). The role of global and local landmarks in virtual environment navigation. *Presence, 9*(1), 69-83.

Stolfo, S., Lee, W., Chan, P., Fan, W., & Eskin, E. (2001). Data mining-based intrusion detectors: An overview of the Columbia IDS project. *Special Interest Group on the Management of Data (SIGMOD) Record, 30* (4), 5-14.

Streeter, L., Vitello, D., & Wonsiewicz, S. (1985). How to tell people where to go: Comparing navigational aids. *International Journal of Man-Machine Studied, 22*, 549-562.

van Dijk, B., op Den, A., Rieks, N., & Zwiers, J. (2003). Navigation assistance in virtual worlds. *Informing Science Journal, 6*, 115-125.

van Luin, J., den Akker, R., & Nijholt, A. (2001). A dialogue agent for navigation support in virtual reality. In *Proceedings of the ACM SIGCHI Conference CHI*, (pp.117-118).

Vinson, N. (1999). Design guidelines for landmarks to support navigation in virtual environments. In *Proceedings of the ACM Conference on Human Factors in Computing Systems*, (pp. 278-285). Pittsburgh, Pennsylvania.

Chapter XVII
Using Grids for Distributed Knowledge Discovery

Antonio Congiusta
DEIS—University of Calabria, Italy

Domenico Talia
DEIS—University of Calabria, Italy

Paolo Trunfio
DEIS—University of Calabria, Italy

ABSTRACT

Knowledge discovery is a compute- and data-intensive process that allows for finding patterns, trends, and models in large datasets. The grid can be effectively exploited for deploying knowledge discovery applications because of the high performance it can offer and its distributed infrastructure. For effective use of grids in knowledge discovery, the development of middleware is critical to support data management, data transfer, data mining and knowledge representation. To such purpose, we designed the Knowledge Grid, a high-level environment providing for grid-based knowledge discovery tools and services. Such services allow users to create and manage complex knowledge discovery applications, composed as workflows that integrate data sources and data-mining tools provided as distributed grid services. This chapter describes the Knowledge Grid architecture and describes how its components can be used to design and implement distributed knowledge discovery applications. Then, the chapter describes how the Knowledge Grid services can be made accessible using the open grid services architecture (OGSA) model.

INTRODUCTION

Knowledge discovery in databases (KDD) is often both a compute- and data-intensive process.

When large datasets are coupled with geographic distribution of data, users, and systems, a variety of technologies must be combined for implementing high-performance distributed knowledge discovery systems. Most of the current off-the-shelf KDD environments require central aggregation of data that, in many cases, is distributed. Data storage in a single site may not always be feasible because of limited network bandwidth, security concerns, scalability problems, and other practical issues.

Data mining in large settings like virtual organization networks, the Internet, corporate intranets, sensor networks, and the emerging world of ubiquitous computing, questions the suitability of centralized KDD architectures for large-scale knowledge discovery in a networked environment. The field of distributed KDD offers an alternative approach. It works by analyzing data in a distributed fashion, and pays particular attention to the trade-off between centralized collection and distributed analysis of data.

When the datasets are large, scaling up the speed of the KDD process is a crucial issue. Distributed knowledge discovery techniques address this problem by using high-performance multicomputer machines and a decentralized approach for mining large datasets that can be used when several interconnected machines are available for running distributed data-mining models. The increasing availability of such machines and networks calls for extensive development of data-analysis algorithms able to scale with datasets, measured in terabytes and petabytes, on distributed and parallel machines with hundreds or thousands of processors. Knowledge discovery is speeded up by executing, in a distributed way, a number of data mining processes on different data subsets, and then combining the results through metalearning. This technology is particularly suitable for applications that typically deal with very large amounts of data (e.g., transaction data, scientific simulation, and telecommunication data) that cannot be analyzed in a single site on traditional machines in acceptable times. Moreover, parallel data-mining algorithms can be a component of distributed data-mining ap-

plications that can exploit either parallelism or data distribution.

Grid technology integrates both distributed and parallel computing; thus, it represents a critical infrastructure for high-performance distributed knowledge discovery. Grid computing is receiving increasing attention both from the research community and from industry and governments, looking to this new computing infrastructure as a key technology for solving complex problems and implementing distributed high-performance applications (Foster, Kesselman, Nick, & Tuecke, 2002). Today there is a large number and variety of grid tools and middleware that allow the user community to use grids for implementing a larger set of applications, with respect to 1 or 2 years ago.

The term "grid" defines a global distributed computing platform through which—like in a power grid—users gain ubiquitous access to a range of services, computing, and data resources. The driving grid applications are traditional high-performance applications, such as high-energy particle physics, and astronomy and environmental modeling, in which experimental devices create large quantities of data that require scientific analysis.

Grid computing differs from conventional distributed computing because it focuses on large-scale resource sharing, offers innovative applications, and, in some cases, it is geared toward high-performance systems. Although originally intended for advanced science and engineering applications, grid computing has emerged as a paradigm for coordinated resource sharing and problem solving in dynamic, multi-institutional virtual organizations in industry and business. Therefore, today's grids can be used as effective infrastructures for distributed high-performance computing and data processing. Grid applications include:

- Intensive simulations on remote supercomputers.

- Cooperative visualization of very large scientific data sets.
- Distributed processing for computationally demanding data analysis.
- Coupling of scientific instruments with remote computers and data archives.

In the last decade, toolkits and software environments for implementing grid applications have become available. These include Condor (http://www.cs.wisc.edu/condor), Legion (http://legion.virginia.edu), Unicore (http://www.unicore.org), and the Globus Toolkit (http://www.globus.org/toolkit). In particular, the Globus Toolkit is the most widely used middleware in scientific and data-intensive grid applications, and is the de facto standard for implementing grid systems. The toolkit addresses security, information/discovery, resource- and data-management, communication, fault-detection, and portability issues. Today, Globus and the other grid tools are used in many projects worldwide. Although most of these projects are in scientific and technical computing, there is a growing number of grid projects in education, industry, and commerce.

Together with the grid shift towards industry and business applications, a parallel shift toward the implementation of data grids has been registered. Data grids are designed to allow large datasets to be stored in repositories and moved about with the same ease as small files can be moved. They represent an enhancement of computational grids, driven by the need to handle large datasets without repeated authentication, and aiming at supporting the implementation of distributed data-intensive applications.

The grid can be effectively exploited for implementing information intensive and knowledge discovery applications. To support this class of applications, tools and services for data mining and knowledge discovery on grids are essential. This objective can be achieved through the development of techniques and tools for supporting data intensive applications, and the integration of data and computational grids with information and Knowledge Grids (see Figure 1). The process of unification of data management and knowledge discovery systems with grid technologies for providing knowledge-based grid services can bring many benefits to science and industry.

Figure 1. Combination of KDD and grid technologies for building a Knowledge Grid

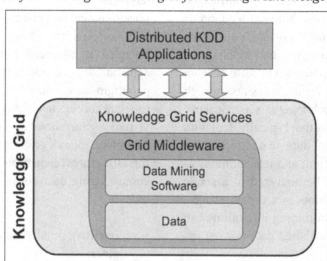

Massive amounts of data are today produced and stored in digital archives. We are able to store petabytes of data in databases and query them at an acceptable rate. However, the extraction of hidden information and knowledge from huge amounts of stored data can make data ownership more competitive.

Grids represent a good opportunity to handle very large datasets distributed over a large number of sites. At the same time, grids can be used as knowledge discovery engines and knowledge management platforms. To effectively use grids for such high-level knowledge-based applications, models, algorithms, and software environments are needed.

The Knowledge Grid is a high-level system developed for providing grid-based knowledge discovery services (Cannataro & Talia, 2003). Services for knowledge discovery on grids have been implemented on top of generic grid services provided by the Globus Toolkit. These services allow researchers, professionals, and scientists to create and manage complex knowledge-discovery applications composed as workflows that integrate data sources, mining tools, computing, and storage provided as distributed services. Knowledge Grid facilities allow users to compose, store, share, and execute these knowledge discovery workflows, as well as publish them as new components and services on the grid.

The knowledge building process in a distributed setting involves collection/generation and distribution of data and information, followed by collective interpretation of processed information into "knowledge." The Knowledge Grid provides high-level abstractions and a set of services based on the use of grid resources to support all the phases of the knowledge discovery process. Therefore, it allows end users to focus on the knowledge discovery process they must develop, without worrying about grid infrastructure and fabric details. This chapter describes knowledge discovery services and features of the Knowledge Grid environment. First, we discuss knowledge discovery services, present the system architecture, and describe how its components can be used to design and implement knowledge discovery applications for science, industry, and commerce. Then we describe how the Knowledge Grid services can be made accessible under the *open grid services architecture* (*OGSA*). Finally, some knowledge discovery applications we implemented on the Knowledge Grid are outlined.

KNOWLEDGE DISCOVERY SERVICES

Today many organizations, industries, and scientific centers produce and manage large amounts of complex data and information. Climate data, astronomic data, and company transaction data are just some examples of massive amounts of digital data repositories that must be stored and analyzed to find useful knowledge in them. This data and information patrimony can be effectively exploited if used as a source to produce the knowledge necessary to support decision making.

Knowledge discovery procedures in all these application areas typically require the creation and management of complex, dynamic, multi-step workflows. At each step, data from various sources can be moved, filtered, integrated, and fed into a data-mining tool. By examining the output results, the analyst chooses which other data sets and mining components can be integrated in the workflow or how to iterate the process to get a knowledge model. Workflows are mapped on a grid by assigning abstract computing nodes to grid hosts and exploiting communication facilities to ensure information/data exchange among the workflow stages.

The knowledge extraction process is both computationally intensive, and collaborative and distributed in nature. Unfortunately, the number of high-level instruments to support the knowledge discovery and management in distributed environments is very low. This is particularly true in grid-based knowledge discovery (Berman, 2001), although some research and development projects and activities in this area have been activated in recent years, such as the Knowledge Grid, Discovery Net, the DataSpace project, and the Datacentric Grid. In particular, the Knowledge Grid (Cannataro & Talia, 2002) we discuss here provides a middleware for knowledge discovery services targeted to a wide range of high-performance distributed applications.

Discovery Net is a project conducted at the Engineering and Physical Sciences Research Council at Imperial College (Curcin, Ghanem, Guo, Kohler, Rowe, Syed, & Wendel, 2002), whose main goal is the design, development, and implementation of an infrastructure for effectively supporting scientific knowledge discovery processes. Within this project, a series of testbeds and demonstrations have been carried out in the areas of life sciences, environmental modeling, and geo-hazard prediction.

The building blocks in Discovery Net are the so-called *knowledge discovery services* (*KDS*), distinguished in *computation services* and *data services*. The former typically comprise algorithms, for example, data preparation and data mining, while the latter define relational tables (as queries) and other data sources. KDS are used to compose moderately complex data-pipelined processes. The composition may be carried out by means of a GUI that provides access to a library of services. An XML-based language called *discovery process markup language* (*DPML*) is used to describe processes.

Among other projects for distributed data analysis, DataSpace proposes a significant system to address efficient data access and transfer over the grid (Grossman & Mazzucco, 2002). DataSpace is a Web-services-based infrastructure for exploring, analyzing, and mining remote and distributed data. DataSpace applications employ a protocol, for working with remote and distributed data, called *DataSpace transfer protocol* (*DSTP*).

DSTP simplifies working with data by providing direct support for common operations, such as working with attributes, keys, and metadata. The DSTP protocol can be layered over specialized high-performance transport protocols such as SABUL (Gu & Grossman, 2003), which allows DataSpace applications to effectively work on wide-area high-performance networks.

Differently from the two environments discussed above, the Datacentric Grid is a system targeted at knowledge discovery on grids designed for mainly dealing with immovable data (Skillicorn, 2002). The nodes at which computations happen are called *data/compute servers* (*DCS*). Besides a compute engine and a data repository, each DCS comprises a *metadata tree*, which is a structure for maintaining relationships among raw data sets and models extracted from them. Furthermore, extracted models become new data sets, potentially useful at subsequent steps and/or for other applications.

The *grid support nodes* (*GSNs*) maintain information about the whole grid. Each GSN contains a directory of DCSs with static and dynamic information about them (e.g., properties and usage), and an execution plan cache containing recent plans along with their achieved performance. Since a computation in the Datacentric Grid is always executed on a single node, execution plans are simple. However, they can start at different places in the model hierarchy because, when they reach a node, they could find, or not, already computed models. The *user support nodes* (*USNs*) carry out execution planning and maintain results.

The Knowledge Grid supports knowledge discovery activities by providing mechanisms and higher-level services for searching resources and representing, creating, and managing knowledge discovery processes, and for composing existing data services and data-mining services in a structured manner, allowing designers to plan, store, document, verify, share, and reexecute their workflows as well as managing their output results. We designed a general framework that, using the generic grid services and middleware, defines a set of tools and services needed to support all the main steps of a KDD process in a distributed environment, from the data selection and data-mining steps to the knowledge storing and interpretation.

In the Knowledge Grid environment, discovery processes are represented as workflows that a user may compose using both concrete and abstract grid resources. Knowledge discovery workflows are defined using a visual interface that shows resources (data, tools, and hosts) to the user, and offers mechanisms for integrating them in a workflow. A formal representation of resources and workflows is stored using an XML-based notation in which a workflow is expressed as a data-flow graph of nodes, each representing either a data-mining service or a data-transfer service. The XML representation allows the workflows for discovery processes to be easily validated, shared, translated in executable scripts, and stored for future executions.

KNOWLEDGE GRID COMPONENTS AND TOOLS

Figure 2 shows the general structure of the Knowledge Grid system and its main components and communication interfaces. The high-level k-grid layer includes services used to compose, validate, and execute a parallel and distributed knowledge-discovery computation. That layer also offers services for storing and analyzing the discovered knowledge. The main services of the high-level k-grid layer are:

- The *data access service* (*DAS*) provides search, selection, transfer, transformation, and delivery of data to be mined.
- The *tools and algorithms access service* (*TAAS*) is responsible for searching, selecting, and downloading data-mining tools and algorithms.
- The *execution plan management service* (*EPMS*). An execution plan is represented by a graph describing interactions and data

Figure 2. The Knowledge Grid general structure and components

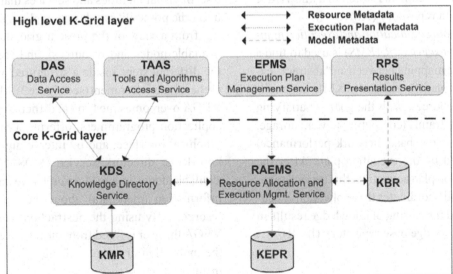

flows between data sources, extraction tools, data mining tools, and visualization tools. The execution plan management service allows for defining the structure of an application by building the corresponding graph and adding a set of constraints about resources. Generated execution plans are stored, through the RAEMS, in the knowledge execution plan repository (KEPR).

- The *results presentation service (RPS)* offers facilities for presenting and visualizing the knowledge models extracted (e.g., association rules, clustering models, classifications). The resulting metadata is stored in the KMR to be managed by the KDS (see Figure 2).

The core k-grid layer includes two main services:

- The *knowledge directory service (KDS)* that manages metadata describing Knowledge Grid resources. Such resources comprise hosts, repositories of data to be mined, tools, and algorithms used to extract, analyze, and manipulate data, distributed knowledge discovery execution plans, and knowledge obtained as a result of the mining process. The metadata information is represented by XML documents stored in a knowledge metadata repository (KMR).

- The *resource allocation and execution management service (RAEMS)* is used to find a suitable mapping between an "abstract" execution plan (formalized in XML) and available resources, with the goal of satisfying the constraints (computing power, storage, memory, database, network performance) imposed by the execution plan. After the execution plan activation, this service manages and coordinates the application execution and the storing of knowledge results in the knowledge base repository (KBR).

All or part of the services provided by the Knowledge Grid can be resident on each grid node where Globus is available. If a grid node offers data sources but does not provides data-mining tools, it does not need to configure tools and algorithms access services. The same principle holds for the other services. For example, the execution plan management service can be configured only on the grid nodes on which an execution plan is produced, whereas the RAEMS must be configured on each grid node that contributes to the execution of a KDD application. In summary, the services designed in the Knowledge Grid architecture are needed to perform all the steps of a KDD process on a grid, but they can be adaptively configured as they are necessary to run a distributed KDD process.

The main components of the Knowledge Grid environment have been implemented and are available through a software prototype, named *VEGA (visual environment for grid applications)*, which embodies services and functionalities ranging from information and discovery services to visual design and execution facilities (Cannataro, Congiusta, Talia & Trunfio, 2002). VEGA offers a simple way to design and execute complex grid applications by exploiting advantages coming from the grid environment. In particular, it offers a set of visual facilities and services that give the users the possibility to design applications starting from a view of the present grid status (i.e., available nodes and resources), and composing the different steps inside a structured and comprehensive environment (see Figure 3). Moreover, VEGA overcomes the typical difficulties of grid application programmers, offering a high-level graphical interface, and by interacting with the knowledge directory service (KDS), to know available nodes in a grid and retrieve additional information (metadata) about their published resources. By using the abstractions offered by VEGA the user is freed from the task of coupling the application structure with the underlying grid infrastructure.

Figure 3. The VEGA visual interface

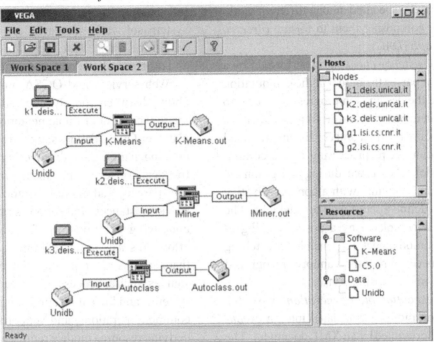

VEGA integrates functionalities of the EPMS and other Knowledge Grid services; in particular, it provides for the following EPMS operations:

- Task composition
- Consistency checking
- Execution plan generation

The *task composition* consists in the definition of the entities involved in the computation and specification of the relationships among them. The task composition facility allows the user to build typical grid applications in an easy, guided, and controlled way, having always a global view of the grid status and the overall building application. Key concepts in the VEGA approach to the design of a grid application are the *visual language* used to describe, in a component-like manner, and through a graphical representation, the jobs constituting an application, and the possibility to group these jobs in workspaces, as to define specific interdependent stages. Structured

applications, composed of multiple sequential stages, exploit both the *workspace* concept and the *virtual resource* abstraction. Thanks to these entities, it is possible to compose applications working on the outcomes of previous phases as if they were available, even if the execution has not been performed yet. For more general applications, comprising a not specified number of nodes or that can be run on different grid deployments, VEGA supports the so-called *abstract resources*. They allow for specifying resources by means of constraints (i.e., required main memory, disk space, CPU speed, operating system, etc.). When abstract resources are employed to define an application, an appropriate matching of abstract resources with physical ones, and a possible optimization phase are performed prior to submitting all the jobs for execution. A computation in the VEGA visual language is organized as a set of workspaces, hosting resources, and specifying relationships among them, so as to define one or more jobs. Jobs belonging to the same workspace are executed concurrently, whereas an arbitrary ordering can be specified among different workspaces, by composing them to form a *directed acyclic graph* (*DAG*).

The *consistency checking* is performed by a module that parses the model of the computation both while the design is in progress and prior to executing it, monitoring and driving user's actions so as to obtain a correct and consistent graphical representation. A preprocessing of the computation model takes place during the graphical composition, allowing, with a context-sensitive control, to define a set of well-formed jobs. The checking is completed by a postprocessing of the computation model, responsible for catching those error occurrences that cannot be recognized during the preprocessing phase.

In the *execution plan generation* phase, the computation model is translated into an *execution plan* represented by an XML document. The execution plan describes a data-mining computation at a high level, neither containing physical information about resources (which are identified by metadata references), nor about status and current availability of such resources. In fact, specific information about the involved resources will be included prior to the allocation process.

KNOWLEDGE GRID AND OGSA

Grid technologies are evolving towards an open grid architecture, called the *open grid services architecture* (*OGSA*), in which a grid provides an extensible set of services that virtual organizations can aggregate in various ways (Talia, 2002).

OGSA defines a uniform-exposed service semantics, the so-called *grid service*, based on concepts and technologies from both the grid computing and Web services communities. Web services define a technique for describing software components to be accessed, methods for accessing these components, and discovery methods that enable the identification of relevant

service providers. Web services are, in principle, independent from programming languages and system software; standards are being defined within the *World Wide Web Consortium* (*W3C*) and other standards bodies.

Web services and OGSA aim at interoperability between loosely coupled services that are independent of implementation, location, or platform. OGSA defines standard mechanisms for creating, naming, and discovering persistent and transient grid service instances, provides location transparency and multiple protocol bindings for service instances, and supports integration with underlying native platform facilities. The OGSA effort aims at defining a common resource model that is an abstract representation of both real resources, such as processors, processes, disks, file systems, and logical resources. It provides some common operations, and supports multiple underlying resource models representing resources as service instances.

In OGSA all services adhere to specified *grid service* interfaces and behaviors required for creating and composing sophisticated distributed systems. Service bindings can support reliable invocation, authentication, authorization, and delegation. To this end, OGSA defines a grid service as a Web service that provides a set of well-defined interfaces, and that follows specific conventions on the use for grid computing. OGSA also defines, in terms of the *Web services description language* (*WSDL*) (Christensen, Curbera, Meredith, & Weerawarana, 2001), mechanisms required for creating and composing sophisticated distributed systems, including lifetime management, change management, and notification. A first specification of the concepts and mechanisms defined in the OGSA is provided by the *open grid services infrastructure* (*OGSI*) (Tuecke et al., 2003).

More recently, the *Web services resource framework* (*WSRF*) was adopted as a refactoring and evolution of OGSI aimed at exploiting new Web services standards, and at evolving OGSI on the base of early implementation and application experiences (Czajkowski et al., 2004). WSRF codifies the relationship between Web services and stateful resources (called *WS-resources*) in terms of a set of conventions on Web services technologies, in particular XML, WSDL, and *WS-Addressing* (Box et al., 2004). The framework describes the WS-resource definition and association with the description of a Web service interface, and describes how to make the properties of a WS-resource accessible through a Web service interface. Despite OGSI and WSRF model stateful resources differently, as a grid service and a WS-resource, respectively, both provide essentially equivalent functionalities. Both grid services and WS-resources, in fact, can be created, addressed, and destroyed in essentially the same ways. The Globus Toolkit 4 is the reference implementation of WSRF.

We are devising an implementation of the Knowledge Grid in terms of the OGSA model. In this implementation, each of the Knowledge Grid services is exposed as a persistent service, using the OGSA conventions and mechanisms. For instance, the EPMS service will implement several interfaces, among which the notification interface that allows the asynchronous delivery to the EPMS of notification messages coming from services invoked as stated in execution plans. At the same time, basic knowledge discovery services can be designed and deployed by using the KDS services for discovering grid resources that could be used in composing knowledge discovery applications.

In the next section, we discuss an example of distributed data mining application to show how its execution can benefit from the Knowledge Grid services provided through the OGSA model.

AN INTRUSION-DETECTION APPLICATION

The goal of the example application is to obtain a classifier for an intrusion-detection system. In particular, the mining process is performed on a dataset containing records generated by network monitoring over a given period of time. Main issues to face with in such applications are the very large size of the used dataset, and the need to extract a number of suitable classification models to be employed into an intrusion detection system. To this end, a number of independent classifiers are first obtained by applying, in parallel, the same learning algorithm over a set of distributed training sets, generated through a random partitioning of the overall dataset. Afterwards, the best classifier is chosen by means of a voting operation by taking into account evaluation criteria like computation time, error rate, confusion matrix, and so forth.

In the scenario of the example, a user application interacts with Knowledge Grid nodes to generate classifiers built from different subsets of a given dataset. The C4.5 data-mining algorithm is used to generate classifiers as decision trees. After the partitioning step, each training set is moved to a Knowledge Grid node providing a C4.5 data-mining service. The induction of the decision trees is performed in parallel on each node, followed by the validation of the models against a testing set. The results are then moved back to the user that may visualize the classifiers to evaluate and select the obtained results.

The application makes use of three types of nodes:

- N_u, the node running the user application that builds the intrusion-detection application and visualizes the results.
- N_0, the node on which the original dataset *DS* is located, and providing a *data-partitioning service*.
- $N_1...N_n$, *n* nodes providing a *C4.5 data-mining service* that perform classification of *n* different subsets of *DS* in parallel. In particular, the C4.5 mining service provides both classification and validation functions exported to remote clients through the corre-

sponding grid service operations "classify" and "validate."

We assume that the user application knows in advance the existence of the initial dataset DS and the partitioning service on node N_0. Moreover, we assume that N_u and $N_0...N_n$ provide Knowledge Grid services, and that $N_0...N_n$ provide services for the reservation of different kinds of resources (e.g., storage, computing cycles, etc.).

Figure 4 shows a possible scenario for the example application. Apart from N_u and N_0, only three computing nodes are represented (N_1, N_2, and N_3). The application can be performed as follows (see Figure 4):

1.　The user application invokes the *TAAS* service on N_u to locate n nodes providing the required C4.5 mining service.

2.　The *TAAS* service of N_u invokes the corresponding services on other Knowledge Grid nodes, in order to obtain information about the needed resources; contacted nodes reply, sending meta-information (only the interaction with the N_1 *TAAS* is shown in dashed line for figure clearness).

3.　The meta-information about nodes $N_1...N_n$ is examined, and such nodes are identified as candidates for the computation. The *TAAS* service on N_u sends this information to the user application.

4.　The user application builds an execution plan for the data process, specifying strategies for data movement and algorithm execution. The execution plan is submitted to the EPMS of N_u.

5.　The *EPMS* invokes the reservation services on $N_0...N_n$ to reserve computing cycles and

Figure 4. A distributed data mining example

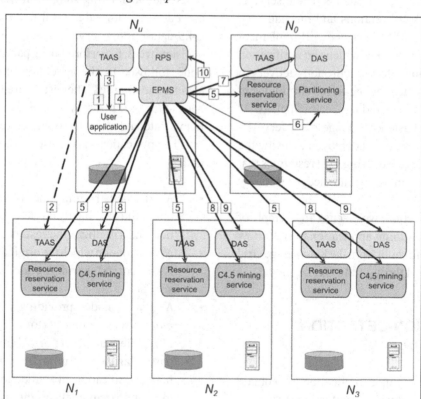

294

storage space. In particular, on N_0 computing, cycles are reserved for executing the partitioning of *DS*, and storage space is used to maintain the extracted subsets. On $N_1...N_n$ cycles are reserved to execute the classification and validation tasks, whereas storage space is used both to maintain the input subsets and the inferred classifiers.

6. The *EPMS* invokes the partitioner service on N_0 to extract n training sets and *n testing sets* from *DS*.

7. The *EPMS* invokes the *DAS* service on N_0 to transfer, to each node $N_1...N_n$, a training/testing set couple.

8. The *EPMS* invokes in sequence: *i*) the "classify" operation of the C4.5 mining service on $N_1...N_n$ to generate the classifiers; *ii*) the "validate" operation of the C4.5 mining service on $N_1...N_n$ to validate the classifiers. As soon as each operation is executed, a notification message is sent to the *EPMS* (not shown in figure).

9. The *EPMS* invokes the *DAS* service on $N_1...N_n$ to transfer the classifiers to N_u.

10. The *EPMS* invokes the *RPS* service on N_u to visualize the classifiers and support the user in evaluating and selecting the results.

KDD APPLICATIONS ON THE KNOWLEDGE GRID

In this section, we describe some significant applications implemented on the Knowledge Grid.

The first one is an implementation of the intrusion detection application described in the previous section (Cannataro, Congiusta, Pugliese, Talia, & Trunfio, 2004). This application is characterized by the employment of a massive dataset (containing millions of records), and by the need of extracting a classification model to be employed in almost real time into the network security system. The Knowledge Grid, thanks to its high-level KDD-oriented features and its

performance, has been a profitable and valuable choice for the application development and execution. This application has been tested on Knowledge Grid deployments including 3 and 8 nodes; the execution times have been compared with those of the sequential execution. The measured speed-up for these configurations has been, respectively, about 2 and 5.

Another application developed on the Knowledge Grid has been focused on bioinformatics, in particular, a "proteomics" application (Cannataro, Comito, Congiusta, & Veltri, 2004). Protein function prediction uses database searches to find proteins similar to a new protein; thus, inferring the protein function. This method is generalized by protein clustering, where databases of proteins are organized into homogeneous families to capture protein similarity. The implemented application carries out the clustering of human proteins sequences using the TribeMCL method. TribeMCL is a clustering method through which it is possible to cluster correlated proteins into groups termed "protein family." This clustering is achieved by analyzing similarity patterns between proteins in a given dataset, and using these patterns to assign proteins into related groups. In many cases, proteins of the same protein family will have similar functional properties. TribeMCL uses the Markov clustering algorithm. The application comprises four phases: *i*) data selection, *ii*) data preprocessing, *iii*) clustering, and *iv*) results visualization. The Data Selection phase extracts sequences from the database. The Data Preprocessing phase prepares the selected data to the clustering operation; in fact, TribeMCL needs a BLAST comparison on its input data. BLAST is a similarity search tool based on string-matching algorithms; given a string, it finds string sequences or subsequences matching with some of the proteins in a given database (this process is called alignment); thus, once the protein sequences have been extracted from the database, a BLAST computation has to be performed. The clustering phase performs the Markov clustering algorithm to obtain a set of protein clusters, and finally, the results visualization phase displays the obtained results. The measurement of application execution times has been done in two different cases: a) only 30 human proteins, and b) all the human proteins in the Swiss-Prot database. Comparing the execution times, we noted that the execution of the clustering phase is a computationally intensive operation and consequently, takes much more time when all the proteins have to be analyzed. However, the grid execution of these phases, using three nodes, has been performed in a time reduced by a factor of about 2.3 with respect to the sequential execution.

The last application we discuss here is concerned with an effort to integrate a query based data mining system into the Knowledge Grid environment (Bueti, Congiusta & Talia, 2004). KDDML-MQL is a system for the execution of complex mining tasks expressed as high level queries through which is possible to combine KDD operators (preprocessing, mining, etc.) to classical database operations such as selection, join, and so forth. A KDDML query has thus the structure of a tree in which each node is a KDDML operator specifying the execution of a KDD task or the logical combination (and/or operators) of results coming from lower levels of the tree. To the end of achieving the integration of such a system (not developed for the grid) into the Knowledge Grid, a slight adaptation of its structure has been needed. We modified KDDML into a distributed application composed of three independent components: *query entering and splitting* (performing the query entering and its subsequent splitting into subqueries to be executed in parallel), *query executor,* and *results visualization.* The distributed execution of KDDML has been modeled according to the master-worker paradigm, a worker being an instance of the query executor. In addition, a proper allocation policy for the subqueries has been implemented. It is based both on optimization criteria (as to balance the subqueries assignment to grid nodes), and on the structure of the tree, in order to correctly reconstruct the final response combining the partial results. Some preliminary experimental results, aimed at testing validity and feasibility of this approach, have been obtained by running some queries on a grid testbed, showing encouraging and satisfactory outcomes.

CONCLUSION

The grid can be effectively exploited for deploying data-driven and knowledge-discovery applications (Berman, Fox, & Hey, 2003). It is a well-suited infrastructure for managing very large data sources and providing high-level mechanisms for extracting valuable knowledge from them. To perform this class of tasks, advanced tools and services for knowledge discovery are vital.

Here we presented the Knowledge Grid: a grid-based software environment that implements grid-enabled knowledge-discovery services built on the Globus Toolkit mechanisms. The Knowledge Grid can be used as a high-level system for providing knowledge discovery services on dispersed resources connected through a grid. These services allow professionals and scientists to create and manage complex knowledge-discovery applications composed as workflows that integrate data sets and mining tools provided as distributed services on a grid. They also allow users to store, share, and execute these knowledge-discovery workflows, as well as publish them as new components and services. The Knowledge Grid provides a higher level of abstraction of the grid resources for knowledge discovery activities; thus, allowing the end-users to concentrate on the knowledge-discovery process without worrying about grid infrastructure details.

In the next years, the grid will be used as a platform for implementing and deploying geographically distributed knowledge discovery (Kargupta, Joshi, Sivakumar, & Yesha, 2004) and knowledge management platforms and applications. Some ongoing efforts in this direction have been recently started. Examples of systems such as Discovery Net, the DataSpace project, the Datacentric Grid, and the Knowledge Grid discussed in this chapter, show the feasibility of the approach, and can represent the first generation of knowledge-based pervasive grids.

The future use of grids is mainly related to the ability to handle large computations and manage worldwide complex-distributed applications. Among those, knowledge-based applications are a major goal. To reach this objective, the grid needs to evolve towards an open decentralized infrastructure based on interoperable high-level services that make use of knowledge both in providing resources and in giving results to end users. Software technologies for the implementation and deployment of Knowledge Grids, as we discussed in this chapter, will provide important elements to build up knowledge-based applications on a small-sized grid or on a worldwide grid. These models, techniques, and tools provide the basic components for developing grid-based complex systems. Examples of such systems include distributed knowledge management environments providing pervasive access, adaptivity, and high performance for virtual organizations in science, engineering, and industry needing to produce knowledge-based applications.

ACKNOWLEDGMENT

This research work is carried out under the FP6 Network of Excellence CoreGRID funded by the European Commission (Contract IST-2002-004265). This work has been also supported by the Italian MIUR FIRB Grid.it project RB-NE01KNFP on high performance grid platforms and tools.

REFERENCES

Berman, F. (2001). From TeraGrid to Knowledge Grid. *Communications of the ACM, 44*(11), 27-28.

Berman, F., Fox, G., & Hey, A. (Eds.). (2003). *Grid computing: Making the global infrastructure a reality.* Chichester, NY: Wiley.

Box, D., et al. (2004). *Web Services Addressing (WS-Addressing).* Retrieved September 14, 2004, from http://www.w3.org/Submission/2004/SUBM-ws-addressing-20040810

Bueti, G., Congiusta, A., & Talia, D. (2004, June). *Developing distributed data mining applications in the Knowledge Grid framework.* Paper presented at the 6th International Conference on High Performance Computing for Computational Science (VECPAR 2004), Valencia, Spain.

Cannataro, M., Comito, C, Congiusta, A., & Veltri, P. (2004). PROTEUS: A bioinformatics problem solving environment on grids. *Parallel Processing Letters, 14*(2), 217-237.

Cannataro, M., Congiusta, A., Pugliese, A., Talia, D., & Trunfio, P. (2004). Distributed data mining on grids: Services, tools, and applications. *IEEE Transactions on Systems, Man, and Cybernetics, Part B, 34*(6), 2451-2465.

Cannataro, M., Congiusta, A., Talia, D., & Trunfio, P. (2002, September). *A data mining toolset for distributed high-performance platforms.* Paper presented at the 3rd International Conference on Data Mining Methods and Databases for Engineering, Finance and Others Fields (Data Mining 2002), Bologna, Italy.

Cannataro, M., & Talia, D. (2003). The Knowledge Grid. *Communications of the ACM, 46*(1), 89-93.

Christensen, E., Curbera, F., Meredith, G., & Weerawarana, S. (2001). *Web Services Description Language (WSDL) 1.1.* Retrieved September 14, 2004, from http://www.w3.org/TR/2001/NOTE-wsdl-20010315

Curcin, V., Ghanem, M., Guo, Y., Kohler, M., Rowe, A., Syed, J., & Wendel, P. (2002, July). *Discovery Net: Towards a grid of knowledge discovery.* Paper presented at the 8th International Conference on Knowledge Discovery and Data Mining (KDD 2002), Edmonton, Canada.

Czajkowski, K., et al. (2004). *From open grid services infrastructure to WS-resource framework: Refactoring & evolution.* Retrieved September 14, 2004, from http://www.globus.org/wsrf/specs/ogsi_to_wsrf_1.0.pdf

Foster, I., Kesselman, C., Nick, J. M., & Tuecke, S. (2002). *The physiology of the grid: An Open Grid Services Architecture for distributed systems integration.* Retrieved August 26, 2006, from http://www.globus.org/alliance/publications/papers/ogsa.pdf

Grossman, R., & Mazzucco, M. (2002). A web infrastructure for the exploratory analysis and mining of data. *IEEE Computing in Science and Engineering, 4*(4), 44-51.

Gu, Y., & Grossman, R. (2003). SABUL: A transport protocol for grid computing. *Journal of Grid Computing, 1*(4), 377-386.

Kargupta, H., Joshi, A., Sivakumar, K., & Yesha, Y. (Eds.). (2004). *Data mining: Next generation challenges and future directions.* Menlo Park, CA: AAAI/MIT Press.

Skillicorn, D. (2002, April). *The case for Datacentric Grids.* Paper presented at the 16th International Parallel and Distributed Processing Symposium (IPDPS 2002), Fort Lauderdale, Florida.

Talia, D. (2002). The Open Grid Services Architecture: Where the grid meets the web. *IEEE Internet Computing, 6*(6), 67-71, 2002.

Tuecke S., et al. (2003). *Open Grid Services Infrastructure (OGSI) Version 1.0.* Retrieved August 26, 2006, from http://www-unix.globus.org/alliance/publications/papers/Final_OGSI_Specification_V1.0.pdf

Chapter XVIII
Fuzzy Miner:
Extracting Fuzzy Rules from Numerical Patterns

Nikos Pelekis
Greece & UMIST Manchester, UK

Babis Theodoulidis
UMIST Manchester, UK

Ioannis Kopanakis
UMIST Manchester, UK

Yannis Theodoridis
University of Piraeus, Greece

INTRODUCTION

Recently, our capabilities of both generating and collecting data have increased rapidly. Consequently, data mining has become a research area with increasing importance. Data mining, also referred to as knowledge discovery in databases (Chen et al., 1996), is the search of relationships and global patterns that exist "hidden" among vast amounts of data. There are various problems that someone has to deal with when extracting knowledge from data, including characterization, comparison, association, classification, predic-

tion, and clustering (Han & Kamber, 2001). This chapter elaborates on the problem of *classification*. Broadly speaking, pattern classification (or recognition) is the science that is concerned with the description or classification of measurements. More technically, pattern classification is the process that finds the common properties among a set of objects in a database and classifies them into different classes according to a classification model.

Classical models usually try to avoid *vague, imprecise,* or *uncertain* information, because it is considered as having a negative influence in the inference process. This chapter accepts the challenge of dealing with such kind of information by introducing a fuzzy system, which deliberately makes use of it. The main idea of fuzzy systems is to extend the classical two-valued modeling of concepts and attributes like *tall, fast*, or *old* in a sense of gradual truth. This means that a person is not just viewed as *tall* or *not tall*, but as tall to a certain degree between 0 and 1. This usually leads to simpler models, which are handled more easily and are more familiar to the human way of thinking.

After providing a brief comparative overview of pattern classification approaches (Section 2) and a short specification of the pattern classification domain in fuzzy systems (Section 3), the chapter follows the above paradigm and describes an effective fuzzy system for the classification of numerical data (Section 4). The initial idea comes from the fact that fuzzy systems are universal approximators (Kosko, 1992; Wang, 1992) of any real continuous function. Such an approximation method (Nozzaki et al., 1997) coming from the domain of fuzzy control systems is appropriately adjusted, extended, and implemented in order to produce a powerful working solution in the domain of pattern classification. An "adaptive" process is also introduced, developed, and incorporated into the previous mechanism for automatically deriving highly accurate linguistic if-then rules. The description of the methodology is combined with the illustration of the design issues of the tool *Fuzzy Miner*. The current work is evaluated (Section 5) by extensive simulation tests and by providing a comparison framework with another tool of the domain that employs a neuro-fuzzy approach, NEFCLASS (Nauck & Kruse, 1995). Finally, the chapter concludes (Section 6) by identifying promising directions for future work pointed to by this research effort.

COMPARATIVE OVERVIEW OF PATTERN CLASSIFICATION APPROACHES

Already, when the field was still in its very infancy, it was realized that the statistics and probability theory (Berger, 1985) had much to offer to pattern classification (Schalkoff, 1992). The question of whether or not a given pattern "belongs" to some pattern class may naturally be treated as a special case of the statistical decision theory problem. Effective, though, as it is, the statistical approach has built-in limitations. For instance, the theory of testing statistical hypotheses entails that a clear-cut yes or no answer should always decide upon the membership of a pattern in a given class. Clearly, not all of the real life patterns admit of such coarse decisions. Sometimes information in a pattern is not simply in the presence or the absence of a set of features, but rather the interconnection of features contains important structural information. Indeed, this relational information is difficult or impossible to be quantified by a feature vector form. This is the underlying basis of structural pattern classification. Structurally based systems assume that pattern structure is quantifiable. As such, complex patterns can be decomposed recursively in simpler subpatterns in almost the same way that a sentence can be decomposed in words. The analogy directed researchers toward the theory of formal languages. The process that results in an answer to a classification question is called syntax analysis or parsing.

Fuzzy Logic and Fuzzy Systems for Pattern Classification

Fuzzy logic (Zimmermann, 1996) is a superset of conventional (Boolean) logic that has been extended to handle the concept of partial truth (values between "completely true" and "completely false"). Fuzzy Pattern Classification is one way to describe systems and the behaviour of systems. Computers always need exact data to process. With Fuzzy Pattern Classification we do not need such exact information. A system can be described by using adjectives like "high," "mid," and "low."

Most applications of fuzzy systems can be found in the area of control engineering (fuzzy control). Fuzzy control applications are based on if-then rules. The antecedent of a rule consists of fuzzy descriptions of measured input values, and the consequent defines a possibly fuzzy output value for the given input. Basically, a fuzzy rule-based system provides an effective way to capture the approximate and inexact nature of the real world. In particular, fuzzy rule-based systems appear useful when the processes are too complex for analysis by conventional quantitative techniques or when the available information from the processes is qualitative, inexact, or uncertain. Fuzzy rule-based systems have the theory of Fuzzy Logic as its theoretical base. Zimmermann states,

"Fuzzy set theory provides a strict mathematical framework in which vague conceptual phenomena can be precisely and rigorously studied" *(Zimmerman, 1996, p. X).*

Fuzzy Sets: A classical (crisp) set is normally defined as a collection of elements or objects $x \, \varepsilon \, X$, which can be finite or countable. Each element can either belong to or not belong to a set A, $A \subseteq X$. Such a classical set can be described in different ways; either one can enumerate the elements that belong to the set, or one can define the member elements by using the characteristic function $1/A$, in which $1/A(x) = 1$ indicates membership of x to A and $1/A(x) = 0$ non-membership. Pattern Classification using fuzzy logic (Cios et al., 1998; Manoranjan et al., 1995) partitions the input

space into categories (pattern classes) $w_1, ..., w_n$ and assigns a given pattern $v = (v_1, v_2, ..., v_n)$ to one of those categories. If v does not fit directly within a category, a "goodness of fit" is reported. By employing fuzzy sets as pattern classes, it is possible to describe the degree to which a pattern belongs to one class or another.

Definition. If X is a collection of objects denoted generically by x, then a fuzzy set A in X is a set of ordered pairs:

$$A = \{(x, \mu_A(x)) \; | x \in X\} \qquad (3.1)$$

μ_A is the membership function that maps X to the membership space M and $\mu_A(x)$ is the grade of membership (also degree of compatibility or degree of truth) of x in A. A widely used function is the so-called triangular membership function

$$\mu_{m,d}(x) = \begin{cases} 1 - \left| \dfrac{m\text{-}x}{d} \right| & \text{if } m - d \leq x \leq m + d \\ 0 & \text{if } x < m - d \text{ or } x > m + d \end{cases}$$

$$(3.2)$$

with $d > 0$ and $m \in \Re$. This function assumes the maximum membership degree of 1 at the value m. It decreases linearly to time left and right of m to membership degree 0. These fuzzy sets are suitable for modelling linguistic terms like *approximately zero*. Of course, the triangular function may be replaced by other functions (e.g., a trapezoidal or a Gaussian function).

Fuzzy Rules: Fuzzy rules are a collection of linguistic statements that describe how a fuzzy inference system should make a decision regarding classifying an input. They combine two or more input fuzzy sets and associate with them an output. Fuzzy rules are always written in the following form:

IF v_1 is A_1 **and** v_2 is A_2 **and** ... v_n is A_n **THEN** *(v_1, v_2, ..., v_n)* belongs to class *w*,

where A_1, A_2, ..., A_n are input fuzzy sets and *w* is output fuzzy set.

For example, one could make up a rule that says:

IF temperature is high **and** humidity is high **THEN** room is hot.

There would have to be membership functions that define what we mean by high temperature (input$_1$), high humidity (input$_2$), and a hot room (output). This process of taking an input such as temperature and processing it through a membership function to determine what we mean by "high" temperature is called *fuzzification*. The purpose of fuzzification is to map the inputs to values from 0 to 1 using a set of input membership functions.

The main advantage of the above connection is its close relation to human thinking. This is also the reason that the knowledge of an expert can easily be incorporated into a fuzzy pattern classification system. But in lack of an expert or in case of a complex system, there is also the possibility of using real information/data from the system to build the fuzzy rules. On the other hand, the disadvantages are the necessity to provide the fuzzy rules, the fact that a fuzzy system cannot learn from data, and that there is no formal method to tune the membership functions.

The approaches introduced so far share some common features and goals. Thus, boundaries between them are not very clear. Each has pitfalls and advantages, and the availability and "shape" of the features often determine the approach chosen to tackle a pattern classification problem. As far as fuzzy, statistical, and structural approaches are concerned, all are valid approaches to the classification problem. The point is that probability (statistical approach) involves crisp set theory and does not allow for an element to be a partial member in a class. Probability is an indicator of the frequency or likelihood that an element is in a class. On the other hand, formal grammars (structural approach) have a difficulty in learning structural rules. Finally, fuzzy set theory deals with the similarity of an element to a class. As such, if we were to classify someone as "senior," fuzzy membership makes much more sense than probability. If we were to classify the outcome of a coin flip, probability makes much more sense.

Neural Networks and Neuro-Fuzzy Systems

The course of argumentation followed so far puts the pattern classification theme into a technical-mathematical framework. Since pattern classification is an ability of intelligent natural systems, it is possible to imitate the neuron (Masters, 1993), the basic unit of the brain, by an analogue logical processing unit, which processes the inputs and produces an output that is either on or off. Thus, by extension, a simple neuron can classify the input in two different classes by setting the output to "1" or "0". The neuron is very good at solving linearly separable problems, but fails completely at solving apparently simple problem such as the XOR one. This issue is easily overcome by multi-layer neurons that use more than one neuron and combine their outputs into other neurons, which would produce a final indication of the class to which the input belongs (Bigus, 1996; Craven & Shavlik, 1997).

Among the previously mentioned solutions, fuzzy logic and neural networks can be an answer to the vast majority of classification problems. Both fuzzy systems and neural networks attempt to determine the transfer function between a feature space and a given class. Both can be automatically adapted by the computer in an attempt to optimize their classification performance. One difference between the two methods is that the membership functions of a fuzzy classifier can be initialized in a state close to the correct solution, while a neural network can only learn from scratch, and, as a result, it can only be initialized in a random state. But their learning capabilities are significant as different learning algorithms are available, and they have great potential for parallelism since the computations of the components are largely independent of each other. Drawbacks still exist, although as shown in the example, there is the impossibility to extract rules from neurons for interpretation. As such, the training of the computer to optimize the classifier is usually much faster with a fuzzy classifier than with a neural network classifier. Consequently, by combining fuzzy logic and neural networks (neuro-fuzzy systems), we can avoid the drawbacks of each method. A neuro-fuzzy classifier of the area, called NEFCLASS, which is also used in the evaluation of our fuzzy system, has been introduced in Nauck and Kruse (1995).

Other Approaches

The necessarily brief overview of the field would be incomplete without mentioning the existence of some alternative approaches, which are neither statistical nor syntactical. For example, the geometrical method (Prabhu, 2003) focuses on finding data representations or organizations that are perceptually meaningful, while the state-space method (Oja, 1983) is concerned with finding ways of searching effectively the hierarchical structures prevalent in many pattern recognition tasks. Furthermore, case-based reasoning methods (Aamodt & Plazas, 1994; Leake, 1996), rough sets techniques (Lenarcik & Piasta, 1997; Pawlak, 1991; Swiniarski, 1998) and clustering methods (Liu et al., 2000) encompass diverse techniques for discovering regularities (or structures or patterns) in complex data sets. They may serve to suggest either hypothetical models for the data-generating mechanism or the existence of previously unknown pattern classes.

Finally, in many applications of fuzzy rule-based systems, fuzzy if-then rules have been obtained from human experts. Recently, various methods were proposed for automatically generating fuzzy if-then rules from numerical data. Most of these methods have involved iterative learning procedures or complicated rule generation mechanisms such as gradient descent learning methods (Nomura et al., 1992), genetic algorithm-based methods (Mitchel, 1996), least-squares methods (Sugeno & Kang, 1998), a fuzzy c-means method (Sugeno & Yasukawa, 1993), and a neuro-fuzzy method (Takagi & Hayashi, 1991). In Wang and Mendel (1992), an efficient rule generation method with no time-consuming iterative procedure is proposed.

FUZZY MINER: A FUZZY SYSTEM FOR SOLVING PATTERN CLASSIFICATION PROBLEMS

In this section, we present a powerful fuzzy system for solving pattern classification problems and we provide the reader with a description of the components of the Fuzzy Miner, their internal processes, and their interrelationships. The reader interested in a more detailed description of the design and implementation issues of Fuzzy Miner is referred to Pelekis (1999). That work has mainly focused on the study and understanding of a method proposed in Nozzaki et al. (1997). Fuzzy if-then rules with non-fuzzy singletons (e.g., real numbers) in the consequent parts are generated by the heuristic method proposed in Nozzaki et al. (1997). In this chapter, we innovatively adjust, extend, and implement this function approximation method in order to produce an effective, working data mining tool in the field of pattern classification. A novel embedded *"adaptive"* process is also introduced, developed, and incorporated into the previous mechanism for automatic deriving highly accurate linguistic if-then rules.

The main advantage of these fuzzy if-then rules is the simplicity of the fuzzy reasoning procedure because no defuzzification step is required. The heuristic method determines the consequent real number of each fuzzy if-then rule as the weighted mean value of given numerical data. Thus, the proposed heuristic method does require neither time-consuming iterative learning procedures nor complicated rule generation mechanisms.

Design & Architecture of the Fuzzy Rule-Based System

Fuzzy rule-based systems are also known as fuzzy inference systems, fuzzy models, fuzzy associative memories, or fuzzy controllers. Basically, such fuzzy rule-based systems are composed of four principal components: a fuzzification interface, a knowledge base, a decision-making logic, and a defuzzification interface. Fuzzy Miner employs the architecture depicted in Figure 1.

In Nozzaki et al. (1997), the authors consider a single-output fuzzy method in the n-dimensional input space $[0, 1]^n$, and we keep for the moment these assumptions just for simplicity reasons. The actual algorithm implemented in Fuzzy Miner introduces a multiple-output fuzzy rule-based system with optional task, the mapping of the input spaces to the $[0, 1]^n$ space (normalization process). Of course, when normalization process

is selected, an appropriate action is performed after the end of the algorithm to reversely map the normalized data to their primitive spaces. Let us assume that the following m input-output pairs are given as training data for constructing a fuzzy rule-based system

$$\{(x_p; y_p) \mid p = 1, 2, \ldots, m\} \qquad (4.1)$$

where $x_p = (x_{p1}, x_{p2}, \ldots, x_{pm})$ is the input vector of the pth input-output pair and y_p is the corresponding output.

The fuzzification interface performs a mapping that converts crisp values of input variables into fuzzy singletons. Basically, a fuzzy singleton is a precise value, and hence, no fuzziness is introduced by fuzzification in this case. This strategy, however, has been widely used in fuzzy system applications because it is easily implemented. Here, we employ fuzzy singletons in the fuzzification interface. On the other end, the defuzzification interface performs a mapping from the fuzzy output of a fuzzy rule-based system to a crisp output. However, the fuzzy rule-based system employed in this chapter does not require a defuzzification interface.

In the following subsections, we present in details the two core modules of the architecture; namely, the knowledge base and the decision making logic.

Figure 1. Architecture of Fuzzy Miner

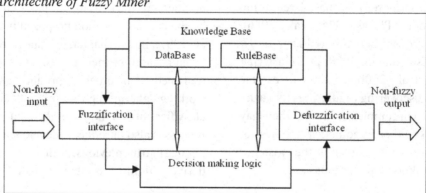

Knowledge Base

The knowledge base of a fuzzy rule-based system consists of two components — a *database* and a *rule base.*

Database: There are two factors that determine a database: a fuzzy partition of the input space and membership functions of antecedent fuzzy sets. In order to develop the appropriate infrastructure, Fuzzy Miner defines three corresponding parametric components—*Database, Fuzzy Partition,* and *Membership Function. Database* provides a complete set of functionalities upon the data (e.g., normalization/denormalization process) that the algorithm needs in order to operate effectively. Someone can think of a *Database* as the realization of a real database, which enables us to store, retrieve, update, and generally manipulate data. The *Database* component is defined as a 2D array, where the first dimension corresponds to the row of a database table, and the second dimension corresponds to the column (input-output space).

We assume that the domain interval of the *i*th input variable x_i is evenly divided into K_i fuzzy sets labelled as $A_{i1}, A_{i2}, ..., A_{iK_i}$ for $i = 1, 2,...,n$. Then the *n*-dimensional input space is divided into $K_1 K_2 ... K_n$ fuzzy subspaces:

$$(A_{1j_1}, A_{2j_2}, ..., A_{nj_n}), j_1 = 1, 2, ..., K_1, \ ... \ j_n = 1, 2, ..., K_n$$

$$(4.2)$$

For example, in the case of a two-dimensional input space, the fuzzy subspace $(A1_{j_1}, A2_{j_2})$ corresponds to the region shown in Figure 2(a). Figure 2(b) shows an example of the fuzzy partition for $K_1 = 5$ and $K_2 = 5$ in the case of a two-input single-output fuzzy rule-based system.

The *Membership Function* component can be perceived as the mean to measure the degree of compatibility of a data value to a fuzzy set, or as the probability that this data value "belongs" to a fuzzy set. In order to be able to use more than one membership function, we adopt a generic representation that enables the definition of different kinds of membership functions. As such, the user of the fuzzy classifier can use not only triangular membership functions, but trapezoidal and bell-shaped ones. In order to represent a triangular fuzzy membership function, three parameters are enough. However, from a practical point of view, to use trapezoidal and/or bell-shaped (Gaussian) membership functions, four parameters are necessary. Figure 3 depicts the three types of membership functions that Fuzzy Miner supports.

Figure 2. (a) Fuzzy subspace and (b) Fuzzy partition for K1 = 5 and K2 = 5

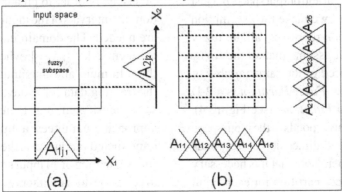

Figure 3. (a) Triangular, (b) Trapezoidal, (c) Bell-shaped

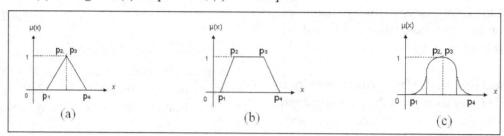

Figure 4. Fuzzy partition structure

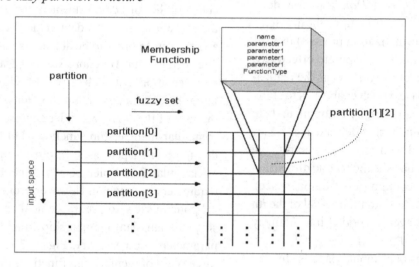

The *Fuzzy Partition* component supports the notion that input and output spaces should be partitioned to a sequence of fuzzy sets. Each of these fuzzy sets has a description of its membership function. Normally, there should be one *Fuzzy Partition* object per input and output space, but just for simplicity reasons, we make the assumption that the *Fuzzy Partition* represents all the fuzzy partitions. We further assume that all the fuzzy partitions are composed of the same number of fuzzy sets N. As such, *Fuzzy Partition* is a 2-D array of *Membership Functions* (see Figure 4). The first dimension corresponds to the input space number, and the second dimension corresponds to the fuzzy set number. Note that it is necessary to use a different fuzzy partition for each input space, because the domain intervals of the input variables may be different.

The main functionality *Fuzzy Partition* offers Fuzzy Miner is the actual fuzzy partitioning taking place at the time of its initialization. More precisely, in order to create *Fuzzy Partition*, the domain intervals of the input and output variables are needed. The domain interval of a variable x_i is taken as $[x_{imin}, x_{imax}]$, where x_{imin} and x_{imax} are the minimum and maximum of the variable in the training data set. Note that the training data set is considered, not the testing data set. This approach guarantees a minimum number of unpredicted outputs. Furthermore, although the fuzzy partition of an input space is only supposed to cover the domain interval of the input variable,

Figure 5. Fuzzy partitioning for triangular membership function

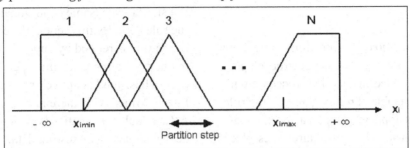

the case of input values lying outside the domain interval must be taken into account. By assigning the value $-\infty$ to the two first parameters of the first fuzzy set and the value $+\infty$ to the two last parameters of the last fuzzy set, the fuzzy partition corresponding to an input variable x covers $\Re$. In Figure 5 we present the partitioning in the case of a triangular membership function.

Rule Base: The rule base consists of a set of fuzzy if-then rules in the form of "IF a set of conditions is satisfied, THEN a set of consequences can be inferred." We assume that the rule base is composed of fuzzy if-then rules of the following form:

Rule $R_{j_i \dots j_n}$:
If x_1 is $A_{1 j_1}$ and ... and x_n is $A_{n j_n}$ then
y is $b_{j_1 \dots j_n}$, $j_1=1, 2,\dots, K_1$; ...; $j_n=1, 2,\dots, K_n$
$$(4.3)$$

where $R_{j_i \dots j_n}$ is the label of each fuzzy if-then rule and $b_{j_i \dots j_n}$ is the consequent real number. These fuzzy if-then rules are referred to as simplified fuzzy if-then rules and have been used in Ichihashi and Watanabe (1990) and Nomura et al. (1992). For determining the consequent real number $b_{j_i \dots j_n}$ of the fuzzy if-then rule $R_{j_i \dots j_n}$ in (4.3), let us define the weight of the p_{th} input-output pair $(x_p; y_p)$ as

$$W_{j_1 \dots j_n}(x_p) = \{\mu_{j_1 \dots j_n}(x_p)\}^a, \qquad (4.4)$$

where a is a positive constant. The role of the positive constant a will be demonstrated by computer

simulations. Using the weight $W_{j_1 \dots j_n}(x_p)$ of each input-output pair, the following heuristic method (the weighted mean value of yp's) determines the consequent real number.

$$b_{j_1 \dots j_n} = \sum_{p=1}^{m} W_{j_1 \dots j_n}(x_p) \cdot y_p \Bigg/ \sum_{p=1}^{m} W_{j_1 \dots j_n}(x_p)$$
$$(4.5)$$

Rulebase is the main component of the application and supports all the functionality that we need in order to implement the various aspects of Fuzzy Miner. It generates fuzzy rules from training data and furthermore is responsible for the decision-making part of the algorithm (see section 4.1.2). An additional task supported by our rule generation method is that of an *adaptive procedure*, which expands a given rulebase during the processing of testing data when the inference engine (decision making) of the algorithm is running. A *Rulebase* is implemented mainly as an array of *Rules* that in its turn is represented as an array of integers corresponding to the conditional part, and an array of *Then Part* objects corresponding to the consequent part, one element per output space. *Then Part* objects are needed in order to calculate the consequent parts of a fuzzy rule (the relatively complex fraction—nominator/denominator—of equation 4.5). The computational development of the above mathematically described process for inferring fuzzy rules, after given learning data and information concerning the number of inputs and outputs of these data, is presented in Figure 6.

Adaptive Procedure. Before illustrating how the decision-making method has been developed, we further introduce an adaptive procedure with which we are capable of refining an existing rule base during its application upon a specific classification scenario. This procedure takes place concurrently with the decision-making process; namely, when testing data are examined, inferred output are calculated and mapped to classes. The idea is based on an advantage of the fuzzy-numerical methods, which is the facility to modify a fuzzy rulebase as new data become available. More specifically, when a new data pair becomes available, one rule is created for this data pair

and is either added to the rule base or updated, if a similar rule (same conditional part) exists in the rule base. By this action, the consequent part of the rule is refined by applying the generation method once more for this specific conditional part. Thus, the "adaptive" procedure enhances Fuzzy Miner with incremental characteristics. All available information is exploited, improving decision making on testing data.

Linguistic Representation. In real-world applications, it may be desirable that linguistic rules are generated from numerical data. In Sugeno and Yasukawa (1993), an approach for deriving linguistic rules from fuzzy if-then rules with fuzzy sets in the consequent parts is proposed. Here, a similar approach is adopted for translating fuzzy if-then

Figure 6. Rule Generation Method

```
Generate rules(numberOfInputs, numberOfOutputs, startOfLearnData, endOfLearnData)
{
        currentRule = 0;
        allocate memory for rulebase[currentRule];
        for all learning data pairs f
                usedData[f] = false;
        create temporary rule;

        for (i = startOfLearnData; i <= endOfLearnData; i++)
        {
                if (!usedData[i])
                {
                        construct rulebase[currentRule];
                        set IF part of rulebase[currentRule];
                        set THEN part of rulebase[currentRule];
                        calculate weight of rulebase[currentRule];
                        for all outputs j
                                numerator[j] = weight * (THEN part of rulebase[currentRule]);
                                denominator[j] = weight;
                        for ( j = i + 1; j <= endOfLearnData; j++)
                        {
                                set IF part of temporary rule;
                                set THEN part of temporary rule;
                                calculate weight of temporary rule;
                                if (!usedData[j] & currentRule has same IF part as temprule)
                                {
                                        for all outputs
                                                update numerator[k];
                                                update denominator[k];
                                        usedData[j] = true;
                                }
                        }
                        for all outputs
                                set THEN part of ruleBase[currentRule];
                        currentRule++;
                }
        }
}
```

rules with consequent real numbers into rules. "Then" part of such rules is a linguistic label and corresponds to the classification of the respective data pairs. This approach can derive classification rules from fuzzy if-then rules with consequent real numbers, which may be generated by other rule generation methods, as well as the described heuristic method. Let us assume that fuzzy if-then rules in (4.3) are given. To translate consequent real numbers into linguistic labels, suppose that the domain interval of an output y is divided into N fuzzy sets (e.g., linguistic labels) $B_1, B_2, ..., B_N$, which are associated with the membership functions $\mu_{B_1}, ..., \mu_{B_N}$, respectively. For example, these fuzzy sets may have linguistic labels such as S: small; MS: medium small; M: medium; ML: medium large; and L: large. In this method, the given fuzzy if-then rules in (4.3) are transformed to the following fuzzy if-then rules:

Rule $R^*_{j_i \cdots j_n}$:
If x_1 is A_{1j_1} and ... and x_n is A_{nj_n} then
y is $B^*_{j_1 \cdots j_n}$, with $CF^*_{j_1 \cdots j_n}$,

$j_1 = 1, 2, ..., K_1; ...; j_n = 1, 2, ..., K_n,$

$$(4.6)$$

where $B^*_{j1..Jn}$ is the consequent fuzzy set characterized by the subsequent membership function:

$$\mu_{B^*_{j_1 \cdots j_n}}(b_{j_1 \cdots j_n}) = \max\{\mu_{B_i}(b_{j_1 \cdots j_n}) \mid i = 1, 2, ..., N\}$$

$$(4.7)$$

and $CF^*_{j1..jn}$ is the degree of certainty defined as

$$CF^*_{j_1 \cdots j_n} = \mu_{B^*_{j_1 \cdots j_n}}(b_{j_1 \cdots j_n})$$

$$(4.8)$$

Decision-Making Logic

The decision-making logic is the kernel of a fuzzy rule-based system that employs fuzzy if-then rules from the rule base to infer the output by a fuzzy reasoning method. In this chapter, we employ the fuzzy reasoning method of equation 4.9 to calculate the inferred output of the fuzzy rule-based

system. Given an input vector $x_p = (x_{p1}, x_{p2}, ..., x_{pn})$, the inferred output $y(x_p)$ is defined by:

$$y(x_p)$$

$$= \sum_{j_1=1}^{K_1} \cdots \sum_{j_n=1}^{K_n} \mu_{j_1 \cdots j_n}(x_p) \cdot b_{j_1 \cdots j_n} \bigg/ \sum_{j_1=1}^{K_1} \cdots \sum_{j_n=1}^{K_n} \mu_{j_1 \cdots j_n}(x_p)$$

$$(4.9)$$

where $\mu_{j_1 \cdots jn}(x_p)$ is the degree of compatibility of the input vector $x_p = (x_{p1}, x_{p2}, ..., x_{pn})$ to the fuzzy if-then rule $R_{ji \cdots jn}$ in (4.6), which is given by:

$$\mu_{j_1 \cdots j_n}(x_p) = \mu_{1j_1}(x_{p1}) \times ... \times \mu_{nj_n}(x_{p_n}).$$

$$(4.10)$$

From (4.9), we can see that the inferred output $y(x_p)$ is the weighted average of the consequent real numbers $b_{j1..jn}$'s of the $K_1 K_2 ... K_n$ fuzzy if-then rules.

Given a testing data set, this method calculates the outputs of the Fuzzy Miner and performs a mapping from the inferred consequent real number to the respective fuzzy set (classification result) to which this real number belongs. Subsequently, this method stores both the original outputs and classifications of the testing data pairs and the inferred outputs with the resulted classifications to an output Database. In order to improve results, we utilize the adaptive procedure, which is an embedded process and not an autonomous one. Finally, in order to evaluate the algorithm for the given testing data, decision-making method estimates the mean square errors between the desired output y_p and the inferred output $y(x_p)$. This **P**erformance **I**ndex (PI) (eq. 5.2) and the number of unpredicted results are returned as the output of the whole process. This procedure is illustrated in Figure 7.

EVALUATION OF THE FUZZY MINER

In this section, we focus on investigating the reliability and the validity of Fuzzy Miner. The

Figure 7. Algorithm for the Decision Making Engine

```
Decision Making ()
{
    construct & initialize an output database DB;
    construct & initialize one numerator & denominator per output;
    unpredictedResults = 0;
    create temporary rule;

    for(i = endOfLearning + 1; i <= endOfTesting; i++)
    {
        set IF part of temporary rule;
        set THEN part of temporary rule;
        calculate weight of temporary rule;
        re-initialize numerators & denominators per output;
        ruleFound = FALSE;

        for(k = 0; k <= NumOfRules; k++)
        {
            estimate degree of compatibility of temporary rule with rulebase[k];

            if(degree > 0)
                for all outputs
                    update numerators & denominators of temprule (using degree);

            if(!ruleFound & temprule has same IF part as currentRule)
            {
                ruleFound = TRUE;
                for all outputs
                    update numerators & denominators of currentRule (using weight);
            }
        }

        for all outputs
        {
            write to output DB      //performing reverse mapping
                1. original output
                2. initial classification
                3. inferred output
                4. resulted classification
        }

        if(denominators != 0)
            calculate performance index;
        else
            unpredicted++;

        if(!ruleFound)
            rulebase[++numOfRules] = temporary rule;
    }

    for all rules
        perform reverse mapping of inferred output to classes &
        set linguistic label for current rule;              // classification result
        set probability of correctness for current rule;    // degree of certainty

    if(writeToDB)
    {
        write inferred data to database;
        if(denormalize)
            if(normalize)      //normalization was performed on load data
                denormalize output data;
    }
}
```

process of classification is deterministic, meaning that the same input data will always produce the same output. As such, in order to measure the performance of the methods implemented in Fuzzy Miner, we have run several experiments that result in interesting conclusions. These experiments have also been compared to the results derived from another classifier, which uses a neuro-fuzzy approach (Nauck & Kruse, 1995). For the experiments, we used the data set from the **Athens Stock Exchange** (ASE, 2004) market.

The ASE data set keeps a vast amount of information concerning the daily transactions of the stock market of Greece. As has been already mentioned, the algorithm works with numerical data, and fuzzy systems are universal approximators of any real continuous function. In order to take advantage of this important feature of fuzzy systems and for the purposes of the evaluation, we designed a classification task based upon the prediction/inference of a function that estimates a real number, which represents the degree of fluctuation of a stock price during a day. The calculation of this real number, as the result of a function, is based upon the following information that the ASE database stores:

- **Max price:** the maximum point in the fluctuation of the price of a stock during a day
- **Min price:** the respective minimum point in the above-mentioned fluctuation
- **Exchanged items:** the number of stocks that were sold/bought during a day
- **Close price:** the ending point in the daily fluctuation of the price of the stock

Our choice for such a function is a formula that takes into account three factors and estimates a real number in the interval [0, 3]. More specifically, each of these factors calculates a normalized number from zero to one, indicating how much a specific stock fluctuated during a day. Based on these sub-formulas, the overall indication of the fluctuation of the stock is derived. Therefore, the function that estimates the consequent (output) real number, which Fuzzy Miner will try to infer/approximate, is described in equation 5.1:

$$f : \Re^4 \rightarrow [0,3] \text{ and } f(x_1, x_2, x_3, x_4)$$
$$= factor1 + factor2 + factor3 \qquad (5.1)$$

$$f(x_1, x_2, x_3, x_4) =$$

$$\frac{(Maxprice - Minprice) - MINdiff}{MAXdiff - MINdiff}$$

$$+ \frac{(Closeprice - Minprice) - MINdiff}{MAXdiff - MINdiff}$$

$$+ \frac{ExchangedItems - MINitems}{MAXitems - MINitems}$$

where x_1, x_2, x_3, x_4 are the four input parameters of the data set.

The first factor indicates the fluctuation of the difference between the maximum price and the

minimum price of a stock. *MAXdiff* and *MINdiff* are the maximum and minimum (respectively) differences in the data set between *Max price* and *Min price* attributes. The second factor models the fluctuation of the difference between the closing price and the minimum price of a stock. This is an indication of how easily a stock maintains its price away from the minimum price. *MAXdiff* and *MINdiff* are the respective maximum and minimum differences between *Close* and *Min price* columns. Finally, the third factor tries to strengthen the two previous factors by adding to them the normalized number of exchanged items. By this, we model the fact that if the fluctuation of a stock is high, then this is more important when the number of exchanged items is also high than when the number of stocks that were sold or bought is low. *MAXitems* and *MINitems* are the maximum and minimum values of *Exchanged items* attribute.

Experimenting with Fuzzy Miner

In order to assess the forecasting ability of Fuzzy Miner, we limited our experiments to the banking sector from which we sampled 3,000 input-output data pairs collected from the daily transactions of eight banking constitutions during a calendrical year (1997). A set of 1,500 tuples was used for learning and the remaining 1,500 tuples for testing. Note that since the fuzzy rule-based system can employ the "adaptive" procedure, test data may be learning data as well, although they do not participate in the creation of the initial fuzzy rule base.

Fitting & Generalization Ability for Training and Testing Data

For the evaluation of the algorithm, the summation of square errors between the desired output y_p and the inferred output $y(x_p)$ for each input-output pair $(x_p; y_p)$ is calculated. This performance index (PI) for Fuzzy Miner is given by the equation 5.2:

$$PI = \sum_{p=1}^{m} \{y(x_p) - y_p\}^2 / 2 \qquad (5.2)$$

The two most important factors of the fuzzy rule-based system are the value of α and the size of the fuzzy partitions. In order to understand the influence of these parameters on the performance index, the algorithm has been invoked with different values of α, varying from 0.1 to 50, and a fuzzy partition size varying from two to 25. The results of the simulations are presented in Table 1, which contains only a subset of the experimentations. Note that for each different value of α, we first present PI without using the adaptive approach, and subsequently PI using the adaptive procedure.

An obvious conclusion that could be inferred from Table 1 is that larger sizes of fuzzy partition lead to a better fitting (smaller PI) to the given input-output data pairs. Using the table containing the complete results of the previously mentioned simulations, the PI for both the original method and the method using the adaptive approach have been plotted against the number of fuzzy sets per fuzzy partition. Figure 8 indicates the strength of the adaptive approach. However, when $\alpha > 1$,

PI might become worse due to the phenomenon of overfitting. Further investigation designates that when the number of fuzzy sets is high, PI decreases very slowly, whereas for a small number of fuzzy sets, PI is much more sensitive to the variation of the fuzzy partition size. Finally, PI asymptotically converges to a specific limit as the fuzzy partition size increases.

An additional observation is that for each specific fuzzy set, PI decreases or increases, depending on the value of α. More specifically, when α is less than five, PI is improving, but when α exceeds that limit, PI starts decreasing. The best fitting is presented when $\alpha = 5$. As a conclusion, we could argue that PI can be improved by choosing the appropriate value of α. Figure 9 illustrates the desired output and two inferred outputs for two different values of α. When $\alpha = 5$, it is self evident that the approximation of the formula is much better than when α is 0.1.

Classification Success

In order to investigate the classification success of Fuzzy Miner with respect to different sizes of the fuzzy partitioning and the values of α, the following table is provided.

Table 1. Performance index against α & number of fuzzy sets

Fuzzy sets \ α	2	3	4	5	6	7
0.1	0.013042	0.005884	0.005064	0.003990	0.003728	0.003467
	0.012900	0.005830	0.005028	0.003912	0.003518	0.003380
0.5	0.011439	0.005519	0.004895	0.003954	0.003716	0.003464
	0.011340	0.005453	0.004866	0.003884	0.003477	0.003378
1	0.009960	0.005212	0.004719	0.003915	0.003713	0.003467
	0.009905	0.005117	0.004702	0.003864	0.003435	0.003392
5	0.006252	0.004578	0.004198	0.003719	0.003249	0.003558
	0.006147	0.004452	0.004268	0.003817	0.003157	0.003688
10	0.005421	0.004470	0.004292	0.003664	0.003294	0.003132
	0.005626	0.004452	0.004212	0.003767	0.004056	0.003988
50	0.005767	0.004848	0.004583	0.003972	0.003937	0.004712
	0.005598	0.005090	0.004789	0.004050	0.004989	0.005109

A conclusion that accords with the conclusion inferred previously is that when the number of fuzzy sets is fixed, then for values of α lower than five, the percentage of classification success increases as α approximates five. When it exceeds five, the trend is either to stabilize or to decrease. This conclusion does not stand strongly, as in the case of PI. This is reasonable due to the vagueness that is introduced by the fuzzy sets. There is the possibility that PI of the classifier can be improved without a corresponding improvement of the classification success. Table 3 presents the trend of the classifier for the case of two fuzzy sets (classes).

The previous reason also explains why Table 2 includes cases where the percentage of success is the same when using the adaptive approach and not. Except for those few situations where the two percentages are identical, the general trend that is followed is that for a number of classes less than

Figure 8. PI against size of fuzzy partitioning

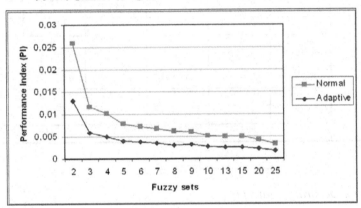

Figure 9. Fluctuation against α

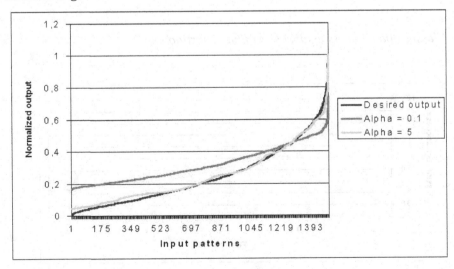

five, the adaptive approach gives higher classification results than the respective approach that is not using it. Unfavourably for fuzzy partition sizes more than five, the phenomenon of overfitting does not allow the adaptive procedure to improve performances. (By overfitting, we mean the situations where the updating of the consequent parts of the fuzzy rules should not be performed if a predefined performance index is reached.)

Table 2. Classification success against α & number of fuzzy sets

Fuzzy sets \ α	2	3	4	5	6	7
0.1	92.8632	88.7925	76.3175	74.5831	70.4470	67.1114
	94.8632	90.4603	76.3175	74.9833	69.5797	66.4443
0.5	93.9300	88.7258	77.1181	75.1835	70.9139	67.3783
	94.9967	90.5270	77.1514	75.3169	69.9133	66.3109
1	95.8301	88.3923	78.0520	75.4503	71.1141	67.5784
	95.9310	90.5935	79.6518	76.3175	70.3135	66.5777
5	96.3302	88.1921	79.9867	77.1181	73.7158	68.4456
	96.3302	90.8606	79.3863	77.3849	70.1134	66.7760
10	94.9967	88.4590	81.0540	77.3849	73.4490	67.5117
	94.9967	90.7272	79.8532	77.9853	69.4463	64.4430
50	93.2234	87.3249	79.6531	75.4503	71.8479	61.7078
	94.5297	89.7265	79.4530	76.0507	66.7111	61.3075

Table 3 Classification accuracy against α

α	0.1	0.5	1	5	10	50
Classification Accuracy	93%	94%	96%	97%	95%	93%

Figure 10. Classification success against size of fuzzy partitioning

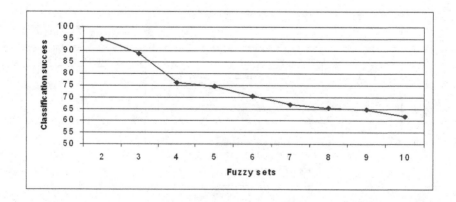

Finally, the easier but also the stronger inference that someone could make from Table 2 is that increasing the number of classes results in decreasing the percentage success of the classifier. This conclusion is illustrated in Figure 10, which shows that low fuzzy partition sizes have as a consequence a rapid reduction of the classification success. On the contrary, for a high number of fuzzy sets, we observe a stabilization in the rate of reduction of the classifier.

Optimizing the Size of the Rule Base

Another interesting observation that someone might notice by running the Fuzzy Miner with the previous parameters, is the variation of PI with respect to the size of the rule base. More specifically, Figure 11 indicates that when PI decreases, the number of the produced fuzzy rules is augmented in a stable rate. Using this graphical representation, it is possible to determine the "optimum" number of rules with respect to a performance requirement. Then, assuming a

linear relation between the number of rules and the fuzzy partition size, the "optimum" number of fuzzy sets can be also determined. The relation between the fuzzy partition size and the inferred fuzzy rules is shown in Figure 12.

The fact that the number of the produced rules is exponential interdependence of the number of fuzzy sets used to partition the input space, could also be inferred from Table 4 where the number of rules for different sizes of fuzzy partition is presented. Table 4 includes an extra row containing the number of rules when the adaptive approach is applied. As expected, the size of the rule base gets larger as new rules are added during the decision-making stage.

Selecting the Right Type of Membership Function

All of the above experiments were performed by selecting the trapezoidal membership function. Legitimately, the inquiry if the other two types of membership function that Fuzzy Miner

Figure 11. Size of rule base against PI

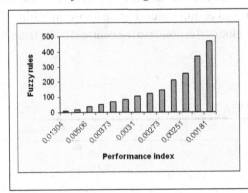

Figure 12. Rule base against fuzzy partition size

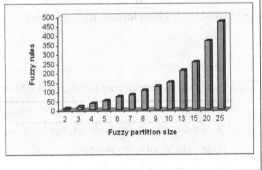

Table 4. Produced rules for different numbers of fuzzy sets

# Fuzzy sets	2	3	4	5	6	7	8	9	10	13	15	20	25
# Rules	9	17	37	49	71	81	106	124	147	209	254	367	470
# Adaptive rules	9	20	41	61	81	94	136	161	190	293	355	532	716

Table . Average performance index for different types of membership function

	Triangular	Trapezoidal	Gaussian
$\alpha = 1$	0.005567	0.005118	0.004069
$\alpha = 5$	0.004462	0.004259	0.003801

Table 6. Average classification success for different types of membership function

	Triangular	Trapezoidal	Gaussian
$\alpha = 1$	76.1704	79.2862	82.3326
$\alpha = 5$	78.3643	80.4647	83.0439

supports provide better performances upon the classification task arises. In order to answer this question, the following two tables are provided, where each column presents the mean value of PI (Table 5) and the classification success (Table 6), respectively. The presented averages are on all possible fuzzy partition sizes and have been calculated for two values of α, where Fuzzy Miner presents relatively stable behaviour.

From these tables, we draw the conclusion that the lowest PI and the higher classification success are derived when using the Gaussian membership function. The second best fitting is accomplished with the trapezoidal function. There is a logical explanation for the differences in the performances of these functions. First, the trapezoidal membership function is better than the triangular because the former gives the maximum degree of compatibility (which is one) in more attribute values than the latter, which gives this maximum membership value just in those in which their value corresponds to the centroid of the triangular

Table 7. Classification accuracy of the NEFCLASS model

When NEFCLASS uses...	Classification success
The default parameterized method (triangular function)	79.7346
Trapezoidal membership function	80.0426
Gaussian membership function	80.9283
Cross validation procedure	92.0244
The *"best"* rule learning method	82.9870
The *"best per class"* rule learning method	81.0765
The training of fuzzy sets approach	85.5647
Pruning strategies after default method	85.7442
Pruning strategies after cross validation	93.8734

shape. As such, trapezoidal membership function gives higher degrees of compatibility in average, so the approximation of the desired output is becoming an easier task. Unfavourably, there is the possibility when using the trapezoidal membership function that some attributes are assigned the maximum degree of compatibility when they should be assigned lower degrees. This problem can be solved either by widening the big base or by narrowing the small base of the trapezoidal shape. Finally, the bell-shaped function has an improved behaviour as it exhibits a smoother transition between its various parts.

Missing Rules

There is the possibility that Fuzzy Miner will not be able to predict an output for all input data pairs. This may occur if there is no rule in the rule base that corresponds to that input data pair. In the simulations performed, this problem occurred only for some specific parameter values, and particularly for large fuzzy partition sizes. The number of unpredicted outputs was very small (rarely more than two). Nevertheless, this is also an additional criterion that must be taken into account when trying to optimize a fuzzy rule-based system.

Comparing Fuzzy Miner with NEFCLASS

Continuing our evaluation, we have chosen NEF-CLASS (Nauck & Kruse, 1995) as the method for contradicting our fuzzy system, as it has many common features and shares the same goals as Fuzzy Miner. We have repeated experiments on NEFCLASS similar to those that were used to explore the validity of Fuzzy Miner. In these experiments, we are interested in the particular characteristics of NEFCLASS, and in this section, we present a summary of the results of the simulations performed. Table 7 reviews the experiments that took place.

Comparing the classification success of NEF-CLASS for its various parameters with the corresponding percentages from the third column of Table 2 (number of fuzzy sets equal to three), we can have a clear picture of the cases that Fuzzy Miner performs better. In general, Fuzzy Miner gives higher classification performances than NEFCLASS. The only case where NEFCLASS classifies patterns with a higher rate is when it uses the cross validation procedure either autonomously or in conjunction with one of the pruning strategies. In the second situation, the success of the classifier is even higher. The explanation for this is that the cross validation procedure uses the whole pattern set both as training and as testing data set, and performs several validation steps in order to create a classifier with the best fitting upon this specific data set. Another conclusion is that both NEFCLASS and Fuzzy Miner have slightly better classification results when using the Gaussian membership function.

The main difference in the philosophy of generating fuzzy rules between NEFCLASS and Fuzzy Miner is not in the algorithmic part of the two approaches. Both algorithms utilize congener methods to produce the antecedent part of a rule, but they diversify in the calculation of the consequent part of the rule, by employing different heuristic methods. The crucial difference that disjoins the two approaches is that NEFCLASS, by using the *"best"* rule and *"best per class"* rule learning method, diminishes the size of rulebase significantly. In addition to this, someone could decrease the number of the produced rules even more by employing one of the supported pruning strategies. On the other hand, our approach searches the whole pattern space and produces rules so every training input pattern is satisfied by some rule. This high number of rules is further increased by introducing the adaptive approach, which creates new rules when test data are processed. This major difference between NEFCLASS and Fuzzy Miner is emanation of the different philosophy and intention of the two applications. NEFCLASS concentrates on the conciseness and the readability of the fuzzy rules, while Fuzzy Miner tries to cover all possible input patterns. That is actually the main explanation why Fuzzy Miner outperforms NEFCLASS in the classification of unseen data.

There is a significant increment in the performance of NEFCLASS when using the method of training the fuzzy sets that were initially constructed by the rule generation method. By training the fuzzy sets, we mean that the base or the height of the initial membership functions is adjusted, so the fuzzy sets can partition the pattern space in a better way. This improved way of fuzzy partitioning means that the degree of membership of a specific input value to a fuzzy set becomes larger or smaller according to some user-defined criterion. Fuzzy Miner also supports a very simple way of tuning the membership functions, which is actually the a parameter. This form of tuning the membership functions after a certain point causes overfitting to data and the classification accuracy of the classifier is worsened. Another problem with improving the shape of the membership functions is that the change that is performed is uniform, so it cannot take into account the uneven distribution that is a common case in real world data. The optimum solution is to try to capture this unevenness of the data set and transmit it to the shape of the membership functions (e.g., scalene triangles).

Both NEFCLASS and Fuzzy Miner employ linguistic representation methods to derive fuzzy if-then rules. The main difference between these rulebases is their size. NEFCLASS produces lesser and smaller rules, but they are more readable and concise. On the other hand, Fuzzy Miner has the disadvantage that could extract many rules for huge data sets, but it has the advantage that, by taking into account the whole pattern set, it can show important exceptions or trends in the data set. From the point of view of the developer, such rules may seem useless, but from the point of view of an expert, these rules may identify exceptional behaviours or problematic cases. As such, NEFCLASS is more likely to lose useful extracted knowledge by excluding whole rules or parts from a rule.

An additional issue that should be mentioned is that the generation of a rulebase in Fuzzy Miner is almost instantaneous (several seconds for large data sets). On the contrary, in order to improve the initial created classifier, NEFCLASS has to train the fuzzy sets, prune the rule base, and repeat the decision-making stage. All these operations result in a more readable rulebase, but the whole process is time consuming. Furthermore, Fuzzy Miner has the advantage that it can classify more than one output concurrently, based on the same set of input attributes. In order to do this in NEF-CLASS, one has to create different input files for each output and to rebuild new classifiers from scratch. Last but not least, in Fuzzy Miner, several configurations can be tried, and the best one can be selected on the basis of the lowest test error. If the number of unpredicted outputs is not zero, a coarser fuzzy partition should be tried. On the other hand, this is not necessary in NEFCLASS, as all these are automatically performed by the cross validation procedure.

CONCLUSION AND FUTURE WORK

This chapter studies the pattern classification problem as it is presented in the context of data mining. More specifically, an efficient fuzzy approach for classification of numerical data is described, followed by the design and the implementation of its corresponding tool called Fuzzy Miner. The approach does not need a defuzzification process; it can be utilized as a function approximator, while by slight changes can be used as a predictor rather than as a classifier. The framework is highly flexible in that its components are configurable to meet various classification objectives. Linguistic representation of the produced fuzzy rules makes the classifier interpretable by native users, whereas the introduction of the adaptive procedure enables expanding and improving the rulebase while examining unseen testing patterns. Fuzzy Miner was evaluated using the Athens Stock Exchange (ASE, 2004) data set. The strategy adopted by Fuzzy Miner was shown to be successful, and the results of the created classifier were presented.

Additional future work is planed in various aspects of Fuzzy Miner. To start with, pruning strategies could be used to improve the interpretability of the classifier. These pruning strategies could be automatic, or some control could be given to the user over the pruning process. Secondly, we have already started designing an algorithm for training the initially created fuzzy sets by changing the length of the base or the height of a membership function, so representing the reality with greater precision. Additionally, in adaptive procedure, we can refine or discard some of the new rules based on the indications of our pruning strategies. As such, it won't be necessary to execute the pruning module for the whole rulebase from scratch every time the adaptive approach is used to improve the classifier. Furthermore, a knowledge expert should be provided with the capability to initialize the rulebase externally, or to change existing rules that do not agree with the expert's domain knowledge. We can further help the expert in the preprocessing stage by providing some statistics on the training data. Another idea is to attach to the system an algorithm to automatically determine the number of fuzzy sets for each variable and a clear criterion of how "good" are the produced fuzzy sets. A new user interface that supports graphical and textual displays (e.g., displays of the fuzzy sets) would be beneficial for interpreting the results of Fuzzy Miner (Kopanakis & Theodoulidis, 2003). Finally, a long-term goal is to integrate Fuzzy Miner with a neural network and to propagate the outcome to a genetic algorithm that would extract the optimum solution upon a specific classification task.

REFERENCES

Aamodt, A., & Plazas, E. (1994). Case-based reasoning: Foundational issues, methodological variations, and system approaches. *AI Comm., 7*, 39-52.

ASE (2004). *The Athens Stock Exchange closing prices*. Retrieved March 26, 2004: http://www.ase.gr/content/en/MarketData/Stocks/Prices/default.asp

Berger, J.O. (1985). *Statistical Decision Theory and Bayesian Analysis* (second edition). Springer-Verlag.

Bigus, J.P. (1996). *Data Mining with Neural Networks: Solving Business Problems*. McGraw-Hill.

Chen, M.S., Han, J., & Yu, P.S. (1996). Data mining: An overview from a database perspective. *IEEE Transactions on Knowledge and Data Engineering, 8*(6), 866-883.

Cios, K., Pedrycz, W., & Swiniarski, R. (1998). *Data Mining Methods for Knowledge Discovery*. Kluwer Academic Publishers.

Craven, M.W., & Shavlik, J.W. (1997). Using neural networks in data mining. *Future Generation Computer Systems, 13*, 211-229.

Han, J., & Kamber, M. (2001). *Data Mining: Concepts and Techniques*. Morgan Kaufmann.

Ichihashi, H., & Watanabe, T. (1990). Learning control system by a simplified fuzzy reasoning model. *Proceedings of IPMU'90*, (pp. 417-419).

Kopanakis, I., & Theodoulidis, B. (2003). Visual data mining modeling techniques for the visualization of mining outcomes. *Journal of Visual Languages and Computing, Special Issue on Visual Data Mining, 14*(6), 543-589.

Kosko, B. (1992). Fuzzy systems as universal approximators. *Proceedings of FUZZ-IEEE '92*, (pp. 1153-1162).

Leake, D.B. (1996). CBR in context: The present and future. In D.B. Leake (Ed.), *Case-Based Reasoning: Experience, Lessons and Future Directions* (pp. 3-30). AAAI Press.

Lenarcik, A., & Piasta, Z. (1997). Probabilistic rough classifiers with mixture of discrete and continuous variables. In T.Y. Lin & N. Cercone (Eds.), *Rough Sets and Data Mining: Analysis for Imprecise Data* (pp. 373-383). Kluwer Academic Publishers.

Liu, B., Xia, Y., & Yu, P. (2000). Clustering through decision tree construction. *Proceedings of ACM CIKM.*

Manoranjan, V.S., Lazaro, A. de Sam, Edwards, D., & Aathalye (1995). A systematic approach to obtaining fuzzy sets for control systems. *IEEE Transactions on Systems, Man and Cybernetics, 25*(1).

Masters, T. (1993). *Practical neural network recipes in C++.* Academic Press.

Mitchel, M. (1996). *An Introduction to genetic algorithms.* MIT Press.

Nauck, D., & Kruse, R. (1995). *NEFCLASS—A neuro-fuzzy approach for the classification of data.* ACM Press.

Nomura, H., Hayashi, I., & Wakami, N. (1992). A learning method of fuzzy inference rules by descent method. In *Proceedings of FUZZ-IEEE '92,* (pp. 203-210).

Nozzaki, K., Ishibuchi, H., & Tanaka, H. (1997). A simple but powerful heuristic method for generating fuzzy rules from numerical data. *Fuzzy Sets and Systems 86*, 251-270.

Oja, E. (1983). *Subspace methods for pattern recognition.* John Wiley.

Pawlak, Z. (1991). *Rough Sets, theoretical aspects of reasoning about data.* Kluwer Academic Publishers.

Pelekis, N. (1999). *Fuzzy m iner: A fuzzy system for solving pattern classification problems.* M.Sc. Thesis, *UMIST.* Retrieved June 15, 2004: http://users.forthnet.gr/ath/pele/HOME_PAGE_NIKOS_PELEKIS/Download/

Prabhu, N. (2003). Gauge groups and data classification. *Applied mathematics and computation, 138*(2-3), 267-289.

Schalkoff, R.J. (1992). *Pattern recognition: Statistical, structural and neuralapproaches.* John Wiley.

Sugeno, M., & Kang, G.T. (1998). Structure identification of fuzzy model. *Fuzzy Sets and Systems, 28*, 15-33.

Sugeno, M., & Yasukawa, T. (1993). A fuzzy-logic-based approach to qualitative modeling. *IEEE Trans. Fuzzy Systems, 1*, 7-31.

Swiniarski, R. (1998). Rough sets and principal component analysis and their applications in future extraction and selection, data model building and classification. In S. Pal & A. Skowron (Eds.), *Fuzzy Sets, Rough Sets and Decision Making Processes.* Springer-Verlag.

Takagi, H., & Hayashi, I. (1991). NN-driven fuzzy reasoning. *Approximate reasoning, 5*, 191-212.

Wang, L.X. (1992). Fuzzy systems as universal approximators. In *Proceedings of FUZZ-IEEE'92,* (pp. 1163-1170).

Wang, L.X., & Mendel, J.M. (1992). Generating fuzzy rules by learning from examples. *IEEE Trans. Systems, Man Cybernet, 22*, 1414-1427.

Zimmermann, H.-J. (1996). *Fuzzy set theory and its applications* (third edition). Kluwer Academic.

ENDNOTE

* A short version appears in the informal Proceedings of the 1st Intermatopmal Workshop on Pattern Representation and Management (PaRMa'04), Heraklion-Crete, Greece, March 2004.

Chapter XIX
Routing Attribute Data Mining Based on Rough Set Theory

Yanbing Liu
UEST of China & Chongqing University of Posts and Telecommunications, China
Shixin Sun, UEST, China

Menghao Wang
Chongqing University of Posts and Telecommunications, China

Jong Tang
Chongqing University of Posts and Telecommunications, China

ABSTRACT

QOSPF (Quality of Service Open Shortest Path First) based on QoS routing has been recognized as a missing piece in the evolution of QoS-based services on the Internet. Data mining has emerged as a tool for data analysis, discovery of new information, and autonomous decision making. This article focuses on routing algorithms and their applications for computing QoS routes in OSPF protocol. The proposed approach is based on a data mining approach using rough set theory, for which the attribute-value system about links of networks is created from network topology. Rough set theory offers a knowledge discovery approach to extracting routing decisions from attribute set. The extracted rules then can be used to select significant routing attributes and to make routing selections in routers. A case study is conducted in order to demonstrate that rough set theory is effective in finding the most significant attribute set. It is shown that the algorithm based on data mining and rough set offers a promising approach to the attribute selection problem in Internet routing.

INTRODUCTION

With the development of high-level applications of data communication networks, routing in net-works needs to satisfy QoS (Quality of Service) demands. Since finding optimal solutions to QoS routing is an NPC (nondeterministic polynomial time completeness) problem, a node would not be able to maintain network information base timely with constant changes of network states (Liu, Xu, Xu, & Cui, 2003; Chickering 1996). IP routing protocols long have been an essential element of internetworking. Recently, this technology has been in the focus of new development, as routing has evolved to handle the needs of next-generation networks. Data connection provides a full range of high-function portable Unicast IP routing software products, including BGP, OSPF, and RIP for both IPv4 and IPv6 networks. Traditional IP routing protocols, therefore, have been extended substantially in a number of areas.

In order to adapt to new demands of computer networks, it is necessary to introduce new feasible and efficient schemes. In order to find an optimal path with only one attribute metric such as bandwidth or number of hops, the traditional OSPF (Open Shortest Path First) uses the cost metric, which is an unsigned 16-bit integer in the range of 1 to 65,535. The default cost for interfaces is calculated, based on the bandwidth by the formula $108/BW$, with BW being the bandwidth of the interface expressed as a full integer of bps (Bruno, 2003). This mechanism may lower the utilization of network resource and cause load imbalance, and it cannot satisfy QoS requirements. The problem of QoS routing is challenging, because selecting paths that meet multiple QoS attribute constraints is a complex algorithmic problem. As current routing protocols already are reaching the limit of feasible complexity, it is important that the complexity introduced by QoS support should not impair the scalability of routing protocols (Li, Zheng, & Nahavandi, 2003). Therefore, the QOSPF based on IP QoS Routing is developed (Crawley, Argon Networks & Nair, 1998). Since then, many researches started to investigate this open problem of routing optimization in QoS extensions to OSPF protocols.

The theory of rough sets recently has emerged as a major mathematical approach for managing uncertainty that arises from inexact, noisy, or incomplete information (Pawlak, 1991). The problem

considered in this article is to decide the link rank by mining a series of link-state attributes based on rough set theory. A case study is conducted in this article in order to show that the reduction algorithm based on rough set theory can offer an attractive method to resolve the attribute selection problem in routing table of IP networks.

QoS-BASED NETWORKS ROUTING PROBLEM

In QoS-based routing, path selection for routing typically is formulated as a shortest-path optimization problem; that is, to select a series of network links connecting source and destination nodes such that particular objective attributes (e.g., cost, bandwidth, delay) are satisfactory (RFC 2386). Because the problem of calculating a path subject to multiple attribute constraints has been proved NP-complete for many common attributes combinations, usually a compromise is made by choosing (mining) a subset QoS parameters. This selection focuses on the mining of an appropriate path based on link attributes information and QoS requirements in the network. As a result, any algorithm that selects any two or more hop counts, time delay, delay jitter, and loss probability as important attributes will try to optimize these paths.

The availability of network routing and network protocols to police network resources suggests a natural solution. The data mining mechanisms that directly use existing QoS mechanisms based on rough set theory could have been developed for constant bit-rate and low-bandwidth unreliable networks. Interactive data mining applications often exhibit busty-traffic patterns and operate on a large routing table. For data mining applications operating on tera-byte-sized data, path optimization is extremely important and largely expected (RFC 2676). Data mining applications driven by humans often have varying needs in terms of link quality and router performance.

However, applications often can adapt to resource constraints by trading off resulting quality for router performance, if network resources are constrained. In other words, the resulting quality can be sacrificed in a controlled manner, when link resources are available with router and response time is critical.

For data mining and rough set applications, user requirements are specified in terms of performance and QoS. One of the main tools of data mining based on rough set is rule induction from raw data represented by a database. Real-life data are frequently imperfect (i.e., erroneous, incomplete, and uncertain). In a routing process, the problem is to determine which next node to select for next hop and which attributes to use for decision at a router. Considering computations at a router, typically while a single attribute is used as the decision variable, one would consider extensions to more than one base link attribute (e.g., *Bandwidth* $> \cdots$, and *Delay* $< \cdots$). Limiting oneself to selecting base attributes, the problem can be formulated in order to determine $f(x_1,...,x_n)$ for all x_i ($i=1,2,...,n$), where $\{x_i\}$ are the attributes and $f(x_1,...,x_n)$ measures the goodness of x_i as decision attributes. Data mining is becoming an interdisciplinary field, drawing work from areas including database technology, knowledge acquisition, and mathematics such as the rough set theory. The rough set theory is a relatively new mathematical approach to problems with data imprecision, vagueness, and uncertainty. The concept of reduction decision table is very useful for feature selection. Because decision table includes condition attributes or features and decision attributes of categories, the procedure of feature selection based on decision table is distinct and effective (Lu & Zhang, 2005).

The importance of attribute selection is its potential for speeding up the processes of routing and improving accuracy of classification. Rough sets can be used for attribute reduction, where attributes that do not contribute to the classification of the given training data can be identified and, consequently, removed. Rough sets also can contribute to the relevance analysis, where the contribution or significance of each attribute is assessed with respect to the given classification task. The problem of finding minimal subsets of a given set of attributes that describe concepts in a given data set is proven to be NP-hard (Chickering, 1996). However, algorithms to reduce computational intensity have been proposed. For example, a method of using a discernability matrix in rough set is proposed to store the differences between attribute values for each pair of data samples (Kusiak, 2001). Rather than searching on the entire data set, the matrix is searched instead to detect redundant attributes. The application of the rough set theory can solve this problem successfully. Usually, the link information of networks is classified by many QoS parameters such as link propagation delay, link available bandwidth, link jitter, possibility of connection and hop counts, and so forth. Then, protocol QOSPF can select the best path with the link rank. We perceive that data mining techniques based on rough set can be applied in order to obtain a reduced representation of routing attribute of data set that is much smaller in volume yet closely maintains the integrity of the original route table.

One of new data mining theories (Kusiak, 2001) is the rough set theory in routing attribute reduction and is used for attributes reduction of routing table, finding hidden route-decision patterns, and generation of new route-decision rules.

RELATED WORK

QoS-Based Routing

QoS-based routing is defined as "a routing mechanism under which paths for flows are determined based on some knowledge of resource availability in the network as well as the QoS requirement of the flows" (RFC 2386) or "a dynamic routing protocol that has expanded its path-selection criteria to include QoS parameters such as available bandwidth, link and end-to-end path utilization, node resources consumption, delay and latency, and induced jitter." In short, it is a dynamic routing scheme with QoS considerations. Routes that can satisfy QoS requirements of a new flow rely on both the knowledge of the flow's requirements and the information about availability of resources in the network. In addition, for the purpose of efficiency, it is important for an algorithm to account for the amount of resources that the network has to allocate to support any new flow. QoS-based routing is supposed to find a path from source to destination that can satisfy a user's requirements on bandwidth, end-to-end delay, and so forth. In addition, this has to be performed dynamically in coping with changes. In case there are several feasible paths available, the path selection can be based on some policy constraints. For example, we can choose a path that costs less in terms of money or the one via the designated service router. Path selections that are based on demand forecast are not accurate. Users may connect to another telecommunications operator, or they may connect to the network earlier or later than forecast. This article presents a method for optimizing a router in the presence of uncertainty, based on rough set theory.

In general, a network prefers to select the cheapest or best path among all paths suitable for a new flow, and it even may decide not to accept a new flow for which a feasible path exists, if the cost is deemed too high. Accounting for these aspects involves several metrics on which the route is based (RFC 2386), which include the following:

- **Possibility of Connection:** Usually, in traditional networks, possibility of link connection is high, and this metric can be ignored. But in wireless network, it is very important.

- **Link-Available Bandwidth:** As mentioned earlier, we currently assume that most QoS requirements are derivable from bandwidth. We further assume that associated with each link is a maximal bandwidth value (e.g., the link physical bandwidth or some fraction thereof that has been set aside for QoS flows). Since in order for a link to be capable of accepting a new flow with given bandwidth requirements, at least that much bandwidth still must be available on the link, then the relevant link metric is, therefore, the (current) amount of available bandwidth.

- **Link Propagation Delay:** This quantity is meant to identify high latency links (e.g., satellite links) that may be unsuitable for real-time requests. This quantity also needs to be advertised as part of extended LSAs (Level Service Agreements), although timely dissemination of this information is not critical, as this parameter is unlikely to change (significantly) over time.

- **Link Jitter:** This quantity is used to measure change of link delay. A path with a smaller jitter is preferable.

- **Hop Count:** This quantity is used to measure path cost in the network. A path with a smaller number of hops is preferable, since it consumes less network resources. As a result, the path selection algorithm will attempt to find the minimum number of hops in a path.

The routing will focus on the selection of an appropriate path, based on link-attribute metrics information and flow requirements on the Internet. Let $m(i, j)$ be a metric of $link(i, j)$. For $P = (i, j, k, ..., m, n)$, a metric $m(p)$ (Jonath & Guo, 2002) is:

- Additive if
  $m(p) = m(i, j) + m(j, k) + ... + m(m, n)$;
- Multiplicative if
  $m(p) = m(i, j)m(j, k)...m(m, n)$;

- Concave if
$$m(p) = \min\{m(i, j), m(j, k), ..., m(m, n)\}.$$

For example, metrics such as propagation delay, link jitter, and hop count are additive, the possibility of connection is multiplicative, and bandwidth is concave. In QoS-based routing, paths of packet flows would be determined based on the previous QoS requirements. The main objective of QoS-based routing is to realize dynamic determination of feasible paths; QoS-based routing can determine a path from among possibly many choices that have a good chance of accommodating QoS of the given flow. Feasible path selection may be subject to policy constraints, such as path cost, provider selection, and so forth (Jonathan & Guo, 2002). It successfully optimizes resource usage. A network state-dependent, QoS-based routing scheme can aid in an efficient utilization of network resources by improving the total network throughput. Such a routing scheme is a basis of efficient network engineering.

QOSPF Routing Algorithm

In traditional data communication networks, routing is concerned primarily with the connectivity. Routing protocols usually characterize networks with a single routing attribute and use shortest-path algorithms for path computations, which are typically transparent to any QoS requirements. As a result, routing decisions are made without any awareness of link-resource availability and relevant user requirements.

OSPF is defined in RFC 2383. It is a link-state routing protocol that uses Dijkstra's shortest paths to destinations. In OSPF, each router sends link-state advertisements about itself and its links to all its adjacent routers. Each router that receives a link-state advertisement records the information in its topology database and sends a copy of the link-state advertisement to each of its adjacency routers. All the link-state advertisements can reach all other routers in the same area, which enables each router in the area to have an identical topology database. A router does not send out routing tables but link-state information about its interfaces. When the topology databases are completed, each router individually will calculate a loop tree and a shortest-path tree. Destinations outside the area also are advertised in link-state advertisements. These, however, do not require that routers run the SPF (Shortest Path First protocol) algorithm before they are added to the routing table. Changes of all metrics need to be advertised as part of extended LSAs so that accurate information is available to the path selection algorithm. OSPF is QoS extensions to OSPF and support for QoS routing, which can be viewed as consisting of three major components in RFC 2386:

1. Obtain the information needed to compute QoS paths and select a path capable of meeting the QoS requirements of a given user request.
2. Establish the path selected to accommodate a new request.
3. Maintain the path assigned for use by a given request.

QOSPF uses a link-state algorithm in order to build and calculate the shortest path to all known destinations. The algorithm by itself is quite complicated. The following is at a very high level and in a simplified way for describing the algorithm:

1. Upon initialization or due to any change in routing information, a router will generate a link-state advertisement, which will represent the collection of all link-states on that router.
2. All routers will exchange link states by means of flooding. Each router that receives a link-state update should store a copy in its link-state database and then propagate the update to other routers.

3. After the database of each router is completed, the router will calculate a Shortest Path Tree to all destinations. The router uses the Dijkstra algorithm to calculate the shortest path tree. The destinations, the associated cost, and the next hop to reach those destinations will form the IP routing table.

4. In case no changes in the QOSPF network occur, such as cost of a link or a network being added or deleted, OSPF should be very quiet. Any changes that occur are communicated via link-state packets, and the Dijkstra algorithm is recalculated to find the shortest path.

In order to reduce communications overhead, routing algorithms based on link status information such as SPF send broadcast messages that contain only information of link status. SPF will broadcast messages that contain only the node's link status instead of the entire routing table. It seems easy to collect information on communication latency of links and to calculate routes with minimal delay; however, this is almost impossible in large networks, because we need to collect information of communication latency of all links frequently by message broadcasting, which leads to extremely heavy communication overheads. In addition, delayed information for the latency may create far from optimal routes. It is not uncommon for an application to transmit data with variable QoS requirements. For example, a video application may require different levels of service, depending upon the content of the connection. Consider a video sequence that consists of a highly dynamic set of action scenes followed by a relatively static sequence. The first part, due to rapid camera movement, is reasonably tolerant with data loss and corruption but intolerant with high-link jitter. In contrast, the static scenes are tolerant with link jitter but require minimal data loss and corruption. In a very large network such as the Internet, it is essential that routing algorithms be scalable. In order to achieve the scalability for adaptive network routing algorithms, it is expected to select (mine) important

QoS attributes, depending on rough set with as little communication overheads as possible.

ROUGH SET METHODOLOGY

Relative Reduction of Rough Set

Rough set theory is a mathematical approach to information analysis introduced by Pawlak (1999).

Rough set based on feature selection is an extension of conventional set theory that supports approximations in making decisions. The rough set itself is an approximation of a vague data set by a pair of precise concepts called lower and upper approximations, which are a classification of the domain of interest into disjoint categories. The lower approximation is a description of domain objects that are known with a certainty belonging to a subset of interests, whereas the upper approximation is a description of objects that possibly would belong to the subset (Parthasarathy, 2001). The main approach to finding rough-set reduction is concerned with the discernability matrix. This section describes the fundamental ideas behind this approach.

An information system S is a quadruple (U, A, V, f), where $U = \{x_1, x_2, ..., x_n\}$ denotes the set of all objects, A is the set of all attributes that are classified further into two disjoint subsets: the condition attributes $C = \{a_i \mid i = 1, ..., m\}$ and decision attribute $D = \{d\}$, such that $A = C \cup D$ and $C \cup D = \phi$. $V = \bigcup_{a \in A} V_a$ is a set of attribute values, where V_a is the domain of attribute a. Notation $a_i(x_j)$ denotes value of x_j on attribute a_i. $f : U \times A \rightarrow V$ is an information function, which appoints the attribute value of every object x in U.

A network can be modeled as a graph $G = (V, E)$. Nodes (V) of the graph represent switches, routers, and hosts (here represent routers). Edges (E) represent communication links. A symmetric link has the same attribute value (bandwidth, propagation delay, etc.). In order to illustrate the operation of these, an example data set of routing attributes (Table 1) is used, where $C = \{a_1, a_2, a_3, a_4\}$ is a routing attribute set in which 'a_1' represents the bandwidth of link: number 1 denotes enough, 2 denotes available, and 3 denotes non-enough. Notation 'a_2' represents the propagation delay of link: 1 denotes low, 2 denotes normal, and 3 denotes high. Notation 'a_3' represents the failure probability of link: 1 denotes low, 2 denotes normal, and 3 denotes under normal. Notation 'a_4' represents the bit-error ratio of link: 1 denotes low, 2 denotes acceptable, and 3 denotes insufferable. Decision attribute $D = \{d\} = \{1, 2, 3\}$: number 1 for good, 2 for normal, and 3 for bad (each of their weights is 1, 2, 3), respectively.

Considering data sets in Table 1 with five objects, four features $a_1 - a_4$ and decision d, features denoting process (route) parameters (e.g., bandwidth, delay) and the decision is the routing performance (good, normal, bad). Some decision rules can be extracted from data sets of Table 1:

- *Rule 1*: if $(a_2 = 1)$ then $(d = 1)$
- *Rule 2*: if $(a_1 \leq 2)$ and $(a_3 = 1)$ and $(a_4 \leq 2)$ then $(d = 1)$
- *Rule 3*: if $(a_1 = 3)$ or $(a_4 = 3)$ or $(a_2 = 3)$ then $(d = 3)$
- *Rule 4*: if $(a_1 \leq 2)$ and $(a_2 = 2)$ and $(a_4 = 2)$ then $(d = 2)$

Table 1. An example data set

$\overset{U}{x}$	a_1	a_2	a_3	a_4	d
e_1	2	1	3	2	1
e_2	2	1	1	1	1
e_3	1	1	1	2	1
e_4	1	2	3	2	2
e_5	2	2	2	2	2
e_6	1	2	3	3	3
e_7	3	2	2	3	3
e_8	3	3	3	1	3

Rough set attributes reduction method can remove redundant conditional attributes (or routing attributes) from a normal data set (Jensen & Shen, 2004). With any $P \subseteq A$, there is an associated equivalence relation $IND(P)$ or U/P:

$$IND(p) = \{(x, y) \in U^2 \mid \forall a \in P, a(x) = a(y)\} \tag{1}$$

Let $X \subseteq U$, the lower and upper approximation of set X, be defined as $\underline{P}X = \{x \mid x \in U, [x]_P \subseteq X\}$ and $\overline{P}X = \{x \mid x \in U, [x]_P \cap X \neq \phi\}$, respectively. Let P and Q be equivalence over U, then the positive region can be defined as:

$$POS_P(Q) = \bigcup_{X \in Q} \underline{P}X \tag{2}$$

A positive region contains all objects in U that can be classified in attributes Q using the information in attribute P. Using this definition of the positive region, a set of attributes Q depends on a set of attributes P. The rough set degree of dependency of a set attributes Q on P is defined in the following:

$$\gamma_P(Q) = \frac{|POS_P(Q)|}{U} \tag{3}$$

Reduction of attributes is achieved by comparing equivalent relations generated by sets of attributes. Attributes are removed so that reduced set can provide decision features that have the same quality as the original. A reduction is defined as a subset X of cardinality R_{min} of conditional attribute set Y such that $\gamma_X(D) = \gamma_Y(D)$, where D is a set of decision attributes. A given data set X may have many attribute-reduction sets, so a set R of all reduction sets is:

$$R = \{I \mid I \subset Y, \gamma_X(D) = \gamma_Y(D)\} \tag{4}$$

The minimal reduction set $R_{min} \subseteq R$ is defined as the set of any reduction set searched in R with cardinality:

$$R_{\min} = \{I \mid I \subset R, \forall X \in R, |I| < |X|\} \qquad (5)$$

These minimal subsets can make decision classes with the same discriminating power as the whole condition attributes.

A discernability matrix is a $n \times n$ matrix in which the classes are diagonal. In the matrix, the (condition) attributes that can be used to discern between the classes in the corresponding row and column are inserted (Li, 2003). Data items of discernability matrix contain attributes used to discern objects. A single element in data items must be a member of reduction; hence, single data items can be included individually through the mining process based on rough set theory. At the same time, data items that include the elements are removed from the discernability matrix. Since the reduction is to find the minimal attribute set discerning all objects, the data items of the discernability matrix contain all discerning information of objects. The intersection of a reduction and data items of the discernability matrix cannot be empty. If there is an empty intersection between some data items with reduction, then corresponding objects would not be discerned by reduction (Li, Tang, Ni, & Yang, 2002).

The information system's discernability matrix $M[C_D(i, j)]_{n \times n}$ where $C_D(i, j)$ is defined as:

$$C_D(i, j)$$
$$= \begin{cases} \{a_k \mid a_k \in C \wedge a_k(x_i) \neq a_k(x_j)\}, d(x_i) \neq d(x_j) \\ 0, \qquad\qquad d(x_i) = d(x_j) \end{cases} \qquad (6)$$

where $i, j = 1, \cdots, n$.

In the definition of the discernability matrix, when $|C_D(i, j)| = 1$, the attribute in $C_D(i, j)$ is one of the core attribute sets. All of attributes in $C_D(i, j)$ where $|C_D(i, j)| = 1$ consist of the core attribute set, which may be null. $C_D(i, j) = 0$ when $C_D(i, j)$ contains a core attribute. Then, a new simple matrix can be obtained.

$$\begin{bmatrix} C_{11} & C_{12} & \cdots & C_{1n} \\ C_{21} & C_{22} & \cdots & C_{2n} \\ \cdots & \cdots & \cdots & \cdots \\ C_{n1} & C_{n2} & \cdots & C_{nn} \end{bmatrix}$$

We can get:

$$L_{ij} = \bigvee_{a_i \in C_{ij}} a_i \qquad (7)$$

$$L = \bigwedge_{C_{ij} \neq 0, C \neq \Phi} L_{ij} \qquad (8)$$

$$L' = \bigvee_i L_i \qquad (9)$$

The reduced attribute set is $L_i \cup Core(A)$. Here, $Core(A)$ denotes the core attribute set. For finding reduces, the decision-relative discernability matrix is of more interest. This only considers those link discernabilities that occur when the corresponding routing attributes differ.

From Table 1, the discernability matrix $M[C_D(i, j)]_{n \times n}$ can be given as follows, (see Box 1).

In this matrix, $|C_D(4, 6)| = 1$, $|C_D(6, 7)| = 1$, we can get $Core(A) = \{a_1, a_4\}$. The decision-relative discernability matrix found in Table 1 is produced. For example, it can be seen from the table that objects 3 and 8 differ in each attribute. Although some attributes in objects 1 and 2 differ, their corresponding decisions are the same, so no entry appears in the decision-relative matrix. Grouping all entries containing single attributes forms the core of the data sets. Here, the core of the set is $\{a_1, a_4\}$. This means that a_1 and a_4 are the most important routing attributes. When $C_D(i, j)$ contains a_1 or a_4, then we set $C_D(i, j) = 0$.

The reduction of the data sets can be derived by converting the previous expression from conjunctive normal form to disjunctive normal form. Although this is guaranteed to discover all minimal subsets, it is a costly operation that

Box 1.

$$\begin{bmatrix} 0 & a_3a_4 & a_1a_3 & a_1a_2 & a_2a_3 & a_1a_2a_4 & a_1a_2a_4 & a_1a_2a_4 \\ & 0 & a_1a_4 & a_1a_2a_3a_4 & a_2a_3a_4 & a_1a_2a_3a_4 & a_1a_2a_3a_4 & a_1a_2a_3a_4 \\ & & 0 & a_2a_3 & a_1a_2a_3 & a_2a_3a_4 & a_1a_2a_3a_4 & a_1a_2a_3a_4 \\ & & & 0 & a_1a_3 & a_4 & a_1a_4 & a_1a_2a_4 \\ & & & & 0 & a_1a_3a_4 & a_1a_3a_4 & a_1a_2a_3a_4 \\ & & & & & 0 & a_1 & a_1a_2a_4 \\ & & & & & & 0 & a_2a_4 \\ & & & & & & & 0 \end{bmatrix}$$

renders the method impractical for even medium-sized data sets. For most applications, a single minimal subset is required for data reduction. This has led to approaches that consider finding individual shortest prime implicates from the discernability function. A common method is to incrementally add those attributes that occur with the highest frequency in the function, removing any clauses containing the attributes, until all clauses are eliminated. However, even this does not ensure that a minimal subset is mined; the search could proceed down to non-minimal paths (Yang & Chiam, 2000).

Algorithm for Selection of QoS Routing Attributes

The following algorithm is proposed to classify link-rank to realize QOSPF.

The first stage is for data preprocessing.

- **Step 1:** From the historical routing record of the subnet and the QoS information about the links, the information system about the links can be built.
- **Step 2:** Then we can draw the discernability matrix about the information system and can conclude the routing-reduction table (Apostolopoulos et al., 1999).

1. $RT = \varphi$, $Core = \varphi$;
2. For every $b \in A$, compute the equivalent class U/P;
3. Construct discernability matrix $M(A) = \{C_D(i,j)\}_{n \times n}$, where $1 < i, j < n$.
4. Do loop, $\forall C_D(i,j) \in M(A)$, if$(| C_D(i,j) | == 1)$ then $Core = Core \cup \{C_D(i,j)\}$;
5. $RT = Core$;
6. $\forall Core$, We set $C_D(i,j) = 0$ when $a \in C_D(i,j)$

The second stage is for the mining process.

- **Step 3:** Do reduction of attribute based on the information system.
- **Step 4:** The logical rules can be concluded from routing-reduction table. Created rules of rough decisions are saved in rule set.
- **Step 5:** With the knowledge of QOSPF, the link with QOS attributes can be mined. The third stage is for obtaining the best path with the knowledge of QOSPF.

The link-state algorithm presented as follows is known as Dijkstra's algorithm based on on-demand computation of QoS paths, which is described to illustrate how the algorithm can select a minimum-delay path with a maximum bandwidth. Some researchers proposed the delay

Exhibit A.

```
Initialization:
for (each destination n in set of nodes in the network) do
begin
DT[n] =infinty;
BW[n] =undefined;
NB[n] =undefined;
end
DT[s] =0; /*the source node s */
BW[s]}=infinty;
Compute QoS routing paths:
S =the set that contains all node in the network;
while (S is not empty) do
begin
u=the node in S whose value in the field DT is minimum;
S =S-{u};
for (each node v adjacent to u) do
begin
if ((b(u, v))>= bandwidth requirement) and
 (DT[v]>DT[u]+d(u, v)) then
begin
DT[v] =DT[v]+d(u, v);
BW[v] =min{BW[u], b(u,v)};
if (the node u is the source node s) then
NB[v] =v;
else
NB[v] =NB[u];
end
end
end
```

and bandwidth attributes to be mined from these steps (Jonathan & Guo, 2002). (See Exhibit A.)

In the previous steps, $b(i, j)$ denotes the available bandwidth on the edge between nodes i and j. Notation $d(i, j)$ denotes the propagation delay on the edge between nodes i and j. Notation BW is the maximum available bandwidth on a path between the source s and destination n. Notation DT is the minimal delay on a path between source s and destination n. Notation NB is the associated routing information.

With the help of data mining based on rough-set theory, the simplification of the routing table is to simplify the condition attributes in the routing table; after that, the routing-attribute reduction possesses the ability of the whole-routing-attribute table before simplification but possesses more important condition attributes (Zhang, 2003).

The approaches previously described are based on Dijkstra's shortest-path algorithms. The Dijkstra algorithm traditionally has been considered more efficient than standard shortest-path computations because of its lower worst-case complexity. The benefit of using Dijkstra's algorithm in QoS path selection has a greater synergy with the existing OSPF implementation. On-demand path computation of Dijkstra-based routing-attribute mining provides advantages in yielding better routes and minimizing the need for storage of data structures, if there are reduced routing attributes for QoS paths (Jonathan & Guo, 2002). The asymptotic worst-case complexity of an implementation of Dijkstra's algorithm is $O(E \log N)$, where N is the number of nodes in the network graph, and E is the number of the edges. The complexity of this rough-set-processing computation is $O(M*N+M^2)$, where M is the number of attributes.

CASE STUDY

Figure 1 shows a weighted graph model as an example of subnet's link to validate the previous algorithm. The node denotes the router. The numbers on the links denote the link-num. Routing is determined by the links' QoS attribute parameters, such as available bandwidth, propagation delay, link jitter, bit-error ratio, and connection possibility. We presume that the standard of classification is the link rank that can be described by I, II, ... VI based on historical routing data. The routing

Figure 1. Subnet Topology

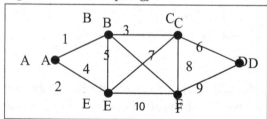

Table 2. Information base of links

algorithm is used to select the best path from A to D with the given QoS attributes.

Building the Information Base

Table 2 shows the information base from Figure 1. All of the values in the table denote the measurements of the attributes (attribute-weight abstract from the real world).

$S = (U, C \cup \{d\}, V, f)$ is an information system, where 'U' is a finite non-null set of link objects, and 'C' is a finite non-null set of link's QoS attributes. A link object has an IP address and participates in the QoS routing graph. Here,

$U = \{1, 2, 3, ..., 10\}$;
$C = \{c_1, c_2, c_3, c_4, c_5\}$;
$D = \{I, II, III, IV, VI\}$;
$V_1 = \{1, 2, 3, 4, 5, 6, 7, 8\}$;
$V_2 = \{1, 2, 3, 4, 5\}$;
$V_3 = \{0, 1, 2, 3, 4, 5\}$;
$V_4 = \{1, 2, 3, 4, 5, 6, 8\}$;
$V_5 = \{1, 2, 3\}$.

Table 2 expresses the function of the information 'f'.

Reduction Information Base

According to Formula (6), discernability matrix $M(C_D(i, j))$ in Table 2, can be given as follows, (see Box 2.)

C \ U	Available bandwidth(C_1)	Propagation delay(C_2)	Link jitter(C_3)	Bit error ratio(C_4)	Connection possibility(C_5)	Link rank(d)
1	4	4	1	5	3	IV
2	5	3	1	8	2	III
3	1	4	3	5	3	IV
4	1	4	3	5	1	VI
5	6	2	2	3	2	II
6	3	4	3	5	1	IV
7	1	5	1	3	1	V
8	3	1	0	3	3	I
9	7	5	4	1	3	III
10	8	1	1	2	3	I

Box 2.

$$\begin{bmatrix}
0 & c_1c_2c_4c_5 & 0 & c_1c_3c_5 & c_1c_2c_3c_4c_5 & 0 & c_1c_2c_4c_5 & c_1c_2c_3c_4 & c_1c_2c_3c_4 & c_1c_2c_4 \\
 & 0 & c_1c_2c_3c_4c_5 & c_1c_2c_3c_4c_5 & c_1c_2c_3c_4 & c_1c_2c_3c_4c_5 & c_1c_2c_4c_5 & c_1c_2c_3c_4c_5 & c_1c_2c_3c_4c_5 & c_1c_2c_4c_5 \\
 & & 0 & c_5 & c_1c_2c_3c_4c_5 & c_1c_5 & c_2c_3c_4c_5 & c_1c_2c_3c_4 & c_1c_2c_3c_4 & c_1c_2c_3c_4 \\
 & & & 0 & c_1c_2c_3c_4c_5 & c_1 & c_2c_3c_4 & c_1c_2c_3c_4c_5 & c_1c_2c_3c_4c_5 & c_1c_2c_4c_5 \\
 & & & & 0 & c_1c_2c_3c_4c_5 & c_1c_2c_3c_5 & c_1c_2c_3c_5 & c_1c_2c_3c_4c_5 & c_1c_2c_3c_4c_5 \\
 & & & & & 0 & c_1c_2c_4 & c_2c_3c_4c_5 & c_1c_2c_3c_4c_5 & c_1c_2c_4c_5 \\
 & & & & & & 0 & c_1c_2c_3c_5 & c_1c_3c_4c_5 & c_1c_2c_4c_5 \\
 & & & & & & & 0 & c_1c_2c_3c_4 & c_1c_3c_4 \\
 & & & & & & & & 0 & c_1c_2c_3c_4 \\
 & & & & & & & & & 0
\end{bmatrix}$$

Discernability matrix is used to the same partition of the data as the whole set of attributes V. To do this, one has to construct the so-called discernability function. This is a Boolean function constructed in the following way (Lu & Zhang, 2005).

For each element $C_{ij} (C_{ij} \neq 0)$, which is not empty, the disjunctive normal form (*DNF*) logic expression Formula (7) can be derived. For example:

$$L_{1,2} = c_1 \vee c_2 \vee c_4 \vee c_5;$$
$$L_{1,4} = c_1 \vee c_3 \vee c_5;$$
$$L_{2,5} = c_1 \vee c_2 \vee c_3 \vee c_4 \vee_5;$$
$$\cdots$$
$$L_{9,10} = c_1 \vee c_2 \vee c_3 \vee c_4$$

Do the conjunctive operation of each *DNF* and get the conjunctive normal form (*CNF*) logic expression Formula (8). In this step, *CNF* of each part in the matrix then can be calculated:

$$CNF_1 = c_1 \vee c_2 \vee c_3 \vee c_4 \vee c_5;$$
$$CNF_2 = c_1 \vee c_2 \vee c_3 \vee c_4$$
$$CNF_3 = c_1 \vee c_2 \vee c_3 \vee c_4 \vee c_5;$$
$$CNF_4 = (c_1 \vee c_2 \vee c_4 \vee c_5) \wedge c_5 = (c_1 \wedge c_5) \vee (c_2 \wedge c_5)$$
$$CNF_5 = c_1 \vee c_2 \vee c_3 \vee c_4;$$

$$CNF_6 = c_1 \vee c_5;$$
$$CNF_7 = c_2 \vee c_5;$$
$$CNF_8 = c_1 \vee c_3;$$
$$CNF_9 = c_1 \vee c_3 \vee c_4;$$
$$CNF_{10} = c_1 \vee c_4.$$

According to Formula (9), the *CNF* of L can be transformed into a new *DNF*, $L' = (c_1 \wedge c_2 \wedge c_5) \vee (c_1 \wedge c_3 \wedge c_5) \vee (c_1 \wedge c_4 \wedge c_5)$. Each *CNF* part in *DNF* that stands for a routing-attribute reduces result, and all the routing attributes in *DNF* are necessary for routing selection. We obtain three reductions of routing attributes: $R_1 = \{c_1, c_2, c_5\}, R_2 = \{c_1, c_3, c_5\}$ and $R_3 = \{c_1, c_4, c_5\}$. The core of the attributes is $Core(A) = R_1 \cap R_2 \cap R_3 = \{c_1, c_2, c_5\} \cap \{c_1, c_3, c_5\} \cap \{c_1, c_4, c_5\} = \{c_1, c_5\}$.

On the other hand, in the previous matrix, $|C_D(3,4)| = 1$, $|C_D(4,6)| = 1$, we also can get $Core(A)$. This means that a_1 and a_4 are the most important routing attributes. When $C_D(i, j)$ contains c_1 or c_5, then we set $C_D(i, j) = 0$.

Now a new simple matrix can be obtained, (see Box 3.)

The reduced routing-attribute set can be $(c_1 \wedge c_2 \wedge c_5) \vee (c_1 \wedge c_3 \wedge c_5) \vee (c_1 \wedge c_4 \wedge c_5)$. It means that the reduction attributes can be $c_1 \wedge c_2 \wedge c_5, c_1 \wedge c_3 \wedge c_5, c_1 \wedge c_4 \wedge c_5$. We now take

Box 3.

$$\begin{bmatrix} 0 & 0 & 0 & 0 & 0 & 0 & 0 & & 0 & & 0 & 0 \\ & 0 & 0 & 0 & 0 & 0 & 0 & & 0 & & 0 & 0 \\ & & 0 & 0 & 0 & 0 & 0 & & 0 & & 0 & 0 \\ & & & 0 & 0 & 0 & 0 & c_2 c_3 c_4 & 0 & 0 \\ & & & & 0 & 0 & 0 & & 0 & & 0 & 0 \\ & & & & & 0 & 0 & & 0 & & 0 & 0 \\ & & & & & & 0 & & 0 & & 0 & 0 \\ & & & & & & & & 0 & & 0 & 0 \\ & & & & & & & & & & 0 & 0 \\ & & & & & & & & & & & 0 \end{bmatrix}$$

only $c_1 \wedge c_2 \wedge c_5$ as an example, and then obtain Table 3.

Logic Rules

We can obtain the associated equivalent relation $IND(P)$ in the result of reduction:

$IND(c_1) = \{\{4,7\},\{6,8\},\{1\},\{2\},\{5\},\{9\},\{10\}\}$
$IND(c_2) = \{\{1,3,4,6\},\{2\},\{7,9\},\{8,10\}\}$
$IND(c_5) = \{\{1,3,8,9,10\},\{2,5\},\{4,6,7\}\}$
$IND(d) = \{\{1,3,6\},\{2,9\},\{4\},\{5\},\{7\},\{8,10\}\}$

In Table 3, we consider the classification of each object on routing attributes, respectively, and then check whether the intersection of arbitrary classifications is empty or not. If intersection is not empty, the value of attribute is the core value in this object; otherwise, there is no core value in this object, which is denoted by "-" (Starzyk, Nelson, & Sturtz, 1999). We can obtain Table 4.

We find that in initial route information, Table 2, there are five attributes besides the decision attribute. In general, the router will record these attributes as historic data, so there are some attributes that are dispensable to "Best Path." After removing the redundant attributes by rough set theory, all the decisions are listed in Table 5. We can derive 12 rules based on this sample.

$(1)(c_1,3) \wedge (c_2,1) \wedge (c_5,3) \to (d,\text{I})$
$(2)(c_1,8) \wedge (c_2,1) \wedge (c_5,3) \to (d,\text{I})$
$(2')(c_2,1) \wedge (c_5,3) \to (d,\text{I})$
$(3)(c_1,6) \wedge (c_2,2) \wedge (c_5,2) \to (d,\text{II})$
$(4)(c_1,5) \wedge (c_2,3) \wedge (c_5,2) \to (d,\text{III})$
$(5)(c_1,7) \wedge (c_2,5) \wedge (c_5,3) \to (d,\text{III})$
$(6)(c_1,4) \wedge (c_2,4) \wedge (c_5,3) \to (d,\text{VI})$
$(7)(c_1,1) \wedge (c_2,4) \wedge (c_5,3) \to (d,\text{VI})$
$(7')(c_2,4) \wedge (c_5,3) \to (d,\text{VI})$
$(8)(c_1,3) \wedge (c_2,4) \wedge (c_5,1) \to (d,\text{IV})$
$(9)(c_1,1) \wedge (c_2,5) \wedge (c_5,1) \to (d,\text{V})$
$(10)(c_1,1) \wedge (c_2,4) \wedge (c_5,1) \to (d,\text{VI})$

Table 3. The decision table

U \ C	Available bandwidth(C_1)	Propagation delay(C_2)	Connection possibility(C_5)	Link rank(d)
1	4	4	3	IV
2	5	3	2	III
3	1	4	3	IV
4	1	4	1	VI
5	6	2	2	II
6	3	4	1	IV
7	1	5	1	V
8	3	1	3	I
9	7	5	3	III
10	8	1	3	I

Table 4. The reduction table

C \ U	Available bandwidth(C_1)	Propagation delay (C_2)	Connection possibility(C_5)	Link rank (d)
1	-	4	3	IV
2	5	3	2	III
3	-	4	3	IV
4	1	4	1	VI
5	6	2	2	II
6	3	4	1	IV
7	1	5	1	V
8	-	1	3	I
9	7	5	3	III
10	-	1	3	I

Table 5. All the possible decisions table

C \ U	Available bandwidth(C_1)	Propagation delay(C_2)	Connection possibility(C_5)	Link rank(d)
1	4	4	3	IV
1'	-	4	3	IV
2	5	3	2	III
3	1	4	3	IV
3'	-	4	3	IV
4	1	4	1	VI
5	6	2	2	II
6	3	4	1	IV
7	1	5	1	V
8	3	1	3	I
8'	-	1	3	I
9	7	5	3	III
10	8	1	3	I
10'	-	1	3	I

The derived rules can be applied to a great number of data in order to distinguish the link into six ranks. Simplification of routing attributes needs not select all routing attributes for keeping the consistence of the routing table. That is to say, some of the link attributes are not to be selected in Dijkstra's algorithm based on on-demand computation of QoS paths. For instance, to the link *Num*.8 and link *Num*.10, route selection only would consider the propagation delay and connection possibility. We found that the link attribute with the same rows could decide the same decision as before. For the compatible data table, the most excellent attribute set can be mined by rough set theory.

The links of the subnet are classified in six ranks. Based on the Dijkstra algorithm (see www.dataconnection.com/network/default.htm) and the mined link-rank attributes, we can find

Figure 2. The rank of link

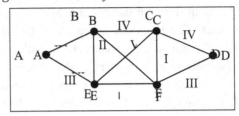

the best path from nodes A to D in Figure 2 as $A \rightarrow E \rightarrow F \rightarrow D$. The routing path is the best on condition that bandwidth is sufficient as well as available, and bit error ratio is acceptable.

In the context of QoS path selection, potential benefits of the previous algorithm are even more apparent. As mentioned before, efficient selection of a best path for flows with QoS requirements usually cannot be handled using a single-attribute optimization routing criterion. While multi-attribute path selection is used to be an intractable problem, the routing algorithm based on attribute mining and rough set to handle an important attribute that is reflective of network resource at no more additional cost of complexity. The corresponding asymptotic worst-case complexity is $O(E \log N) + O(M * N + M^2)$.

CONCLUSION

The complication of traditional solution schemes to QoS Routing is NP completeness. A data mining method for QoS routing based on rough set theory has been presented in this article. Based on QoS routing concepts and rough set theory, we studied rule-mining algorithms for selecting routing attributes and routing data reduction. The case study has shown that the method is sound in selecting the best route in networks. In this article, link data are classified into different ranks with rough set theory according to QoS attributes, and QOSPF based on QoS routing can be realized. Rough-set theory can be applied to deal with link data with given QoS attributes. Rough-set theory also can be applied to rapidly rank QoS attributes in terms of their significance in path selections. The proposed approach of QOSPF based on rough-set theory can select a path with QoS parameters and can offer QoS-based services in Internet routing. For further work, run-time analysis with changes of connectivity and diameter of connection graphs should be considered. Moreover, the tests in a real network environment also may be considered.

ACKNOWLEDGMENT

The authors thank Dr. Y.Y. Yao and Dr. J.T. Yao for their help during the first author's visit in Canada. Y. Liu also wishes to express his sincere thanks to all those who have worked or are currently working with him and for those helpful discussions with Dr. Wang Guoyin and Dr. Wu Yu. The authors thank the referees for their useful suggestions.

The Natural Science Foundation of CQUPT, CSTC under Grant No.2005BB2060, and the Natural Science Foundation of CQMEC under Grant No.KJ050507 supported the work.

REFERENCES

Apostolopoulos, G., et al. (1999). *RFC 2676*. IETF. Retrieved from ftp://ftp.chu.edu.tw/pub/Documents/RFC/rfc2676.txt

Bruno, A. (2003). CCIE #2738. CCIE routing and switching exam certification guide.

Chickering, D.M. (1996). *Learning Bayesian networks is NP-complete*. Berlin: Springer-Verlag.

Crawley, E., et al. (1998). *RFC 2386: A framework for QoS-based routing in the Internet*. IETF. Retrieved from ftp://ftp.chu.edu.tw/pub/Documents/RFC/rfc2386.txt

Jensen, R., & Shen, Q. (2004). Semantics-preserving dimensionality reduction: Rough and fuzzy-rough based approaches. *IEEE Transactions on Knowledge and Data Engineering, 16,* 1457-1471.

Jonathan, H., & Guo, X. (2002). *Quality of service control in high-speed network*. Beijing: Tsinghua University Press.

Kusiak, A. (2000). Decomposition in data mining: An industrial case study. *IEEE Trans, Electron, Packag, and Manufact., 23,* 345-353.

Kusiak, A. (2001). Rough set theory: A data mining tool for semiconductor manufacturing. *IEEE Transactions on Electronics Packaging Manufacturing, 24,* 44-50.

Li, L., Tang, B., Ni, Z., & Yang, T. (2002). A study on spatial attribute data mining based on rough set. In *Proceedings of the International Conference on Machine Learning and Cybernetics* (pp. 743-746).

Li, P., Zheng, H., & Nahavandi S. (2003). The application of rough set and Kohonen network to feature selection for object extraction. In *Proceedings of the Second International Conference on Machine Learning and Cybernetics* (pp. 1185-1189).

Li, Y. (2003). Classification of clients in client relationship management base on rough set theory. In *Proceedings of the 2nd International Conference on Machine Learning and Cybernetics* (pp. 242-246).

Liu, H., Xu, M., Xu, K., & Cui, Y. (2003). Research on Internetwork routing protocol: A survey. *Science of Telecommunications, 19,* 28-32.

Lu, X., & Zhang, J. (2005). A new method of decision rule mining for IT project risk management based on rough set. In *Proceedings of the International Conference on Services Systems and Services Management* (pp. 1022-1026).

Parthasarathy, S. (2001). Towards network-aware data mining. In *Proceedings of the 15ᵗʰ International Parallel and Distributed Processing Symposium* (pp. 157-161).

Pawlak, Z. (1991). *Theoretical aspects of reasoning about data.* Boston: Kluwer Academic Publishers.

Starzyk, J.A., Nelson, D.E., & Sturtz, K. (1999). Reduct generation in information systems. *Bulletin of the International Rough Set Society, 3,* 19-22.

Yang, Y., & Chiam, T.C. (2000). Rule discovery based on rough set theory. In *Proceedings of the Third International Conference.*

Zhang, Y. (2003). Rough set and genetic algorithms in path planning of robot. In *Proceedings of the Second International Conference on Machine Learning and Cybernetics* (pp. 698-701).

This work was previously published in International Journal of Data Warehousing and Mining, Vol. 2, Issue 3, edited by D. Taniar, pp. 27-41, copyright 2006 by Idea Group Publishing (an imprint of IGI Global).

338

Compilation of References

Aamodt, A., & Plazas, E. (1994). Case-based reasoning: Foundational issues, methodological variations, and system approaches. *AI Comm., 7*, 39-52.

Abello, J., Pardalos, P. M., & Resende, M. G. C. (Eds). (2002). *Handbook of massive data sets.* Kluwer Academic Publishers.

Agraval, R., & Srikant, R. (1994). Fast algorithms for mining association rules. In *Proceedings of the 20th VLDB Conference.* Santiago, Chile.

Agrawal, R., & Srikant, R. (1995). Mining sequential patterns. In *Proceedings of the International Conference on Data Engineering,* (pp. 3-14). Taipei, Taiwan.

Agrawal, R., Mannila, H., Srikant, R., Toivonen, H., & Verkamo, A. I. (1995). Fast discovery of association rules. In *Advances in knowledge discovery and data mining.* Cambridge: AAAI/MIT Press.

Ali, S. & Smith, K.A. (2003). Automatic parameter selection for polynomial kernel. In *Proceedings of the IEEE International Conference on Information Reuse and Integration,* USA (pp. 243-249).

Ali, S. & Smith, K.A. (2004a). Laplace kernel with automatic smoothing parameter estimation for support vector machine. *Computational Management Science (submitted).*

Ali, S. & Smith, K.A. (2004b). Automatic kernel selection for support vector machines. *Neurocomputing (submitted).*

Allwein, E. L., Schapire, R. E., & Singer, Y. (2000). Reducing multiclass to binary: A unifying approach for margin classifiers. *Journal of machine learning research, 1*, 113-141.

Altschul, S. F., Gish, W., Miller, W., Myers, E., & Lipman, D. J. (1990). Basic local alignment search tool. *Journal of Molecular Biology, 215*, 403-410.

Altschul, S. F., Madden, T. L., Schäffer, A. A., Zhang, J., Zhang, Z., Miller, W., et al. (1997). Gapped BLAST and PSI-BLAST: A new generation of protein database search programs. *Nucleic Acids Res., 25*, 3389-3402.

Amari, S.-I. & Wu, S. (1999). Improving support vector machine classifiers by modifying kernel functions. *Neural Networks, 12*, 783-789.

Amato, R., Ciaramella, A., Deniskina, N., Del Mondo, C., di Bernardo, D., Donalek, C., Longo, G., Mangano, G., Miele, G., Raiconi, G., Staiano, A., & Tagliaferri, R. (2006). A multi-step approach to time series analysis and gene expression clustering. *Bioinformatics, 22*(5), 589-596.

An, A., & Cercone, N. (1999). Discretization of continuous attributes for learning classification rules. In *Proceedings of the Third Pacific-Asia Conference on Methodologies for Knowledge Discovery and Data Mining* (pp. 509-514).

Andersen, S. K., Olesen, K. G., & Jensen, F. V. (1990). HUGIN—A shell for building Bayesian belief universes for expert systems. In G. Shafer & J. Pearl (Eds.), *Readings in uncertain reasoning* (pp. 332-337). San Francisco, CA: Morgan Kaufmann Publishers.

Andreeva, A., Howorth, D., Brenner, S. E., Hubbard, T. J., Chothia, C., & Murzin, A. G. (2004). SCOP database in 2004: Refinements integrate structure and sequence family data. *Nucleic Acids Res., 32*, D226–D229.

Anshakov, O.M., Finn, V.K., & Skvortsov, D.P. (1989). On axiomatization of many-valued logics associated with formalization of plausible reasoning. *Studia Logica, 42*(4), 423-447.

Apostolopoulos, G., et al. (1999). *RFC 2676*. IETF. Retrieved from ftp://ftp.chu.edu.tw/pub/Documents/RFC/rfc2676.txt

Archer, N., & Wang, S. (1993). Application of the back-propagation neural network algorithm with monotonicity constraints for two-group classification problems. *Decision Sciences, 24*, 60-75.

Arning, A., Agrawal, R., & Raghavan, P. (1996). A linear method for deviation detection in large databases. In *Proceeding of the Second International Conference on Knowledge Discovery and Data Mining* (pp. 164–169, Portland, Oregon.

ASE (2004). *The Athens Stock Exchange closing prices.* Retrieved March 26, 2004: http://www.ase.gr/content/en/MarketData/Stocks/Prices/default.asp

Auer, P., Holte, R.C., & Maass, W. (1995). Theory and applications of agnostic PAC-learning with small decision trees. In *Proceedings of the Eighth European Conference on Machine Learning* (pp. 21-29).

Bäck, T., Fogel, D. B., & Michalewicz, Z. (1997). *Handbook of evolutionary computation.* Oxford: Oxford University Press.

Bailey, T. L., & Elkan, C. P. (1995). Unsupervised learning of multiple motifs in biopolymers using EM. *Machine Learning, 21*(1-2), 51-80.

Bakirtzis, A., Biskas, P., Zoumas, C., & Petridis, V. (2002). Optimal power flow by enhanced genetic algorithm. *IEEE Transactions on Power Systems, 15*(1), 229-236.

Baldi, P., & Brunak, S. (1998). *Bioinformatics: The machine learning approach.* Cambridge, MA: MIT Press.

Baldi, P., Frasconi, P., & Smyth, P. (2003). *Modeling the internet and the WEB: Probabilistic methods and algorithms.* Chichester: Wiley.

Bandyopadhyay, S., & Maulik, U. (2002). Genetic clustering for automatic evolution of clusters and application to image classification. *Pattern Recognition, 35*, 1197-1208.

Baragona, R. (2001). A simulation study on clustering time series with metaheuristic methods. *Quaderni di Statistica, 3*, 1-26.

Bartnikowski, S., Granberry, M., Mugan, J., & Truemper, K. (2006). Transformation of rational and set data to logic data. In E. Triantaphyllou & G. Felici (Eds.), *Data mining and knowledge discovery approaches based on rule induction techniques.* Berlin: Springer-Verlag.

Bay, S.D. (2000). Multivariate discretization of continuous variables for set mining. In *Proceedings of the Sixth ACM SIGKDD International Conference on Knowledge Discovery and Data Mining* (pp. 315-319).

Bay, S.D., & Pazzani, M.J. (1999). Detecting change in categorical data: Mining contrast sets. In *Proceedings of the Fifth ACM SIGKDD International Conference on Knowledge Discovery and Data Mining* (pp. 302-306).

Bennett, K., & Mangasarian, O.L. (1992). Robust linear programming discrimination of two linearly inseparable sets. *Optimization Methods and Software, 1*, 23-34.

Bennett, K.P., Wu, S., & Auslender, L. (1999). On support vector decision trees for database marketing. In *Proceedings of the IEEE International Joint Conference on Neural Networks, IJCNN'99* (pp. 904-909).

Berger, J.O. (1985). *Statistical Decision Theory and Bayesian Analysis* (second edition). Springer-Verlag.

Berman, F. (2001). From TeraGrid to knowledge grid. *Communications of the ACM, 44*(11), 27-28.

Berman, F., Fox, G., & Hey, A., (Eds.). (2003). *Grid computing: Making the global infrastructure a reality.* Chichester, NY: Wiley.

Berry, M., & Linoff, G. (1997). *Data mining techniques for marketing, sales, and customer support.* New York: Wiley.

Bezdek, J.C. (1987). *Pattern recognition with fuzzy objective function algorithms.* New York: Plenum Press.

Bezdek, J. C., Keller, J., Krisnapuram, R., & Pal, N. R. (1999). *Fuzzy models and algorithms for pattern recognition and image processing.* Kluwer Academic Publisher.

Bigus, J.P. (1996). *Data Mining with Neural Networks: Solving Business Problems.* McGraw-Hill.

Bishop, C. (1995). *Neural networks for pattern recognition.* Oxford: Oxford University.

Bishop, C. M. (1999). Latent variable models. In M. I. Jordan (Ed.), *Learning in graphical models* (pp. 371-403). MIT Press.

Bishop, C. M., Svensen, M., & Williams, C. K. I. (1998). GTM: The generative topographic mapping. *Neural Computation, 10*(1), 215-235.

Bishop, C.M. (Ed.). (2004). *Neural network in pattern recognition.* New York: Oxford University Press.

Blaheta, D. (2002). Handling noisy training and testing data. In *Proceedings of the 7th Conference on Empirical Methods in Natural Language Processing* (pp. 111-116). Philadelphia, PA.

Blake, C. & Merz, C.J. (2002). UCI repository of machine learning databases. Retrieved from *http://www.ics.uci.edu/~mlearn/MLRepository.html*

Blake, C., & Merz, C. (1998). *UCI repository of machine learning databases.* Department of Information and Computer Sciences, University of California, Irvine Retrieved from http://www.ics.uci.edu/~mlearn/MLRepository.html

Bodt, B., Forester, J., Hansen, C., Heilman, E., Kaste, R., & O'May, J. (2002). Data mining combat simulations: a new approach for battlefield parameterization. *Army Research Conference*, Orlando FL, December 2002.

Boldyrev, N.G. (1974). Minimization of Boolean partial functions with a large number of "Don't Care" conditions and the problem of feature extraction. *Proceedings of International Symposium "Discrete Systems"* (pp.101-109). Riga, Latvia.

Boom 2.3. Retrieved from http://vlsicad.eecs.umich.edu

Booth, R. C., & Roland, W. B. (1991). Neural network-based combustion optimization reduces NOx emissions while improving performance. In *Proceedings of the 1991 IEEE*

Workshop on Dynamic Modeling Control Applications for Industry Applications (pp. 1-6).

Boros, E., Hammer, P. L., Ibaraki, T., & Kogan, A. (1997). A logical analysis of numerical data. *Mathematical Programming. 79*, 163-190.

Boros, E., Hammer, P. L., Ibaraki, T., Kogan, A., Mayoraz, E., & Muchnick, I. (2000). An implementation of Logical Analysis of Data. *IEEE Transactions on Knowledge and Data Engineering. 12*, 292-306.

Boros, E., Ibaraki, T., & Mikino, K. (1998). Error-free and best-fit extensions of partially defined Boolean Functions. *Information and Computation, 140*, 254-283.

Boser, B.E., Guyon, I., & Vapnik, V.N. (1992). A training algorithm for optimal margin classifiers. In *Proceedings of the Fifth Annual Workshop of Computational Learning Theory*, Pittsburgh, Pennsylvania (pp. 144-152).

Boulle, A., Chandramohan, D., & Weller, O. (2001). A case study of using artificial neural networks for classifying cause of death from autopsy. *International Journal of Epidemiology, 30*, 515-520.

Bowman, D. (1999). Maintaining spatial orientation during travel in an immersive virtual environment. *Presence, 8*(6), 618-631.

Box, D., et al. (2004). *Web Services Addressing (WS-Addressing).* Retrieved September 14, 2004, from http://www.w3.org/Submission/2004/SUBM-ws-addressing-20040810

Bredensteiner, E.J., & Bennett, K. P. (1999). Multicategory classification by support vector machines. *Computational Optimization and Application, 12*, 53-79.

Breiman, L. (1996). Bagging predictors. *Machine Learning, 26*(2), 123-140.

Breiman, L., Friedman J.H., Olshen R.A., & Stone, C.J. (1984). *Classification and regression trees.* Wadsworth International.

Brodley, C. E., & Friedl, M. A. (1996). Identifying and eliminating mislabeled training instances. In *Proceedings* of the 13th *National Conference on Artificial Intelligence* (pp. 799-805). Portland, OR: AAAI Press.

Brodley, C. E., & Friedl, M. A. (1999). Identifying mislabeled training data. *Journal of Artificial Intelligence Research, 11*, 131-167.

Brown, M., Grundy, W., Lin, D., Cristianini, N., Sugnet, C., Furey, T., Ares, M. Jr. & Haussler, D. (2000). Knowledge-based analysis of microarray gene expression data by using support vector machines. *Proceedings of the National Academy of Science USA, 97*(1), 262–267.

Bruno, A. (2003). CCIE #2738. CCIE routing and switching exam certification guide.

Bueti, G., Congiusta, A., & Talia, D. (2004, June). *Developing distributed data mining applications in the Knowledge Grid framework.* Paper presented at the 6th International Conference on High Performance Computing for Computational Science (VECPAR 2004), Valencia, Spain.

Burdea, G., & Coiffet, P. (2003). *Virtual reality technology.* NJ: John Wiley and Sons.

Burges, C. J. C. (1998). A tutorial on support vector machines for pattern recognition. *Data Mining and Knowledge Discovery, 2*, 121-167.

Cannataro, M., & Talia, D. (2003). The Knowledge Grid. *Communications of the ACM, 46*(1), 89-93.

Cannataro, M., Comito, C, Congiusta, A., & Veltri, P. (2004). PROTEUS: A bioinformatics problem solving environment on grids. *Parallel Processing Letters, 14*(2), 217-237.

Cannataro, M., Congiusta, A., Pugliese, A., Talia, D., & Trunfio, P. (2004). Distributed data mining on grids: Services, tools, and applications. *IEEE Transactions on Systems, Man, and Cybernetics, Part B, 34*(6), 2451-2465.

Cannataro, M., Congiusta, A., Talia, D., & Trunfio, P. (2002, September). *A data mining toolset for distributed high-performance platforms.* Paper presented at the 3rd International Conference on Data Mining Methods and Databases for Engineering, Finance and Others Fields (Data Mining 2002), Bologna, Italy.

Carlos, S., Pavel, B., & Brazdil, P. (2004). A meta-learning method to select the kernel width in support vector regression. *Machine Learning, 54*, 195-209.

Carpineto, C., & Romano, G. (1996). A lattice conceptual clustering system and its application to browsing retrieval. *Machine Learning, 24*, 95-122.

Castelli, V., Hutchins, S. T., Li, C.-S., & Turek, J. J. E. (2001). *Modifying an unreliable training set for supervised classification.* United States Patent 6,298,351.

Ceri, C., Gotlob, G., & Tanca, L. (1990). *Logic programming and databases.* Springer.

Chakrabarti, S. (2003). *Mining the web: Discovering knowledge from hypertext data.* New York: Morgan Kaufmann.

Chan, P., Fan, W., Prodromidis, A., & Stolfo, S. (1999). Distributed data mining in credit card fraud detection. *IEEE Intelligent Systems, 14*(6), 67-74.

Chang, C.-C., & Lin, C.-J. (2001). *LIBSVM: A library for support vector machines* [Computer software and manual]. Retrieved May 11, 2005, from http://www.csie.ntu.edu. tw/~cjlin/libsvm

Chang, K. (2000). *Nonlinear dimensionality reduction using probabilistic principal surfaces.* Unpublished PhD Dissertation. The University of Texas at Austin, USA.

Chang, K., & Ghosh, J. (2001). A unified model for probabilistic principal surfaces. *IEEE Transac tions on Pattern Analysis and Machine Intelligence, 23*(1), 22-41.

Chapelle, O., Vapnik, V., Bousquet, O., & Mukherjee, S. (2002). Choosing multiple parameters for support vector machines. *Machine Learning, 46*(1), 131-159.

Charitos, D. (1997). Designing space in VE's for aiding wayfinding behaviour. In *Proceedings of the 4th UK Virtual Reality Special Interest Group Conference, Brunel University.*

Chen, J., & Stanney, K. (1999). A theoretical model of wayfinding in virtual environments: Proposed strategies for navigational aiding. *Presence, 8*(6), 671-685.

Chen, M.S., Han, J., & Yu, P.S. (1996). Data mining: An overview from a database perspective. *IEEE Transactions on Knowledge and Data Engineering, 8*(6), 866-883.

Cheung, K.-W., Kwok, J. T., Law, M. H., & Tsui, K.-C. (2003). Mining customer product ratings for personalized marketing. *Decision Support Systems, 35*(2), 231-243.

Chickering, D.M. (1996). *Learning Bayesian networks is NP-complete.* Berlin: Springer-Verlag.

Chiş, M. (2000). A new evolutionary hierarchical clustering technique. In Babeş-Bolyai University, *Research Seminars, Seminar on Computer Science* (pp. 13-20).

Chiş, M., & Dumitrescu, D. (2002). Evolutionary hierarchical clustering for data mining. In *Proceedings of the Symposium "Zilele Academice Clujene", Computer Science Section, 14-22 June 2002, Seminar on Computer Science* (pp. 12-18).

Chong, A. Z. S., Wilcox, S. J., & Ward, J. (2000). Neural network models of the combustion derivatives emanating from a chain grate stoker fired boiler plant. In *Proceedings of the. 2000 IEEE Seminar Advanced Sensors and Instrumentation Systems for Combustion Processes* (pp. 61-64).

Christensen, E., Curbera, F., Meredith, G., & Weerawarana, S. (2001). *Web Services Description Language (WSDL) 1.1.* Retrieved September 14, 2004, from http://www.w3.org/TR/2001/NOTE-wsdl-20010315

Cios, K. J., Pedrycz, W., & Swiniarski, R. (1998). *Data mining methods for knowledge discovery.* The Netherlands: Kluwer, Dordrecht.

Cios, K., Pedrycz, W., & Swiniarski, R. (1998). *Data Mining Methods for Knowledge Discovery.* Kluwer Academic Publishers.

Coakley, J., & Brown, C. (2000). Artificial neural networks in accounting and finance: Modeling issues. *International Journal of Intelligent Systems in Accounting, Finance and Management, 9*, 119-144.

Conroy, R. (2001). *Spatial navigation in immersive virtual environments.* Ph.D. Thesis, University College London.

Cooper, G. F. (1984). *NESTOR: A computer-based medical diagnostic aid that integrates causal and probabilistic knowledge.* Ph.D. Thesis, Medical Information Sciences, Stanford, CA: Stanford University.

Cortes C. & Vapnik, V. (1995). Support vector networks. *Machine Learning, 20*, 273-297.

Cortes, C., & Vapnik, V. (1995). Support vector networks. *Machine Learning, 20*, 273-297.

Cosmadakis, S., Kanellakis, P.C., & Spyratos, N. (1986). Partition semantics for relations. *Journal of Computer and System Sciences, 33*(2), 203-233.

Cover, T. M., & Hart, P. E. (1967). Nearest neighbor pattern classification. *IEEE Transactions on Information Theory,* IT-13, 21-27.

Crama, Y., & Hammer, P. L. (2002). *Boolean functions— Theory, algorithms, and applications.* Retrieved from http://www.rogp.hec.ulg.ac.be/Crama/

Crama, Y., Hammer, P.L., & Ibaraki, T. (1988). Cause-effect relationships and partially defined Boolean functions. *Annals of Operations Research. 16*, 299-326.

Craven, M. W., Mural, R. J., Hauser, L. J., & Uberbacher, E. C. (1995). Predicting protein folding classes without overly relying on homology. *ISMB, 3*, 98–106.

Craven, M.W. & Shavlik, J.W. (1997). Using neural networks in data mining. *Future Generation Computer Systems, 13*, 211-229.

Crawley, E., et al. (1998). *RFC 2386: A framework for QoS-based routing in the Internet.* IETF. Retrieved from ftp://ftp.chu.edu.tw/pub/Documents/RFC/rfc2386.txt

Cristianini, N., & Shawe-Taylor, J. (2000). *An introduction to support vector machines and other kernel-based learning methods.* Cambridge, UK: Cambridge University Press.

Cronin, J.J,. & Taylor, A. (1992). Measuring service quality: A reexamination and extension. *Journal of Marketing, 56*(3), 55-68.

Curcin, V., Ghanem, M., Guo, Y., Kohler, M., Rowe, A., Syed, J., & Wendel, P. (2002, July). *Discovery Net: Towards a grid of knowledge discovery*. Paper presented at the 8th International Conference on Knowledge Discovery and Data Mining (KDD 2002), Edmonton, Canada.

Czajkowski, K., et al. (2004). *From open grid services infrastructure to WS-resource framework: Refactoring & evolution*. Retrieved September 14, 2004, from http://www.globus.org/wsrf/specs/ogsi_to_wsrf_1.0.pdf

D'Andrade, R. (1978). U-statistic hierarchical clustering. *Psychometrika, 4*, 58-67.

Dabholkar, P.A., Thorpe, D.I., & Rentz, J.O. (1996). A measure of service quality for retail stores: Scale development and validation. *Journal of the Academy of Marketing Science, 24*(1), 3-16.

Dandekar, T., Schuster, S., Snel, B., Huynen, M., & Bork, P. (1999). Pathway alignment: application to the comparative analysis of glycolytic enzymes. *Biochemical Journal, 343*, 115-124.

Darken, R., & Peterson, B. (2002). Spatial orientation, wayfinding, and representation. In K. Stanney (Ed.), *Handbook of virtual environments: Design, implementation, and applications* (pp. 493-518). Mahwah, NJ: Lawrence Erlbaum Associates.

Darken, R., & Sibert, J. (1996). Navigating large virtual spaces. *The International Journal of Human-Computer Interaction, 8*(1), 49-72.

Dasu, T., & Johnson, T. (2003). *Exploratory data mining and data cleaning*. Hoboken, NJ: Wiley Series in Probability and Statistics.

Davies, D. L., & Bouldin, D. W. (1979). A cluster separation measure. *IEEE Trans. Pattern Analysis and Machine Intelligence, 1*, 224-227.

Dawid, A. P., & Skene, A. M. (1979). Maximum likelihood estimation of observer error-rates using the EM algorithm. *Applied Statistics, 28*(1), 20-28.

DeCoste, D., & Schölkopf, B (2002). Training invariant support vector machines. *Machine Learning, 45*(1-3), 161-290.

Demetrovics J., & Vu, D.T. (1993). Generating Armstrong relation schemes and inferring functional dependencies from relations. *International Journal on Information Theory & Applications, 1*(4), 3-12.

Dempster, A. P., Laird, N. M., & Rubin, D. B. (1977). Maximum likelihood from incomplete data via the EM algorithm. *Journal of the Royal Statistical Society, Series B, 39*, 1-38.

Deshpande, M., & Karypis, G. (2000). *Selective Markov models for predicting Web-page accesses*. Retrieved from http://www.cs.umn.edu/~karypis

Dickinson, M. (2002). The Great Observatories Origins Deep Survey: An overview. Retrieved from http://www.stsci.edu/science/goods/abstract.html

Dietterich, T., & Bakiri, G. (1995). Solving multiclass learning problems via error-correcting output codes. *Journal of Artificial Intelligence Research, 2*, 263-286.

Dimeo, R., & Lee, K.Y. (1995). Boiler-turbine control system design using a genetic algorithm. *IEEE Transactions on Energy Conversion, 10*(4), 752-759.

Ding, C. H., & Dubchak, I. (2001). Multi-class protein fold recognition using support vector machines and neural networks. *Bioinformatics, 17*, 349-358.

Domencich, T., & Mc Fadden, D. (1975). *Urban travel demand, a behavioural analysis*. Amsterdam, Holland: North-Holland Publishing Company.

Dougherty, J., Kohavi, R., & Sahami, M. (1995). Supervised and unsupervised discretization of continuous features. In *Machine Learning: Proceedings of the Twelfth International Conference* (pp. 194-202).

Dowling, C.E. (1993). On the irredundant generation of knowledge spaces. *Journal of Math. Psych., 37*(1), 49-62.

Dubchak, I., Muchnik, I., Holbrook, S. R., & Kim, S. H. (1995). Prediction of protein folding class using global description of amino acid sequence. *Proc. Natl Acad. Sci. USA, 92*, 8700-8704.

Dubchak, I., Muchnik, I., Mayor, C., Dralyuk, I., & Kim, S. H. (1999). Recognition of a protein fold in the context of the Structural Classification of Proteins (SCOP) classification. *Proteins, 35*, 401-407.

Duda, R. O., Hart, P. E., & Stork, D. G. (2001). *Pattern classification (2nd ed.).* John Wiley & Sons.

Duda, R. O., Hart, P. E., & Stork, D. G. (2001). *Pattern classification.* New York: Wiley Interscience.

Duin, R.P.W. (1996). A note on comparing classifier. *Pattern Recognition Letters, 1*, 529-536.

Dumitrescu, D. (1999). *Mathematical foundations of the classification theory.* Bucharest, Romania: Romanian Academy Press.

Dumitrescu, D., Lazzerini, B., & Hui, L. (2000). Hierarchical data structure detection using evolutionary algorithms. *Studia Babeş-Bolyai University, Ser.Informatica.*

Dumitrescu, D., Lazzerini, B., Jain, L., & Dumitrescu, A. (1999). *Evolutionary computation.* Boca Raton, FL: C.R.C. Press.

Durbin, R., Eddy, S., Krogh, A., & Mitchison, G. (1998). *Biological sequence analysis: Probabilistic models of proteins and nucleic acids.* Cambridge, UK: Cambridge University Press.

Edwards, A.L. (1957). *Techniques of attitude scale construction.* New York: Appleton-Century-Crofts Inc.

Eiben, A. E., Hinterding, R., & Mizhalewicz, Z. (1999). Parameter control in evolutionary algorithms. *IEEE Transactions on Evolutionary Computation, 3*(2), 124-141.

Eisen, J. A. (2000). Horizontal gene transfer among microbial genomes: New insights from complete genome analysis. *Current Opinion in Genetics & Development, 10*, 606-611.

Eisen, M. B., Spellman, P. T., Brown, P. O., & Botstein, D. (1998). Cluster analysis and display of genome-wide expression patterns. *PNAS, 95,*14863–14868.

Elvins, T., Nadeau, D., & Kirsh, D. (2001). Worldlets: 3D thumbnails for wayfinding in large virtual worlds. *Presence, 10*(6), 565-582.

Eskin, E. (2000). Detecting errors within a corpus using anomaly detection. In *Proceedings of the 1st Conference of the North American Association for Computational Linguistics.* Seattle, WA.

Espresso II. Retrieved from http://www-cad.eecs.berkeley.edu

Evgeniou, T., Pontil, M., & Poggio, T. (2000). Regularization networks and support vector machines. *Advances in Computational Mathematics, 13*(1), 1-50.

Fagiuoli, E., & Zaffalon, M. (1998). 2U: An exact interval propagation algorithm for polytrees with binary variables. *Artificial Intelligence, 106*(1), 77-107.

Fayyad, U., Piatetsky-Shapiro, G., Smyth, P., & Uthurusamy, R. (1995). *Advances in knowledge discovery and data mining.* Cambridge, MA: AAAI/MIT Press.

Fayyad, U.M. & Irani, K.B. (1992). On the Handling of Continuous-Valued Attributes in Decision Tree Generation. *Machine Learning* 8, 87-102.

Fayyad, U.M., & Irani, K.B. (1993). Multi-interval discretization of continuous-valued attributes for classification learning. In *Proceedings of the Thirteenth International Joint Conference on Artificial Intelligence* (pp. 1022-1027).

Felici, G., & Truemper K. (2005). The Lsquare system for mining logic data. In *Encyclopedia of Data Warehousing and Mining* (vol. 2.) IGI Information Science Reference

Felici, G., & Truemper, K. (2000). A MINSAT approach for learning in logic domains. *INFORMS Journal on Computing, 13*, 1-17.

Felici, G., & Truemper, K. (2002). A minsat approach for learning in logic domains, *INFORMS Journal on Computing, 14*(1), 20-36.

Felici, G., Sun, F., & Truemper, K. (2004). Learning logic formulas and related error distributions. In E. Triantaphyllou & G. Felici (Eds.), *Data mining and knowledge discovery approaches based on rule induction techniques.* Berlin: Springer-Verlag.

Felici, G., Sun, K-S., & Truemper. K. (2006). Learning logic formulas and related error distributions. In G. Felici and E. Trintaphyllou (Eds.), *Data mining and knowledge discovery approaches based on rule induction techniques.* Springer.

Felsenstein, J. (1989). PHYLIP—Phylogeny Inference Package (Version 3.2). *Cladistics, 5,* 164-166.

Finn, V. K. (1984). Inductive models of knowledge representation in man-machine and robotics systems. *Proceedings of VINITI, Vol. A,* 58-76.

Finn, V. K. (1988). Commonsense inference and commonsense reasoning. *Review of Science and Technique (Itogi Nauki i Tekhniki), Series "The Theory of Probability. Mathematical Statistics. Technical Cybernetics," 28,* 3-84.

Finn, V. K. (1991). Plausible reasoning in systems of JSM type. *Review of Science and Technique (Itogi Nauki i Tekhniki), Series "Informatika," 15,* 54-101.

Finn, V. K. (1999). The synthesis of cognitive procedures and the problem of induction. *NTI, Series 2*(1-2), 8-44. Moscow, Russia: VINITI.

Fisher, R. A. (1936). The use of multiple measurements in taxonomic problems. *Annals of Eugenics, 7,* 179-188.

Forst, C. V., & Schulten, K. (2001). Phylogenetic analysis of metabolic pathways. *Journal of Molecular Evolution, 52,* 471-489.

Forsyth, D. A., & Ponce, J. (2003). *Computer vision: A modern approach.* Englewood Cliffs, NJ: Prentice Hall.

Foster, I., Kesselman, C., Nick, J. M., & Tuecke, S. (2002). *The physiology of the grid: An Open Grid Services Architecture for distributed systems integration.* Retrieved August 26, 2006, from http://www.globus.org/alliance/publications/papers/ogsa.pdf

Franceschini, F. (2001). *Dai prodotti ai servizi. Le nuove frontiere per la misura della qualità.* Torino, Italy: UTET.

Freed, N., & Glover, F. (1986). Evaluating alternative linear programming models to solve the two-group discriminant problem. *Decision Sciences, 17,* 151-162.

Fu, T.-C., Chung F.-L., Ng, V., & Luk, R. (2001). Pattern discovery from stock time series using self-organizing maps.

Paper presented at the *KDD 2001 Workshop on Temporal Data Mining,* August 26-29, San Francisco.

Fu, Z., Golden, B., Lele, S., Raghavan, S., & Wasil, E. (2003). A genetic algorithm-based approach for building accurate decision trees. *Informs Journal of Computing, 5,* 3-22.

Gaasterland, T., & Selkov, E. (1995). Reconstruction of metabolic networks using incomplete information. In *Proceedings of the Third International Conference on Intelligent Systems for Molecular Biology* (pp. 127-135). Menlo Park, CA: AAAI Press.

Gagneur, J., Jackson, D., & Casar, G. (2003). Hierarchical analysis of dependence in metabolic networks. *Bioinformatics, 19,* 1027-1034.

Gaivoronski, A., & Stella, F. (1998). Stochastic optimization with structured distributions: The case of Bayesian nets. *Annals of Operations Research, 81,* 189-211.

Galitsky, B. A., Kuznetsov, S. O., & Vinogradov, D. V. (2005). JASMINE: A hybrid reasoning tool for discovering causal links in biological data. Retrieved from http://www.dcs.bbk.ac.uk/~galitsky/*Jasmine*

Gama, J., Torgo, L., & Soares, C. (1998). Dynamic discretization of continuous attributes. In *Proceedings of the Sixth Ibero-American Conference on Artificial Intelligence* (pp. 160-169).

Gamberger, D., Lavrač, N., & Grošelj, C. (1999). Experiments with noise filtering in a medical domain. In *Proceedings of the 16th International Conference on Machine Learning* (pp. 143-151). Bled, Slovenia.

Ganascia, J. - Gabriel. (1989). EKAW - 89 tutorial notes: Machine learning. *Third European Workshop on Knowledge Acquisition for Knowledge-Based Systems* (pp. 287-296). Paris, France.

Ganter, B. (1984). *Two basic algorithms in concepts analysis* (FB4-Preprint, No. 831). TH Darmstadt.

Geman, S., & Geman, D. (1984). Stochastic relaxation, gibbs distribution, and the Bayesian restoration of images. *IEEE Transactions on Pattern Analysis and Machine Intelligence, 6*(6), 721-741.

Gentle, J.E. (2002). *Elements of computational statistics.* New York: Springer-Verlag.

Ghezelayagh, H., & Lee, K. Y. (2002). Intelligent predictive control of a power plant with evolutionary programming optimizer and neuro-fuzzy identifier. In *Proceedings of the 2002 IEEE World Congress on Computational Intelligence* (pp. 1-6).

Gimlin, D. R., & Ferrell, D. R. (1974). A *k-k'* error correcting procedure for nonparametric imperfectly supervised learning. *IEEE Transactions on Systems, Man, and Cybernetics, SMC-4*(3), 304-306.

Giraud-Carrier, C., & Martinez, T. (1994). An incremental learning model for commonsense reasoning. *Proceedings of the Seventh International Symposium on Artificial Intelligence (ISAI'94), ITESM* (pp. 134-141).

Giudici, P. (2003). *Applied data mining, statistical methods for business and industry.* London:Wiley.

Glover, F. (1990). Improved linear programming models for discriminant analysis. *Decision Sciences, 21,* 771-785.

Golay, X., Kollias, S., Stoll, G., Meier, D., Valavanis, A., & Boesiger, P. (1998). A new correlation-based fuzzy logic clustering algorithm for fMRI. *Magnetic Resonance in Medicine, 40,* 249-260.

Goldberg, D. E. (1989). *Genetic algorithms.* New York: Addison Wesley Longman, Inc,.

Goodrich, M., & Tamassia, R. (2001). *Algorithm design: Foundations, analysis, and Internet examples.* New York: Wiley.

Goutte, C., Toft, P., & Rostrup, E. (1999). On clustering fMRI time series. *Neuroimage, 9*(3), 298-310.

Gowda, K. C., & Krishna, G. (1979). Editing and error correction using the concept of mutual nearest neighborhood. In *Proceedings of the International Conference on Cybernetics and Society* (pp. 222-226). IEEE Press.

Grady, S. (2003). *Virtual reality: Simulating and enhancing the world with computers.* Facts on File Inc.

Grefenstette, J. J. (1986). Optimization of control parameters for genetic algorithms. *IEEE Transactions on Systems, Man, and Cybernetics,* SMC-16, 122-128.

Gribskov, M., McLachlan, A. D., & Eisenberg, D. (1987). Profile analysis: Detection of distantly related proteins. *Proc. Natl. Acad. Sci. USA, 84* (pp. 4355-4358).

Grossman, R., & Mazzucco, M. (2002). A web infrastructure for the exploratory analysis and mining of data. *IEEE Computing in Science and Engineering, 4*(4), 44-51.

Grundy,W. N., Bailey, T. L., Elkan, C. P., & Baker, M. E. (1997). Meta-MEME: Motif-based hidden Markov Models of biological sequences. *Computer Applications in the Biosciences, 13*(4), 397-406.

Gu, Y., & Grossman, R. (2003). SABUL: A transport protocol for grid computing. *Journal of Grid Computing, 1*(4), 377-386.

Guha, S., Rastogi, R., & Shim, K. (1998). CURE: An efficient clustering algorithm for large databases. In *Proccedings of ACG SIGMOD International Conference on Management of Data* (pp. 73-82).

Guha, S., Rastogi, R., & Shim, K. (2000). ROCK: A robust clustering algorithm for categorical attributes. *Information Systems, 25*(5), 345-366.

Gunn, S.R. (1998). Support vector machines for classification and regression. Southampton, UK: University of Southampton.

Gunther, R., Kazman, R., & Macgregor, C. (2004). Using 3D sound as a navigational aid in virtual environments. *Behaviour and Information Technology, 23*(6), 435-446.

Guruswami, V., & Sahai, A. (1999). Multiclass learning, boosting, and error-correcting codes. In *12th Annual Conference on Computational Learning Theory* (pp. 145-155).

Guyon, I., Matic, N., & Vapnik, V. (1996). Discovering informative patterns and data cleaning. In U. M. Fayyad, G. Piatetsky-Shapiro, P. Smyth, & R. Uthurusamy (Eds.), *Advances in Knowledge Discovery and Data Mining* (pp. 181-203). Menlo Park, CA: AAAI Press.

Guyon, I., Matić, N., & Vapnik, V. (1996). Discovering informative patterns and data cleaning. In U. M. Fayyad, G. Piatetsky-Shapiro, P. Smyth, & R. Uthurusamy (Eds.), *Advances in knowledge discovery and data mining* (pp. 181-203). AAAI/MIT Press.

Guyon, I., Weston, J., Barnhill, S., & Vapnik, V. (2002). Gene selection for cancer classification using support vector machines. *Machine Learning, 46*(1/3), 389-422.

Hall, L. O., Özyurt, B., & Bezdek, J. C. (1999). Clustering with a genetically optimized approach. *IEEE Transactions on Evolutionary Computation, 3*(2), 103-112.

Hammer, P.L. (1986). *Partially defined Boolean functions and cause-effect relationships.* Paper presented at the International Conference on Multi-Attribute Decision Making Via OR-Based Expert Sytems, University of Passau, Passau Germany.

Hampel, F. R., Rousseeuw, P. J., Ronchetti, E. M. & Stahel, W. A. (1986). *Robust statistics: The approach based on influence functions.* New York: Wiley.

Han J., & Kamber M. (2000). *Data mining: Concepts and techniques.* New York: Morgan Kaufmann.

Han, J. & Kamber, M. (2001). *Data Mining: Concepts and Techniques.* Morgan Kaufmann.

Hand, D. J. (1981). *Discrimination and classification.* New York: John Wiley.

Hand, D. J., Heikki, M., & Smyth, P. (2001). *Principles of data mining.* Cambridge: MIT Press.
Hastie, T., Tibshirani, R., & Friedman, J. (2001). *The elements of statistical learning: Data mining, inference and prediction.* New York: Springer-Verlag.

Hanisch, D., Zien, A., Zimmer, R., & Lengauer, T. (2002). Co-clustering of biological networks and gene expression data. *Bioinformatics, 18,* S145-S154.

Haussler, D. (1999). *Convolution kernels on discrete structures* (Technical Report UCSC-CRL-99-10). Santa Cruz: University of California.

Heckerman, D. (1999). A tutorial on learning with Bayesian networks. In M. Jordan (Ed.), *Learning in Graphical Models* (pp. 301-354). Cambridge, MA: The MIT Press.

Heckerman, D., Geiger, D., & Chickering, D. M. (1995). Learning Bayesian networks: The combination of knowledge and statistical data. *Machine Learnin,. 20,* 197-243.

Heilman, E., *et al.* (2002) Identifying battlefield metrics through experimentation. *Proceedings of the 7th International Command & Control Research & Technology Symposium.*

Henikoff, S., & Henikoff, J.G. (1994). Protein family classification based on search a database of blocks. *Genomics, 19,* 97-107.

Herrera, F., & Lozano, M. (2003). Fuzzy adaptive genetic algorithms: design, taxonomy, and future directions. *Soft Computing, 7,* 545-562.

Heymans, M., & Singh, A. J. (2003). Deriving phylogenetic trees from the similarity analysis of metabolic pathways. *Bioinformatics, 19,* i138-i146.

Ho, T. K., & Baird, H. S. (1997). Large-scale simulation studies in image pattern recognition. *IEEE Transactions on Pattern Analysis and Machine Intelligence, 19*(10), 1067-1079.

Holland, J. H. (1975) *Adaptation in natural and artificial systems.* Cambridge, MA: MIT Press.

Holm, L., & Sander, C. (1999). Protein folds and families: Sequence and structure alignments. *Nucleic Acids Res., 27,* 244–247.

Holme, P., Huss, M., & Jeong, H. (2003). Subnetwork hierarchies of biochemical pathways. *Bioinformatics, 19,* 532-538.

Holte, R.C. (1993). Very simple classification rules perform well on most commonly used datasets. *Machine Learning, 11,* 63-91.

Hsu, C.-W., Chang, C.-C., & Lin, C.-J. (2003). *A practical guide to support vector classification.* Retrieved May 11, 2005, from http://www.csie.ntu.edu.tw/~cjlin/papers/guide/guide.pdf

Huang, Z., Chen, H., Hsu, C.-J., Chen, W.-H., & Wu, S. (2004). Credit rating analysis with support vector machines and neural networks: A market comparative study. *Decision Support Systems, 37*(4), 543-558.

Huber, P. J. (1981). *Robust statistics.* New York,: John Wiley & Sons.

Huntala, Y., Karkkainen, J., Porkka, P., & Toivonen, H. (1999). TANE: An efficient algorithm for discovering functional and approximate dependencies. *The Computer Journal, 42*(2), 100-111.

Hyun-Chul, K., Shaoning, P., Hong-Mo, J., Daijin, K., & Sung-Yang, B. (2002). Pattern classification using support vector machine ensemble. In *Proceedings of the IEEE 16th International Conference on Pattern Recognition* (pp. 160-163).

Ichihashi, H. & Watanabe, T. (1990). Learning control system by a simplified fuzzy reasoning model. *Proceedings of IPMU'90*, (pp. 417-419).

Ingram, R., & Benford, S. (1995). Legibility enhancement for information visualization. In *Proceedings of IEEE Conference on Visualization (IEEE VIZ'95)*, Atlanta, GA.

Ingram, R., & Benford, S. (1996). The application of legibility techniques to enhance 3-D information visualizations. *The Computer Journal, 39*(10), 819-836.

Ingram, R., Benford, S., & Bowers, J. (1996). Building virtual cities: Applying urban planning principles to the design of virtual environments. In *Proceedings of the ACM Symposium on Virtual Reality Software and Technology (VRST96)*, (pp. 83-91). Hong Kong, China.

Jaakkola, T., Diekhans, M., & Haussler, D. (1999). Using the Fisher Kernel Method to detect remote protein homologies. In *Proceedings of the Seventh International Conference on Intelligent Systems for Molecular Biology* (pp. 149-158). Menlo Park, CA: AAAI Press.

Jaakkola, T., Diekhans, M., & Haussler, D. (2000). A discriminative framework for detecting remote protein homologies. *Journal of Computational Biology, 7*, 95-114.

Jain, A. K., & Dubes, R. C. (1998). *Algorithms for clustering data.* Englewood Cliffs, NJ: Prentice Hall.

Jang, G.S., Lai, F., & Parng, T.M. (1993). Intelligent stock trading decision support system using dual adaptive-structure neural networks. *Journal of Information Science and Engineering, 9*, 271-297.

Jefferys, W.H., & Berger, J.O. (1992). Ockhams razor and Bayesian analysis. *American Science, 80*, 64-72.

Jensen, F.V. (1996). *An introduction to Bayesian networks.* London: UCL Press.

Jensen, R., & Shen, Q. (2004). Semantics-preserving dimensionality reduction: Rough and fuzzy-rough based approaches. *IEEE Transactions on Knowledge and Data Engineering, 16*, 1457-1471.

Joachims, T. (1998). Text categorization with support vector machines: Learning with many relevant features. In *Proceedings of the ECML'98, 10th European Conference on Machine Learning* (pp. 137-142).

Joachims, T. (2002). *Learning to classify text using support vector machines: Methods, theory and algorithms.* Norwell, MA: Kluwer Academic.

Johnson, S. C. (1967). Hierarchical clustering schemes. *Psychometrika, 2*, 241-254.

Jonathan, H., & Guo, X. (2002). *Quality of service control in high-speed network.* Beijing: Tsinghua University Press.

Kalpakis, K., Gada, D., & Puttagunta, V. (2001). Distance measures for effective clustering of ARIMA time-series. *Proceedings of the 2001 IEEE International Conference on Data Mining* (pp. 273-280)., San Jose, CA, Nov. 29 – Dec. 2, 2001.

Kamath, A.P., Karmarkar, N.K., Ramakrishnan, K.J., & Resende, M.G.C. (1992). A continuous approach to inductive inference. *Mathematical Programming, 57*, 215-238.

Kantardzic, M. (2002). *Data mining: Concepts, models, methods, and algorithms.* Piscataway, NJ: IEEE Press.

Kantardzic, M., Rashad, S. & Sadeghian, P. (2004). Spatial navigation assistance system for large virtual environments: The data mining approach. In *Proceedings of the Mathematical Methods for Learning*, (pp. 44-46). Villa Geno, Como, Italy.

Kargupta, H., Joshi, A., Sivakumar, K., & Yesha, Y. (Eds.). (2004). *Data mining: Next generation challenges and future directions*. Menlo Park, CA: AAAI/MIT Press.

Karp, P.D., Riley, M., Saier, M., Paulsen, I. T., Paley, S. M., & Pellegrini-Toole, A. (2000). The EcoCyc and MetaCyc databases. *Nucleic Acids Research, 28*, 56-59.

Karypis, G., Han, E., & Kumar, V. (1999). CHAMELEON: A hierarchical clustering algorithm using dynamic modeling. *IEEE Computer, 32*(8), 68-75.

Kaufman, L., & Rousseeuw, P. J. (1990). *Finding groups in data: An introduction to cluster analysis*. New York: John Wiley & Sons.

Kohavi, R., & Sahami, M. (1996). Error-based and entropy-based discretization of continuous features. In *Proceedings of the Second International Conference on Knowledge Discovery and Data Mining* (pp. 114-119).

Kohonen, T. (1995). *Self-organizing maps*. Berlin: Springer-Verlag.

Kopanakis, I. & Theodoulidis, B. (2003). Visual data mining modeling techniques for the visualization of mining outcomes. *Journal of Visual Languages and Computing, Special Issue on Visual Data Mining, 14*(6), 543-589.

Kosko, B. (1992). Fuzzy systems as universal approximators. *Proceedings of FUZZ-IEEE '92,* (pp. 1153-1162).

Kreßel, U. (1999). Advances in kernel methods—Support vector learning. In B. Scholkopf, C. Burges, & A. J. Smola (Eds.), *Pairwise classification and support vector machines*. Cambridge, MA: MIT Press.

Krishna, K., & Murty, M. N. (1999). Genetic k-means algorithms. *IEEE Transactions on Systems Man and Cybernetics - Part B: Cybernetics, 29*(3), 433-439.

Krishnakumar, K., & Goldberg, D. E. (1992). Control system optimization using genetic algorithms. *Journal of Guidance, Control, and Dynamics, 15*(3), 735-740.

Krishnapuram, R., Joshi, A., Nasraoui, O., & Yi, L. (2001). Low-complexity fuzzy relational clustering algorithms for web mining. *IEEE Transactions on Fuzzy Systems, 9*(4), 595-607.

Krogh, A., Brown, M., Mian, I. S., Sjolander, K., & Haussler, D. (1994). Hidden Markov models in computational biology: Applications to protein modeling. *Journal of Molecular Biology, 235*, 1501-1531.

Kubika, J., & Moore, A. (2003). Probabilistic noise identification and data cleaning. In X. Wu, A. Tuzhilin, & J. Shavlik (Eds.), *The Third IEEE International Conference on Data Mining* (pp. 131-138). IEEE Computer Society.

Kusiak, A. (2000). Decomposition in data mining: An industrial case study. *IEEE Trans, Electron, Packag, and Manufact., 23*, 345-353.

Kusiak, A. (2001). Rough set theory: A data mining tool for semiconductor manufacturing. *IEEE Transactions on Electronics Packaging Manufacturing, 24*, 44-50.

Kusiak, A. (2001a). Feature transformation methods in data mining. *IEEE Transactions on Electronics Packaging Manufacturing, 24*(3), 214-221.

Kusiak, A. (2002). A data mining approach for generation of control signatures. *ASME Transactions: Journal of Manufacturing Science and Engineering, 124*(4), 923-926.

Kusiak, A., Shah, S., & Dixon, B. (2003). Data mining enhanced decision-making approach for kidney dialysis survival time. In *Proceedings of the 5th IFAC Symposium on Modeling and Control in Biomedical Systems*, Melbourne, Australia, August 2003 (pp. 35-39).

Kuznetsov, S. O. (1993). Fast algorithm of constructing all the intersections of finite semi-lattice objects. *NTI, Series 2*(1), 17-20. Moscow, Russia: VINITI.

Kuznetsov, S. O., & Obiedkov, S. A. (2001). Comparing performance of algorithms for generating concept lattices. *J. Exp. Theor. Artif. Intell. 14*(2-3), 183-216.

Kwedlo, W., & Krętowski, M. (1999). An evolutionary algorithm using multivariate discretization for decision rule induction. In *Proceedings of the European Conference on Principles of Data Mining and Knowledge Discovery* (pp. 392-397).

Lagarias, J.C., Reeds, J.A., Wright, M.H., & Wright, P.E. (1998). Convergence properties of the Nelder-Mead simplex method in low dimensions. *SIAM Journal of Optimisation, 9*, 112-147.

Lam W., & Bacchus, F. (1994). Learning bayesian belief networks. An approach based on the MDL principle. *Computational Intelligence, 10*, 269-293.

Lam, C. P., & Stork, D. G. (2003). Evaluating classifiers by means of test data with noisy labels. In *Proceedings of the 18th International Joint Conference on Artificial Intelligence* (pp. 513-518). Acapulco, Mexico.

Lancaster, K. (1966). A new approach to consumer theory. *Journal of Political Economy, 74*(2), 132-157.

Lavraĉ, N., & Džeroski, S. (1994). *Inductive logic programming: Techniques and applications.* Chichester: Ellis Horwood.

Lavraĉ, N., & Flash, P. (2000). *An extended transformation approach to inductive logic programming. CSTR – 00 -002,* March, 2000 (pp. 1-42).University of Bristol, Department of Computer Science.

Lavraĉ, N., Gamberger, D., & Jovanoski, V. (1999). A study of relevance for learning in deductive databases. *Journal of Logic Programming, 40*(2/3), 215-249.

Lawrence, D. (1987). Genetic algorithms in search, optimization, and machine learning. Addison-Wesley.

Lawrence, S., & Giles, L. (2000). Overfitting and neural networks: Conjugate gradient and back-propagation. *IEEE Computer Society,* 114-119.

Leake, D.B. (1996). CBR in context: The present and future. In D.B. Leake (Ed.), *Case-Based Reasoning: Experience, Lessons and Future Directions* (pp. 3-30). AAAI Press.

Lee, K. Y., Perakis, M., Sevcik, D., Santoso, N. I., Lausterer, G., & Samad, T. (2000). Intelligent distributed simulation and control of power plants. *IEEE Transactions on Energy Conversion, 15*(1), 116-123.

Lenarcik, A. & Piasta, Z. (1997). Probabilistic rough classifiers with mixture of discrete and continuous variables. In T.Y. Lin & N. Cercone (Eds.), *Rough Sets and Data Mining: Analysis for Imprecise Data* (pp. 373-383). Kluwer Academic Publishers.

Li, C., & Biswas, G. (1999). Temporal pattern generation using hidden Markov model based unsupervised classification. In D. J. Hand, J. N. Kok, M. R. Berthold (Eds.), IDA '99, *LNCS 164,* Springer-Verlag: Berlin, 245-256.

Li, L., Tang, B., Ni, Z., & Yang, T. (2002). A study on spatial attribute data mining based on rough set. In *Proceedings of the International Conference on Machine Learning and Cybernetics* (pp. 743-746).

Li, P., Zheng, H., & Nahavandi S. (2003). The application of rough set and Kohonen network to feature selection for object extraction. In *Proceedings of the Second International Conference on Machine Learning and Cybernetics* (pp. 1185-1189).

Li, Y. (2003). Classification of clients in client relationship management base on rough set theory. In *Proceedings of the 2nd International Conference on Machine Learning and Cybernetics* (pp. 242-246).

Liao, L., & Noble, W. S. (2003). Combining pairwise sequence similarity and support vector machines for detecting remote protein evolutionary and structural relationships. *Journal of Computational Biology, 10*, 857-868.

Liao, L., Kim, S., & Tomb, J-F. (2002). Genome comparisons based on profiles of metabolic pathways", In *The Proceedings of The Sixth International Conference on Knowledge-Based Intelligent Information & Engineering Systems* (pp. 469-476). Crema, Italy: IOS Press.

Liao, T. W. (2005). Clustering of time series data - a survey. *Pattern Recognition, 38*(11), 1857-1874.

Liao, T. W. (2007). A clustering procedure for exploratory mining of vector time series. *Pattern Recognition,* doi:10.1016/j.patcog.2007.01.005.

Liao, T. W., Bodt, B., Forester, J., Hansen, C., Heilman, E., Kaste, R., & O'May, J. (2002). *Army Research Conference,* Orlando FL, December 2002.

Liberles, D. A., Thoren, A., von Heijne, G., & Elofsson, A. (2002). The use of phylogenetic profiles for gene predictions. *Current Genomics, 3*, 131-137.

Lim, T.-S. (2002). Knowledge discovery central, datasets. Retrieved from *http://www.KDCentral.com*

Lin, J., & Gerstein, M. (2002). Whole-genome trees based on the occurrence of folds and orthologs: Implications for comparing genomes on different levels. *Genome Research, 10*, 808-818.

Lin, Y. (2002). Support vector machines and the bayes rule in classification. *Data Mining and Knowledge Discovery, 6*, 259-275.

Lisi, F., & Malerba, D. (2004). Inducing multi-level association rules from multiple relations. *Machine Leaning, 55*, 175-210.

Liu, B. (1996). Intelligent route finding: Combining knowledge, cases and an efficient search algorithm. In *Proceedings of the 12th European Conference on Artificial Intelligence (ECAI-96)*, (pp. 380-384). Budapest, Hungary.

Liu, B., Xia, Y., & Yu, P. (2000). Clustering through decision tree construction. *Proceedings of ACM CIKM*.

Liu, H., Hussain, F., Tan, C.L., & Dash, M. (2002). Discretization: An enabling technique. *Data Mining and Knowledge Discovery, 6*, 393-423.

Liu, H., Xu, M., Xu, K., & Cui, Y. (2003). Research on Internetwork routing protocol: A survey. *Science of Telecommunications, 19*, 28-32.

Lorena, L. A. N., & Furtado, J. C. (2001). Constructive genetic algorithms for clustering problems. *Evolutionary Computation, 9*(3), 309-327.

Louviere, J. (1988). Conjoint analysis modelling of stated preferences. A review of theory, methods, recent developments and external validity. *Journal of Transport Economics ad Policy, 22*(1), 93-119.

Louviere, J.J., Hensher D.A., & Swait J. (2000). *Stated choice methods. Analysis and application*. Cambridge, England: Cambridge University Press.

Lu, X., & Zhang, J. (2005). A new method of decision rule mining for IT project risk management based on rough set. In *Proceedings of the International Conference on Services Systems and Services Management* (pp. 1022-1026).

Lynch, K. (1960). *The image of the city*. Cambridge, MA: MIT Press.

Maass, W. (1994). Efficient agnostic PAC-learning with simple hypotheses. In *Proceedings of the Seventh Annual ACM Conference on Computerized Learning Theory* (pp. 67-75).

MacQueen, J. (1967). Some methods for classification and analysis of multivariate observations. In L. M. LeCam & J. Neyman (Eds.), *Proceedings of the 5th Berkeley Symposium on Mathematical Statistics and Probability*, Vol. 1 (pp. 281-297), University of California Press, Berkeley.

Magliano, J., Cohen, R., Allen, G., & Rodrigue, J. (1995). The impact of a wayfinder's goal on learning a new environment: Different types of spatial knowledge as goals. *Journal of Environmental Psychology, 15*, 65-75.

Mandenhall, W. & Sincich, T. (1995). *Statistics for engineering and the sciences* (4th ed.). Prentice Hall.

Mangasarian, O.L. (1965). Linear and nonlinear separation of patterns by linear programming. *Operations Research, 13*, 444-452.

Mangasarian, O.L. (1968). Multi-surface method of pattern separation. *IEEE Transactions on Information Theory, 14*, 801–807.

Mangasarian, O.L., Setiono, R., & Wolberg, W. H. (1990). Pattern recognition via linear programming: Theory and application to medical diagnosis. In T. H. Coleman & Y. Li (Eds.), *Large-scale numerical optimization* (pp. 22-30). Philadelphia: SIAM Publication.

Mangasarian, O.L., Street, W.N., & Wolberg, W.H. (1995). Breast cancer diagnosis and prognosis via linear programming. *Operations Research, 43*, 570-577.

Mannila, H., & Räihä, K.–J. (1992). On the complexity of inferring functional dependencies. *Discrete Applied Mathematics, 40*, 237-243.

Mannila, H., & Räihä, K.–J. (1994). Algorithm for inferring functional dependencies. *Data & Knowledge Engineering, 12*, 83-99.

Manoranjan, V.S., Lazaro, A. de Sam, Edwards, D., & Aathalye (1995). A systematic approach to obtaining fuzzy sets for control systems. *IEEE Transactions on Systems, Man and Cybernetics, 25*(1).

Manski, C. (1973). *The analysis of quantitative choice*. Unpublished doctoral dissertation, Department of Economics, Massachusetts Institute of Technology, Cambridge.

Marcotte, E. M., Xenarios, I., van Der Bliek, A. M., & Eisenberg, D. (2000). Localizing proteins in the cell from their phylogenetic profiles. *Proc. Natl. Acad. Sci. USA*, 97, (pp. 12115-12120).

Mas-Colell, A., Whinston M.D., & Green J.R. (1995). *Microeconomic theory*. New York: Oxford University Press.

Masters, T. (1993). *Practical neural network recipes in C++*. Academic Press.

Maulik, U., & Bandyopadhyay, S. (2000). Genetic algorithm-based clustering technique. *Pattern Recognition*, *33*, 1455-1465.

Mc Fadden, D. (1974). Conditional logit analysis of qualitative choice behaviour. In P. Zarembka (Ed.), *Frontiers in econometrics* (pp. 105-142). New York: Academic Press.

McCluskey, E. J. (Ed.). (1986). *Logic design principle*. NJ: Prentice Hall, Inc.

McNeill, M., Sayers, H., Wilson, S., & McKevitt, P. (2002). A spoken dialogue system for navigation in non-immersive virtual environments. *Computer Graphics*, *21*(4), 713-722.

Megretskaya, I. A. (1988). Construction of natural classification tests for knowledge base generation. In *The Problem of the Expert System Application in the National Economy* (pp. 89-93). Kishinev, Moldavia.

Mephu Nguifo, E., & Njiwoua, P. (1998). Using lattice based framework as a tool for feature extraction. In H. Lui & H. Motoda (Eds.), *Feature extraction, construction, and selection: A data mining perspective*. Kluwer.

Michalewicz, Z. (1992). Genetic algorithms + data structures = evolution programs. Berlin: Springer-Verlag.

Michalski, R. S. (1983). A theory and methodology of inductive learning. *Artificial Intelligence*, *20*, 111-161.

Michalski, R. S., & Larsen, I. B. (1978). *Selection of most representative training examples and incremental generation of VL1 Hypotheses: The Underlying methodology and the description of programs ESEL and AQII*. (Report No.

78-867). Dep. of Comp. Science, Univ. of Illinois at Urbana-Champaign, IL, USA.

Michalski, R. S., & Ram, A. (1995). Learning as goal-driven inference. In A. Ram & D. B. Leake (Eds), *Goal-driven learning*. Cambridge, MA: MIT Press/Bradford Books.

Mill, J. S. (1900). *The system of logic*. Moscow, Russia: Russian Publishing Company "Book Affair."

Miller, R.J., & Yang, Y. (1997). Association rules over interval data. In *Proceedings of the ACM SIGMOD International Conference on Management of Data* (pp. 452-461).

Mitchel, M. (1996). *An Introduction to genetic algorithms*. MIT Press.

Mitchell, T. (1997). *Machine learning*. Boston, MA: McGraw Hill.

Mitra, S., Pal, S. K., & Mitra, P. (2002). Data mining in soft computing framework: A survey. *IEEE Transactions on Neural Networks*, *13*(1), 3-14.

Mittelman, D., Sadreyev, R., & Grishin, N. (2003). Probabilistic scoring measures for profile-profile comparison yield more accurate shore seed alignments. *Bioinformatics*, *19*, 1531-1539.

Modjeska, D., & Waterworth, J. (2000). Effects of desktop 3D world design on user navigation and search performance. In *Proceedings of IEEE Information Visualization*, (pp. 215-220). Salt Lake City, Utah.

Moeser, S. (1988). Cognitive mapping in a complex building. *Environment and Behavior*, *20*, 21-49.

Morik, K., Brockhausen, P., & Joachims, T. (1999). Combining statistical learning with a knowledge-based approach: A case study in intensive care monitoring. In *Proceedings of the 16th International Conference on Machine Learning* (pp. 268-277).

Muller, K.-R., Mika, S., Ratsch, G., Tsuda, K., & Scholkopf, B. (2001). An introduction to kernel-based learning algorithms. *IEEE Transactions on Neural Networks, 12*(2), 181-201.

Murphy, J. J. (Ed.). (1997). *Analisi Tecnica dei Mercati Finanziari*. New York Institute of Finance.

Murthy, S. K. (1998). Automatic construction of decision trees from data: A multi-disciplinary survey. *Data Mining and Knowledge Discovery, 2*, 345-389.

Murzin, A. G., Brenner, S. E., Hubbard, T., & Chothia, C. (1995). SCOP: A structural classification of protein database for the investigation of sequence and structures. *J. Mol. Biol., 247*, 536-540.

Nabney, I. T. (2002). *Netlab: Algorithms for pattern recognition.* Springer-Verlag.

Naidenova, X. A. (1992). Machine learning as a diagnostic task. In I. Arefiev (Ed.), *Knowledge-Dialogue-Solution* (pp.26-36). Materials of the short-term scientific seminar. Saint-Petersburg, Russia.

Naidenova, X. A. (1996). Reducing machine learning tasks to the approximation of a given classification on a given set of examples. In *Proceedings of the 5th National Conference at Artificial Intelligence (Kazan, Tatarstan), 1*, 275-279.

Naidenova, X. A. (1999). The data-knowledge transformation. *Text processing and cognitive technologies (Pushchino, Russia), 3*, 130-151.

Naidenova, X. A. (2001). Inferring good diagnostic tests as a model of common sense reasoning. *Proceedings of the International Conference "Knowledge-Dialog-Solution", 2*, 501-506. Saint-Petersburg, Russia: State North-West Technical University, Publishing House "Lan".

Naidenova, X. A., & Ermakov, A. E. (2001). The decomposition of algorithms of inferring good diagnostic tests. In A. Zakrevskij (Ed.), *Proceedings of the 4th International Conference "Computer – Aided Design of Discrete Devices (CAD DD'2001)" (Vol. 3*, pp. 61-69), Institute of Engineering Cybernetics, National Academy of Sciences of Belarus. Minsk, Belarus.

Naidenova, X. A., & Polegaeva, J. G. (1986). An algorithm of finding the best diagnostic tests. In G. E. Mintz & P. P. Lorents (Eds), *The 4th All Union Conference "Application of mathematical logic methods"* (pp. 63-67). Tallinn, Estonia: Institute of Cybernetics, National Acad. of Sciences of Estonia.

Naidenova, X. A., & Polegaeva, J. G. (1991). SISIF—The system of knowledge acquisition from experimental facts. In J. L. Alty & L. I. Mikulich (Eds.), *Proceedings of the IFIP TC5/WG5.3 Conference "Industrial applications of artificial intelligence"* (pp. 87-92). North-Holland, Amsterdam, the Netherlands.

Naidenova, X. A., Plaksin, M. V., & Shagalov, V. L. (1995b). Inductive inferring all good classification tests. *Proceedings of International Conference "Knowledge-Dialog-Solution" (Jalta, Ukraine), 1*, 79-84.

Naidenova, X. A., Polegaeva, J. G., & Iserlis, J. E. (1995a). The system of knowledge acquisition based on constructing the best diagnostic classification tests. In *Proceedings of International Conference "Knowledge-Dialog-Solution" (Jalta, Ukraine), 1*, 85-95.

Narra, K., & Liao, L. (2004). Using extended phylogenetic profiles and support vector machines for protein family classification. In *The Proceedings of the Fifth International Conference on Software Engineering, Artificial Intelligence, Networking, and Parallel/Distributed Computing* (pp. 152-157). Beijing, China: ACIS Publication.

Narra, K., & Liao, L. (2005). Use of extended phylogenetic profiles with E-values and support vector machines for protein family classification. *International Journal of Computer and Information Science, 6*(1).

Nauck, D. & Kruse, R. (1995). *NEFCLASS—A neuro-fuzzy approach for the classification of data.* ACM Press.

Neapolitan, R.E. (1990). *Probabilistic reasoning in expert systems: Theory and algorithms.* New York: John Wiley & Sons.

Negro, G. (1995). *Organizzare la qualità nei servizi.* Milano, Italy: Edizioni del Sole 24 Ore.

Nevill-Manning, C. G., Wu, T. D., & Brutlag, D. L.(1998). Highly specific protein sequence motifs for genome analysis. *Proc. Natl. Acad. Sci. USA, 95*(11), 5865-5871.

Ng, A. Y. (1997). Preventing "overfitting" of cross-validation data. In *Proceedings of the 14th International Conference on Machine Learning* (pp. 245-253). Nashville, TN.

Niimi, A., & Tazaki, E. (2000). *Genetic programming combined with association rule algorithm for decision tree construction*. Paper presented at the fourth international conference on knowledge-based intelligent engeneering systems & allied technologies, Brighton, UK.

Nijholt, A., Zwiers, J., & van Dijk, B. (2001). Maps, agents and dialogue for exploring a virtual world. In *Proceedings of the 5th World Multiconference on Systemics, Cybernetics and Informatics (SCI 2001)*, (pp. 94-99). Orlando, Florida.

Noble, W. (2004). Support vector machine applications in computational biology. In B. Scholkopf, K. Tsuda, & J-P. Vert. (Eds.), *Kernel methods in computational biology* (pp. 71-92). Cambridge, MA: The MIT Press.

Nomura, H., Hayashi, I., & Wakami, N. (1992). A learning method of fuzzy inference rules by descent method. In *Proceedings of FUZZ-IEEE '92*, (pp. 203-210).

Nourine, L., & Raynaud, O. (1999). A fast algorithm for building lattices. *Information Processing Letters, 71*, 199-204.

Nozzaki, K., Ishibuchi, H., & Tanaka, H. (1997). A simple but powerful heuristic method for generating fuzzy rules from numerical data. *Fuzzy Sets and Systems 86*, 251-270.

O'May, J., & Hailman, E. (2002). OneSAF killer/victim scoreboard capability. ARL-TR-2829.

O'Neill, M. (1992). Effects of familiarity and plan complexity on wayfinding in simulated buildings. *Journal of Environmental Psychology, 12*, 319-327.

Obitko, M. (2004). *Introduction to genetic algorithms*, 1998. Retrieved from http://cs.felk.cvut.cz/~xobitko/ga/, Accessed on 02/03/2004

Oja, E. (1983). *Subspace methods for pattern recognition*. John Wiley.

Onoda, T., Murata, H., Ratsch, G., & Muller, K.-R. (2002). Experimental analysis of support vector machines with different kernels based on non-intrusive monitoring data. In *Proceedings of the IEEE International Joint Conference on Neural Networks* (pp. 2186-2191).

Ooyen, A., & Nienhuis, B. (1992). Improving the convergence of the back-propagation algorithm. *Neural Networks, 5*, 465-471.

Ore, O. (1944). Galois connexions. *Trans. Amer. Math. Society, 55*(1), 493-513.

Ormoneit, D., & Tresp, V. (1998). Averaging, maximum likelihood and Bayesian estimation for improving gaussian mixture probability density estimates. *IEEE Transaction on Neural Networks, 9*(4), 639-650.

Orsenigo, C., & Vercellis, C. (2003). Multivariate classification trees based on minimum features discrete support vector machines. *IMA Journal of Management Mathematics, 14*, 221-234.

Orsenigo, C., & Vercellis, C. (2004). Discrete support vector decision trees via tabu-search. *Journal of Computational Statistics and Data Analysis, 47*, 311-322.

Orsenigo, C., & Vercellis, C. (2007a). Multicategory classification via discrete support vector machines. *Computational Management Science*.

Orsenigo, C., & Vercellis, C. (2007b). Accurately learning from few examples with a polyhedral classifier. *Computational Optimization and Applications*.

Osuna, E., Freund, R., & Girosi, F. (1997). Training support vector machines: An application to face detection. In *Proceedings of the 1997 Conference on Computer Vision and Pattern Recognition—CVPR'97* (pp. 130-138). Washington, DC: IEEE Computer Society.

Ou, Y.-Y., Chen, C.-Y., Hwang, S.-C., & Oyang, Y.-J. (2003). Expediting model selection for support vector machines based on data reduction. In *Proceedings of the IEEE International Conference on Systems, Man and Cybernetics* (pp. 786-791).

Overbeek, R., Larsen, N., Pusch, G. D., D'Souza, M., Selkov Jr., E., Kyrpides, N., Fonstein, M., Maltsev, N., & Selkov, E. (2000). WIT: Integrated system for high throughput genome sequence analysis and metabolic reconstruction. *Nucleic Acids Res., 28*, 123-125.

Paab, G., Leopold, E., Larson, M., Kindermann, J., & Eickeler, S. (2002). SVM classification using sequences of phonemes and syllables. In *Proceedings of the European Conference on Machine Learning, ECML*, Helsinki.

Parasuraman, A., Zeithaml, V.A., & Berry, L.L. (1988). SERVQUAL: A multiple item scale for measuring consumer perceptions of service quality. *Journal of Retailing, 64*(1), 12-37.

Park, J. B., Park, Y. M., Won, J. R., & Lee, K. (2000). An improved genetic algorithm for generation expansion planning. *IEEE Transactions on Energy Conversion, 15*(3), 913-922.

Parrado-Hernandez, E., Mora-Jimenez, I., Arenas-Garca, J., Figueiras-Vidal, A.R., & Navia-Vazquez, A. (2003). Growing support vector classifiers with controlled complexity. Pattern Recognition, 36, 1479-1488.

Parthasarathy, S. (2001). Towards network-aware data mining. In *Proceedings of the 15th International Parallel and Distributed Processing Symposium* (pp. 157-161).

Pawlak, Z. (1991). *Rough Sets, theoretical aspects of reasoning about data*. Kluwer Academic Publishers.

Pawlak, Z. (1991). *Theoretical aspects of reasoning about data*. Boston: Kluwer Academic Publishers.

Pearl, J. (1988). *Probabilistic reasoning in intelligent systems: Networks of plausible inference*. San Mateo, CA: Morgan Kaufmann Publishers.

Pearson, W. (1990). Rapid and sensitive sequence comparison with FASTP and FASTA. *Meth. Enzymol., 183*, 63-98.

Pelekis, N. (*1999*). *Fuzzy m iner: A fuzzy system for solving pattern classification problems*. M.Sc. Thesis, *UMIST*. Retrieved June 15, 2004: http://users.forthnet.gr/ath/pele/HOME_PAGE_NIKOS_PELEKIS/Download/

Pellegrini, M., Marcotte, E. M., Thompson, M. J., Eisenberg, D., & Yeates, T. O. (1999). Assigning protein functions by comparative genome analysis: Protein phylogenetic profiles. *Proc. Natl. Acad. Sci. USA, 96*, (pp. 4285-4288).

Pernkopf, F. (2005). Bayesian network classifiers versus selective k-NN classifier. *Pattern Recognition, 38*, 1-10.

Peterson, B., Stine, J., & Darken, R. (2000). A process and representation for modeling expert navigators. In *Proceedings of the 9th Conference on Computer Generated Forces and Behavioral Representation*, (pp. 459-470). Orlando, Florida.

Peterson, B., Wells, M., Furness, T., & Hunt, E. (1998). The effects of the interface on navigation in virtual environments. In *Proceedings of Human Factors and Ergonomics Society 1998 Annual Meeting*, (pp. 1496-1505). Chicago, IL.

Pham, D. T., & Chan, A. B. (1998). Control chart pattern recognition using a new type of self-organizing neural network. *Proc. Instn. Mech. Engrs., 212*(1), 115-127.

Pham, D., & Karaboga, D. (2000). *Intelligent optimization techniques*. London: Springer-Verlag.

Piaget, J. (1959). *La genèse des structures logiques élémentaires*. Neuchâtel.

Pitkow, J., & Pirolli, P. (1999). Mining longest repeating subsequence to predict world wide web surfing. In *Second USENIX Symposium on Internet Technologies and Systems*, Boulder, CO.

Policker, S., & Geva, A. B. (2000). Nonstationary time series analysis by temporal clustering. *IEEE Transactions on Systems, Man, and Cybernetics—Part B: Cybernetics, 30*(2), 339-343.

Prabhu, N. (2003). Gauge groups and data classification. *Applied mathematics and computation, 138*(2-3), 267-289.

Quinlan, J. R. (1986). Induction of decision trees. *Machine Learning, 1*(1), 81-106.

Quinlan, J. R. (1993). *C4.5: Programs for machine learning*. San Mateo: Morgan Kaufmann.

Quinlan, J. R. (Ed.). (1993). *C4.5: Programs for machine learning*. San Mateo, CA: Morgan-Kaufmann.

Quinlan, J.R. (1986). Induction of decision trees. *Machine Learning, 1*, 81-106.

Quinlan, R. (1993). *C4.5: Programs for machine learning*. San Francisco: Morgan Kaufmann.

Rabiner, L. R. (1989). A tutorial on hidden Markov models and selected applications in speech recognition. *Proc. IEEE, 77*, 257-286.

Rahm, E., & Hai Do, H. (2000). Data cleaning: Problems and current approaches. *IEEE Data Engineering Bulletin, 23*(4), 3-13.

Ramakrishnan, R., & Gehrke, J. (2003). *Database management systems* (3rd ed.). New York: McGraw-Hill.

Rasiova, H. (1974). *An algebraic approach to non-classical logic (Studies in Logic, Vol. 78)*. Amsterdam-London: North-Holland Publishing Company.

Riguet, J. (1948). Relations binaires, fermetures, correspondences de Galois. *Bull. Soc. Math., 76*(3), 114-155.

Rojas, Ignacio, González, J., Pomares, H., Merelo, J. J., Castillo, P. A., & Romero, G. (2002). Statistical analysis of the main parameters involved in the design of a genetic algorithm. *IEEE Transactions on Systems, Man, and Cybernetics-Part C: Applications and Reviews, 32*(1), 31-37.

Rosenblatt, F. (1958, November). The Perceptron: A probabilistic model for information storage and organization in the brain, *Psychological Review, 65*, 386-408.

Ross, S.M. (2000). *Introduction to probability and statistics for engineers and scientists* (2nd ed.). London: Academic Press.

Ruddle, R., Payne, S., & Jones, D. (1998). Navigating large-scale "desk-top" virtual buildings: Effects of orientation aids and familiarity. *Presence: Teleoperators and Virtual Environments, 7*(2), 179-192.

Ruddle, R., Randall, S., & Jones, D. (1996). Navigation and spatial knowledge acquisition in large-scale virtual buildings: An experimental comparison of immersive and "desk-top" displays. In *Proceedings of the 2nd International FIVE Conference*, (pp. 125-136).

Rulequest Research. Gritbot. Retrieved from http://www.rulequest.com

Sadeghian, P., Kantardzic, M., & Rashad, S. (2006b). Knowledgeable navigation in virtual environments. In C.

Ghaoui (Ed.), *Encyclopedia of human computer interaction* (pp. 389-395). Hershey, PA: Idea Group Inc.

Sadeghian, P., Kantardzic, M., Lozitskiy, O., & Sheta, W. (2005). Route recommendations: Navigation assistance in complex virtual environments. In *Proceedings of the 20th International Conference on Computers and their Applications*, (pp. 220-225). New Orleans, LA.

Sadeghian, P., Kantardzic, M., Lozitskiy, O., & Sheta, W. (2006a). The frequent wayfinding-sequence (FWS) methodology: Finding preferred routes in complex VEs. *The International Journal of Human-Computer Studies, 64*(4), 356-374.

Sadreyev, R., & Grishin, N. (2003). Compass: A tool for comparison of multiple protein alignments with assessment of statistical significance. *Journal of Molecular Biology, 326*, 317-336.

Salzberg, S. (1991). A nearest hyper rectangle learning method. *Machine Learning, 6*, 277-309.

Sansom, D. C., Downs, T., & Saha, T. K. (2002). Evaluation of support vector machine based forecasting tool in electricity price forecasting for Australian National Electricity Market participants. *Journal of Electrical and Electronics Engineering Australia, 22*(3), 227-233.

Satalich, G. (1995). *Navigation and wayfinding in virtual reality: Finding proper tools and cues to enhance navigation awareness*. Master Thesis, University of Washington, Seattle, WA.

Scannapieco, M., Missier, P., & Batini, C. (2005). Data quality at a glance. *Datenbank-Spektrum, 14*, 6-14.

Schalkoff, R.J. (1992). *Pattern recognition: Statistical, structural and neural approaches*. John Wiley.

Schmidt-Schauss, M., & Smolka, G. (1991). Attributive concept descriptions with complements. *Artificial Intelligence, 48*(1), 1-26.

Schneeweiss, W.G. (Ed.). (1989). *Boolean funcions with engineering applications and computer programs*. Berlin: Springer-Verlag.

Schölkopf, B., & Smola, A. (2002). *Learning with kernels: Support vector machines, regularization, optimization, and beyond.* Cambridge, MA: MIT.

Scholkopf, B., & Smola, A. J. (2001). *Learning with kernels: Support vector machines, learning).* Cambridge, MA: The MIT Press.

Schwarm, S., & Wolfman S. (2000). Cleaning data with Bayesian methods. Final project report for CSE574, University of Washington, Retrieved the 23rd April 2002, from http://www.cs.washington.edu/homes/wolfwork/

Shanmugam, K., & Breipohl, A. M. (1971). An error correcting procedure for learning with an imperfect teacher. *IEEE Transactions on Systems, Man, and Cybernetics,* SMC-*1*(3), 223-229.

Sherman, W., & Craig, A. (2002). *Understanding virtual reality.* San Francisco: Morgan Kaufmann.

Shreider, J. (1974). Algebra of classification. *Proceedings of VINITI, Series 2*(9), 3-6.

Siepel, A., & Haussler, D. (2004). Combining phylogenetic and hidden Markov Models in biosequence analysis. *J. Comput. Biol., 11*(2-3), 413-428.

Skillicorn, D. (2002, April). *The case for Datacentric Grids.* Paper presented at the 16th International Parallel and Distributed Processing Symposium (IPDPS 2002), Fort Lauderdale, Florida.

Smith, K.A., Woo, F., Ciesielski, V., & Ibrahim, R. (2001). Modelling the relationship between problem characteristics and data mining algorithm performance using neural networks. In C. Dagli et al. (Eds.), *Smart engineering system design: Neural networks, fuzzy logic, evolutionary programming, data mining, and complex systems* (pp. 357-362). ASME Press.

Smith, K.A., Woo, F., Ciesielski, V., & Ibrahim, R. (2002). Matching data mining algorithm suitability to data characteristics using a self-organising map. In A. Abraham & M. Koppen (Eds.), *Hybrid information systems* (pp. 169-180). Heidelberg: Physica-Verlag.

Smith, T. F., & Waterman, M. S.(1981). Identification of common molecular subsequences. *Journal of Molecular Biology, 147,* 195-197.

Smyth, B., & McGinty, L. (2002). The route to personalization: A case-based reasoning perspective. *Expert Update, 5*(2), 27-36.

Smyth, P., & Wolpert, D. H. (1999). An evaluation of linearly combining density estimators via stacking. *Machine Learning Journal, 36,* 59-83.

Smyth, P., Fayyad, U. M., Burl, M. C., & Perona, P. (1996). Modeling subjective uncertainty in image annotation. In U. M. Fayyad, G. Piatetsky-Shapiro, P. Smyth, & R. Uthurusamy (Eds.), *Advances in knowledge discovery and data mining* (pp. 517-539). AAAI/MIT Press.

Soliman, M. (2004). *A model for mining distributed frequent sequences.* Ph.D. Dissertation, University of Louisville, Louisville, KY.

Spears, W. M. (1995). Adapting crossover in evolutionary algorithms. In J. R. McDonnell, R. G. Reynolds, & D. B. Fogel (Eds.), *Proceedings of the Fourth Annual Conference on Evolutionary Programming* (pp.367-384). Cambridge: MIT Press.

Spears, W. M., & De Jong, K. A. (1991). On the virtues of parameterized uniform crossover. In R. Belew & L. Booker (Eds.), *Proceedings of the Fourth International Conference on Genetic Algorithms* (pp. 230-236). San Mateo, CA: Morgan Kaufmann.

Spellman, P. T., Sherlock, G., Zhang, M. Q., Iyer, V. R., Anders, K., Eisen, M. B., Brown, P. O., Botstein, D., & Futcher, B.(1998). Comprehensive identification of cell cycle-regulated genes of the yeast saccharomyces cerevisiae by microarray hybridization. *Molecular Biology of the Cell, 9,* 3273–3297.

Sperner, E. (1928). Eine Satz uber untermengen einer endlichen menge. *Mat. Z, 27*(11), 544-548.

Spiegelhalter, D. J., Dawid, A. P., Lauritzen, S. L., & Cowell, R.G. (1993). Bayesian analysis in expert systems. *Statistical Science, 8*(3), 219-283.

Srikant, R., & Agrawal, R. (1996). Mining quantitative association rules in large relational tables. In *Proceedings of the ACM SIGMOD International Conference on Management of Data* (pp. 1-12).

Srinivas, M., & Patnaik, L. M. (1994). Adaptive probabilities of crossover and mutation in genetic algorithms. *IEEE Transactions on Systems, Man, and Cybernetics, 24*(4), 656-666.

Staiano, A. (2004). *Unsupervised neural networks for the extraction of scientific information from astronomical data.* Unpublished PhD Dissertation, University of Salerno, Italy.

Starzyk, J.A., Nelson, D.E., & Sturtz, K. (1999). Reduct generation in information systems. *Bulletin of the International Rough Set Society, 3*, 19-22.

Steck, S., & Mallot, H. (2000). The role of global and local landmarks in virtual environment navigation. *Presence, 9*(1), 69-83.

Stephan, V., Debes, K., Gross, H. M., Wintrich, F., & Wintrich, H. (2001). A new control scheme for combustion process using reinforcement learning based on neural networks. *International Journal of Computational Intelligence and Applications, 1*(2), 121-136.

Stolfo, S., Lee, W., Chan, P., Fan, W., & Eskin, E. (2001). Data mining-based intrusion detectors: An overview of the Columbia IDS project. *Special Interest Group on the Management of Data (SIGMOD) Record, 30* (4), 5-14.

Stone, M. (1974). Cross-validatory choice and assessment of statistical predictions. *Journal of the Royal Statistical Society, 36*, 111-147.

Street, W.N. (in press). Multicategory decision trees using nonlinear programming. *Informs Journal on Computing.*

Streeter, L., Vitello, D., & Wonsiewicz, S. (1985). How to tell people where to go: Comparing navigational aids. *International Journal of Man-Machine Studied, 22*, 549-562.

Stumme, G., Taouil, R., Bastide, Y., Pasquier, N., & Lakhal, L. (2000). Fast computation of concept lattices using data mining techniques. In *Proceeding the 7th International*

Workshop on Knowledge Representation Meets Databases (KRDB 2000) (pp. 129-139).

Stumme, G., Wille, R., & Wille, U. (1998). Conceptual knowledge discovery in databases using formal concept analysis methods. *Proceeding the 2nd European Symposium on Principles of Data Mining and Knowledge Discovery (PKDD'98).*

Sugeno, M. & Yasukawa, T. (1993). A fuzzy-logic-based approach to qualitative modeling. *IEEE Trans. Fuzzy Systems, 1*, 7-31.

Sugeno, M.& Kang, G.T. (1998). Structure identification of fuzzy model. *Fuzzy Sets and Systems, 28*, 15-33.

Swiniarski, R. (1998). Rough sets and principal component analysis and their applications in future extraction and selection, data model building and classification. In S. Pal & A. Skowron (Eds.), *Fuzzy Sets, Rough Sets and Decision Making Processes.* Springer-Verlag.

Syswerda, G. (1989). Uniform crossover in genetic algorithms. In J. Schaffer (Ed.), *Proceedings of the Third International Conference on Genetic Algorithms* (pp. 2-9), Los Altos: Morgan Kaufmann Publishers.

Tagliaferri, R., Ciaramella, A., Milano, L., Barone, F., & Longo, G. (1999). Spectral analysis of stellar light curves by means of neural networks. *Astronomy and Astrophysics Supplement Series, 137*, 391–405.

Takagi, H. & Hayashi, I. (1991). NN-driven fuzzy reasoning. *Approximate reasoning, 5*, 191-212.

Talia, D. (2002). The Open Grid Services Architecture: Where the grid meets the web. *IEEE Internet Computing, 6*(6), 67-71, 2002.

Tamayo, P., Slonim, D., Mesirov, J., Zhu, Q., Kitareewan, S., Dmitrovsky, E., Lander, E. S., & Golub, T. R. (1999). Interpreting patterns of gene expression with self-organizing maps: Methods and application to hematopoietic differentiation. *PNAS, 96*, 2907–2912.

Tamhane, A.C. & Dunlop, D.D. (2000). *Statistics and data analysis.* Prentice Hall.

Teas, R.K. (1993). Expectations, performance evaluation, and consumers' perceptions of quality. *Journal of Marketing, 57*(4), 18-34.

Teng, C. M. (1999). Correcting noisy data. In *Proceedings of the 16th International Conference on Machine Learning* (pp. 239-248). Bled, Slovenia.

The Data Warehouse Institute, (2003). Data quality and the bottom line: Achieving business success through a commitment to high quality data, ZD Net UK, Retrieved the 13th April 2003, from http://whitepapers.zdnet.co.uk/0,39025945,60096183p-39000581q,00.htm.

Thurstone, L. (1927). A law of comparative judgment. *Psychological Review, 34*, 273-286.

Toronen, P., Kolehmainen, M., Wong, G., & Castren, E.(1999). Analysis of gene expression data using self-organizing maps. *FEBS Letters, 451*, 142–146.

Triantaphyllou, E., Allen, L., Soyster, L., & Kumara, S.R.T. (1994). Generating logical expressions from positive and negative examples via a branch-and-bound approach. *Computer and Operations Research, 21*, 185-197.

Truemper, K. (1996). *The Leibniz system for logic programming*. Version 4.2, Leibniz Plano, Texas 75023, U.S.A.

Truemper, K. (2004). *Design of logic-based intelligent systems*. New York: Wiley.

Tuecke S., et al. (2003). *Open Grid Services Infrastructure (OGSI) Version 1.0*. Retrieved August 26, 2006, from

van Dijk, B., op Den, A., Rieks, N., & Zwiers, J. (2003). Navigation assistance in virtual worlds. *Informing Science Journal, 6*, 115-125.

van Luin, J., den Akker, R., & Nijholt, A. (2001). A dialogue agent for navigation support in virtual reality. In *Proceedings of the ACM SIGCHI Conference CHI*, (pp.117-118).

Vapnik, V. (1995). *The nature of statistical learning theory*. New York: Springer.

Vapnik, V. (1998). *Statistical Learning Theory: Adaptive and learning systems for signal processing, communications, and control*. New York: Wiley.

Vapnik, V. (Ed.). (1995). *The nature of statistical learning theory*. Springer-Verlag.

Vapnik, V. N. (1998). *Statistical learning theory*. New York: Wiley.

Vapnik, V., & Lerner, A. (1963). Pattern recognition using generalized portrait method. *Automation and Remote Control, 24*, 774-780.

Vert, J-P. (2002). A tree kernel to analyze phylogenetic profiles. *Bioinformatics, 18*, S276-S284.

Vinogradov, D. V. (1999). Logic programs for quasi-axiomatic theories, *NTI, Series 2*(1-2), 61-64. Moscow, Russia: VINITI.

Vinson, N. (1999). Design guidelines for landmarks to support navigation in virtual environments. In *Proceedings of the ACM Conference on Human Factors in Computing Systems*, (pp. 278-285). Pittsburgh, Pennsylvania.

Wan, V. & Renals, S. (2002). Evaluation of kernel methods for speaker verification and identification. In *Proceedings of IEEE International Conference on Acoustics, Speech, and Signal Processing, ICASSP'02*, (pp. 669-672).

Wang, K., & Goh, H.C. (1997). Minimum splits based discretization for continuous features. In *Proceedings of the Fifteenth International Joint Conference on Artificial Intelligence* (pp. 942-951).

Wang, L.X. & Mendel, J.M. (1992). Generating fuzzy rules by learning from examples. *IEEE Trans. Systems, Man Cybernet, 22*, 1414-1427.

Wang, L.X. (1992). Fuzzy systems as universal approximators. In *Proceedings of FUZZ-IEEE'92*, (pp. 1163-1170).

Watkins, C. (1999). Dynamic alignment kernels. In A. J. Smola, P. Bartlett, B. SchÄolkopf, & C. Schuurmans (Ed.), *Advances in large margin classifiers*. Cambridge, MA: The MIT Press.

Weiss, S. A., & Kulikowski, C. A. (1991). *Computer systems that learn: Classification and prediction methods from statistics, neural nets, machine learning, and expert systems*. San Mateo, CA: Morgan Kaufmann.

Weston, J. & Watkins, C. (1999). Multi-class support vector machines. In M. Verleysen (Ed.), In *Proceedings of the Seventh European Symposium on Artificial Neural Networks*, Bruges, Belgium.

Wille, R. (1992). Concept lattices and conceptual knowledge system. *Computer Math. Appl.*, *23*(6-9), 493-515.

Wilson, D. L. (1972). Asymptotic properties of nearest neighbor rules using edited data. *IEEE Transactions on Systems, Man, and Cybernetics*, *2*(3), 408-421.

Witten, I.H., & Frank E. (2005). *Data mining: Practical machine learning tools and techniques.* Morgan Kaufmann.

Witten, I.H., & Frank, E. (2000). *Data mining.* San Diego: Academic Press.

Witten, I.H., & Frank, E. (Ed.). (2005). *Data mining: Practical machine learning tools and techniques* (2nd ed.). San Francisco: Morgan Kaufmann.

Woese, C. (1987). Bacterial evolution. *Microbial Rev.*, *51*, 221-271.

Wolberg W. H., & Mangasarian, O. L. (1990). Multisurface method of pattern separation for medical diagnosis applied to breast cytology, In *Proceedings of the National Academy of Sciences, 87* (pp. 9193-9196).

Wolberg, W.H., & Mangasarian, O.L. (1990). Multisurface method of pattern separation for medical diagnosis applied to breast cytology. In *Proceedings of the National Academy of Sciences U.S.A, 87*, 9193-9196.

Wolpert, D. H. (1992). Stacked generalization. *Neural Networks*, *5*(2), 241-259.

Wu, X. (1996). A Bayesian discretizer for real-valued attributes. *The Computer Journal*, *39*, 688-691.

Yang, Y., & Chiam, T.C. (2000). Rule discovery based on rough set theory. In *Proceedings of the Third International Conference.*

Yeung, D.Y. (1993). Constructive neural networks as estimators of Bayesian discriminant functions. *Pattern Recognition, 26*, 189-204.

Zaffalon, M. (2002). Exact credal treatment of missing data. *Journal of Statistical Planning and Inference*, *105*(1), 105-122.

Zaffalon, M., & Fagiuoli, E. (2003). Tree-based credal networks for classification. *Reliable computing*, *9*(6), 487-509.

Zaki, M. J., & Hsiao C.-J. (2002). CHARM: An efficient algorithm for closed Itemset mining. In *2nd SIAM International Conference on Data Mining.*

Zakrevskij, A. D. (1982). Revealing implicative regularities in the Boolean space of attributes and pattern recognition, *Kibernetika*, *1*, 1-6.

Zakrevskij, A. D. (1987). Implicative regularities in formal cognition models. *LMPS'87 Abstracts*, *1*, 373-375.

Zakrevskij, A. D. (2001). A logical approach to the pattern recognition problem. *Proceedings of the International Conference "Knowledge-Dialog-Solution" (KDS'2001)*, *2*, 238-245. *Saint-Petersburg, Russia: State North-West Technical University, Publishing House "Lan").*

Zakrevskij, A. D., & Vasylkova, I, V. (1997). Inductive inference systems in logical recognition in case of partial data. *Proceedings of the Fourth International Conference on Pattern Recognition and Information Processing (Minsk – Szczecin)*, *1*, 322-326.

Zhang, K., Wang, J. T. L., & Shasha, D. (1996). On the editing distance between undirected acyclic graphs. *International Journal of Foundations of Computer Science*, *7*, 43-58.

Zhang, S., Liao, L., Tomb, J-F., Wang, J. T. L. (2002). Clustering and classifying enzymes in metabolic pathways: Some preliminary results. In *ACM SIGKDD Workshop on Data Mining in Bioinformatics*, Edmonton, Canada (pp. 19-24).

Zhang, Y. (2003). Rough set and genetic algorithms in path planning of robot. In *Proceedings of the Second International Conference on Machine Learning and Cybernetics* (pp. 698-701).

Zhu, X., Wu, X., & Chen, Q. (2003). Eliminating class noise in large datasets. In *Proceedings of the 20th International Conference on Machine Learning* (pp. 920-927). Washington, DC.

Zimmermann, H.-J. (1996). *Fuzzy set theory and its applications* (third edition). Kluwer Academic.

About the Contributors

Giovanni Felici graduated in statistics at the University of Rome "La Sapienza." He received his MSc in operations research and operations management at the University of Lancaster, UK, in 1990, and his PhD in operations research at the University of Rome "La Sapienza." He is presently a permanent researcher at IASI, the Istituto di Analisi dei Sistemi ed Informatica of the National Research Council (CNR), Italy, where he started his research activity in 1994 working on research projects in logic programming and mathematical optimization. His current research activity is mainly devoted to to the application of optimization techniques to data-mining problems, with particular focus on integer programming algorithms for learning in logic and expert systems.

Carlo Vercellis is full professor at the Politecnico di Milano, where he teaches courses in optimization and business intelligence. He is also director of the research group MOLD—mathematical modeling, optimization, learning from data. Previously, after his graduation in Mathematics at the Università di Milano, he had been with the National Research Council (CNR), the Bocconi University, the Università di Milano. He has coordinated national and international research programs funded by EEC, CNR, and MIUR. His current research interests include mathematical models for learning, such as support vector machines and classification trees; data mining and machine learning, and their applications to relational marketing and biolife sciences; optimization models and methods, in particular with applications to supply chain and revenue management. In the past he was involved in research on design and analysis of algorithms for combinatorial optimization, project management, transportation models. He is author of several books and more than 70 papers, mostly appearing in refereed international journals and edited books.

* * * * *

Paolo Baldini graduated in computer science engineering at University of Pavia in 2004 with Professor Paolo Giudici: thesis on associative algorithms for Web-usage mining. His interest for data mining was born during a seminar for the artificial intelligence course at the engineering faculty, and thanks to professor Giudici, he had the possibility to write a thesis on this subject. His research interests are, in particular, Web mining, and risk management. Now he is an expert consultant of ERM at SMOUSE company of Piacenza.

Monica Chis was born in Cluj-Napoca, Romania. She studied computer science in Babes-Bolyai University, Cluj-Napoca, Romania. Since November 1999, she has been a PhD student at Department of Computer Science, Faculty of Mathematics and Computer Science, Babes-Bolyai University, Cluj-

Napoca, Romania. Since 2000 she has been Senior Lecturer at Avram Iancu University, Cluj-Napoca, Romania. She is working in the field of software development. Her research interests are in the field of evolutionary data analysis.

Antonio Congiusta is a research fellow in computer engineering at the University of Calabria, Italy. He received his PhD in systems and computer engineering from the University of Calabria in 2006. From 2002 to 2003 he was a research fellow at ICAR-CNR, Institute of the Italian National Research Council, working on grid-based data-mining systems and applications. His research interests are focused on grid programming environments, grid services, workflow-based grid systems, and distributed knowledge discovery.

Lara De Vinco got her MSc in computer science at Department of Mathematics and Computer Science, University of Salerno in 2004, with a thesis concerning the application of nonlinear latent variable models to astronomical data visualization. Her scientific interests are in neural networks and data mining. Since 2005, she has been at Nexera S.c.p.a.—Napoli, as project manager involved in the development of intelligent video surveillance systems.

Enrico Fagiuoli graduated in physics at the Università degli Studi di Milano in 1969. Since 1983, he has been associate professor of operation research, first at the Università degli Studi di Milano, and later at the Dipartimento di Informatica, Sistemistica e Comunicazione of the Università degli Studi di Milano-Bicocca, where he teaches data mining. His research interests include stochastic orderings between random variables, and computational methods and algorithms for Bayesian belief networks and their applications.

Valerio Gatta is a postdoctoral researcher at Department of National Contability and Analisys of the Social Processes, Faculty of Statistics, University of Rome "La Sapienza", Italy. He received his doctorate (PhD) in economics and statistics in 2005, and he got his diploma (MSc) in econometrics & management science at Erasmus University of Rotterdam. His research interests include quantitative marketing, service quality estimation, discrete choice models, stated preferences surveys, and statistical techniques, in particular for transport sector.

Paolo Giudici (BS in Economics, Bocconi University, 1989; MsC in Statistics, University of Minnesota, 1990; PhD in Statistics, University of Trento, 1993) is professor of statistics at the University of Pavia where he teaches statistics, business statistics, and risk management (at Borromeo College). He is director of the data mining laboratory (http://www.datamininglab.it), member of the University Assessment Board (http://www.unipv.it/nuv), and coordinator of the Institute of Advanced Studies school on "methods for the management of complex systems" (http://www.unipv.it/complexity). He is the author of 77 publications, among which are two research books. He has spent research periods abroad; in particular at the University of Bristol, the University of Cambridge, and at the Fields Institute for research in the mathematical sciences (Toronto). He is the coordinator of several national and European integrated research projects on data mining. He is responsible for the risk management interest group of the European network for business and industrial statistics (http://www.enbis.org). He is also a member of the Italian Statistical Society, the Italian Association for Financial Risk management (http://www.aifirm.com), and the Royal Statistical Society.

Dr. **Mehmed Kantardzic** is a professor in the computer engineering and Computer Science Department at the University of Louisville, and is the director of the data mining lab. He is the author of several books including *Data Mining: Concepts, Models, Methods, and Algorithms*, and an editor of the recently published book, *Next Generation of Data-Mining Applications*.

Dr. **Andrew Kusiak** is a professor in the Department of Mechanical and Industrial Engineering at the University of Iowa in Iowa City, Iowa. He is interested in applications of computational intelligence in automation, manufacturing, product development, and healthcare. Dr. Kusiak has published numerous books and technical papers in journals sponsored by professional societies, such as AAAI, ASME, IEEE, IIE, ESOR, IFIP, IFAC, INFORMS, ISPE, and SME. He speaks frequently at international meetings, conducts professional seminars, and consults for industrial corporations. Dr. Kusiak is IIE Fellow, serves on editorial boards of over 30 journals, edits book series, and is the editor-in-chief of the *Journal of Intelligent Manufacturing*.

Dr. **Chuck P. Lam** is an expert in statistical pattern recognition and a data strategist for business organizations. He graduated in electrical engineering from San Jose State University and received his MSc and PhD in electrical engineering from Stanford University. Dr. Lam has published refereed papers and book chapters on data acquisition, neural networks, collaborative filtering, and social networks. His consulting clients include Sun Microsystems, Hitachi, IBM, and numerous Internet startups.

T. Warren Liao received his PhD in Industrial Engineering from Lehigh University in 1990, and is currently a professor with Industrial Engineering Department, Louisiana State University. His research interests include soft computing, pattern recognition, data mining, and their applications in manufacturing. He has more than 60 refereed journal publications, and was the guest editor for several journals including *J. of Intelligent Manufacturing, Computers & Industrial Engineering, Applied Soft Computing*, and *International Journal of Industrial Engineering*.

Massimo Liquori graduated in 2001 cum laude at the University of Rome "La Sapienza," faculty of Economics. He also received his PhD in applied mathematics at the University of Rome "La Sapienza" with a thesis on classification techniques with application in finance. He is currently assistant researcher at the University of Rome "La Sapienza," Department of Mathematics, Faculty of Economics.

Giuseppe Longo (MSc 1980 in Physics) is full professor of astrophysics at Department of Physical Sciences (DSF), University "Federico II" of Napoli. His research interests cover extragalactic astronomy (cluster of galaxies and individual galaxies), cosmic rays, and neural networks applied to astrophysics. He leads the Astroneural collaboration, a joint enterprise among researchers of DSF and of the Department of Mathematics and Computer Science of the University of Salerno, aimed to explore the applications of neural networks and advanced soft computing tools to astronomical data mining. He is responsible for the Research Area of Astrophysics and Plasma Physics at the DSF; P.I. of the DSF Research Unit in the 6-th European Framework Research Infrastructure Project VO-Tech and a member of the Science Advisory Board for the VO-Tech Project.

Jonathan Mugan is currently a PhD student in the Computer Sciences Department at the University of Texas at Austin. He received an MS degree in computer science from the University of Texas at Dallas, and an MBA and a BA from Texas A&M University. His research interests include machine learning and intelligent robotics.

Xenia Naidenova was born in Leningrad (Saint-Petersburg) in 1940. She graduated from Lenin Electro-technical Institute of Leningrad (now Saint-Petersburg Electrotechnical University) in 1963, and received the specialty of computer engineering. From this institute, she obtained her doctoral degree (PhD) in technical sciences in 1979. In 1995, she started to work as senior researcher at the Scientific Research Centre of Saint-Petersburg Military Medical Academy, where she is engaged in developing knowledge discovery and data-mining systems to support solving medicine and psychological diagnostic tasks. Under Xenia Naidenova, some advanced knowledge acquisition systems based on machine learning original algorithms have been developed, including a tool for adaptive programming applied diagnostic medical systems. She is also the author of more than 100 scientific articles, and a member of the Russia Association for Artificial Intelligence, founded in 1989.

Sara Omerino graduated in computer science, at the Università degli Studi di Milano-Bicocca, in 2003. Her research interests concern methods and algorithms for neural networks, algorithms for solving optimization problems on Bayesian belief networks, and quantitative methods for data mining. She developed a software prototype for data cleaning that exploits Bayesian belief networks. Since 2004, she has been working at Etnoteam, and is responsible for the customer relationship management system developed for Banca Intesa.

Carlotta Orsenigo is researcher at the Università di Milano, where she teaches courses in mathematics and data mining. She also teaches courses in optimization and business intelligence at the Politecnico di Milano, where she graduated in management and production engineering. Her current research interests include optimization models and algorithms for data mining and knowledge discovery, such as support vector machines and classification trees, and applications of these models to marketing and biolife sciences. She is author of papers appearing in international journals and books.

Dr. **Pedram Sadeghian** is a recent graduate of the Computer Engineering and Computer Science Department at the University of Louisville. He completed his studies under the guidance of Dr. Kantardzic, and his thesis was titled "Navigation Assistance in Large-Scale Virtual Environments: The Data Mining Approach." His research interests include data mining, fractal theory, and virtual environments.

Andrea Scozzari graduated in 1997 cum laude in statistics at the University of Rome "La Sapienza." In 2001, he received his PhD in operations research at the University of Rome "La Sapienza." He is currently assistant researcher at the University of Rome "La Sapienza" at the department of mathematics, faculty of economic. His research activity is mainly focused on network facility location. Some results related to location problems were published by, or are under review for important international journals. Recently, his research interest has moved to classification problems with applications in economics and finance.

Shital Shah is a PhD student in industrial engineering at The University of Iowa, Iowa City. He completed his MS in IE at The University of Alabama, Tuscaloosa, and BE in production engineering at VJTI, Bombay, India. His area of interest is in computational intelligence, data mining, informatics, operations research, simulation, and decision support systems. He is in the process of publishing various data mining and informatics related research papers. His is the member of ALPHA-PI-MU and INFORMS.

Dr. **Walaa M. Sheta** has been an associate professor of computer graphics at the Informatics Research Institute at the Mubarak City for Scientific Research (MUCSAT) since 2006. He holds a visiting researcher position at the University of Louisville in USA and the University of Salford in the UK. He received his MSc and PhD in information technology from the Institute of Graduate Studies and Research, University of Alexandria, in 1992 and 2000, respectively. He received BSc from the Faculty of Science, University of Alexandria in 1989. His research interests include real-time computer graphics, virtual reality, human-computer interaction, and visual data mining.

Antonino Staiano (MSc 1997, PhD 2003 in computer science) research interests cover machine learning, statistical pattern recognition, evolutionary algorithms, data mining, and bioinformatics. Since October 2000, he has been developing neural-based models and tools for data visualization and clustering applied to astrophysics and genetic data. In 2006, he was granted a position at Italian National Institute of Astrophysics (INAF) for the European VO-Tech project: an international coordinated program designed to provide the European astronomical community with tools, systems, research support, and data interoperability standards necessary to explore the international astronomical and astrophysical data archives. Since February 2007, he has been assistant professor at Department of Applied Science, University of Napoli "Parthenope."

Fabio Stella graduated in computer science at the Università degli Studi di Milano in 1990. In 1995, he received the PhD in computational mathematics and operation research, with a dissertation concerning Bayesian belief networks and stochastic optimization. Since 2001, he has been associate professor of operation research at the Dipartimento di Informatica, Sistemistica e Comunicazione of the Università degli studi di Milano-Bicocca. His research interests include computational methods and algorithms for artificial neural networks architecture selection, algorithms for solving optimization problems on Bayesian belief networks, quantitative methods for adverse drug reactions, data, Web e-text mining, as well as online algorithms for portfolio selection.

Dr. **David G. Stork** is chief scientist at Ricoh Innovations and Visiting Lecturer at Stanford University. A graduate of MIT (BS) and the University of Maryland (PhD), he has been on the faculties of Wellesley College, Swarthmore College, Clark University, Boston University, and Stanford University. Dr. Stork holds 33 patents, and has published numerous peer-reviewed papers and book chapters. His deepest interests are in adaptive pattern recognition by machines and humans, and novel uses of the Internet.

Roberto Tagliaferri (MSc '84 in computer science) is full professor of computer science and neural networks at the University of Salerno. His research covers the area of neural nets: neural dynamics, fuzzy neural nets, applications to signal and image processing with astronomical and geological data, industrial and medical computer-aided diagnosis. He is the author or the coauthor of more than 100 publications in the area of neural networks. Since 1995, coeditor of the *Proceedings of the Italian Workshops on Neural Nets (WIRN)*, cochair of the Bioinformatics SIG of the INNS, Member of the Director Council of the IIASS (International Institute for Advanced Scientific Studies) "E. R. Caianiello." Senior member of the IEEE, member of INFN, INFM, and AIIA.

Domenico Talia is a professor of computer engineering at the Faculty of Engineering, University of Calabria, Italy, a research associate at ICAR-CNR, and a partner at Exeura s.r.l. His research interests include grid computing, distributed knowledge discovery, parallel data mining, parallel programming languages, and peer-to-peer systems. He published three books and about 200 papers in international journals and conference proceedings. He is a member of the editorial boards of *IEEE TKDE*, the *Future Generation Computer Systems Journal*, the *International Journal on Web and Grid Services*, and the *Web Intelligence and Agent Systems International Journal*. He is a member of the executive committee of the CoreGRID Network of Excellence.

Dr. **Evangelos Triantaphyllou** did all his graduate studies at Penn State University from 1984 to 1990. While at Penn State, he earned a Dual MS in environment and operations research (OR) (in 1985), an MS in computer science (in 1988) and a dual PhD in industrial engineering and operations research (in 1990). His PhD dissertation was related to data mining by means of optimization approaches. Since the spring of 2005, he has been a Professor in the Computer Science Department at the Louisiana State University (LSU) in Baton Rouge, LA. Before that, he had served for 11 years as an assistant, associate, and full professor in the Industrial Engineering Department at the same university. He has also served for one year as an interim associate dean for the College of Engineering at LSU. His research is focused on decision-making theory and applications, data mining and knowledge discovery, and the interface of operations research and computer science. In 1999 he received the prestigious IIE (Institute of Industrial Engineers), OR Division, research award for his research contributions in those fields. In 2005 he received an LSU Distinguished Faculty Award as recognition of his research, teaching, and service accomplishments. In 2000, Dr. Triantaphyllou published a bestseller book on multicriteria decision making. Besides the previous monograph on decision making, he has published another monograph on data mining by means of a logic-based approach (Springer, 2007). He has also coedited two books related to data mining, one on data mining by means of rule induction (2006) and another one on the mining of enterprise data (2007).

Klaus Truemper, PhD, is Professor Emeritus of Computer Science at the University of Texas, Dallas. Dr. Truemper received his doctorate in operations research from Case Western Reserve University in 1973. In 1988 he received the prestigious Senior Distinguished U.S. Scientist Award from the Alexander von Humboldt Foundation (Germany). Dr. Truemper's work includes the books *Matroid Decomposition*, *Effective Logic Computation*, and *Design of Logic-based Intelligent Systems and the Leibniz Sy*stem software.

Paolo Trunfio is an assistant professor of computer engineering at the Faculty of Engineering, University of Calabria, Italy. He received his PhD in systems and computer engineering from the same university in 2005. From 2001 to 2002 he collaborated with the ICAR-CNR Institute, working on parallel and distributed systems, and data mining. His current research interests include grid computing, service-oriented architectures, peer-to-peer systems, and distributed knowledge discovery.

Index

A

adaptive
  genetic algorithm (AGA) 161
  procedure 308
ad hoc 67, 141
agglomerative hierarchical clustering (HAC) 148
analysis of the variance (ANOVA) 161
a priori 131
ARMA models 159
artificial neural network (ANN) 25, 93, 187
association and sequence rules 234, 240
attribute language (AVL) 42

B

Bayesian
  belief networks (BBNs) 205, 206
  decision rule 209
Bayes Risk 209
binarization procedure 27
binary
  classification 83
  representation 27
Bioinoformatics 132
BLAST 132
BLOCKs 132
Boolean functions 24, 28, 29, 37
Bottom-up methods 5
branch node 112

C

circulating fluidized boiler (CFB) 187
classification method 1
clustering 146, 147, 148, 153
  hierarchy 150
commonsense reasoning 41, 44, 46, 58
Comparative Genomics 133
consensus filtering 222

control signatures 179, 180, 186
cumulative distribution function (CDF) 111
CURE (Clustering Using REpresentatives) 149
curse of dimensionality 90, 94
customer satisfaction survey (CSS) 66, 69
Cutpoint 1

D

data
  access service (DAS) 289
  analysis 204
  cleaning 204, 205, 206, 218
  mining 83, 128, 157, 159, 174, 179, 180, 285, 290, 294, 297
  quality 205, 208, 218
data/compute servers (DCS) 288
Datacentric Grid 288
DataSpace transfer protocol (DSTP) 288
decision trees 34, 65, 66, 73, 93
deductive
  hierarchical database (DHDB) 42
  inference 42, 44
diagnostic test 41, 44
directed acyclic graph (DAG) 206, 292
disconfirmation theory 67
discovery process markup language (DPML) 288
discrete choice models 69, 71
discrete support vector machines 116, 128
Discrete SVM 116
discretization 1, 2, 5, 6, 16
discriminant analysis 35
disjunctive normal form (DNF) 26, 29
dynamic time warping (DTW) 162

E

eMotif 133
empirical risk 119
empirical risk minimization (ERM) 101

enzyme commission (EC) 134
error correcting output codes (ECOC) 122
Euclidean
  distance 131, 168
  metric 131
evolutionary
  algorithm 146, 147, 149, 150, 156
  hierarchical clustering algorithm (EvHiCA) 152
execution plan management service (EPMS) 289
expectation-maximization (EM) 208, 246
exploratory data mining 205
Extended Phylogenetic Profiles 141

F

FASTA 132
feature space 90
first-order logic language (FOL) 42
fitness function 151, 152
formal concept analysis (FCA) 43
frequent wayfinding-sequence (FWS) 265
functional annotations 132
functional magnetic resonance imaging (fMRI) 158
fuzzy
  adaptive genetic algorithms (FAGA) 162
  partition 306
  rule-based system 304
  rules 301
  sets 301
fuzzy c-means (FCM) 157
  algorithm 159

G

GA mechanism 180
generative topographic mapping (GTM) 246
genetic-algorithm(GA)-based methods 157
genetic algorithm (GA) 158, 159, 175, 179, 180
genetic clustering 158, 159, 163, 176
good irredundant tests (GIRTs) 50
good maximally redundant tests (GMRTs) 50
Great Observatories Origins Deep Survey (GOODS) 251
grid
  computing 285, 292, 298
  projections 250
  support nodes (GSNs) 288

H

Hidden Markov Model 94
hierarchal Profilings 132

hierarchical clustering 146, 147, 153
hierarchical profile 132

I

inductive inference 42, 64
inductive logic programming (ILP) 42
inheritance pattern 140
instance-based (IB) 223
interval cut-point method 27
iterative case filtering algorithm (ICF) 222

J

JSM-method of hypotheses 43

K

k-medoids-based genetic algorithms 159
kernel
  functions 139
  trick 91
Killer-Victim Scoreboard (KVS) method 168
knowledge
  based expert systems 24
  base repository (KBR) 290
  directory service (KDS) 290
  discovery in databases (KDD) 284
  discovery services (KDS) 288
  execution plan repository (KEPR) 290
  grid 287, 289
  metadata repository (KMR) 290

L

Lagrangian dual problem 91
lattice theory 41, 43, 44, 58
leaf node 112
Level of Significance 209
Linear Discriminant Analysis 92
linear optimization (LO) 121
Linguistic Representation 308
log file 234
logical analysis of data (LAD) 26
logic programming 78, 81

M

Machine Learning 82, 90, 97
majority filtering 222
Markov model 241
Master tree 135
Mathematical Statistics 82

maximal frequent itemsets (MFI) 237
Maximum Likelihood (ML) 102, 104
Membership Functions 306
MEME 132
metabolic pathway profiles (MPP) 137
minimum description length (MDL) 5
missing values 206, 209, 210
multicategory classification 117, 118, 121, 128
multinominal logit (ML) 69

N

Naïve Bayes 34
National Science Foundation (NSF) 281
navigation assistance 265, 282
Nearest Neighbor 93, 250
Nearest Triangulation 250
neural networks 27, 34, 37
nonlinear programming 25
Nrnodes 150

O

open grid services architecture (OGSA) 284, 287,
    292
open grid services infrastructure (OGSI) 292
outliers detection 206, 214

P

P-Tree Approach 135
pairwise decomposition 117
Partition around Medoids (PAM) 158
pattern recognition 82
phylogenetic profiles 132
power plant 179, 180
priori knowledge 268
probabilistic principal surfaces (PPS) 244
probability density function (PDF) 111
Profile-Profile 132
profile similarity 131
Protein Folding Classification 116
PSI-BLAST 132

Q

quadratic problem (QSVM) 119
Quadratic Programs (QPs) 92

R

radial basis function (RBF) 33, 92, 102
    kernel 102, 106

parameter 102
random utility theory (RUT) 68
rate of change (ROC) 26
rational data 1, 2, 6, 8
reduction techniques (RT) 223
requent wayfinding-sequence (FWS) 266
results presentation service (RPS) 290
Retail Service Quality Scale 67
ROCK (RObust Clustering using linKs) 149
Rule Base 307
rule intersection 180

S

saturation filter 222
self-organizing maps (SOM) 246
sequence
    mining 266, 268
    rules 234, 235, 241
    to-model 158
service quality 66, 67, 69, 73
servqual scale 67
single cut-point method 27
stated preference models (SP) 65, 66
stochastic oscillators (SO 26
structural
    classes 117
    risk minimization (SRM) 101
support vector machines (SVM) 25, 33, 82, 83, 101,
    139
    Fisher 132
    literature 102
support vectors 92

T

target variable 73
task composition 291
time series 157, 158, 159, 160, 168
tools and algorithms access service (TAAS) 289
tree distance 137
tree kernel approach 138, 141
truncated datasets 206, 212, 218

U

UCR Time Series Data Mining Archive 167
user support nodes (USNs) 288

V

virtual
  environments (VEs) 265, 281
  resource
    abstraction 291

W

wayfinding 265, 267, 268
Web services description language (WSDL) 292
Web services resource framework (WSRF) 292
World Wide Web Consortium (W3C) 292